Lecture Notes in Computer Science　　　9947

Commenced Publication in 1973
Founding and Former Series Editors:
Gerhard Goos, Juris Hartmanis, and Jan van Leeuwen

More information about this series at http://www.springer.com/series/7407

Akira Hirose · Seiichi Ozawa
Kenji Doya · Kazushi Ikeda
Minho Lee · Derong Liu (Eds.)

Neural Information Processing

23rd International Conference, ICONIP 2016
Kyoto, Japan, October 16–21, 2016
Proceedings, Part I

 Springer

Editors
Akira Hirose
The University of Tokyo
Tokyo
Japan

Seiichi Ozawa
Kobe University
Kobe
Japan

Kenji Doya
Okinawa Institute of Science and
   Technology Graduate University
Onna
Japan

Kazushi Ikeda
Nara Institute of Science and Technology
Ikoma
Japan

Minho Lee
Kyungpook National University
Daegu
Korea (Republic of)

Derong Liu
Chinese Academy of Sciences
Beijing
China

ISSN 0302-9743   ISSN 1611-3349 (electronic)
Lecture Notes in Computer Science
ISBN 978-3-319-46686-6   ISBN 978-3-319-46687-3   (eBook)
DOI 10.1007/978-3-319-46687-3

Library of Congress Control Number: 2016953319

LNCS Sublibrary: SL1 – Theoretical Computer Science and General Issues

Printed on acid-free paper

This Springer imprint is published by Springer Nature
The registered company is Springer International Publishing AG
The registered company address is: Gewerbestrasse 11, 6330 Cham, Switzerland

Preface

This volume is part of the four-volume proceedings of the 23rd International Conference on Neural Information Processing (ICONIP 2016) held in Kyoto, Japan, during October 16–21, 2016, which was organized by the Asia-Pacific Neural Network Society (APNNS, http://www.apnns.org/) and the Japanese Neural Network Society (JNNS, http://www.jnns.org/). ICONIP 2016 Kyoto was the first annual conference of APNNS, which started in January 2016 as a new society succeeding the Asia-Pacific Neural Network Assembly (APNNA). APNNS aims at the local and global promotion of neural network research and education with an emphasis on diversity in members and cultures, transparency in its operation, and continuity in event organization. The ICONIP 2016 Organizing Committee consists of JNNS board members and international researchers, who plan and run the conference.

Currently, neural networks are attracting the attention of many people, not only from scientific and technological communities but also the general public in relation to the so-called Big Data, TrueNorth (IBM), Deep Learning, AlphaGo (Google DeepMind), as well as major projects such as the SyNAPSE Project (USA, 2008), the Human Brain Project (EU, 2012), and the AIP Project (Japan, 2016). The APNNS's predecessor, APNNA, promoted fields that were active but also others that were leveling off. APNNS has taken over this function, and further enhances the aim of holding technical and scientific events for interaction where even those who have extended the continuing fields and moved into new/neighboring areas rejoin and participate in lively discussions to generate and cultivate novel ideas in neural networks and related fields.

The ICONIP 2016 Kyoto Organizing Committee received 431 submissions from 38 countries and regions worldwide. Among them, 296 (68.7 %) were accepted for presentation. The first authors of papers that were presented came from Japan (100), China (78), Australia (22), India (13), Korea (12), France (7), Hong Kong (7), Taiwan (7), Malaysia (6), United Kingdom (6), Germany (5), New Zealand (5) and other countries/regions worldwide.

Besides the papers published in these four volumes of the Proceedings, the conference technical program includes

- Four plenary talks by Kunihiko Fukushima, Mitsuo Kawato, Irwin King, and Sebastian Seung
- Four tutorials by Aapo Hyvarinen, Nikola Kazabov, Stephen Scott, and Okito Yamashita,
- One Student Best Paper Award evaluation session
- Five special sessions, namely, bio-inspired/energy-efficient information processing, whole-brain architecture, data-driven approach for extracting latent features from multidimensional data, topological and graph-based clustering methods, and deep and reinforcement learning
- Two workshops: Data Mining and Cybersecurity Workshop 2016 and Workshop on Novel Approaches of Systems Neuroscience to Sports and Rehabilitation

The event also included exhibitions and a technical tour.

Kyoto is located in the central part of Honshu, the main island of Japan. Kyoto formerly flourished as the imperial capital of Japan for 1,000 years after 794 A.D., and is presently known as "The City of Ten Thousand Shrines." There are 17 sites (13 temples, three shrines, and one castle) in Kyoto that form part of the UNESCO World Heritage Listing, named the "Historic Monuments of Ancient Kyoto (Kyoto, Uji and Otsu Cities)." In addition, there are three popular, major festivals (Matsuri) in Kyoto, one of which, "Jidai Matsuri" (The Festival of Ages), was held on October 22, just after ICONIP 2016.

We, the general chair, co-chair, and Program Committee co-chairs, would like to express our sincere gratitude to everyone involved in making the conference a success. We wish to acknowledge the support of all the sponsors and supporters of ICONIP 2016, namely, APNNS, JNNS, KDDI, NICT, Ogasawara Foundation, SCAT, as well as Kyoto Prefecture, Kyoto Convention and Visitors Bureau, and Springer. We also thank the keynote, plenary, and invited speakers, the exhibitors, the student paper award evaluation committee members, the special session and workshop organizers, as well as all the Organizing Committee members, the reviewers, the conference participants, and the contributing authors.

October 2016

Akira Hirose
Seiichi Ozawa
Kenji Doya
Kazushi Ikeda
Minho Lee
Derong Liu

Organization

General Organizing Board

JNNS Board Members

Honorary Chairs

Shun-ichi Amari   RIKEN
Kunihiko Fukushima   Fuzzy Logic Systems Institute

Organizing Committee

General Chair

Akira Hirose   The University of Tokyo, Japan

General Co-chair

Seiichi Ozawa   Kobe University, Japan

Program Committee Chairs

Kenji Doya   OIST, Japan
Kazushi Ikeda   NAIST, Japan
Minho Lee   Kyungpook National University, Korea
Derong Liu   Chinese Academy of Science, China

Local Arrangements Chairs

Hiroaki Nakanishi   Kyoto University, Japan
Ikuko Nishikawa   Ritsumeikan University, Japan

Members

Toshio Aoyagi   Kyoto University, Japan
Naoki Honda   Kyoto University, Japan
Kazushi Ikeda   NAIST, Japan

Shin Ishii   Kyoto University, Japan
Katsunori Kitano   Ritsumeikan University, Japan
Hiroaki Mizuhara   Kyoto University, Japan
Yoshio Sakurai   Doshisha University, Japan
Yasuhiro Tsubo   Ritsumeikan University, Japan

Financial Chair

Seiichi Ozawa   Kobe University, Japan

Member

Toshiaki Omori   Kobe University, Japan

Special Session Chair

Kazushi Ikeda   NAIST, Japan

Workshop/Tutorial Chair

Hiroaki Gomi   NTT Communication Science Laboratories, Japan

Publication Chair

Koichiro Yamauchi   Chubu University, Japan

Members

Yutaka Hirata   Chubu University, Japan
Kay Inagaki   Chubu University, Japan
Akito Ishihara   Chukyo University, Japan

Exhibition Chair

Tomohiro Shibata   Kyushu Institute of Technology, Japan

Members

Hiroshi Kage   Mitsubishi Electric Corporation, Japan
Daiju Nakano   IBM Research - Tokyo, Japan
Takashi Shinozaki   NICT, Japan

Publicity Chair

Yutaka Sakai   Tamagawa University, Japan

Industry Relations

Ken-ichi Tanaka	Mitsubishi Electric Corporation, Japan
Toshiyuki Yamane	IBM Research - Tokyo, Japan

Sponsorship Chair

Ko Sakai	University of Tsukuba, Japan

Member

Susumu Kuroyanagi	Nagoya Institute of Technology, Japan

General Secretaries

Hiroaki Mizuhara	Kyoto University, Japan
Gouhei Tanaka	The University of Tokyo, Japan

International Advisory Committee

Igor Aizenberg	Texas A&M University-Texarkana, USA
Sabri Arik	Istanbul University, Turkey
P. Balasubramaniam	Gandhigram Rural Institute, India
Eduardo Bayro-Corrochano	CINVESTAV, Mexico
Jinde Cao	Southeast University, China
Jonathan Chan	King Mongkut's University of Technology, Thailand
Sung-Bae Cho	Yonsei University, Korea
Wlodzislaw Duch	Nicolaus Copernicus University, Poland
Tom Gedeon	Australian National University, Australia
Tingwen Huang	Texas A&M University at Qatar, Qatar
Nik Kasabov	Auckland University of Technology, New Zealand
Rhee Man Kil	Sungkyunkwan University (SKKU), Korea
Irwin King	Chinese University of Hong Kong, SAR China
James Kwok	Hong Kong University of Science and Technology, SAR China
Weng Kin Lai	Tunku Abdul Rahman University College, Malaysia
James Lam	The University of Hong Kong, SAR Hong Kong
Kittichai Lavangnananda	King Mongkut's University of Technology, Thailand
Min-Ho Lee	Kyungpoor National University, Korea
Soo-Young Lee	Korea Advanced Institute of Science and Technology, Korea
Andrew Chi-Sing Leung	City University of Hong Kong, SAR China
Chee Peng Lim	University Sains Malaysia, Malaysia
Chin-Teng Lin	National Chiao Tung University, Taiwan
Derong Liu	The Institute of Automation of the Chinese Academy of Sciences (CASIA), China

Chu Kiong Loo	University of Malaya, Malaysia
Bao-Liang Lu	Shanghai Jiao Tong University, China
Aamir Saeed Malik	Petronas University of Technology, Malaysia
Danilo P. Mandic	Imperial College London, UK
Nikhil R. Pal	Indian Statistical Institute, India
Hyeyoung Park	Kyungpook National University, Korea
Ju. H. Park	Yeungnam University, Republic of Korea
John Sum	National Chung Hsing University, Taiwan
DeLiang Wang	Ohio State University, USA
Jun Wang	Chinese University of Hong Kong, SAR Hong Kong
Lipo Wang	Nanyang Technological University, Singapore
Zidong Wang	Brunel University, UK
Kevin Wong	Murdoch University, Australia
Xin Yao	University of Birmingham, UK
Li-Qing Zhang	Shanghai Jiao Tong University, China

Advisory Committee Members

Masumi Ishikawa	Kyushu Institute of Technology
Noboru Ohnishi	Nagoya University
Shiro Usui	Toyohashi University of Technology
Takeshi Yamakawa	Fuzzy Logic Systems Institute

Technical Program Committee

Abdulrahman Altahhan	Tetsuo Furukawa
Sabri Arik	Kuntal Ghosh
Sang-Woo Ban	Anupriya Gogna
Tao Ban	Hiroaki Gomi
Matei Basarab	Shanqing Guo
Younes Bennani	Masafumi Hagiwara
Ivo Bukovsky	Isao Hayashi
Bin Cao	Shan He
Jonathan Chan	Akira Hirose
Rohitash Chandra	Jin Hu
Chung-Cheng Chen	Jinglu Hu
Gang Chen	Kaizhu Huang
Jun Cheng	Jun Igarashi
Long Cheng	Kazushi Ikeda
Zunshui Cheng	Ryoichi Isawa
Sung-Bae Cho	Shin Ishii
Justin Dauwels	Teijiro Isokawa
Mingcong Deng	Wisnu Jatmiko
Kenji Doya	Sungmoon Jeong
Issam Falih	Youki Kadobayashi

Juyang Weng
Bin Xu
Tetsuya Yagi
Nobuhiko Yamaguchi
Hiroshi Yamakawa
Toshiyuki Yamane
Koichiro Yamauchi
Tadashi Yamazaki
Pengfei Yan
Qinmin Yang
Xiong Yang

Zhanyu Yang
Junichiro Yoshimoto
Zhigang Zeng
Dehua Zhang
Li Zhang
Nian Zhang
Ruibin Zhang
Bo Zhao
Jinghui Zhong
Ding-Xuan Zhou
Lei Zhu

Contents – Part I

Neural Data Analysis

Robotics and Control

Bio-Inspired/Energy-Efficient Information Processing: Theory, Systems, Devices

Neurodynamics

Bioinformatics

Biomedical Engineering

Data Mining and Cybersecurity Workshop

Deep and Reinforcement Learning

Emotion Prediction from User-Generated Videos by Emotion Wheel Guided Deep Learning

Che-Ting Ho, Yu-Hsun Lin, and Ja-Ling Wu$^{(\boxtimes)}$

Department of GINM, National Taiwan University, Taipei, Taiwan
{murasaki0110,lymanblue,wjl}@cmlab.csie.ntu.edu.tw

Abstract. To build a robust system for predicting emotions from user-generated videos is a challenging problem due to the diverse contents and the high level abstraction of human emotions. Evidenced by the recent success of deep learning (e.g. Convolutional Neural Networks, CNN) in several visual competitions, CNN is expected to be a possible solution to conquer certain challenges in human cognitive processing, such as emotion prediction. The emotion wheel (a widely used emotion categorization in psychology) may provide a guidance on building basic cognitive structure for CNN feature learning. In this work, we try to predict emotions from user-generated videos with the aid of emotion wheel guided CNN feature extractors. Experimental results show that the emotion wheel guided and CNN learned features improved the average emotion prediction accuracy rate to 54.2 %, which is better than that of the related state-of-the-art approaches.

Keywords: Deep learning · Emotion wheel · Emotion prediction

1 Introduction

As the bandwidth of Internet broadens and the ability to capture videos from various consumer devices increases, users can easily record and share videos on social networks and/or websites to express their feelings. Moreover, understanding user emotions expressed in these contents is beneficial for many applications (e.g. predicting users' opinions for a certain event or topic). Therefore, how to robustly predict emotions from user-generated videos (UGV) became an interesting research topic in AI and machine learning communities, very recently. However, predicting emotions in UGV is a very challenging problem since UGV usually have diversified content and without pre-defined script and professional post-editing. This explains why the average prediction accuracy achieved in the state-of-the-art work (e.g. [20]) is still less than 50 %.

Building an image-based computational framework for conducting emotion prediction has been studied extensively over the last decade. A comprehensive survey [19] on previous approaches depicts several systems on the basis of low-level visual features. Considering the emotional cues could be carried in high-level abstractions that beyond visual or audio features, such as attributes in images,

© Springer International Publishing AG 2016
A. Hirose et al. (Eds.): ICONIP 2016, Part I, LNCS 9947, pp. 3–12, 2016.
DOI: 10.1007/978-3-319-46687-3_1

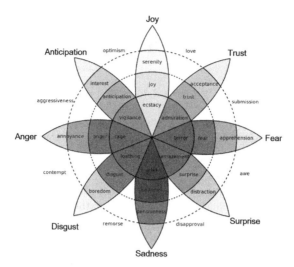

Fig. 1. The psychological emotion wheel which consists of 8 basic emotions.

the research work [2] proposed a framework using mid-level attribute features for emotion detection. They built a large scale visual sentiment ontology based on different emotional concepts and images collected from Flickr. The detector was trained for each concept to construct a detector bank for visual sentiments.

In contrast to the progress made in image sentiment analysis, the efforts for emotion prediction in videos are mainly focusing on movies. The research work [11] proposed a mean-shift based clustering framework to classify movie clips into genres such as action, comedy, horror or drama, according to several low-level visual features. In [18], they utilized multiple modalities existed in videos (such as visual and audio) to formulate a number of audiovisual features for bridging the cognitive gap. This approach revealed the advantages of fusing different modalities for emotion analysis in movies.

The research work [20] first introduced the emotion analysis problem and constructed a UGVEmotion dataset based on the 8 basic emotions as shown in Fig. 1. This type of videos has several unique characteristics as compared with movies. The first one is that UGV usually contain a single clear emotion in a short time period. The second one is that UGV are often taken by amateur users, resulting diverse environments and unstable qualities.

The design of high performance handcraft features for emotion prediction in UGV remains a challenging problem since numerous existing features have been evaluated in the research work [20] and the obtained prediction accuracy is still less than 50 %.

On the other hand, the deep learning technique (e.g. CNN) learns the features from the data directly and has demonstrating significant performance gain in numerous visual tasks (e.g. ImageNet challenge [6,15], object recognition [3], face verification [16], semantic segmentation [7] and video classification [5]) and drawing considerable attention in computer vision and machine learning societies.

Inspired by the success of CNN, we are interested in leveraging deep learning framework to resolve the challenging emotion prediction problem in UGV.

2   Related Work

2.1   Emotion Wheel

In emotion wheel, human emotion responses can be divided into eight primary classes with bipolar relationships [10]: joy vs. sadness, trust vs. disgust, fear vs. anger, and surprise vs. anticipation as demonstrated in Fig. 1. The theory of emotion wheel [10] argues these basic emotions are important to increase the fitness for surviving of human beings (e.g. fear inspires fight-or-flight response). The theory of emotion wheel suggests that the other emotions (e.g. optimism) are the combination of these 8 primary emotions. In other words, these 8 basic emotions laid the foundations for emotion analysis and prediction.

2.2   Deep Learning

Convolutional neural network (CNN) is a learning framework that aims to learn features from data. Due to the progress on GPU programming, one has the ability to train CNN on a large scale image dataset like ImageNet [12] with an affordable time cost.

2.3   Emotion Prediction

The research of emotion analysis focused on the challenging UGV scenarios in recent years. The latest research works about emotion prediction in UGV can be summarized as follows:

- UGVEmotion [20] A comprehensive computational framework [20] has been built including several modalities, such as visual, audio, and other high-level attributes.
- Multimodal DBM [8] The deep boltzmann machine (DBM) method is adopted by [8] for emotion prediction in UGV. In contrast to extract features from each modality (e.g. visual, audio), they proposed to learn a joint density model over the multimodal inputs for DBM. The multimodal DBM improves the prediction accuracy as compared with that of the original UGVEmotion.

3   Proposed Method

3.1   Emotion Wheel Guided CNN Feature Extractor

Since the outputs of the last fully connected layer in CNN are commonly treated as effective features for video classification [21], we construct a feature learning hierarchy through combination and concatenation of CNN-based feature

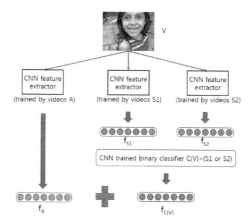

Fig. 2. The proposed emotion wheel guided CNN learning hierarchy for feature extraction. The final concatenated features is denoted by the square bracket $[f_A, f_{C(V)}]$.

Table 1. The proposed subset divisions of the emotion wheel.

Subset	Emotion
S1	Anger, Fear, Joy, Sadness
S2	Anticipation, Disgust, Surprise, Trust

extractors and classifier, as shown in Fig. 2. The obtained concatenating feature $[f_A, f_{C(V)}]$ is expected to perform well in emotion prediction since CNN has very good credit for learning appropriate features from data, directly.

In Fig. 1, the angular distance between two emotions of the emotion wheel represents the level of emotion similarity. Therefore, the bipolar emotion pair (e.g. joy vs. sadness) locates in the opposite positions of the emotion wheel. In order to obtain a more discriminative feature extractor, we divide the UGVEmotion training videos into 2 disjoint subsets (as shown in Table 1) with maximal emotion diversity in each subset.

As illustrated in Fig. 2, the proposed emotion wheel guided CNN feature extractor consists of 3 feature extractors and 1 binary classifier for constructing the final representative feature. The functions and parameters of the proposed feature learning hierarchy can be summarized as follows.

- f_A: f_A denotes the CNN feature extractor which is trained by using all the training videos in UGVEmotion dataset.
- f_{S1} or f_{S2}: f_{S1}(or f_{S2}) represents the CNN feature extractor that is trained by the training videos in UGVEmotion dataset whose emotion category belongs to subset S1 (or S2).
- $C(V)$: A CNN trained binary classifier that predicts the emotion category, $C(V)$, of video V is either S1 or S2.

The final representative feature is the concatenated result of the features extracted by f_A and $f_{C(V)}$. As pre-described, the concatenation operation (or result) is represented by a square bracket, $[f_A, f_{C(V)}]$, in this work.

4 Frame-Level Feature Extraction

4.1 Preprocessing

UGV usually have some post-production frames inserted by users (e.g. additional captions like "thanks for watching"). We want to detect and remove these frames because these additional materials contribute very little visual information in regard to emotion expression. Therefore, we manually labeled 1000 frames for training a detector. And then we compute a $16 \times 16 \times 16$ color histogram for each frame. Since those post-production frames typically have monotonous colors, the variance of color histogram will be a good feature for distinction. Then, we utilize color histogram variance as the feature to train an SVM model to detect and remove the above-mentioned emotion unrelated frames from the original UGVEmotion dataset.

4.2 Visual Feature Extraction

The visual CNN architecture is optimized for single-image inputs, while the input of UGV is, in general, a video. In order to train a CNN with image-based architecture but applicable to UGVEmotion dataset, we subsample the input video with 3 FPS (frames-per-second) sampling rate and rescale each image to a fixed size of 256×256 pixels with 3 (i.e. RGB) color channels. The visual feature extraction of each frame follows the procedure illustrated in Fig. 2.

4.3 Audio Feature Extraction

In the previous subsection, a CNN feature extraction pipeline has been built for each video frame. Audio waves are naturally one-dimensional signals; however, a two-dimensional image-like data representation is a must for applying the pre-described transfer learning technique to learn effective CNN based audio features. Therefore, we transform the audio waves into spectrograms to meet this requirement. Each spectrogram has a window of size one second with 0.5 s overlap. The same feature extraction pipeline, as shown in Fig. 2, is applied to the spectrograms to extract audio features.

4.4 Motion Feature Extraction

The video classification task would be challenge for CNN since the original architecture is tailored to suit single-image inputs. There is research work [14] addressing video classification by CNN incorporating with spatial and temporal streams. However, the video classification accuracy of state-of-the-art CNN based approach [14] is still less than that of the improved dense trajectories (IDT) feature [17] with stacked Fisher encoding [9,13]. Therefore, we choose IDT plus fisher vector (IDT+FV) encoding as our motion feature for UGV emotion prediction.

5   Video-Level Feature Encoding

Figure 3 illustrates our proposed system flow for fusing multimodal features. The frame-level features are post-processed and encoded as a video level feature. First, the IDT feature, addressed by Fisher vector encoding, is treated as a video-level motion feature. Second, in contrast to the common averaging of frame-level features, the video-level feature encoding for the CNN-based frame-level visual and audio features is resolved by bags of words (BoW) approach in this paper. Finally, the early fusion method presented in [1] is adopted to fuse visual, audio and motion video-level features together to train our SVM classifier for emotion prediction.

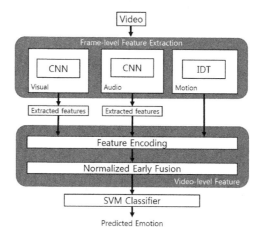

Fig. 3. The operational flow and system block diagram of the proposed multimodal feature fusion. The emotion wheel guided CNN feature extractor is applied to both visual data and audio data. The motion activity is captured by IDT feature plus Fisher vector (IDT+FV) encoding.

6   Experiments

In order to evaluate the performance of the proposed emotion wheel guided feature engineering and the overall prediction accuracy of the multimodal feature fusion, we did a series of tests of our system on UGVEmotion dataset [20], which is the largest UGV dataset with emotion annotations, publicly available up to now. The emotion UGV dataset is composed of videos downloaded from YouTube and Flickr. The released dataset consists of video clips annotated with one of the 8 basic emotion categories, depicted in Fig. 1.

The "VGG" CNN model [15] is chosen as our deep learning testbed for emotion prediction, because of its top performance in ImageNet 2014 competition.

Table 2. Performance comparison of various tested feature engineering approaches. The notations V., A., Att. and Mot. represent visual, audio, attribute and motion, respectively.

Features	V	A	Att	Mot
f_A (from scratch)	9.5	8	-	-
f_A	11.7	12.6	-	-
f_A (from scratch) + SVM	33.6	20.1	-	-
f_A + SVM	48	37.8	-	-
$[f_{S1}, f_{S2}]$ + SVM	47.2	35.6	-	-
$f_{C(V)}$ + SVM	47.8	36.9	-	-
$[f_A, f_{S1}, f_{S2}]$ + SVM	48.5	37.3	-	-
$[f_A, f_{C(V)}]$ + SVM	50.1	38.5	-	-
UGVEmotion	41.9	28.8	40	-
IDT-FV (motion) + SVM	-	-	-	40.8

Table 3. Performance comparison of different video-level encoding methods.

Algorithms	Encoding	Visual	Audio
$[f_A, f_{C(V)}]$	Avg.	50.1	38.5
$[f_A, f_{C(V)}]$	BoW	51.3	40.2

Table 4. Overall system performance comparisons. The notations V., A., Att. and Mot. represent visual, audio, attribute and motion, respectively.

Algorithms	UGVEmotion		Multimodal DBM	Proposed	
Category	Visual	V. + A. + Att	V. + A. + Att	Visual	V. + A. + Mot.
Anger	49.4	53	50.9	41	**64**
Anticipation	3	7.6	3.4	13	**18**
Disgust	35.1	44.6	39.9	38	**47**
Fear	45	47.3	54.5	63	**70**
Joy	44.8	48.3	59	**65**	56
Sadness	23.5	20	21.7	**58**	50
Surprise	75.6	76.9	**82.8**	66	69
Trust	10.3	28.5	31.2	**41**	37
Overall	41.9	46.1	49.9	51.3	**54.2**

The model provided by [15] was trained on ImageNet 2014 dataset using Caffe learning toolkit [4].

We examined the performance (in terms of prediction precision) of different feature engineering approaches for CNN based visual and audio feature extraction. The overall system performance is also compared with the state-of-the-art results reported in **UGVEmotion** [20] and **Multimodal DBM** [8].

6.1 Performance of the Proposed Feature Engineering

Table 2 illustrates the performances of different feature engineering approaches for emotion prediction. All CNN based feature extractors (including both visual and audio) are incorporated the transfer learning technique from ImageNet into the tests, except the results reported in the first row (i.e. the "from scratch" approach). The naive application of CNN model gives the lowest prediction accuracy without applying the transfer learning technique from ImageNet (as shown in the row indicated as f_A (from scratch) of Table 2). The incorporation of SVM classifier provides much better prediction accuracy than that of CNN built in soft-max classifier for the extracted features. Therefore, SVM is chosen as the final feature classifier in our system.

The integration of CNN feature extractor f_A (with transfer learning from ImageNet) and SVM has already outperformed UGVEmotion in prediction precision. The proposed emotion wheel guided hierarchical feature extractor shows its outstanding performance (prediction accuracy goes up to 50.1 % by visual feature only and 38.5 % by audio feature only) among the various tested feature engineering configurations.

6.2 Performance of Video-Level Feature Encoding

Table 3 shows BoW encoding method provides a little bit better prediction performance as compared with the common averaging approach for constructing video-level features. Empirically, BoW encoding improves the prediction accuracy in both visual and audio features.

6.3 Performance of Overall System

The overall emotion prediction accuracy is demonstrated in Table 4, in which the highest prediction accuracy result of each one of the comparisons is denoted in boldface. The proposed system (by visual feature or by fused multimodal feature) provides the highest prediction results in 7 out of the 8 emotion categories. The work Multimodal DBM [8] shows the highest precision in 'surprise' emotion.

From Table 4, the proposed visual feature has already outperformed the two afore-cited state-of-the-art approaches in overall emotion prediction accuracy. The visual feature achieves the highest prediction accuracy in 3 out of the 8 emotion categories even compared with the proposed multimodal approach. This phenomenon implies that the emotion prediction is dominated mainly by visual features. The fact that those emotion categories "fear, joy, sadness, surprise" can be well-predicted (i.e. accuracy > 55 %) by using visual feature only gives a strong support for the arguments of emotion wheel theory. The emotion wheel theory suggests that these basic emotions are important for human survival

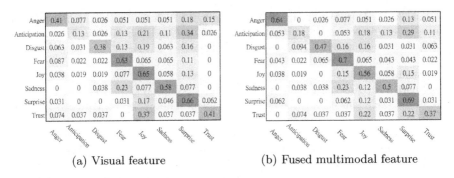

(a) Visual feature   (b) Fused multimodal feature

Fig. 4. The confusion matrices of our emotion prediction results based on (a) the visual feature and (b) the fused multimodal feature.

(e.g. fear inspires fight-or-flight response), because human brain is equipped with superior surviving ability through rapid visual information processing.

The overall system performance achieves 54.2 % prediction accuracy which outperforms the corresponding state-of-the-art results. The fused multimodal feature, in most of cases, provides better prediction accuracy than the situation using visual feature alone.

Figure 4 illustrates the confusion matrices of our emotion prediction results based on (a) the visual feature and (b) the fused multimodal feature. The visual feature based confusion matrix shows that the emotion 'anticipation' will easily be mis-predicted as the emotion 'surprise'. The emotion 'trust' will wrongly be predicted as the emotion 'joy'. However, as evidenced by Fig. 4(b), if the fused multimodal feature is used, the occurrences of the above-mentioned mis-prediction will be largely reduced.

7   Conclusion

UGV emotion prediction is an emerging and challenging problem. In order to resolve this challenge, we leverage the psychological emotion wheel theory to guide the design of CNN based feature extractors. The prediction accuracy is improved up to 54.2 % which outperforms the state-of-the-art related research results.

References

1. Ayache, S., Quénot, G., Gensel, J.: Classifier fusion for SVM-based multimedia semantic indexing. In: Amati, G., Carpineto, C., Romano, G. (eds.) ECIR 2007. LNCS, vol. 4425, pp. 494–504. Springer, Heidelberg (2007). doi:10.1007/978-3-540-71496-5_44
2. Borth, D., Ji, R., Chen, T., Breuel, T., Chang, S.-F.: Large-scale visual sentiment ontology and detectors using adjective noun pairs. In: ACM MM 2013, pp. 223–232. ACM, New York (2013)

3. Girshick, R., Donahue, J., Darrell, T., Malik, J.: Rich feature hierarchies foraccurate object detection and semantic segmentation. In: IEEE CVPR 2014 (2014)
4. Jia, Y., Shelhamer, E., Donahue, J., Karayev, S., Long, J., Girshick, R., Guadarrama, S., Darrell, T.: Caffe: convolutional architecture for fast feature embedding, pp. 675–678 (2014)
5. Karpathy, A., Toderici, G., Shetty, S., Leung, T., Sukthankar, R., Fei-Fei, L.: Large-scale video classification with convolutional neural networks. In: IEEE CVPR 2014 (2014)
6. Krizhevsky, A., Sutskever, I., Hinton, G.E.: Imagenet classification with deep convolutional neural networks. In: Pereira, F., Burges, C., Bottou, L., Weinberger, K. (eds.) NIPS 2012, pp. 1097–1105 (2012)
7. Long, J., Shelhamer, E., Darrell, T.: Fully convolutional networks for semanticsegmentation. In: IEEE CVPR 2015 (2015)
8. Pang, L., Ngo, C.-W.: Mutlimodal learning with deep boltzmann machine for emotion prediction in user generated videos. In: ICMR 2015, ICMR 2015, pp. 619–622. ACM, New York (2015)
9. Peng, X., Zou, C., Qiao, Y., Peng, Q.: Action recognition with stacked fisher vectors. In: Fleet, D., Pajdla, T., Schiele, B., Tuytelaars, T. (eds.) ECCV 2014. LNCS, vol. 8693, pp. 581–595. Springer, Heidelberg (2014). doi:10.1007/978-3-319-10602-1_38
10. Plutchik, R.: Emotion: Theory, Research and Experience, vol. 1. Academic Press, New York (1980)
11. Rasheed, Z., Sheikh, Y., Shah, M.: On the use of computable features for film classification. IEEE TCSVT 15(1), 52–64 (2005)
12. Russakovsky, O., Deng, J., Su, H., Krause, J., Satheesh, S., Ma, S., Huang, Z., Karpathy, A., Khosla, A., Bernstein, M., Berg, A.C., Fei-Fei, L.: Imagenet large scale visual recognition challenge. IJCV, 1–42, April 2015
13. Simonyan, K., Vedaldi, A., Zisserman, A.: Deep fisher networks for large-scale image classification. In: Burges, C., Bottou, L., Welling, M., Ghahramani, Z., Weinberger, K. (eds.) NIPS 2013, pp. 163–171 (2013)
14. Simonyan, K., Zisserman, A.: Two-stream convolutional networks for action recognition in videos. In: Ghahramani, Z., Welling, M., Cortes, C., Lawrence, N., Weinberger, K. (eds.) NIPS 2014, pp. 568–576 (2014)
15. Simonyan, K., Zisserman, A.: Very deep convolutional networks for large-scaleimage recognition. In: ICLR 2015 (2015)
16. Taigman, Y., Yang, M., Ranzato, M., Wolf, L.: Deepface: closing the gap to human-level performance in face verification. In: IEEE CVPR 2014, pp. 1701–1708 (2014)
17. Wang, H., Schmid, C.: Action recognition with improved trajectories. In: ICCV 2013 (2013)
18. Wang, H.L., Cheong, L.-F.: Affective understanding in film. IEEE TCSVT 16(6), 689–704 (2006)
19. Wang, W., He, Q.: A survey on emotional semantic image retrieval. In: IEEE ICIP 2008, pp. 117–120, October 2008
20. Jiang, Y.-G., Xue, X., Baohan, X.: Predicting emotions in user-generated videos. In: AAAI 2014, Canada (2014)
21. Zha, S., Luisier, F., Andrews, W., Srivastava, N., Salakhutdinov, R.: Exploiting image-trained CNN architectures for unconstrained video classification. In: BMVC 2015 (2015)

Deep Q-Learning with Prioritized Sampling

Jianwei Zhai, Quan Liu$^{(\boxtimes)}$, Zongzhang Zhang, Shan Zhong,
Haijun Zhu, Peng Zhang, and Cijia Sun

School of Computer Science and Technology, Soochow University,
Suzhou 215000, China
quanliu@suda.edu.cn, 20144227023@stu.suda.edu.cn

Abstract. The combination of modern reinforcement learning and deep learning approaches brings significant breakthroughs to a variety of domains requiring both rich perception of high-dimensional sensory inputs and policy selection. A recent significant breakthrough in using deep neural networks as function approximators, termed Deep Q-Networks (DQN), proves to be very powerful for solving problems approaching real-world complexities such as Atari 2600 games. To remove temporal correlation between the observed transitions, DQN uses a sampling mechanism called experience reply which simply replays transitions at random from the memory buffer. However, such a mechanism does not exploit the importance of transitions in the memory buffer. In this paper, we use prioritized sampling into DQN as an alternative. Our experimental results demonstrate that DQN with prioritized sampling achieves a better performance, in terms of both average score and learning rate on four Atari 2600 games.

Keywords: Reinforcement Learning · Deep Learning · Deep Reinforcement Learning · Policy control · Prioritized sampling

1 Introduction

Recent breakthroughs in both modern reinforcement learning (RL) and deep learning (DL) have given rise to a new research direction called deep reinforcement learning (DRL) which combines advances in DL for sensory inputs processing with RL [1–3]. One of the most notable DRL methods called Deep Q-Networks (DQN) which combines a deep convolutional neural network with a variant Q-learning algorithm in RL, has been shown to be capable of learning control policies in complex environments with high-dimensional sensory inputs. DQN outperformed previous algorithms based on handcrafted features and achieved or even surpassed a level comparable to that of a skilled human player in some Atari 2600 games, using the same network architecture and hyperparameters [2].

On-line RL agents incrementally update the parameters of value functions when they encounter a sequence of highly correlated transitions [4]. However, most DL methods have two main requirements: (a) the training samples are

© Springer International Publishing AG 2016
A. Hirose et al. (Eds.): ICONIP 2016, Part I, LNCS 9947, pp. 13–22, 2016.
DOI: 10.1007/978-3-319-46687-3_2

independent of each other; (b) the samples can be reused many times during training. A technique called experience replay [5] can be utilized to meet these requirements. Therefore, this sampling mechanism was applied to the DQN method [1,2], which stabilized the training of the algorithm. At each time step, the agent stores every observed transition into the memory buffer and then samples uniformly to get a number of mini-batch transitions for updating the parameters.

However, this approach of uniform sampling is in some respects limited because the memory buffer does not differentiate the importance of distinct transitions and always overwrites with recent transitions owing to the finite memory size [2]. So in this paper, we propose a novel sampling mechanism termed prioritized sampling which is more effective and efficient than the case of all transitions are sampled uniformly. Specifically, we set two priority levels on sampling transitions. On one hand, a more efficient sampling method should emphasize the transitions from which the agent can learn the most. And we all know that transitions with positive rewards are more informative and valuable for learning. So we assign these transitions with higher priority to be sampled during training. This modification of sampling can make transitions with positive rewards be sampled more frequently so as to learn optimal policies faster. On the other hand, we add an explicit penalty term to every transition being accessed to reduce the probability of being sampled again. The degrees of punishment will be higher with the increase of sampled times in order to make parts of the transitions never been sampled recently have chances to be reused in time. More importantly, we set the priority of sampling by measuring the magnitude of rewards higher than the latter.

This paper presents a new model-free, off-policy RL algorithm, called PS-DQN. PS-DQN makes two improvements based on the DQN algorithm. First, its network is trained with samples obtained by prioritized sampling to eliminate correlations between observed transitions. Second, it uses a soft target network to give consistent target Q-values during temporal difference backups. On four Atari 2600 games, PS-DQN appears better than DQN empirically, in terms of both average score and learning rate.

2 Background

2.1 Reinforcement Learning

In reinforcement learning, the agent interacts sequentially with an unknown environment, with the goal of maximizing cumulative rewards [4]. The environment is often formalized as a sequence of state transitions (s_t, a_t, r_t, s_t+1) of a Markov Decision Process (MDP). The action-value function is used in many RL algorithms. It describes the expected return after taking an action a_t in state s_t and thereafter following policy π:

$$Q^\pi(s,a) = \mathbb{E}_\pi[R_t|s_t = s, a_t = a, \pi] \tag{1}$$

The action-value function obeys a fundamental recursion known as the Bellman equation:

$$Q^*(s, a) = \mathbb{E}[r + \gamma \max_{a'} Q^*(s', a')|s, a] \tag{2}$$

We generally use the Bellman equation as an iterative update to estimate the action-value function:

$$Q_{i+1}(s, a) = \mathbb{E}[r + \gamma \max_{a'} Q_i(s', a')|s, a] \tag{3}$$

As the number of iterations approaches to infinity, value iteration algorithms are guaranteed to converge to the optimal action-value functions, $Q_i \longrightarrow Q^*$. However, it is impractical to estimate the optimal Q-value of every single state-action pair without any generalization because of the high complexity of state-action space. One of the core ideas to alleviate the computational challenge is to represent the Q-value using a function approximator such as a neural network, $Q(s, a) = Q(s, a; \theta)$, although some RL algorithms (e.g.,Q-learning) appear to be highly unstable when being combined with non-linear function approximators [6].

2.2   Deep Q-Networks

Deep Q-Networks (DQN) uses two main innovations in order to alleviate the unstability of learning when combining traditional RL algorithms with a deep convolutional neural network [2]. One key innovation is that DQN uses the experience replay mechanism. At each time-step t, the agent stores the transition tuple $e_t = (s_t, a_t, r_t, s_{t+1})$ into a memory buffer $D = \{e_1, e_2, \ldots, e_t\}$ and then samples transitions uniformly for training. Another key innovation behind the success of DQN is the use of two separate Q-networks $Q(s, a; \theta)$ and $Q(s, a; \theta^-)$ with current parameters θ and old parameters θ^- respectively. At every updating iteration i, the current parameters θ are updated so as to minimize the mean-squared Bellman error with respect to old parameters θ^-, by optimizing the following loss function:

$$L_i(\theta_i) = \mathbb{E}\left[\left(r + \gamma \max_{a'} Q(s', a'; \theta_i^-) - Q(s, a; \theta_i)\right)^2\right] \tag{4}$$

Differentiating the loss function with respect to the current parameters, we arrive at the following gradient:

$$\nabla_{\theta_i} L_i(\theta_i) = \left(r + \gamma \max_{a'} Q(s', a'; \theta_i^-) - Q(s, a; \theta_i)\right) \nabla_{\theta_i} Q(s, a; \theta) \tag{5}$$

Then, the parameters of the network are adjusted in the direction of the gradient descent of the loss function:

$$\theta_{i+1} = \theta_i + \alpha \nabla_{\theta_i} L_i(\theta_i) \tag{6}$$

DQN is off-policy, that is to say, it estimates the optimal Q-values by the greedy strategy $a = \max_a Q(s, a; \theta)$, while selecting actions by an ε-greedy strategy with the purpose of ensuring the balance between exploration and exploitation. In brief, DQN follows the greedy action with probability $1 - \varepsilon$ and selects a random action with probability ε.

3   Model Architecture and Algorithm

3.1   Details of Deep CNN Architecture

We use a deep convolutional neural network architecture in which there is a separate output unit for predicting Q-values of discrete actions. The main advantage of this architecture is its ability to compute Q-values for all possible actions in a given state representation with only a single forward pass along the network. The exact architecture used for most deep reinforcement learning tasks, demonstrated in Fig. 1, is as follows. The model is similar to the DQN's architecture except that the full-connected layer is followed by a dropout operation [7] in order to handle the problem of over-fitting. The input to our network is a $84 \times 84 \times 4$ image produced by the preprocessing procedure [2]. The first convolution layer convolves 32 filters of 8×8 with stride 4 with the input image and followed by a rectifier nonlinearity (ReLU). The second hidden layer convolves 64 filters of 4×4 with stride 2, again applies a rectifier nonlinearity. Then the final convolution layer of our network convolves 64 filters of 3×3 with stride 1 followed by a rectifier. This is followed by a fully-connected hidden layer consisting of 512 rectifier units and then a dropout operation. Finally, the output layer is a fully-connected linear layer with a single output for each valid action. The number of valid actions varied from 4 to 18 on the Atari 2600 games.

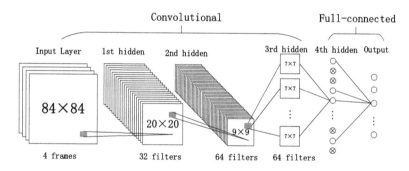

Fig. 1. The exact architecture of the deep neural network.

3.2   Prioritized Sampling

As in DQN, uniform sampling has some limitations. This is because this method does not make fully use of some valuable transitions from which we can learn the most, and always overwrites with recent transitions owing to the finite memory size. To alleviate this problem, here we use our prioritized sampling method instead of uniform sampling.

Generally, we improve the efficiency of sampling by making the transitions with positive rewards or high magnitude of TD errors to be sampled more frequently. On the one hand, the transitions that provide positive rewards are extremely rare at the primary stage of learning. However, these transitions are more valuable and informative than others for agents to learn from. So in our algorithm, we use two separate memory buffers to improve the utilization of transitions with positive rewards. The higher priority buffer D_1 is used to store transitions with positive rewards, and naturally the lower priority buffer D_2 stores transitions with non-positive rewards. Then we use a similar stratified sampling method to sample transitions from the higher priority buffer with probability ρ and from the other with probability $1 - \rho$.

On the other hand, some fraction of transitions are never sampled before they drop out of the buffers. To alleviate this problem, we add a term v_t for tracking the sampled times in the transition tuple $e_t = (s_t, a_t, r_t, v_t, s_{t+1})$. We can assume that the frequently replayed transitions will have low TD errors because of more opportunities to modify the Q-values so as to approximate the target action-value functions. Naturally, we prefer to sample the transitions that have not been sampled for a while, because they have relatively larger TD errors. So our innovation is to use a prioritized sampling based on the sampled times of each transition for ensuring every sample is replayed from time to time. We define the priority of transition j as:

$$p_j = \frac{1}{(v_j + 1)} \tag{7}$$

As we can see from this definition, the priority of transition j monotonously decreases with the increase of sampled times.

However, greedy sampling solely based on the priorities of transitions may make the training data lack of diversity. Specifically, to avoid expensive sweeps over the entire memory buffers, greedy sampling is prone to sample the transitions with higher priorities, meaning that a transition that has a lower priority after a replay may not be sampled anymore. One consequence is that the TD errors used for updating Q-values shrink slowly, especially when using the deep neural network as a function approximator. It is necessary to propose a method that can take full advantage of the transitions' priorities and ensure the diversity of sampling at the same time.

So we introduce a stochastic sampling method that falls in between pure greedy sampling by priority and sampling uniformly. We make a guarantee that the probability of being sampled to be monotonically increased in a transition's priority, while ensuring a non-zero probability of being sampled even for a transition with the lowest priority. Specifically, we define the probability of sampling transition j as:

$$P(j) = \frac{p_j^\alpha}{\sum_{i=1}^{i=size(D_1)} p_i^\alpha} \tag{8}$$

where $p_j > 0$ is the priority of transition j. The exponent α determines the degree of prioritization when sampling transitions. It is obvious that setting $\alpha = 0$ corresponds to the uniform sampling, while $\alpha = 1$ corresponds to the pure greedy sampling case.

We combine our prioritized sampling method with a deep reinforcement learning agent known as DQN. Our main modification is to replace the uniform sampling used by DQN with our prioritized sampling method. The full algorithm is presented in the next section.

Algorithm 1. deep Q-learning with prioritized sampling

1: Randomly initialize Q-Network Q with weights θ and soft target Q-Network $\hat{Q}$ with weights $\theta^- \leftarrow \theta$; memory buffers D_1 and D_2 to capacity N; mini-batch M, $p_1 = 1$.
2: **for** episode $1, M$ **do**
3:     Initialize sequence $s_1 = \{x_1\}$ and preprocessed sequence $\Phi_1 = \Phi(s_1)$.
4:     **for** $t = 1, T$ **do**
5:         With probability ϵ select a random action a_t.
6:         Otherwise select $a_t = \arg\max_a Q(\Phi(s_t), a; \theta)$.
7:         Execute action a_t and observe reward r_t and new image x_{t+1}.
8:         Set $s_{t+1} = \{s_t, a_t, x_{t+1}\}$, $v_t = 0$ and preprocess $\Phi_{t+1} = \Phi(s_{t+1})$.
9:         **if** $r_t > 0$ **then**
10:             Store transition $(\Phi_t, a_t, r_t, v_t, \Phi_{t+1})$ in D_1.
11:         **else**
12:             Store transition $(\Phi_t, a_t, r_t, v_t, \Phi_{t+1})$ in D_2.
13:         **for** $m = 1, M$ **do**
14:             **if** random() $< \rho$ **then**
15:                 Sample a transition $(\Phi_j, a_j, r_j, v_j, \Phi_{j+1})$ from D_1 according to a
16:                 probability distribution $P(j) = (p_j)^\alpha / \sum_i (p_i)^\alpha$.
17:             **else**
18:                 Sample a transition $(\Phi_j, a_j, r_j, v_j, \Phi_{j+1})$ from D_2 according to a
19:                 probability distribution $P(j) = (p_j)^\alpha / \sum_i (p_i)^\alpha$.
20:             Update replayed times: $v_j = v_j + 1$.
21:             Update transition priority: $p_j = 1/(v_j + 1)$.
22:         **end for**
23:         Set $y_j = \begin{cases} r_j & \text{if } s_{j+1} \text{ is terminal} \\ r_j + \gamma \max_{a'} \hat{Q}(\Phi_{j+1}, a'; \theta^-) & \text{otherwise} \end{cases}$
24:         Perform gradient descent step on the loss $L(\theta) = (y_j - Q(\Phi_j, a_j; \theta))^2$ with
25:         respect to the network parameters θ.
26:         Update the target networks: $\theta^- \leftarrow \tau\theta + (1 - \tau)\theta^-$.
27:     **end for**
28: **end for**

3.3 Algorithm

Directly implementing Q-learning with a deep neural network proved to be unstable in many environments. However, such non-linear function approximators appear to be necessary to learn more abstract and valuable feature representation when confronting with large state space. Therefore, we need to make two improvements to ensure the stability of our algorithm.

As in DQN, we firstly use the experience replay mechanism to address the problem of highly correlation between samples. The transitions in the form of $(s_t, a_t, r_t, v_t, s_{t+1})$ are stored into different buffers according to the magnitude of r_t. Parameters of the network are updated by performing stochastic gradient descent using a mini-batch of transitions obtained by prioritized sampling from the buffers.

The second modification aiming at ensuring the stability of our algorithm is to use a separate target network $Q(s, a; \theta^-)$ to generate the target Q-values: $y_i = r + \gamma \max_{a'} Q(s', a'; \theta_i^-)$. We use a soft target update, rather than directly copying the weights of the current network to the target network. Instead, the weights of these target networks are then updated by slowly tracking the current network: $\theta^- \longleftarrow \tau\theta + (1 - \tau)\theta^-$ with $\tau \ll 1$. Generating θ^- in this way makes the target Q-values change slowly, greatly improving the stability of learning the optimal action-value functions. The full algorithm for training the network is presented in Algorithm 1.

4   Experiments

4.1   Experimental Set up

We perform an evaluation of our proposed PS-DQN agent by conducting experiments on four Atari 2600 games using the Arcade Learning Environment [8] (ALE). ALE provides a challenging and diverse set of RL problems where an agent must learn to play the games directly from the high-dimensional sensory video inputs. In our experiments, all hyper-parameters are identical to DQN unless stated differently.

Firstly, the "soft" factor τ is set to be 0.05 for having the target Q-network slowly track the current network. The gradients are clipped to fall within [-5,5] to guarantee the stability of learning. In addition, the parameter ρ starts at 0.5 and decays linearly to 0.25 over the first million frames because smarter agent will get more transitions with positive rewards through incremental learning. For the hyper-parameter α that are utilized to ensure the diversity of transitions, we did a coarse grid search (evaluated on the game of Breakout), and found that the setting $\alpha = 0.6$ appears best in our algorithm. On each game, the network is trained on a single GPU for 50 million frames consuming one week and utilized two memory buffers with the capacity of one million most recent frames.

To summarize, our experiments only use a minimal prior knowledge consisting of the input sensory images, the game-specific scores and a single set of hyper-parameters across all games, resulting in an artificial agent with the capability of learning to being expert in a diverse of challenging tasks.

4.2   Main Evaluation

We select the following four games for evaluation: Breakout, Boxing, Pong and Space_invaders as tested problems. The deep Q-Network described in [2] is used as a baseline. A random number of frames were skipped by repeatedly taking the null or do nothing action before giving control to the agent for ensuring variation in the initial conditions. The learned policies are then evaluated after every 250000 steps (an epoch) based on the average reward per episode obtained by running an ϵ-greedy policy with $\epsilon = 0.05$ for 125000 steps.

In RL, we usually set an evaluation metric which is the total reward the agent collects in an episode. Naturally, our first metric is the best average reward per episode of 50 epochs for the two agents. The comparisons of training processes of two agents on the four Atari games are depicted in Fig. 2.

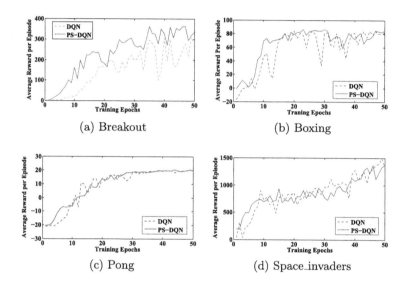

(a) Breakout

(b) Boxing

(c) Pong

(d) Space_invaders

Fig. 2. The average reward per episode for the two agents on four Atari 2600 games as a function of the number of training epochs.

We find that adding prioritized sampling to DQN gives rise to a significant improvement on four games embodied in higher average scores especially at the early stage of training in most of the games. This improvement can be ascribed to the increase of utilization rate of valuable and informative transitions with positive rewards. Furthermore, we find that training agents by PS-DQN are more stable in all games with the exception of Space_invaders. This behavior caused by our prioritized sampling which makes every transition be sampled with a certain probability, rather than biasing toward out-of-date transitions which have been sampled hundreds of times. However, the average reward per episode metric tends to be noisy because small changes to the weights of a policy can lead to large changes in the distribution of states the policy visits [1]. So we used a more stable metric which is the average maximum predicted action-value function. According to the results depicted in Fig. 3, we make a conclusion that our PS-DQN algorithm leads to a faster convergence speed than DQN. On

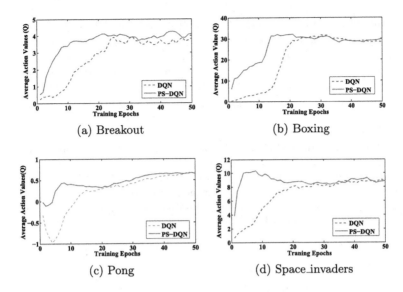

Fig. 3. The average maximum predicted Q-value per episode for the two agents on four Atari 2600 games as a function of the number of training epochs.

Breakout and Boxing, negative reward doesn't exist in the game and positive reward is rare at the early stage of training. So with our sampling mechanism, Q-values increase smoothly until the network converges. On pong, negative rewards appear frequently during the early stage of playing. So the Q-values curve which represents the learning process of DQN has a low peak as depicted in Fig. 3(c). This fluctuation of the Q-value function is adverse to the learning of the agent. Fortunately, our preference to transitions with positive rewards avoids a local minimum of average Q-values, resulting in a performance boost on the stability of learning. However, we do not lack the positive rewards at the beginning of the agent's learning on Space_invaders. Our sampling mechanism inevitably leads to an overuse of the samples with positive-reward. So it is not difficult to explain the overestimation of Q-values at the beginning of training as depicted in Fig. 3(d).

Nevertheless, our PS-DQN algorithm also has some limitations. We can see that there is almost no improvement in the curve of average reward of Space_invaders reflecting in lower average rewards for some epochs and the instability of the training process induced by overusing of the transitions with positive-rewards. The experimental results indicate that our PS-DQN agent is considerably appropriate for these sensing and controlling tasks which could generate a mass of zero rewards except for some scattered positive rewards at the primary stage of learning, such as Breakout and Boxing.

5   Conclusion and Future Work

We presented a novel algorithm, called PS-DQN, by combining the deep Q-network with a prioritized sampling strategy. According to the analysis of our experimental results on four Atari 2600 games, we arrive at a conclusion that using prioritized sampling may lead to a faster and more stable learning process, and a better performance of scoring on some tested games. These preliminary results might provide empirical clues for further research, in particular developing an automatic way to adapt the hyper-parameter ρ on-line based on the distribution of transitions distinguished by the magnitude of rewards.

Acknowledgements. This work was funded by National Natural Science Foundation (61272005, 61303108, 61373094, 61502323, 61472262). We would also like to thank he reviewers for their helpful comments. Natural Science Foundation of Jiangsu (BK2012616), High School Natural Foundation of Jiangsu (13KJB520020), Key Laboratory of Symbolic Computation and Knowledge Engineering of Ministry of Education, Jilin University (93K172014K04), Suzhou Industrial application of basic research program part (SYG201308, SYG201422).

References

1. Mnih, V., Kavukcuoglu, K., Silver, D., et al.: Playing atari with deep reinforcement learning. In: Deep Learning Workshop of the 27th Advances in Neural Information Processing Systems, NIPS, Lake Tahoe (2013)
2. Mnih, V., Kavukcuoglu, K., Silver, D., et al.: Human-level control through deep reinforcement learning. Nature **518**(7540), 529–533 (2015)
3. Silver, D., Huang, A., Maddison, C.J., et al.: Mastering the game of Go with deep neural networks and tree search. Nature **529**(7587), 484–489 (2016)
4. Sutton, R.S., Barto, A.G.: Reinforcement Learning: An Introduction. MIT Press, Cambridge (1998)
5. Lin, L.J.: Reinforcement learning for robots using neural networks. Technical report, DTIC Document (1993)
6. Tsitsiklis, J.N., Van, R.B.: An analysis of temporal-difference learning with function approximation. IEEE Trans. Autom. Control **42**(5), 674–690 (1997)
7. Hinton, G.E., Srivastava, N., Krizhevsky, A., et al.: Improving neural networks by preventing co-adaptation of feature detectors. Comput. Sci. **3**(4), 212–223 (2012)
8. Bellemare, M.G., Naddaf, Y., Veness, J., et al.: The arcade learning environment: an evaluation platform for general agents. J. Artif. Intell. Res. **47**(1), 253–279 (2012)

Deep Inverse Reinforcement Learning by Logistic Regression

Eiji Uchibe[1,2(✉)]

[1] Department of Brain Robot Interface,
ATR Computational Neuroscience Laboratories,
2-2-2 Hikaridai, Seika-cho, Soraku-gun, Kyoto 619-0288, Japan
uchibe@atr.jp
[2] Neural Computation Unit, Okinawa Institute of Science and Technology
Graduate University, 1919-1 Tancha, Onna-son, Okinawa 904-0495, Japan

Abstract. This study proposes model-free deep inverse reinforcement learning to find nonlinear reward function structures. It is based on our previous method that exploits the fact that the log of the ratio between an optimal state transition and a baseline one is given by a part of reward and the difference of the value functions under linearly solvable Markov decision processes and reward and value functions are estimated by logistic regression. However, reward is assumed to be a linear function whose basis functions are prepared in advance. To overcome this limitation, we employ deep neural network frameworks to implement logistic regression. Simulation results show our method is comparable to model-based previous methods with less computing effort in the Objectworld benchmark. In addition, we show the optimal policy, which is trained with the shaping reward using the estimated reward and value functions, outperforms the policies that are used to collect data in the game of Reversi.

Keywords: Inverse Reinforcement Learning · Deep learning · Density ratio estimation · Logistic regression

1 Introduction

Inverse Reinforcement Learning (IRL), which is a method of estimating a reward function that can explain a given agent's behavior [8,16], provides a computational scheme to implement imitation learning. It is also a promising approach for understanding the learning processes of biological systems such as driving a vehicle [4,9] and playing table tennis [6] because the reward specifies the goal of the behavior.

Previously we developed IRL under the Linearly solvable Markov Decision Process (LMDP) [13] that directly estimates the state-dependent reward and the value function [14]. This method exploits the fact that the logarithm of the ratio between an optimal and a baseline state transition is represented by a state-dependent reward and the difference of the value functions under the LMDP framework and they are efficiently estimated by logistic regression to classify

© Springer International Publishing AG 2016
A. Hirose et al. (Eds.): ICONIP 2016, Part I, LNCS 9947, pp. 23–31, 2016.
DOI: 10.1007/978-3-319-46687-3_3

whether the data are sampled from the optimal transition probability. Unlike most previous IRL methods such as Maximum Entropy-based IRL (MaxEnt-IRL) [16], our IRL does not need to find an optimal policy for every iteration. However, most previous studies (including our method) use linear function approximators in which a set of basis functions are prepared manually.

Recently, Wulfmeier et al. proposed DeepIRL, which combined MaxEnt-IRL with a deep neural network architecture to find nonlinear reward functions [15]. However, their method suffers from the same two problems as MaxEnt-IRL. One is that their method is a model-based approach, and an environmental state transition probability is assumed to be known in advance. The other is that the optimal policy should be computed for every iteration step and it is computationally expensive. Finn et al. proposed a similar deep inverse optimal control method [3] based on MaxEnt-IRL and relative entropy-based IRL [2].

This paper extends our previous method by introducing deep learning frameworks to identify the nonlinear representation of reward and value functions. The application of deep learning frameworks is straightforward, and the network structure of binary classifiers is derived from the simplified Bellman equation under LMDP. In the Objectworld benchmark, our simulation results show that the performance of our model-free method resembles that of Wulfmeier's model-based method with less computing time. In addition, we show that appropriate reward can be successfully retrieved and that the optimal policy trained with the estimated reward outperforms policies that are used to collect data in the game of Reversi. Furthermore, learning speed can be improved by the estimated value function by the shaping reward theory [7].

2   Linearly Solvable Markov Decision Process

Let $\mathcal{X}$ and $\mathcal{U}$ respectively be continuous state and continuous action spaces. At time step t, a learning agent observes environmental current state $\boldsymbol{x}_t \in \mathcal{X}$ and executes action $\boldsymbol{u}_t \in \mathcal{U}$ that is sampled according to a stochastic policy $\pi(\boldsymbol{u}_t \mid \boldsymbol{x}_t)$. Consequently, an immediate reward $r(\boldsymbol{x}_t, \boldsymbol{u}_t)$ is given by the environment and the environment makes a state transition based on state transition probability $P_T(\boldsymbol{y}_t \mid \boldsymbol{x}_t, \boldsymbol{u}_t)$ from $\boldsymbol{x}_t$ to $\boldsymbol{y}_t = \boldsymbol{x}_{t+1} \in \mathcal{X}$ by executing the action $\boldsymbol{u}_t$. The goal of (forward) reinforcement learning is to construct an optimal policy $\pi(\boldsymbol{u} \mid \boldsymbol{x})$ that maximizes the given objective function. Several objective functions exist, and the most widely used one is a discounted sum of rewards given by

$$V(\boldsymbol{x}) = \mathbb{E}\left[\sum_{t=0}^{\infty} \gamma^t r(\boldsymbol{x}_t, \boldsymbol{u}_t)\right],$$

where $\gamma \in [0, 1)$ is called the discount factor. The optimal state value function for the discounted reward setting satisfies the following Bellman equation:

$$V(\boldsymbol{x}) = \max_{\boldsymbol{u}} \left[r(\boldsymbol{x}, \boldsymbol{u}) + \gamma \mathbb{E}_{\boldsymbol{y} \sim P_T(\cdot|\boldsymbol{x}, \boldsymbol{u})}\left[V(\boldsymbol{y})\right]\right]. \tag{1}$$

Eq. (1) is the nonlinear equation due to the max operator.

The Linearly solvable Markov Decision Process (LMDP), also known as KL-control, simplifies Eq. (1) under some assumptions [13]. LMDP's key trick is to directly optimize the state transition probability instead of the policy. More specifically, two conditional probability density functions are introduced. One is the controlled probability denoted by $\pi(y \mid x)$, which can be interpreted as an optimal state transition. The other is the uncontrolled probability denoted by $b(y \mid x)$, which can be regarded as an innate state transition of the target system. Theoretically, $b(y \mid x)$ is arbitrary and can be constructed by $b(y \mid x) = \int P_T(y \mid x, u)b(u \mid x)du$, where $b(u \mid x)$ is a uniformly random policy.

Then the reward function is restricted to the following form:

$$r(x, u) = q(x) - \frac{1}{\beta}\mathrm{KL}(\pi(\cdot \mid x) \parallel b(\cdot \mid x)), \tag{2}$$

where $q(x)$, β, and $\mathrm{KL}(\pi(\cdot \mid x) \parallel b(\cdot \mid x))$ respectively denote a state-dependent reward function, a positive inverse temperature, and the Kullback Leibler (KL) divergence between the controlled and uncontrolled state transition densities. In this case, the Bellman equation (1) is written as

$$V(x) = q(x) + \max_{\pi} \int \pi(y \mid x) \left[-\frac{1}{\beta} \log \frac{\pi(y \mid x)}{b(y \mid x)} + \gamma V(y) \right] dy.$$

We can maximize the right hand side of the above equation by applying the Lagrangian method [13] to obtain the following solution:

$$\exp(\beta V(x)) = \exp(\beta q(x)) \int b(y \mid x) \exp(\beta\gamma V(y))dy. \tag{3}$$

The optimal controlled probability for the discounted reward setting is given by

$$\pi(y \mid x) = \frac{b(y \mid x) \exp(\beta\gamma V(y))}{\int b(y' \mid x) \exp(\beta\gamma V(y'))dy'}. \tag{4}$$

Note that Eq. (3) remains nonlinear even though desirability function $Z(x) = \exp(\beta V(x))$ is introduced because of the existence of discount factor γ.

3   Deep Inverse Reinforcement Learning

3.1   Bellman Equation for IRL

From Eqs. (3) and (4), we derive the following critical relation for the discounted reward setting:

$$\frac{1}{\beta} \log \frac{\pi(y \mid x)}{b(y \mid x)} = q(x) + \gamma V(y) - V(x). \tag{5}$$

Equation (5) plays an important role in our IRL algorithms. Similar equations can be derived for average-reward, first-exit, and finite-horizon problems. Note that the right hand side of Eq. (5) is not a temporal difference error because $q(x)$ is the state-dependent part of the reward function in Eq. (2).

Applying the Bayes rule into Eq. (5) yields the following:

$$\log \frac{\pi(\boldsymbol{x}, \boldsymbol{y})}{b(\boldsymbol{x}, \boldsymbol{y})} = \log \frac{\pi(\boldsymbol{x})}{b(\boldsymbol{x})} + \beta q(\boldsymbol{x}) + \gamma \beta V(\boldsymbol{y}) - \beta V(\boldsymbol{x}). \tag{6}$$

Our goal is to estimate $\beta q(\boldsymbol{x})$ and $\beta V(\boldsymbol{x})$ from the observed data, and we assume two datasets of state transitions. One is $\mathcal{D}^\pi$ from the controlled probability:

$$\mathcal{D}^\pi = \{(\boldsymbol{x}_i^\pi, \boldsymbol{y}_i^\pi)\}_{i=1}^{N^\pi}, \quad \boldsymbol{y}_i^\pi \sim \pi(\cdot \mid \boldsymbol{x}_i^\pi),$$

where N^π denotes the number of data points. In the standard IRL setting, $\mathcal{D}^\pi$ is interpreted as data from experts to be investigated. The other is a dataset from the uncontrolled probability:

$$\mathcal{D}^b = \{(\boldsymbol{x}_j^b, \boldsymbol{y}_j^b)\}_{j=1}^{N^b}, \quad \boldsymbol{y}_j^b \sim b(\cdot \mid \boldsymbol{x}_j^b),$$

where N^b denotes the number of data points. We are interested in estimating ratios $\pi(\boldsymbol{x})/b(\boldsymbol{x})$ and $\pi(\boldsymbol{x}, \boldsymbol{y})/b(\boldsymbol{x}, \boldsymbol{y})$ from $\mathcal{D}^b$ and $\mathcal{D}^\pi$.

3.2   LogReg-IRL: Logistic Regression-Based IRL

This subsection shows how Eq. (6) is used to estimate $\beta q(\boldsymbol{x})$ and $\beta V(\boldsymbol{x})$. LogReg, which is a density estimation method using logistic regression [1,11], is appropriate to estimate the log ratio of the following two densities: $\log \pi(\boldsymbol{x})/b(\boldsymbol{x})$ and $\log \pi(\boldsymbol{x}, \boldsymbol{y})/b(\boldsymbol{x}, \boldsymbol{y})$. First, to estimate $\log \pi(\boldsymbol{x})/b(\boldsymbol{x})$, assign a selector variable $\eta = -1$ to the samples from the uncontrolled probability and $\eta = 1$ to the samples from the controlled probability:

$$b(\boldsymbol{x}) = \Pr(\boldsymbol{x} \mid \eta = -1), \quad \pi(\boldsymbol{x}) = \Pr(\boldsymbol{x} \mid \eta = 1).$$

The density ratio can be represented by applying the Bayes rule:

$$\begin{aligned}
\frac{\pi(\boldsymbol{x})}{b(\boldsymbol{x})} &= \left(\frac{\Pr(\eta = 1 \mid \boldsymbol{x}) \Pr(\boldsymbol{x})}{\Pr(\eta = 1)} \right) \left(\frac{\Pr(\eta = -1 \mid \boldsymbol{x}) \Pr(\boldsymbol{x})}{\Pr(\eta = -1)} \right)^{-1} \\
&= \frac{\Pr(\eta = -1)}{\Pr(\eta = 1)} \frac{\Pr(\eta = 1 \mid \boldsymbol{x})}{\Pr(\eta = -1 \mid \boldsymbol{x})}.
\end{aligned}$$

The first ratio, $\Pr(\eta = -1)/\Pr(\eta = 1)$, is estimated by N^b/N^π, and the second ratio is computed after estimating the conditional probability $\Pr(\eta \mid \boldsymbol{x})$ by a logistic regression classifier:

$$\Pr(\eta \mid \boldsymbol{x}) = \sigma\left(\eta f_x(\boldsymbol{x}; \boldsymbol{w}_x)\right),$$

where $\sigma(x) = 1/(1 + \exp(-x))$ is a sigmoid function and $f_x(\boldsymbol{x}; \boldsymbol{w}_x)$ denotes a deep neural network function parameterized by the weight vector $\boldsymbol{w}_x$. Note that the logarithm of the density ratio is given by

$$\log \frac{\pi(\boldsymbol{x})}{b(\boldsymbol{x})} = f_x(\boldsymbol{x}; \boldsymbol{w}_x) + \log \frac{N^b}{N^\pi}. \tag{7}$$

Network weights $\boldsymbol{w}_x$ can be estimated by the backpropagation whose objective function is given by the negative regularized log-likelihood. The closed-form solution is not derived, but it is possible to minimize it efficiently by standard nonlinear optimization methods such as backpropagation.

Next, $\log \pi(\boldsymbol{x}, \boldsymbol{y})/b(\boldsymbol{x}, \boldsymbol{y})$ is estimated by Eq. (6) in the same way. Nonlinear function approximators for $\beta q(\boldsymbol{x})$ and $\beta V(\boldsymbol{x})$ are respectively introduced by

$$\beta q(\boldsymbol{x}) \approx f_q(\boldsymbol{x}; \boldsymbol{w}_q), \quad \beta V(\boldsymbol{x}) \approx f_V(\boldsymbol{x}; \boldsymbol{w}_V), \tag{8}$$

where $f.(\boldsymbol{x}; \boldsymbol{w}.)$ denotes the deep neural network function parameterized by the weights $\boldsymbol{w}.$ and subscripts q and V respectively represent the reward and state value functions. By substituting Eqs. (7) and (8) into Eq. (6), we obtain the following relationship:

$$\log \frac{\pi(\boldsymbol{x}, \boldsymbol{y})}{b(\boldsymbol{x}, \boldsymbol{y})} = f_x(\boldsymbol{x}; \boldsymbol{w}_x) + f_q(\boldsymbol{x}; \boldsymbol{w}_q) + \gamma f_V(\boldsymbol{y}; \boldsymbol{w}_V) - f_V(\boldsymbol{x}; \boldsymbol{w}_V) + \log \frac{N^b}{N^\pi}.$$

The above equation is also interpreted as a density ratio estimation problem, and network parameters $\boldsymbol{w}_q$ and $\boldsymbol{w}_V$ are estimated by logistic regression in which the classifier is given by

$$\Pr(\eta \mid \boldsymbol{x}, \boldsymbol{y}) = \sigma \left(\eta \left(f_x(\boldsymbol{x}; \boldsymbol{w}_x) + f_q(\boldsymbol{x}; \boldsymbol{w}_q) + \gamma f_V(\boldsymbol{y}; \boldsymbol{w}_V) - f_V(\boldsymbol{x}; \boldsymbol{w}_V) \right) \right). \tag{9}$$

Note that the inverse temperature β is not estimated as an independent parameter. The parameter vectors can be optimized by standard logistic regression algorithms.

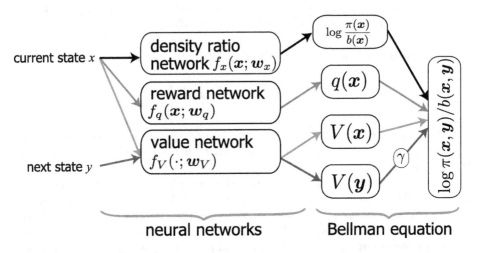

Fig. 1. Proposed network architecture for inverse reinforcement learning that consists of three networks: density ratio, reward, and value function. Then the Bellman equation (6) is computed from the outputs of the three networks.

Figure 1 shows our proposed deep neural network. The input consists of current state x and next state y, and the output consists of label η that addresses whether (x, y) are given from $\mathcal{D}^\pi$ or $\mathcal{D}^b$. The network has three sub-networks: density-ratio network $f_x(x; w_x)$, reward network $f_q(x; w_q)$, and value network $f_V(w; w_V)$. Note that state y is given only to the value network. After the sub-networks compute the log of the density ratio, the reward, and the value functions, the log of the density ratio is computed by the Bellman equation (6) and used to compute the classifier's probability (9).

4   Experiments

4.1   Objectworld Benchmark

To validate our method, we select the ObjectWorld benchmark [5] because it is also used by Wulfmeier *et al.* [15]. It is a 32×32 grid of states with five actions per state: motions in all four directions and staying in place. State x is described by a two-dimensional vector where each dimension represents the minimum distance to an object of one of two colors. Each action has a 30 % chance of moving in a different random direction. The reward function is positive for cells that are both within a distance of 3 of color 1 and a distance of 2 of color 2, negative if only within a distance of 3 of color 1 and zero otherwise. See task description [5] for more details.

We created two environments: training and transfer. The optimal policy is computed under the true reward in the training environment and its policy including 30 % random actions is used to collect optimal dataset $\mathcal{D}^\pi$, while a random policy is used to collect baseline dataset $\mathcal{D}^b$. In this experiment, our method is implemented by a feedforward neural network with two hidden layers with rectified linear units and one linear output layer for the density ratio, reward, and value networks. As a comparison, the original deep IRL by Wulfmeier *et al.* and the model-free Wulfmeier's method are evaluated where the state transition probability is estimated by $\mathcal{D}^b$.

In accordance with our previous study, we evaluated the performances of the proposed method and DeepIRL by the expected value difference scores, which are measures of the sub-optimality of the learned policy under the true reward. It is the difference between the expected sum of the rewards obtained for the optimal policy given the true reward and that for the optimal policy based on the estimated reward.

Figures 2(a) and (b) respectively show the scores in the training and test environments. The score of our model-free proposed method was almost the same as that of the model-based Wulfmeier's method in the training and transfer environments, while the model-free Wulfmeier's method performed poorly because the estimated environmental state transition probability was less accurate. Note that our method is model-free and does not require the state transition probability of the environment. Figure 2(c) compares the computing time, and our method found a solution much faster than the other two methods because it does not need to solve forward reinforcement learning problems.

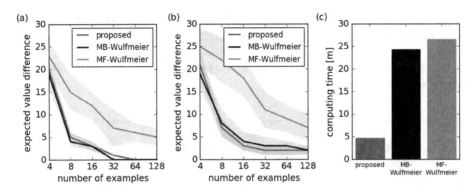

Fig. 2. Results of Objectworld benchmark: (a) Training case where identical environment is used to collect test data; (b) Transfer case where a new environment with different object configurations is created to collect test data; (c) Computing time.

4.2 Reversi

Reversi (a.k.a. Othello), which is a deterministic, perfect information, zero-sum game for two players, has been studied by the AI community. The game's goal is to control a majority of the pieces at the end of the game by forcing as many of your opponent's pieces to be turned over on an 8×8 board as possible. A single move might change up to 20 pieces, and an average of 60 moves are needed to complete the game. Although reinforcement learning has been successfully applied to the game of GO, which is much more complicated than Reversi, it took a huge amount of computing time to find an optimal policy because a sparse reward was used [10]. This provides motivation for finding a dense reward structure by IRL.

To collect optimal dataset $\mathcal{D}^\pi$, we prepared three stationary policies (RANDOM, HEUR, and COEV) used in previous studies [12], and every policy repeatedly plays against every other. Then the state transitions are retrieved from the game trajectories of the winners. On the other hand, baseline dataset $\mathcal{D}^b$ is constructed by retrieving the state transitions from the trajectories in which two RANDOM policies play the game. The numbers of samples are $|\mathcal{D}^\pi| = |\mathcal{D}^b| = 10^5$.

In this experiment, we used the same type of neural networks used in Sect. 4.1. The input is given by a 64-node vector $x \in \{-1, 0, +1\}^{64}$, where 0 represents an empty square, and $+1$ and -1 respectively represent the black and white pieces. Every hidden layer has 100 nodes.

We evaluated the Q-learning method trained by the following rewards and the three stationary policies against all other methods using the winning rate as performance measures: (1) $q(x)$, estimated by the proposed method, (2) shaping reward $q(x) + \gamma V(y) - V(x)$, estimated by the proposed method, (3) reward estimated by the model-free Wulfmeier's method, and (4) sparse reward, where $1, 0.5$, and -1 respectively stand for win, tie, and lose. Figures 3(a) and (c) show that Q-learning trained with the estimated reward outperformed RANDOM and Q-learning trained with the sparse reward at the beginning and took about

3×10^4 plays to defeat HEUR and COEV. Figure 3(b) shows that the learning speed was significantly improved by the shaping reward. For example, the Q-learning agent trained with the reward estimated by Wulfmeier's method failed to collect samples to win the game because the agent trained with the shaping reward learned faster.

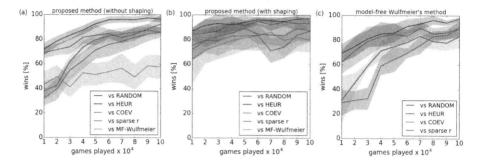

Fig. 3. Learning performance with estimated reward: (a) Agent trained with reward $q(\boldsymbol{x})$ estimated by our method plays with three stationary policies and two learning agents trained by sparse reward and reward estimated by model-free Wulfmeier's method; (b) Agent trained with shaping reward $q(\boldsymbol{x}) + \gamma V(\boldsymbol{y}) - V(\boldsymbol{x})$ estimated by our method plays with other agents; (c) Agent trained with reward estimated by model-free Wulfmeier's method plays with other agents.

5   Conclusion

We integrated our IRL method with deep neural networks for the automatic extraction of features. Our model-free method found rewards that produced comparable performance to the model-based Wulfmeier's method with less computing time in the ObjectWorld benchmark. Next, our method and the model-free Wulfmeier's method were utilized to estimate the reward function in the game of Reversi. Since our method estimates the value function at the same time, it is used as the potential function for reward shaping to accelerate speed of learning. The optimal policy trained with the shaping reward outperformed three stationary policies and optimal policies trained by Wulfmeier's method. More systematic investigations on such deep networks as activation functions and learning algorithms are needed for future study.

Acknowledgements. This paper is based on results obtained from a project commissioned by the New Energy and Industrial Technology Development Organization (NEDO).

References

1. Bickel, S., Brückner, M., Scheffer, T.: Discriminative learning under covariate shift. J. Mach. Learn. Res. **10**, 2137–2155 (2009)
2. Boularias, A., Kober, J., Peters, J.: Relative entropy inverse reinforcement learning. In: Proceedings of the 14th International Conference on Artificial Intelligence and Statistics, vol. 15 (2011)
3. Finn, C., Levine, S., Abbeel, P.: Guided cost learning: deep inverse optimal control via policy optimization. In: Proceedings of the 33rd International Conference on Machine Learning, pp. 49–58 (2016)
4. Kuderer, M., Gulati, S., Burgard, W.: Learning driving styles for autonomous vehicles from demonstration. In: Proceedings of IEEE International Conference on Robotics and Automation, pp. 2641–2646 (2015)
5. Levine, S., Popović, Z., Koltun, V.: Nonlinear inverse reinforcement learning with gaussian processes. Adv. Neural Inf. Process. Syst. **24**, 19–27 (2011)
6. Muelling, K., Boularias, A., Mohler, B., Schölkopf, B., Peters, J.: Learning strategies in table tennis using inverse reinforcement learning. Biol. Cybern. **108**(5), 603–619 (2014)
7. Ng, A.Y., Harada, D., Russel, S.: Policy invariance under reward transformations: theory and application to reward shaping. In: Proceedings of the 16th International Conference on Machine Learning (1999)
8. Ng, A.Y., Russell, S.: Algorithms for inverse reinforcement learning. In: Proceedings of the 17th International Conference on Machine Learning, pp. 663–670 (2000)
9. Shimosaka, M., Nishi, K., Sato, J., Kataoka, H.: Predicting driving behavior using inverse reinforcement learning with multiple reward functions towards environmental diversity. In: Proceedings of IEEE Intelligent Vehicles Symposium, pp. 567–572 (2015)
10. Silver, D., et al.: Mastering the game of go with deep neural networks and tree search. Nature **529**(7587), 484–489 (2016)
11. Sugiyama, M., Suzuki, T., Kanamori, T.: Density Ratio Estimation in Machine Learning. Cambridge University Press, Cambridge (2012)
12. Szubert, M., Jaśkowski, W., Krawiec, K.: On scalability, generalization, and hybridization of coevolutionary learning: a case study for Othello. IEEE Trans. Comput. Intell. AI Games **5**(3), 214–226 (2013)
13. Todorov, E.: Efficient computation of optimal actions. PNAS **106**(28), 11478–11483 (2009)
14. Uchibe, E., Doya, K.: Inverse reinforcement learning using dynamic policy programming. In: Proceedings of the 4th IEEE International Conference on Development and Learning and on Epigenetic Robotics, pp. 222–228 (2014)
15. Wulfmeier, M., Ondrúška, P., Posner, I.: Maximum entropy deep inverse reinforcement learning. In: NIPS Deep Reinforcement Learning Workshop (2015)
16. Ziebart, B.D., Maas, A., Bagnell, J.A., Dey, A.K.: Maximum entropy inverse reinforcement learning. In: Proceedings of the 23rd AAAI Conference on Artificial Intelligence (2008)

Parallel Learning for Combined Knowledge Acquisition Model

Kohei Henmi$^{(\boxtimes)}$ and Motonobu Hattori

Interdisciplinary Graduate School of Medicine, Engineering and Agriculture,
University of Yamanashi, Kofu, Yamanashi, Japan
{g16tk012,m-hattori}@yamanashi.ac.jp

Abstract. In this paper, we propose a novel learning method for the combined knowledge acquisition model. The combined knowledge acquisition model is a model for knowledge acquisition in which an agent heuristically find new knowledge by integrating existing plural knowledge. In the conventional model, there are two separate phases for combined knowledge acquisition: (a) solving a task with existing knowledge by trial and error and (b) learning new knowledge based on the experience in solving the task. However, since these two phases are carried out serially, the efficiency of learning was poor. In this paper, in order to improve this problem, we propose a novel knowledge acquisition method which realizes two phases simultaneously. Computer simulation results show that the proposed method much improves the efficiency of learning new knowledge.

Keywords: Direct-Vision-Based reinforcement learning · Combined knowledge acquisition model

1 Introduction

Given a task which has been never experienced before, we humans can manage to solve it by combining existing knowledge by trial and error. Then, once the task is successfully solved, the experience becomes new knowledge and will be used for other tasks later. Such ability to acquire knowledge is indispensable for intelligent systems like a robot. We have already proposed such a knowledge acquisition model based on neural networks and reinforcement learning. The learning of the model consists of two phases. In the first phase, it uses plural knowledge in corporation in order to solve an inexperienced task. After the task has been solved, new knowledge is constructed by integrating plural knowledge used into one in the second phase. We have shown that our model gradually becomes able to solve complicated tasks as it acquires new integrated knowledge, and it can be applied to real environment [1–3]. However, since the above two phases are carried out serially, it takes a long time to build new knowledge. Above all things, it is likely that we humans carry out one phase in parallel with the other. So, the objective of our research is to develop a learning method for

© Springer International Publishing AG 2016
A. Hirose et al. (Eds.): ICONIP 2016, Part I, LNCS 9947, pp. 32–39, 2016.
DOI: 10.1007/978-3-319-46687-3_4

the combined knowledge acquisition model which realizes learning of two phases simultaneously.

The rest of this paper is organized as follows. In Sect. 2, we briefly review our combined knowledge acquisition model. In Sect. 3, we propose parallel learning method for the model. Then, computer simulation results are shown in Sect. 4. Finally, we make some conclusions in Sect. 5.

2   Combined Knowledge Acquisition Model

Here, we explain the Direct-Vision-Based (DVB) reinforcement learning [4] proposed by Shibata et al. first, which is used in the learning of the combined knowledge acquisition model. Then, we explain the architecture and learning algorithm of the combined knowledge acquisition model.

2.1   Direct-Vision-Based Reinforcement Learning

Shibata et al. have proposed a reinforcement learning called *Direct-Vision-Based (DVB) reinforcement learning* for the box pushing task by a mobile robot [4]. In the DVB reinforcement learning, only raw visual sensor signals are given to a multilayer neural network, and which is trained by Back Propagation using training signals that is generated based on reinforcement learning. They have shown that the mobile robot could learn to go and push a box directly from only visual image without any image pre-processing, control methods and task knowledge. That is, the DVB reinforcement learning is inherently a very strong technique which can learn the whole process from sensors to motors including recognition of environment, attention, memory, planning, control and so on. So, we have employed the DVB reinforcement learning in our combined knowledge acquisition model.

The learning of the DVB reinforcement learning is based on actor-critic architecture where actor (action command generator) and critic (state value generator) are composed of the output layer of a multilayer neural network. Namely, the hidden layers are commonly used by both actor and critic.

In the learning of the critic, Temporal Difference (TD) error $\hat{r}_t$ is used

$$\hat{r}_t = r_t + \gamma P(s_t) - P(s_{t-1}) \tag{1}$$

where r_t is a reward, γ is a discount factor, s_t is a state vector, and $P(s_t)$ denotes a state value. The state value at $t-1$, $P(s_{t-1})$ is trained by

$$P^T(s_{t-1}) = P(s_{t-1}) + \hat{r}_t = r_t + \gamma P(s_t) \tag{2}$$

where $P^T(s_{t-1})$ denotes the training signal for the state value. On the other hand, the actor output vector $a(s_{t-1})$ is trained by

$$a^T(s_{t-1}) = a(s_{t-1}) + \hat{r}_t \mathbf{rnd}_{t-1} \tag{3}$$

where $\boldsymbol{a}^T(\boldsymbol{s}_{t-1})$ denotes the training signal for the actor vector, and $\mathbf{rnd}_{t-1}$ is a random vector from the uniform distribution for trial and error. By using Eqs. (2) and (3), the multilayer neural network is trained by Back Propagation algorithm. As the learning progresses, actor vector becomes to gain more state value.

2.2 Architecture and Learning Algorithm

Figure 1 shows the architecture of the conventional combined knowledge acquisition model. Assume that there are N knowledge, each of which is represented by a multilayer neural network learned by DVB reinforcement learning. That is, each multilayer neural network has been constructed to solve a particular task, and we regard it as a piece of knowledge.

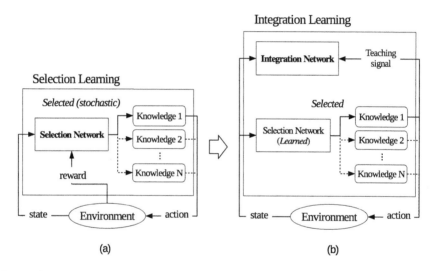

(a)   (b)

Fig. 1. Architecture of the conventional combined knowledge acquisition model. (a) Selection learning is followed by (b) integration learning.

In the conventional combined knowledge acquisition model, given an inexperienced task, selection learning is carried out first, in which the selection network is learned by DVB reinforcement learning so that appropriate knowledge can be selected for the state of the environment (Fig. 1(a)). The selection network selects the knowledge i which takes the maximum value of the sum of the output value $a_i(\boldsymbol{s}_t)$ and the random value $\mathrm{rnd}_{t,i}$, where $\mathrm{rnd}_{t,i}$ is the ith element of $\mathbf{rnd}_t$. The actor output $a_i(\boldsymbol{s}_{t-1})$ is trained by

$$a_i^T(\boldsymbol{s}_{t-1}) = \begin{cases} a_i(\boldsymbol{s}_{t-1}) + \hat{r}_t \mathrm{rnd}_{t-1,i} & \text{if} \quad i = selected\_action \\ a_i(\boldsymbol{s}_{t-1}) & \text{otherwise} \end{cases} \qquad (4)$$

where $a_i^T(s_{t-1})$ denotes the training signal for the actor output. The selection network is trained by Back Propagation based on the training signals, Eqs. (2) and (4).

After the task success rate by the selection network has become sufficiently high, integration learning is carried out (Fig. 1(b)). In this phase, the action produced by the selection network is learned and integrated in the integration network as follows:

(1) Observe the state of the environment, s_t.
(2) Input the state of the environment s_t to the selection network which has been already learned, and then select a piece of knowledge among N knowledge according to the output of the selection network.
(3) Execute the corresponding network for selected knowledge, that is, s_t is given to the network and actor vector $a(s_t)$ is obtained as output.
(4) Input s_t to the integration network, and let it learn with $a(s_t)$ as teaching vector by Back Propagation.
(5) Act against the environment with $a(s_t)$. This causes the change of the state of the environment. Go to (1).

Repeating (1) to (5) until a certain criterion is satisfied, combined knowledge is integrated into the integration network. The obtained integration network is added to the pool of knowledge as the $N + 1$th knowledge for the future use. One of advantages to integrate plural knowledge into a single multilayer neural network like this is that its action becomes much more seamless than that by the selection network [1–3]. Furthermore, accumulating knowledge in this manner, it becomes able to solve more complicated tasks which can not be solved without combination of knowledge [2,3].

3   Parallel Learning for Combined Knowledge Acquisition Model

Since the learning of the conventional combined knowledge acquisition model requires two phases, it takes a long time to build new integrated knowledge. From engineering point of view, the learning should be much more efficient. Moreover, it is unlikely that such a serial manner of knowledge acquisition is executed in our brain. So, we propose a new learning method for the model which enables both selection learning and integration learning simultaneously.

Figure 2 shows the architecture of the proposed parallel learning. Although it is almost identical to Fig. 1(b), the fundamental difference between Figs. 2 and 1(b) is that the selection network of the proposed method has not been learned yet. Therefore, there is one big concern about parallel learning, which is that teacher signals coming from the selection network is not reliable especially in the beginning of learning.

In order to deal with this problem, we propose a learning method for the integration network, which is based on TD error $\hat{r}_t$ (Eq. (1)). TD error $\hat{r}_t$ is a value indicating good or bad of state transition from s_{t-1} to s_t. If $\hat{r}_t$ is positive,

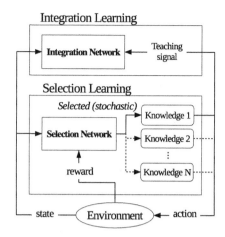

Fig. 2. Architecture of the proposed parallel learning.

it shows that the selection of knowledge by the selection network at $t - 1$ was good. Conversely, if $\hat{r}_t$ is negative, the selection was bad. Here, we propose two learning method for the integration network:

(1) The integration network is learned only when $\hat{r}_t > 0$.
(2) The integration network is learned with positive learning rate when $\hat{r}_t > 0$, and with negative learning rate when $\hat{r}_t \leq 0$.

Hereafter, we indicate (1) as *TD-p learning* and (2) as *TD-pn learning*.

4   Computer Simulation Results

4.1   Experimental Setup

In order to examine the effectiveness of the proposed parallel learning, we used a small mobile robot called *Khepera* in a robot simulator, Webots Ver.7.2.0. Khepera is mounted eight light sensors, eight infrared sensors, and 64-dimensional visual sensor. Each of these 80 dimensional sensor values was normalized into a real number between 0.0 and 1.0, and applied to neural networks as a state vector. Khepera is controlled by giving the rotation speed of the left and right wheels. Therefore, the output vector of the integration network is 2 dimensions.

In order to evaluate the performance of the proposed parallel learning, we set *adaptive obstacle avoidance task*, in which the robot has to go straight as much as possible while avoiding obstacles and walls to the left or to the right adaptively in the environment. Before performing this task, *left-turn obstacle avoidance task* and *right-turn obstacle avoidance task* were separately carried out. For example, in the left-turn obstacle avoidance task, the robot has to go straight as much as possible while avoiding obstacles and wall to the left. Then, obtained two pieces of integrated knowledge were stored in the pool of knowledge. In addition, we prepared twelve primitive actions shown in Fig. 3 as fundamental knowledge (In each fundamental knowledge, the rotation speed of wheels was fixed to perform specific action. That is, it is not represented by a neural network). Therefore, fourteen pieces of knowledge in total were available for the adaptive obstacle avoidance task. The experimental environment is shown in Fig. 4. The action area is 200 cm × 200 cm which is surrounded by a height of 10 cm white wall, and in which 40 obstacles were placed randomly in each trial except around the initial position of Khepera (indicated by slanted lines in Fig. 4). The size of each obstacle is 10 cm × 10 cm × 10 cm. The initial position of Khepera was fixed at the center of the environment, and the initial angle was chosen randomly in the range of 360°.

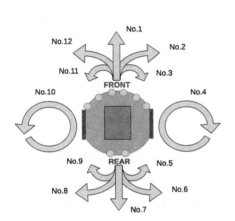

Fig. 3. Khepera and 12 pieces of primitive knowledge.

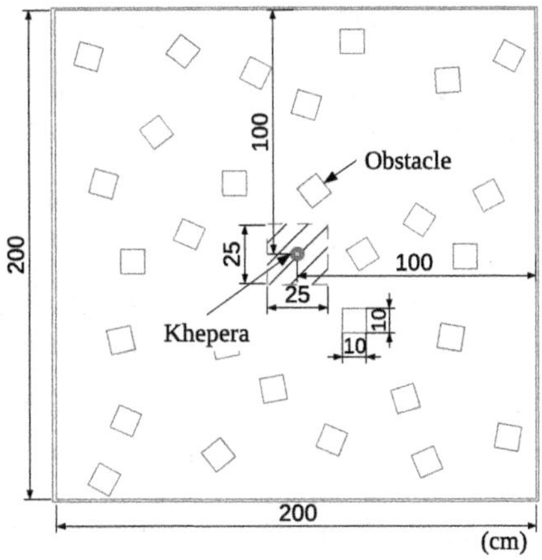

Fig. 4. Experimental environment.

Table 1 shows parameters and conditions used in this experiment.

In each simulation, we carried out 100 times of test trial every 10 times of learning trials, and regarded as the learning had been completed when the success rate and going-straight rate became more than 95 % and 0.9, respectively. The going-straight rate β was defined by

$$\beta = \frac{\min\left(\sum_{i=1}^{100} w_i^L, \sum_{i=1}^{100} w_i^R\right)}{\max\left(\sum_{i=1}^{100} w_i^L, \sum_{i=1}^{100} w_i^R\right)} \tag{5}$$

where $w_i^{L(R)}$ denotes the rotation distance of the left (right) wheel at the ith trial. So, the going-straight rate value of 1 indicates the robot went straight during test trials, and 0 means that the robot was rotating by centering one of the wheels.

Table 1. Parameters and conditions used in the adaptive obstacle avoidance task.

No. of hidden layers	3
No. of hidden neurons	40
Learning rate for BP	0.1(TD-p, TD-pn), -0.005 (TD-pn)
No. of trials in learning	10
No. of trials in test	100
Reward	0.05
Penalty	-0.9
Conditions for reward	Knowledge of going-straight is selected
	Knowledge of right-turn is selected when obstacle/wall is in the left
	Knowledge of left-turn is selected when obstacle/wall is in the right
Condition for successful trial	Elapsed 2000 steps without failure
Condition for failure	Collision with obstacle/wall
Required success rate (α)	95 %
Required going-straight rate (β)	0.9

Table 2. Averaged trails required for learning to satisfy criteria (α and β) based on 10 simulations.

	Conventional	TD-p	TD-pn	without TD
Selection & Integration	-	239	181	2205
Selection	246	-	-	-
Integration	365	-	-	-
Total	611 ± 235.4	239 ± 78.8	181 ± 46.1	2205 ± 1151.9

Table 2 shows the averaged trials required for learning until the criteria (α and β) were satisfied based on 10 simulations.

There was significant difference between the total trials required for the conventional model and those for the proposed TD-p and TD-pn learning ($p < 0.05$), and also between those for the parallel learning without using TD error and proposed methods ($p < 0.05$). This result shows that the integration network can acquire integrated knowledge rapidly by using the proposed parallel learning. Surprisingly, the learning of the integration network by the proposed methods tends to be faster than that of the selection network by the conventional method. This result shows that the sign of TD error is an excellent guide for the integration learning.

Although there is not significant difference between TD-p and TD-pn, it seems that TD-pn is faster than TD-p. This is because in a certain situation an appropriate action and inappropriate one exist at opposite poles in this task. That is, if turning to the left doesn't yield good state transition, it means that the robot should have turned to the right. So, TD-pn learning worked very well in obstacle avoidance task. We think the effectiveness of the TD-pn learning depends on the characteristics of a task.

5   Conclusions

In this paper, in order to improve the efficiency of learning and validity of the combined knowledge acquisition model, we have proposed parallel learning methods. The proposed methods enable the learning of selection network and integration network simultaneously by using TD error. We have applied the proposed methods to obstacle avoidance task and shown that they could significantly reduce required trials for learning in comparison with the conventional learning method.

In the future research, we will apply the proposed parallel learning method to more complicated tasks.

References

1. Yabe, T., Hattori, M.: Combined knowledge acquisition model by integration of existing knowledge (in Japanese). In: Proceedings of Forum on Information Technology, H-005, pp. 399–400 (2006)
2. Yabe, T., Hattori, M.: Research on characteristic and real environment applicability of combined knowledge acquisition model by integration of knowledge. In: Proceedings of 70th National Convention of Information Processing Society of Japan, 5V–6, 2, pp. 283–284 (2008)
3. Shikina, S., Hattori, M.: Learning for selection of existing knowledge in combined tasks (in Japanese). In: Proceedings of 72th National Convention of Information Processing Society of Japan, 2U–8, 2, pp. 239–340 (2010)
4. Shibata, K., Iida, M.: Acquisition of box pushing by direct-vision-based reinforcement learning. In: SICE 2003 Annual Conference, vol. 3, pp. 2322–2327 (2003)

Emergence of Higher Exploration in Reinforcement Learning Using a Chaotic Neural Network

Yuki Goto[✉] and Katsunari Shibata

Department of Electrical and Electronic Engineering, Oita University,
700 Dannoharu, Oita 870-1192, Japan
iwishdayss@gmail.com, shibata@oita-u.ac.jp

Abstract. Aiming for the emergence of higher functions such as "logical thinking", our group has proposed completely novel reinforcement learning where exploration is performed based on the internal dynamics of a chaotic neural network. In this paper, in the learning of an obstacle avoidance task, it was examined that in the process of growing the dynamics through learning, the level of exploration changes from "lower" to "higher", in other words, from "motor level" to "more abstract level". It was shown that the agent learned to reach the goal while avoiding the obstacle and there is an area where the agent looks to pass through the right side or left side of the obstacle randomly. The result shows the possibility of the "higher exploration" though the agent sometimes collided with the obstacle and was trapped for a while as learning progressed.

Keywords: Reinforcement learning · Chaotic neural network · Higher exploration · Emergence of intelligence · Obstacle avoidance

1 Introduction

Our group has pointed out the difficulty of developing a program by hand for such massively parallel and highly flexible computation that our brain is doing, and proposed the approach that a Neural Network (NN) is responsible for the whole process from sensors to motors and various functions emerge in the NN through Reinforcement Learning (RL) [1,2]. Recent excellent performance of "Deep Learning" especially in the area of recognition [3] and the surprising result in the TV games by combining it with RL [4] are thanks to its emergence ability of useful internal representations, and support the significance of our approach.

Because higher functions such as "memory", "prediction", "logical thinking" and so on, need to cope with dynamics, a Recurrent Neural Network (RNN) is used on behalf of a layered NN. The emergence of "memory" or "prediction" has been confirmed in a simple task [5,6]. However, the learning of a task requiring multiple state transitions is not easy [7], and the emergence of what we can call "logical thinking" has not been shown yet.

© Springer International Publishing AG 2016
A. Hirose et al. (Eds.): ICONIP 2016, Part I, LNCS 9947, pp. 40–48, 2016.
DOI: 10.1007/978-3-319-46687-3_5

Therefore, we have felt the need of another approach in which desired dynamics is not obtained from scratch in a non-chaotic "silent" NN, but is reformed from chaotic dynamics through learning in a chaotic NN. We have also thought that "exploration" should be considered as a function based on internal dynamics as well as "memory" or "prediction", and random-like "exploration" is expected to grow up in "logical thinking" through learning. According to the hypothesis, we have proposed a completely novel RL where exploration is performed based on the internal chaotic dynamics without adding external random numbers [8].

On the other hand, recently, the ability of reservoir computing has been unveiled, and it was surprising that complicated dynamic patterns are easily learned using a chaotic NN by FORCE Learning [9]. In addition, it was shown that by adding exploration noises from the outside, a chaotic NN can learn complicated dynamic patterns based on a reward signal without giving any target signal directly [10]. From the above ability of chaotic NNs, RL using a chaotic NN is expected to develop greatly hereafter.

Authors thought that in the process of growing from "exploration" to "logical thinking", the level of exploration changes from "lower", which is motor-level, to "higher", which is more abstract level. For example in a forked road, we don't move our each muscle randomly, but choose whether to go the right way or left way in more abstract action space. That is because we have already learned that though we go on a non-road area, we cannot get a good result.

Therefore in this paper, aiming to show the possibility of emergence of the higher exploration, we replace the forked road situation with an obstacle avoidance task in which an agent learns to reach a goal while avoiding an obstacle, and whether the agent passes the right side or the left side of the obstacle is focused on. The task refers to [11], in which an agent learned appropriate actions based on regular RL using a layered NN, but there was a place where the agent could not move before the obstacle when no random number for exploration is added.

2 Reinforcement Learning Using a Chaotic Neural Network

Reinforcement learning is autonomous and purposive learning of appropriate actions to get more reward and less punishment. Generally, an agent explores stochastically based on random numbers. However here, as mentioned in Introduction, an agent explores by chaotic dynamics that a chaotic NN produces without adding noises or random numbers. In this paper, for continuous input-output mappings, Actor-Critic is used as a RL architecture. A chaotic NN and a non-chaotic layered NN are used for actor and critic respectively as shown in Fig. 1, and the sensor signals are the input for both NNs. Here, the neuron model used in both NNs is static that is different from [9] or [10] as

$$u_{j,t}^{(l)} = \sum_{i=1}^{N^{(l-1)}} w_{j,i}^{(l)} o_{i,t}^{(l-1)} \left(+ \sum_{i=1}^{N^{(l)}} w_{j,i}^{\mathrm{FB}} o_{i,t-1}^{(l)} \right) \tag{1}$$

where $u_{j,t}^{(l)}$ and $o_{j,t}^{(l)}$ are the internal state and the output of the j-th neuron in the l-th layer at time t, $w_{j,i}^{(l)}$ is the synaptic weight from the i-th neuron in the $(l\text{-}1)$-th layer to the j-th neuron in the l-th layer. The second term in the right-hand side is only for the hidden layer in the chaotic NN, and $w_{j,i}^{\mathrm{FB}}$ is the weight for the recurrent connection from the i-th neuron in the hidden layer. The activation function is the sigmoid (tanh) function $f(\ )$ whose value ranges from -0.5 to 0.5, and the output is $o_{j,t}^{(l)} = f(u_{j,t}^{(l)})$. The chaotic NN has two actor outputs $\mathbf{A}(\mathbf{S}_t)$ that are used as motion signals, and the non-chaotic NN has a critic output $V(\mathbf{S}_t)$ where $\mathbf{S}_t$ is the sensor inputs at time t.

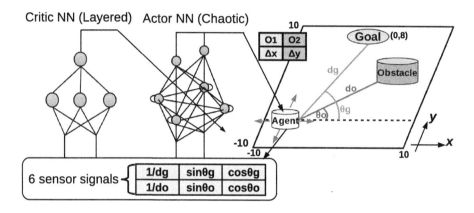

Fig. 1. Reinforcement learning system and the obstacle avoidance task in this paper

For learning, TD-error $\hat{r}_t$ is represented as

$$\hat{r}_t = r_{t+1} + \gamma V(\mathbf{S}_{t+1}) - V(\mathbf{S}_t) \tag{2}$$

where r_{t+1} is the reward given at time $t+1$, γ is a discount factor. T_{V_t} is the target for the critic output at time t and is computed as

$$T_{V_t} = V(\mathbf{S}_t) + \hat{r}_t = r_{t+1} + \gamma V(\mathbf{S}_{t+1}). \tag{3}$$

The critic NN is trained once according to Error Back Propagation using this target signal. To adjust the value range, 0.5 is added to the output of the critic NN and 0.5 is subtracted from the target T_{V_t} before using them actually.

In the chaotic NN in this paper, chaotic dynamics is produced by strong feedback connections between hidden neurons, and there is no feedback connections from the output. For the input-hidden and hidden-output connection (synaptic) weight $w_{j,i}^{(l)}$ in the chaotic NN, is modified using the trace $c_{j,i,t}^{(l)}$ as

$$\Delta w_{j,i,t}^{(l)} = \eta_A^{(l)} \hat{r}_t c_{j,i,t}^{(l)} \tag{4}$$

where $\Delta w_{j,i,t}^{(l)}$ is the modification of the weight $w_{j,i}^{(l)}$ at time t, $\eta_A^{(l)}$ is a learning rate for the l-th layer of the actor chaotic NN. The trace $c_{j,i,t}^{(l)}$ holds the past contribution of the pre-synaptic signal to the output increase in the post-synaptic neuron, and at each time step, it takes in the pre-synaptic signal $o_{i,t}^{(l-1)}$ and forgets the past trace value according to the change in the post-synaptic neuron $\Delta o_{j,t}^{(l)} = o_{j,t}^{(l)} - o_{j,t-1}^{(l)}$ as

$$c_{j,i,t}^{(l)} = (1 - |\Delta o_{j,t}^{(l)}|) \cdot c_{j,i,t-1}^{(l)} + \Delta o_{j,t}^{(l)} \cdot o_{i,t}^{(l-1)}. \tag{5}$$

The feedback connection weights $w_{j,i}^{FB}$ are not modified here.

3   Simulation

In this paper, to examine the acquisition of higher exploration, an obstacle avoidance task is simulated referring to the task in [11]. In this simulation, as shown in Fig. 1, there is a 20×20 field, and a goal is fixed at the upper center area $(0, 8)$. An obstacle and an agent are located randomly at the beginning of every episode. The agent moves according to the outputs of the actor chaotic NN, and when it reaches the circle with a radius of 1.0 around the goal, 1.0 is given as a reward. When it reaches the circle with a radius of 1.5 around the obstacle or it collides with a wall at the boundary of the field, -0.01 is given as a punishment. The episode is terminated when the agent either reaches the goal or fails to do so in 1,000 steps. 6 sensor signals as shown in Fig. 1 are sent to the both networks as input. Each of the two actor outputs decides the one-step move in x or y direction. The parameters used in the simulation are shown in Table 1.

At first, critic (state value) and actions when the obstacle is put at $(0, 0)$ are observed in the two cases after 100,000 episodes (a) and after 1,000,000 episodes (b) of learning. The agent was located at $x = -2, -1, 0, 1, 2$, $y = -8$ and the trajectories and the change in the critic values along the trajectories are shown in Fig. 2. It can be seen that after 1,000,000 episodes (b), the trajectories are smoother and the agent reaches the goal in smaller steps than in the case after 100,000 episodes (a). However after 1,000,000 episodes (b) when the agent starts from $(2, -8)$ (red trajectory), the agent collided with the obstacle and could not move for 8 steps. Therefore, the number of steps to the goal when the agent moved along the red trajectory is larger than the others.

Figure 3 shows the distribution of the critic output as a function of the agent location when the obstacle is put as the above. In both cases, the critic value is larger as the agent location is closer to the goal and lower around $(0, -2)$ where the obstacle disturbs the agent to go to the goal. This result shows that the agent learned that when the agent is close to the goal, the state is good, and when the obstacle exists around the line segment from the agent to the goal, the state is not good. In (b) after 1,000,000 episodes, the critic value is higher in total than (a) after 100,000 episodes, and that shows the agent can reach the goal in smaller number of steps in the case of (b).

Table 1. The parameters used in the simulation

Name		Actor net	Critic net
Step limit in each episode		1,000	
Number of layers		3	
Number of inputs		6	
Number of hidden neurons		100	10
Number of outputs		2	1
Value range of sigmoid function		$-0.5 - 0.5$	
Gain of sigmoid	Output	1	
Function	Hidden	2	1
Learning rate η	Output <- Hidden	0.00001	1
	Hidden <- Input	0.001	1
	Hidden <- Hidden	0.0	—
Range of initial weights (uniformly random)	Hidden <- Hidden (feedback)	±20	—
	others	±1	
Discount factor γ		—	0.95

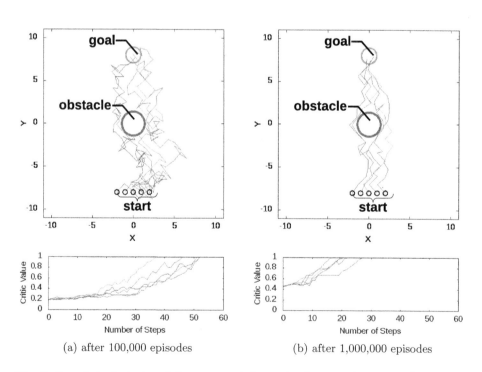

(a) after 100,000 episodes   (b) after 1,000,000 episodes

Fig. 2. Sample trajectories of the agent and change in the critic (state value) along the trajectories (Color figure online)

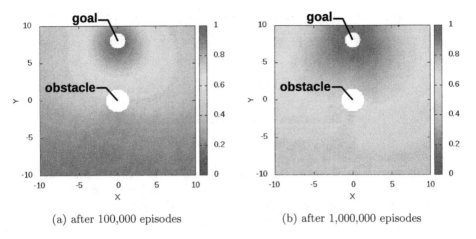

(a) after 100,000 episodes   (b) after 1,000,000 episodes

Fig. 3. Distribution of critic (state value) output as a function of the agent location (Color figure online)

The learning curve is shown in Fig. 4. The red trace shows the number of steps from the initial location of the agent to the goal for each episode, and the blue trace shows the average number of steps over every 100 episodes. Since the agent learns how to go to the goal and avoid the obstacle, the number of steps is decreased. However, after 200,000 episodes, although the average number of steps (blue trace) still continues to decrease, the number of steps looks to increase. This mean that, when the agent collided with the obstacle, it was sometimes trapped at the place for a while such as the red trajectory in Fig. 2(b).

In this paper, as an index of chaotic property, Lyapunov exponent, which shows the sensitivity to small perturbations, is computed. When the Lyapunov exponent is positive, the dynamics is chaotic. Here, every 1,000 episodes, a random vector whose size is normalized to 0.001 is added to the internal state of the hidden neurons in the chaotic NN. After one-step action according to the actor outputs, the Euclidean distance d of the hidden states from the case when no perturbation is added was compared between before and after the action. The above is performed in 400 situations in which the agent's location varies as $x = -9, -7, \cdots, 9, y = -2, -8$ and the obstacle location varies as $x = -9, -7, \cdots, 9, y = 0, 5$, and the Lyapunov exponent λ is calculated by

$$\lambda = \frac{1}{400} \sum_{p=1}^{400} ln \frac{d_{after}^{(p)}}{d_{before}^{(p)}} = \frac{1}{400} \sum_{p=1}^{400} ln \frac{d_{after}^{(p)}}{10^{-3}}. \tag{6}$$

The change in Lyapunov exponent according to the learning progress is shown in Fig. 5. The Lyapunov exponent is decreased quickly before the 100,000th episode and slowly after the 100,000th episode keeping the value positive. As shown in Fig. 2, in the case of after 100,000 episodes (a), the influence of the chaotic dynamics looks large, but around the end of the learning (after 1,000,000 episodes (b)), it looks smaller though still some irregularities can be seen.

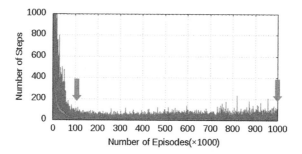

Fig. 4. Learning curve: change in the number of steps to the goal (red trace: steps at every episode, blue trace: average steps for every 100 episodes, pink arrows: the detail performances are shown in Figs. 2, 3 and 6) (Color figure online)

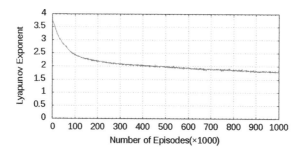

Fig. 5. Change in the Lyapunov exponent during learning

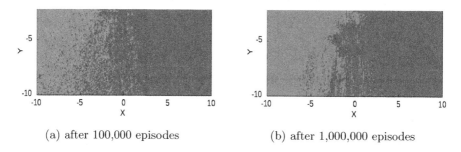

(a) after 100,000 episodes   (b) after 1,000,000 episodes

Fig. 6. Distribution of the agent initial location from which the agent passed the right side or left side of the obstacle to reach the goal (blue: left side, red: right side) (Color figure online)

In order to discuss whether the "higher exploration" emerges or not, Fig. 6 shows how the side of the obstacle through which the agent passed to avoid it varies depending on the initial location in the area $y < -2$ where the agent is located farther than the obstacle from the goal. In the both cases, after 100,000 episodes (a) and after 1,000,000 episodes (b), the agent is likely to pass through the right side of the obstacle when the initial location is in the right part of the field, and vice versa. Around the boundary of the two areas, especially in (a), the side the agent passed varies frequently depending on the initial location and so the agent looks to choose the side randomly. In (b), the distribution of these two areas is more symmetrical and reasonable than in (a). The result also shows that the agent is not trapped completely in front of the obstacle even without adding any external random numbers to the actor output, and that is different from the result in [11]. It is thought that the possibility of the emergence of higher exploration in which learning is reflected could be shown although it is ideal not to collide with the obstacle.

4 Conclusion

It was shown that by RL using a chaotic NN, the agent learned to go to the goal while avoiding a randomly-located obstacle. The distribution of the agent initial location where the agent passed the right side or left side of the obstacle did not have a clear boundary and the agent looks to choose the side to pass randomly. There was no place where the agent could not move to the right or left to avoid the obstacle. These results suggest the emergence of higher exploration, which would appear on the way to the emergence of "thinking", we expect. In the latter half of learning, Lyapunov exponent was decreased, and the agent sometimes collided with the obstacle and was trapped at the place for a while. Since the sensor inputs in this task are different from in [11], it is necessary to think about the solution of the problem from both sides of task setting and control of the chaotic property.

Acknowledgement. This work was supported by JSPS KAKENHI Grant Number 15K00360.

References

1. Shibata, K., Okabe, Y.: Reinforcement learning when visual signals are directly given as inputs. In: Proceedings of ICNN 1997, vol. 3, pp. 1716–1720 (1997)
2. Shibata, K.: Emergence of intelligence through reinforcement learning with a neural network. In: Mellouk, A. (ed.) Advances in Reinforcement Learning, pp. 99–120. InTech (2011)
3. Krizhevsky, A., et al.: ImageNet classification with deep convolutional neural networks. Adv. NIPS **25**, 1097–1105 (2012)
4. Mnih, V., et al.: Playing Atari with deep reinforcement learning. In: NIPS Deep Learning Workshop 2013 (2013)
5. Shibata, K., Utsunomiya, H.: Discovery of pattern meaning from delayed rewards by reinforcement learning with a recurrent neural network. In: Proceedings of IJCNN 2011, pp. 1445–1452 (2011)
6. Shibata, K., Goto, K.: Emergence of flexible prediction-based discrete decision making and continuous motion generation through actor-Q-learning. In: Proceedings of ICDL-Epirob 2013, ID 15 (2013)

7. Sawatsubashi, Y., et al.: Emergence of discrete and abstract state representation through reinforcement learning in a continuous input task. In: Kim, J.-H., Matson, E.T., Myung, H., Xu, P. (eds.) Robot Intelligence Technology and Applications 2012. AISC, vol. 208, pp. 13–22. Springer, Heidelberg (2012)
8. Shibata, K., Sakashita, Y.: Reinforcement learning with internal-dynamics-based exploration using a chaotic neural network. In: Proceedings of IJCNN 2015, #15231 (2015)
9. Sussillo, D.C.: Learning in Chaotic Recurrent Neural Networks. Columbia University, Ph.D. thesis (2009)
10. Hoerzer, G.M., et al.: Emergence of complex computational structures from chaotic neural networks through reward-modulated Hebbian learning. Cereb. Cortex **24**(3), 677–690 (2014)
11. Shibata, K., et al.: Direct-vision-based reinforcement learning in "Going a Target" task with an obstacle and with a variety of target sizes. In: Proceedings of NEURAP 1998, pp. 95–102 (1998)

Big Data Analysis

Establishing Mechanism of Warning for River Dust Event Based on an Artificial Neural Network

Yen Hsun Chuang, Ho Wen Chen[✉], Wei Yea Chen,
and Ya Chin Teng

Department of Environmental Science and Engineering, Tunghai University,
No. 181, Sec. 3, Taichung Port Road, Xitun District,
Taichung City 407, Taiwan (R.O.C.)
hwchen@thu.edu.tw

Abstract. PM_{10} is one of contributors to air pollution. One cause of increases in PM_{10} concentration in ambient air is the dust of bare land from rivers in drought season. The Taan and Tachia river are this study area, and data on PM_{10} concentration, $PM_{2.5}$ concentration and meteorological condition at air monitoring site are used to establish a model for predicting next PM_{10} concentration ($PM_{10}(T + 1)$) based on an artificial neural network (ANN) and to establish a mechanism for warning about $PM_{10}(T + 1)$ concentration exceed $150 \ \mu g/m^3$ from rivers in drought season. The optimal architecture of an ANN for predicting $PM_{10}(T + 1)$ concentration has six input factors include PM_{10}, $PM_{2.5}$ and meteorological condition. The train and test R was 0.8392 and 0.7900. $PM_{10}(T)$ was the most important factor in predicting $PM_{10}(T + 1)$ by sensitivity analysis. Finally, mechanism constraints were established for warning of high $PM_{10}(T + 1)$ concentrations in river basins.

Keywords: Artificial neural network · Dust · Warning mechanism · Predictive model

1 Introduction

Four major rivers (the Taan River, the Tachia River, the Wu River and the Choshui River) flow in central Taiwan. Owing to the extreme steepness of the riverbed upstream segments of all four rivers, the annual sediment yields of fine sludge in the downstream segments of the four rivers are extremely huge [1]. *Moreover, there have been happened bare-soil downstream during the drought season when dams were constructed upstream of the rivers obstruct most of the water in the upstream portions.* Farmers turn over bare soil, exposing it to insolation for around one month before growing watermelons. This action destroys the condensation surface layer of the soil that prevents the underlying dust from becoming suspended in the wind, exacerbating the phenomenon of river dust during strong monsoon seasons [2–4]. Under weather conditions, the Asian dust storms (ADS) that are caused by frontal activities in northern China in mid-October to mid-May can affect the air quality all over Taiwan [5, 6]. During these ADS, high PM_{10} concentrations are detected at most of the air-quality

© Springer International Publishing AG 2016
A. Hirose et al. (Eds.): ICONIP 2016, Part I, LNCS 9947, pp. 51–60, 2016.
DOI: 10.1007/978-3-319-46687-3_6

monitoring stations in Taiwan. Not only does pollution arrive by long-range transport, but also local area suffers from local pollution [7, 8]. Lin et al. [5] analyzed aerosol concentrations in Taiwan from 2002 to 2008 and found that major transport paths for the dry type (DT) dust cases passed through anthropogenic source areas in the low boundary when the major path for dust cases passed over the ocean in wet type (WT). Lin et al. [9] found that, during dust events, some anthropogenic chemicals, such as CO and SO_2, precede dust particles over northern Taiwan following frontal passage. Although chemical pollution that accompanies dust has been observed throughout northern Taiwan, the developing of river dust in local area and the long-range transport of aerosol particles over the island and the contributions of these two forms aerosols from continental East Asia in concert with local aerosol emissions have not been quantified.

Environmental epidemiologic studies have demonstrated the impact of ADS on many human health measures, such as mortality, hospitalizations, emergency room visits and clinic visits [10–12]. Most relevant investigations have examined the health impact of ADS events by identifying temporal changes in these health measures under either the ambidirectional framework or the lag framework. The ambidirectional framework compares health measures both prospectively and retrospectively based upon an arbitrarily chosen reference period with respect to ADS events and the seven days after an ADS event, and investigates the varying health measures before, during, and after such events [13]. Lag framework studies of delayed health impacts that occur over time support the hypothesis that the most significant impact on human health as a direct result of ADS events may actually occur many days after the storm has ended [12]. Nonetheless, to date, the geographic variation of health risks from exposure to ADS events has not been thoroughly investigated.

To protect human health, real-time information about air quality is required. Reliable forecasts of air quality should be provided by different air quality predictive models not only to predict the occurrence of severe pollution episodes, but also to abate emissions when the probability of the occurrence of such episodes is heightened. Recently, many researchers tested NN-based methods to forecast airborne PM concentrations [14, 15]. de Gennaro et al. [16] developed and tested to forecast PM_{10} daily concentration in two contrasted environments in NE Spain by using artificial neural network (ANN) and hourly PM concentration. The best forecasted performance indexes for the regional background site in Montseny ($R^2 = 0.86$, SI = 0.75), influenced by local and sometimes unexpected sources. Hooyberghs, Mensink, Dumont, Fierens and Brasseur [17] presented an ANN for forecasting one day ahead the daily average PM_{10} concentrations in Belgium, in which the most important input variable was the boundary layer height.

In this study, an ANN was used to forecast the PM_{10} concentration in the central Taiwan's river basin. ANN–based model that uses hourly PM_{10} concentrations provided an alert for high dust concentrations. The model input data for predicting 24 h average PM_{10} concentrations one day in advance were the hourly PM_{10} and $PM_{2.5}$ concentrations, and meteorological data such as wind speed, wind direction, rainfall, solar radiation, temperature and relative humidity. The main goal of this work is the protection of the alerted population based on accurate and timely information.

2   Methodology

2.1   Study Area

Two major rivers (the Taan River and the Tachia River) flow through Taichung in central Taiwan. In recent years, the Taiwan EPA has reported that a particular type of local air pollutant event is suspected to be affected by the river dust in monsoon season. This type of episode is referred to herein a "river dust episode". To manage the problem, the Taiwan EPA established monitoring stations (Shalue) to detect the concentration of PM_{10} and meteorological conditions in the two corresponding river basins. Figure 1 is a site map of the two river basins with hot spots of dust. Based on historical information from the Shalue monitoring station obtained from 2005 to 2013, the sources of dust were north and northeast of the site. Table 1 present the concentration of PM_{10} and meteorological conditions there.

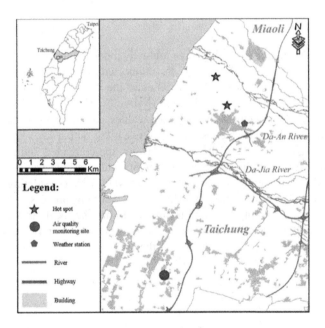

Fig. 1. Study area

2.2   Model for Predicting PM_{10} Concentration Based on ANN

Data-driven approaches to determining the nonlinear relationships between input and output variables include those based on ANN. An ANN imitates the behaviors of the human brain, recognizing the patterns of the relationships between input and output human brain, recognizing the patterns of the relationships between input and output variables after a period of learning from a set of training data. The basic structure of an

Table 1. Concentrations of coarse and fine particles and meteorological parameters in Shalue monitoring station

Year	PM$_{10}$(μg/m^3)	PM$_{2.5}$ (μg/m^3)	Temp.[a] (°C)	RH[b] (%)	Rainfall (mm)	Pressure (hpa)	Sunlight (hr)	Wind speed (m/s)
2005	56.87	34.03	23.32	73.62	0.22	1012.88	0.28	3.89
2006	54.59	31.90	23.78	74.97	0.18	1012.69	0.29	3.78
2007	62.51	31.73	24.37	73.73	0.24	1012.36	0.31	3.65
2008	61.91	31.29	23.38	73.80	0.20	1012.78	0.28	3.60
2009	56.28	32.87	23.41	73.78	0.12	1012.41	0.26	3.66
2010	49.87	32.70	23.34	77.44	0.14	1013.17	0.23	3.55
2011	51.97	34.47	23.04	73.63	0.08	1012.95	0.22	3.79
2012	45.57	29.70	23.45	75.79	0.19	1011.81	0.22	3.49
2013	49.43	35.12	23.47	75.59	0.23	1012.80	0.23	3.45

[a]Temperature
[b]Relative Humidity

ANN can be described as comprising three mutually independent layers - input, hidden, and output layers (Fig. 2). The layers are connected with neural synaptic weight coefficients (w_{ij}). Typically, when inputs (x_i) enter the system, they are multiplied by synaptic weights and summed at each node ($\sum(w_{ij})(x_i)$). When these values exceed corresponding threshold values, the summed values can be passed throughout the network, activating all hidden nodes until they reach the output layer. After the input has been processed in all layers, the errors between the computed and non-computed outputs are presented as well. The synaptic weights and threshold values are uncertain numbers. However, the network can adjust these values by considering a particular example case in the learning process. The overall learning process is composed of repeat calculating the training data and testing data until the local error close to zero.

The hidden nodes are activated by using the sigmoid activation function. When the networks finished the training at first epoch, the computed errors are sent back to the

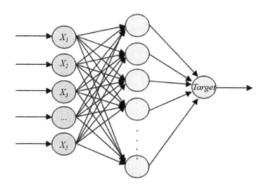

Fig. 2. Basic structure of ANN

first step again for correcting the error and weight adjustment of each layer. The network recognizes the optimal network pattern the target or an expected output. This process of data-based training is called back-propagation. The back-propagation algorithm works as follows [18, 19].

(1) Define the input data set for training the neural network
(2) Calculate the actual outputs of each hidden layer (j) using the following mathematical formula, where n is the number of inputs of neuron j, and θ is the threshold value.

$$y_i = f\left(\sum\nolimits_{i=1}^{n} w_{ij}x_i' - \theta_j\right) \qquad (3)$$

(3) Transform the output values (y_j) in the hidden layer using the sigmoid function as follows.

$$f(x) = \frac{1}{(1 - \exp(-y_j))} \qquad (4)$$

(4) Calculate the actual outputs of the output layer (k), where m denotes the number of inputs of neuron k, and transform y_k by applying the sigmoid function,

$$y_k = f\left(\sum\nolimits_{j=1}^{m} w_{jk}x_{jk} - \theta_{jk}\right) \qquad (5)$$

(5) Calculate the error in the output layer (ε_k), where $y_{d,k}$ is the desired output of neuron k in the output layer.

$$\varepsilon_k = y_k(1 - y_k)(y_{d,k} - y_k) \qquad (6)$$

(6) Change the weights in the output layer (w_{ij}^*), where η is a constant, called the learning rate.

$$w_{ij}^* = w_{ij} + (\eta)(y_j)(\varepsilon_k) \qquad (7)$$

(7) Calculate (back-propagate) the error in the hidden layer (ε_j), where l denotes the number of outputs of neuron k in the hidden layer.

$$\varepsilon_j = (y_i)(1 - y_j)\sum\nolimits_{k=1}^{l} \varepsilon_k w_{jk} \qquad (8)$$

(8) Change the weights of the relationship between input variables with the hidden neurons (w_{ij}^*).

$$w_{ij}^* = w_{ij} + (\eta)(x_i)(\varepsilon_j) \qquad (9)$$

(9) Repeat the above calculations and allow the system to learn how the error is related to the inputs, outputs, and weights. Finally, the optimal condition among the parameters will be identified.

3   Result and Discussion

3.1   Optimal Model Selection and the Results of Network Training and Testing

To identify the optimal model, eight situations of $PM_{10}(T + 1)$ concentration prediction are considered (Table 2) *and the mean of $PM_{10}(T + 1)$ is presented predict PM_{10} value for next day.* Table 2 presents the results of training and testing in the eight situations. In eight situations, the result show that the VI model is the best situation which training R and testing R were 0.8352 and 0.7900. The optimal solution involves $PM_{10}(T)$ concentration, temperature (T), $PM_{2.5}(T)$ concentration, sunlight per hour (T), pressure (T) and wind speed (T). Figure 3 displays the MAPE that compares the predicted and actual concentrations of PM_{10} obtained using the training data. Figure 4 displays the MAPE that compares the predicted and actual PM_{10} concentration obtained using the testing data. *Compare with [20], our results show when actual value over then 75 µg/m³, the model can predict high concentration in MAPE < 20 %.*

Table 2. The predictive program of PM_{10} concentration

Type	Input variable									Train-R	Test-R
	PM_{10}	$PM_{2.5}$	Temp.	Sun-light	Pres-sure	Wind speed	Rain fall	RH			
I	✓									0.7359	0.7361
II	✓		✓							0.7813	0.7561
III	✓	✓	✓							0.7680	0.7623
IV	✓	✓	✓	✓						0.8101	0.7574
V	✓	✓	✓	✓	✓					0.8208	0.7735
VI	✓	✓	✓	✓	✓	✓				0.8352	0.7900
VII	✓	✓	✓	✓	✓	✓	✓			0.8412	0.7835
VIII	✓	✓	✓	✓	✓	✓	✓	✓	✓	0.8544	0.7700

3.2   Accuracy Analysis and Sensitivity Analysis

In this study, 75 µg/m³ was used as the critical standard to classify $PM_{10}(T + 1)$ concentration in two type ($PM_{10}(T + 1) < 75$ µg/m³ and $PM_{10}(T + 1) \geq 75$ µg/m³). The results in Table 3 demonstrate that the overall accuracy of the VI model was 85.99 % when the correct classification rate exceeded 80 %. Compared actual concentration of $PM_{10}(T + 1) \geq 75$ ug/m³ with predictive concentration of $PM_{10}(T + 1) < 75$ µg/m³ was about 6.32 %. To evaluate the relative importance approach, the

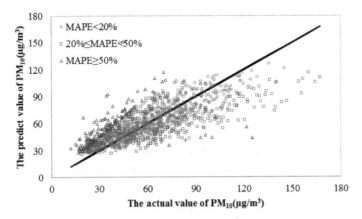

Fig. 3. Comparison between predict and actual concentrations of PM_{10} obtained using training data

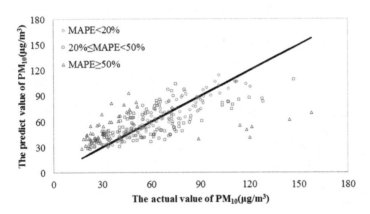

Fig. 4. Comparison between predicted and actual concentrations of PM_{10} obtained using testing data

connection and the actual concentration of $PM_{10}(T + 1) < 75$ μg/m^3 with predictive concentration of $PM_{10}(T + 1) \geq 75$ ug/m^3 was about 7.71 % when the incorrect classification rate was lower than 20 %.

Sensitivity analysis can be applied to investigate the relative importance among the various input factors on $PM_{10}(T + 1)$ emission by means of the weights method proposed weights between the input, hidden, and output layers were used to identify which input factor is the most important in determining $PM_{10}(T + 1)$. The six considered input factors were the actual concentration of $PM_{10}(T)$, temperature, concentration of $PM_{2.5}(T)$, wind speed (T), sunlight per hour (T), and pressure (T). Table 4 presents the relative importance of the factors. The actual concentration of $PM_{10}(T)$ is the most significant factor in the

Table 3. Confusion matrix of predicted and actual data

		Predictive concentration		
		$PM_{10}(T + 1) < 75$ μg/m^3	$PM_{10}(T + 1) \geq 75$ ug/m^3	Total
Actual concentration	$PM_{10}(T + 1) < 75$ μg/m^3	272(74.72 %)	28(7.70 %)	300(82.42 %)
	$PM_{10}(T + 1) \geq 75$ ug/m^3	23(6.32 %)	41(11.26 %)	64(17.58 %)
	Total	295(81.04 %)	69(18.96 %)	364(100.00 %)
Accuracy		85.99 %		

Table 4. The impact factor results of VI model from weight method

Input variable	U$^+$	U$^-$	$|$ U$^+$–U$^-$ $|$	Ranking
$PM_{10}(T)$	−0.5215	−0.8342	0.3127	1
Temp. (T)	−0.8007	−0.5565	0.2441	2
$PM_{2.5}(T)$	−0.8299	−0.6623	0.1676	3
Sunlight (T)	−0.7223	−0.7079	0.0144	6
Pressure (T)	−0.8034	−0.8908	0.0874	5
Wind speed (T)	−0.8455	−0.7234	0.1221	4

Table 5. The mechanism constraints of warning on river dust event and forecast rate

Constraints	Result	$PM_{10} \geq$ 100 μg/m^3	N	The forecast rate	The forecast rate in efficiency
The predict concentration of $PM_{10} \geq 75$ μg/m^3	69	17	364	18.96 %	24.64 %
The actual concentration of $PM_{10} \geqq 76.28$ μg/m^3 Temp. < 24.06°C $PM_{2.5} \geqq 45.52$ μg/m^3 The wind direction from North Rainfall < 0.019 mm Wind speed $\geqq 2.17$ m/s	25	10	364	6.87 %	40.00 %

prediction of $PM_{10}(T + 1)$, followed by temperature and concentration of $PM_{2.5}$. *The mean show when the high concentration of actual PM_{10} was happened, the high concentration of PM_{10} might be happened next day.*

3.3   Mechanism for Warning About River Dust Event

To establish a mechanism to warn about a river dust event, three values of the most significant factor, concentration of $PM_{10}(T + 1)$ - include $PM_{10}(T + 1) < 75$ μg/m^3, 75 μg/m$^3 \leqq PM_{10}(T + 1) < 100$ μg/m^3 and $PM_{10}(T + 1) \geqq 100$ μg/m^3 are identified, *what the level was definition by monitoring site at local area.*

In predicting process, we found the two constraints for warning were predicted $PM_{10} (T + 1) \geq 75 \ \mu g/m^3$ *which is the first constraints, the second constraint involved actual* $PM_{10}(T) \geq 76.28 \ \mu g/m^3$, *temperature* $< 24.06°C$, $PM_{2.5} \geq 45.52 \ \mu g/m^3$, *northerly wind direction, rainfall* $< 0.019 \ mm$ *and wind speed* $\geq 2.17 \ m/s$ *by using basic statistical analysis.* Table 5 presents the warning forecasts in efficiency. The result of first constraint demonstrates the efficiency of warning forecast rate (24.64 %), and the result of second constraint demonstrates the efficiency of warning forecast rate (40.00 %).

4 Conclusion

The study applied a prediction model and warning mechanism to prevent the event of the high dust concentration. The input variable of the prediction model was PM_{10} concentration. The prediction of the dust concentration ignored both the area of bare land and the source of the pollutant. This prediction model had high accuracy (80 %). Step 1 of the warning mechanism examines whether the concentration of PM_{10} prediction exceeds 75 $\mu g/m^3$, and step 2 of warning mechanism examines whether the concentration of PM_{10} prediction exceeds 100 $\mu g/m^3$ and must conform the second constraints.

References

1. Kuo, C.Y., Lin, C.Y., Huang, L.M., Wang, S.Z., Shieh, P.F., Lin, Y.R., Wang, J.Y.: Spatial variations of the aerosols in river-dust episodes in central Taiwan. J. Hazard. Mater. **179**, 1022–1030 (2010)
2. Sharratt, B., Auvermann, B.: Dust Pollution from Agriculture. Reference Module in Food Science. In: Encyclopedia of Agriculture and Food Systems, pp. 487–504 (2014)
3. Wang, R.M., You, C.F., Chu, H.Y., Hung, J.J.: Seasonal variability of dissolved major and trace elements in the Gaoping (Kaoping) River Estuary. Southwestern Taiwan J. Mar. Syst. **76**, 444–456 (2009)
4. Chang, S.Y., Fang, G.C., Chou, C.C.K., Chen, W.N.: Chemical compositions and radiative properties of dust and anthropogenic air masses study in Taipei Basin, Taiwan, during spring of 2004. Atmos. Environ. **40**, 7796–7809 (2006)
5. Lin, C.Y., Chou, C.C.K., Wang, Z.F., Lung, S.C., Lee, C.T., Yuan, C.S., Chen, W.N., Chang, S.Y., Hsu, S.C., Chen, W.C., Liu, S.C.: Impact of different transport mechanisms of Asian dust and anthropogenic pollutants to Taiwan. Atmos. Environ. **60**, 403–418 (2012)
6. Hsu, C.Y., Chiang, H.C., Lin, S.L., Chen, M.J., Lin, T.Y., Chen, Y.C.: Elemental characterization and source apportionment of PM10 and PM2.5 in the western coastal area of central Taiwan. Sci. Total Environ. **541**, 1139–1150 (2016)
7. Xue, M., Ma, J.Z., Yan, P., Pan, X.L.: Impacts of pollution and dust aerosols on the atmospheric optical properties over a polluted rural area near Beijing city. Atmos. Res. **101**, 835–843 (2011)
8. Cheng, M.C., You, C.F., Cao, J.J., Jin, Z.D.: Spatial and seasonal variability of water-soluble ions in PM2.5 aerosols in 14 major cities in China. Atmos. Environ. **60**, 182–192 (2012)

9. Lin, C.Y., Wang, Z., Chen, W.N., Chang, S.Y., Chou, C.C.K., Sugimoto, N., Zhao, X.: Long-range transport of Asian dust and air pollutants to Taiwan: observed evidence and model simulation. Atmos. Chem. Phys. **7**, 423–434 (2007)

10. Wang, Y.C., Lin, Y.K.: Mortality associated with particulate concentration and Asian dust storms in Metropolitan Taipei. Atmosp. Environ. **117**, 32–40 (2015)

11. Yu, H.L., Yang, C.H., Chien, L.C.: Spatial vulnerability under extreme events: a case of Asian dust storm's effects on children's respiratory health. Environ. Int. **54**, 35–44 (2013)

12. Aili, A.S.J., Oanh, N.T.K.: Effects of dust storm on public health in desert fringe area: case study of northeast edge of Taklimakan Desert. China. Atmos. Pollut. Res. **6**, 805–814 (2015)

13. Chien, L.C., Yang, C.H., Yu, H.L.: Estimated effects of Asian dust storms on spatiotemporal distributions of clinic visits for respiratory diseases in Taipei children (Taiwan). Environ. Health Perspect. **120**, 1215–1220 (2012)

14. Antanasijevic, D.Z., Pocajt, V.V., Povrenovic, D.S., Ristic, M.D., Peric-Grujic, A.A.: PM10 emission forecasting using artificial neural networks and genetic algorithm input variable optimization. Sci. Total Environ. **443**, 511–519 (2013)

15. Fernando, H.J.S., Mammarella, M.C., Grandoni, G., Fedele, P., Di Marco, R., Dimitrova, R., Hyde, P.: Forecasting PM10 in metropolitan areas: efficacy of neural networks. Environ. Pollut. **163**, 62–67 (2012)

16. de Gennaro, G., Trizio, L., Di Gilio, A., Pey, J., Perez, N., Cusack, M., Alastuey, A., Querol, X.: Neural network model for the prediction of PM10 daily concentrations in two sites in the Western Mediterranean. Sci. Total Environ. **463**, 875–883 (2013)

17. Hooyberghs, J., Mensink, C., Dumont, G., Fierens, F., Brasseur, O.: A neural network forecast for daily average PM10 concentrations in Belgium. Atmosp. Environ. **39**, 3279–3289 (2005)

18. Wang, L., Zeng, Y., Chen, T.: Back propagation neural network with adaptive differential evolution algorithm for time series forecasting. Expert Syst. Appl. **42**, 855–863 (2015)

19. Bai, Y., Li, Y., Wang, X.X., Xie, J.J., Li, C.: Air pollutants concentrations forecasting using back propagation neural network based on wavelet decomposition with meteorological conditions. Atmos. Pollut. Res. **7**, 557–566 (2016)

20. Perez, P., Reyes, J.: An integrated neural network model for PM10 forecasting. Atmos. Environ. **40**, 2845–2851 (2006)

Harvesting Multiple Resources for Software as a Service Offers: A Big Data Study

Asma Musabah Alkalbani[1], Ahmed Mohamed Ghamry[2,3]($\boxtimes$),
Farookh Khadeer Hussain[1]($\boxtimes$), and Omar Khadeer Hussain[2,3]($\boxtimes$)

[1] Decision Support and e-Service Intelligence Lab, School of Software,
Center of Quantum Computation and Intelligent Systems, University of Technology,
Sydney, NSW 2007, Australia
Asma.M.Alkalbani@student.uts.edu.au, Farookh.Hussain@uts.edu.au
[2] School of Business, University of New South Wales Canberra (UNSW Canberra),
Campbell, Australia
a.ghamry@unsw.edu.au
[3] Australian Defence Force Academy, Canberra, ACT 2602, Australia
O.Hussain@adfa.edu.au

Abstract. Currently, the World Wide Web (WWW) is the primary resource for cloud services information, including offers and providers. Cloud applications (Software as a Service), such as Google App, are one of the most popular and commonly used types of cloud services. Having access to a large amount of information on SaaS offers is critical for the potential cloud client to select and purchase an appropriate service. Web harvesting has become a primary tool for discovering knowledge from the Web source. This paper describes the design and development of Web scraper to collect information on SaaS offers from target Digital cloud services advertisement portals, namely www.getApp.com, and www.cloudreviews.com. The collected data were used to establish two datasets: a SaaS provider's dataset and a SaaS reviews/feedback dataset. Further, we applied sentiment analysis on the reviews dataset to establish a third dataset called the SaaS sentiment polarity dataset. The significance of this study is that the first work focuses on Web harvesting for cloud computing domain, and it also establishes the first SaaS services datasets. Furthermore, we present statistical data that can be helpful to determine the current status of SaaS services and the number of services offered on the Web. In our conclusion, we provide further insight into improving Web scraping for SaaS service information. Our datasets are available online through www.bluepagesdataset.com.

Keywords: Software as a Service · Service offer · Web harvesting · SaaS dataset

1 Introduction

Over the past few years, with the continuous and rapid growth of cloud computing technologies, Software-as-a-Service (SaaS) has become one of the world's

© Springer International Publishing AG 2016
A. Hirose et al. (Eds.): ICONIP 2016, Part I, LNCS 9947, pp. 61–71, 2016.
DOI: 10.1007/978-3-319-46687-3_7

largest digital business industries. SaaS shows a hybrid year-to-year increase, and several reports indicate that SaaS is becoming widely accepted. For instance, Gartner stated that in 2014, SaaS achieved 48.8 billion dollars in revenue [1]. One prediction indicates that by 2020, sales of SaaS will be more than 132 billion dollars. Another prediction by the International Data Corporation (IDC) is that by 2017, the SaaS market will be worth $107 billion, more than twice as much as its 2013 estimate of $47.4 billion [2]. Hence, the SaaS market has become highly competitive for SaaS service providers all over the world.

The Internet is the primary resource and the only distribution channel for the SaaS industry, transforming the Internet into a global SaaS marketplace. For example, there is a vast amount of SaaS information provided by SaaS-related websites containing SaaS offers, SaaS prices and details on SaaS providers. In addition, there has been a growth in Web-based portals, such as cloud reviews [3] and getApp [4], which provide a list of service offers collected from multiple sources on the Web.

Generally, publicly available search engines, such as Google, Yahoo, and Bing, are used to search for SaaS service offers on the World Wide Web (WWW). The results of these search engines show the potential that exists for extracting SaaS offers from the Web. However, the key issues lie within the quality of the results, as these search engines do not recognize SaaS offers. Usually, the obtained results comprise both relevant and irrelevant web sources. Consequently, accessing information on SaaS offers remains a problem as there is a lack of an available and efficient searching and information retrieval tools to find SaaS offers on the Web.

Therefore, our research concept is to utilize the existing content and structures of SaaS offers used from multiple sources to investigate SaaS offers on the Web in order to provide a complete view of the available SaaS offers. In other words, this study attempts to discover the SaaS offers which are available on the Web today.

The majority of the research to date, however, has focused on enhance SaaS discovery by using semantic technology to enhance Cloud information retrieval such as in [5,6]. The results of these research studies have shown the potential that lies in using semantic technologies to enhance data retrieval results from existing text-based search engines. Research by [7] proposed semantic information filtering of a search engine's results. Basically, the filter identifies the similarities between the cloud ontology concepts and the search engine's results, and then based on the specific threshold, it identifies if the information retrieved is relevant or irrelevant to the cloud domain. The cloud domain ontology comprises 424 concepts, which present the information. The semantic filter has been evaluated using virtual websites with up to 15700 web pages, including irrelevant and relevant cloud service providers' virtual sites.

In 2013 Noor et al. [8] consulted a cloud ontology to crawl Web-based resources, and then stored the crawling result in a local repository in order to obtain a cloud dataset. This study is considered to be the first effort toward obtaining a cloud dataset, but the dataset has several limitations, including a lack of primary service information, such as service name and service URLs, and the data values do not have the semantic meaning associated with them. Even though there have been numerous efforts to enhance the discovery of cloud services over the Web, the main limitation of these studies is that they fail to address the issue of investigating and discovering cloud service offers across multiple Web resources.

Therefore, to address this issue, this work introduces a framework for harvesting multiple SaaS offer resources to build the first SaaS repository. In this paper, we propose a SaaS Web scraper which crawls across several publicly available web portals to establish SaaS datasets. Our proposed method shows better results compared with existing approaches in regard to the provision of details on how many SaaS services are available today on the Web. We successfully collected around 5294 existing SaaS offers accessible on the Web today. Our dataset on SaaS offers comprises the main attributes that are needed for service selection. Moreover, this dataset assists in drawing a statistical distribution of SaaS offers, therefore providing accurate conclusions.

Web harvesting (Web scraping) is a computer technique to extract information and data from Web sources [9]. In other words, it is the transformation of unstructured data (HTML format) into structured data, also called Web data extraction. Although there has been much research on the subject of Web harvesting, no previous study so far has used the Web harvesting technique to investigate, collect, and gather information about cloud computing from the Web. In this work, we apply the Web harvesting task that targets Web sources to extract data about SaaS offers and SaaS consumers' reviews. This paper aims to obtain SaaS data from multiple sources. For this study, the target is restricted: to extract SaaS offers and consumers' reviews and feedback on the services from multiple sources on the WWW.

In this work, we establish three datasets: a SaaS offers dataset, a SaaS reviews dataset, and a SaaS polarity dataset, which can be potentially used as a resource for SaaS service discovery, selection, and composition. Moreover, this data could be used for SaaS knowledge discovery which plays a vital role in the construction of a SaaS knowledge base. This research makes the following contributions: we examine the potential of using the Web harvesting technique to extract information on SaaS services offers and SaaS consumers' reviews from multiple sources on the WWW; we introduce the notion of a SaaS services dataset to collect SaaS service offers that can be potentially used as a base for SaaS service discovery, selection, and composition; a SaaS dataset can also be used for SaaS knowledge discovery in order to construct a SaaS knowledge base; by continually scraping the existing SaaS service sources available on the Web, the dataset is capable of providing up-to-date data on SaaS services, hence this dataset is effective for service discovery, we collect and analyse the results and present various statistics including how many SaaS are accessible and what different categories of SaaS are accessible and we apply sentiment analysis and run several machine learning experiments on the SaaS reviews dataset containing the SaaS polarity dataset that can be accessed on the Web today.

To the best of our knowledge, this is the first study to do the following:

1. to investigate the Web to discover the amount of SaaS offers available today on the Web;
2. to establish a SaaS offers dataset;
3. to collect SaaS consumers' reviews to analyse and investigate overall satisfaction of SaaS consumers. Such analysis could provide useful information to improve the quality of SaaS provided;
4. to establish a SaaS polarity dataset that can possibly be used for applying some machine learning prediction techniques and for deep learning as well;
5. to provide ongoing research which aims at establishing the largest cloud services dataset and knowledge base.

The rest of this paper is organized as follows: Sect. 2 describes some of the related work; Sect. 3 discusses resources to find SaaS service offers; Sect. 4 describes the architecture of the dataset; Sect. 5 describes the methodology of our research; Sect. 6 discusses the results and the evaluation of harvesting SaaS web sources; Sect. 7 describes some of the challenges in the discovery of services; Sect. 8 discussed conclusion and future work.

2   Related Work

Recently, researchers have shown an increased interest in the discovery of cloud service issues, whereas previously many had focused on the discovery of cloud services on the Web using semantic technologies. A considerable amount of literature has been published on building a cloud ontology to enhance the dynamic discovery of cloud services over the WWW. A recent study by Afify et al. [5] developed a system for cloud service discovery containing a business ontology that assists service registry, service discovery, filtering and ranking the final result. This study does not support the dynamic discovery of cloud services, hence the information of the service offers need to be provided manually.

Research by Magesh et al. [10] proposed a semantic description for cloud service offers including service name, service level of agreement, service price, and service features. The study suggested representing each cloud offer as a single ontology and then combining all of them to construct a global ontology. The constructed global ontology has 64 entities and 128 properties. Unfortunately, this study neglects the need for quality of service information and rating attributes in selecting the services. In another effort by Kang et al. [11] a cloud service ontology was introduced to enhance the dynamic discovery of cloud services on the Web. This research used several reasoning methods to find semantic similarities between the user's request and the search engine results. The selected services are ranked, based on the price in the time slots that were determined by the consumers.

A different approach taken by Tahamtan et al. [12] is to assist a business organization to find an appropriate cloud business service, and a cloud business functions ontology was proposed to achieve this goal. The ontology includes most of the business function concepts and classifications outlined in [13]. In order to locate the right service with the right provider, the ontology is designed to map between the cloud service concepts and business concepts. In addition, the ontology includes some other important service attributes, such as service characteristics and service delivery model. Unfortunately, as with other existing work, their work, too, fails to account for the model's QoS parameters.

Although semantic searching methods may partially support the discovery of cloud services, they do not provide users with efficient ways to find proper services. Additionally, a scarcity of contributions in the current literature to determining the current status and distribution of cloud services. Very little research has been conducted on investigating cloud services on the Web. In [8] the authors' details on cloud services were collected throughout the Web by crawling Web sources. However, the study does not provide a complete view of the cloud services on the Web, and also the dataset provided in this study lacks primary service information, such as service name and service URL, and the data values do not have the semantic meaning associated with it. Therefore, it may provide inaccurate or misleading conclusions.

A recent study by Alkalbani et al. [14] proposed establishing a central repository for SaaS services. The study makes use of an open source, namely the Nutch-Hadoop crawler, to crawl details on SaaS offer from the Web and then stores the result in a local repository. The key shortcoming of this research is that the study only provides a service URL and service name.

3   SaaS Services Offer Resources

Finding information on SaaS service offers is not an easy task, especially since SaaS offers do not have standards to support service publishing, service description, discovery of service providers and their services offered, as is the case for Web services. For Web services, a service registry has been developed which plays a vital role as a publicly available, central access point to describe and publish Web services using semantic annotations. However, generally speaking, cloud service discovery is strictly tied to publicly available general search engines, such as Google, Yahoo, and Bing. These engines search text to find information related to service offers. However, they usually retrieve both relevant and irrelevant information. The following briefly describes the range of possible Web resources for finding SaaS services offers on the Web.

3.1   Cloud Service Portals or Directories

Web-based service directories or portals, such as getApp, cloudreviews, and others, is one possible method for finding SaaS services. The majority of services listed in these portals and directories have been collected from different cloud providers. Capturing SaaS services from these portals requires access to their repositories which are not publicly accessible. Another way to capture service offer data is by building a custom web scraper designed to capture and collect the service offer data from each portal independently, which is the focus of our research.

4   SaaS Datasets

Service offer information is a business structure for publishing service and business information which should be carefully considered when selecting the service. In the case of SaaS service offers, the service offer is simply meta-information or in other words, a HTML document. Therefore, for the purpose of this study, we target only meta-information for each service offer, which usually includes: service name, service description, service provider, service URL, service rating, service price, etc. In addition, this research considers collecting the services' rating as a part of the service offer that needs to be known when making a service selection decision. Also, this study considers collecting service reviews and feedback which could provide a useful summary about SaaS users' satisfaction. Thus, this research establishes three SaaS datasets: (1) SaaS offers dataset, (2) SaaS reviews dataset, and (3) SaaS sentiment polarity dataset. The next section details our procedure to construct these datasets.

5   Methodology

The mechanism implemented to achieve our research objectives is as follows: (1) defining the accessible Web-based sources from which SaaS information can be obtained, including publicly available web portals such as getApp, (2) designing and building a Web scraper to automatically crawl and collect information about SaaS including SaaS offers and SaaS consumers' reviews/feedback, and finally storing the results in a local repository to establish SaaS datasets, which is explained in more detail in the results section. Our research framework, as shown in Fig. 1, consists of the following stages:

- **Meta-collector:** To collect the meta-resources, first we download the Web source for each Web portal home page (HTML document), and then we extract the service offer links (URLs) from the home page sources. Then, we obtain the meta-information for each service link (URL).
- **Meta-validator:** We verify and validate the collected URLs and ensure that we retrieve all the offered URLs.
- **Meta-storage:** We store the meta-information on each URL (service offer page source) as a *"Meta-Source object"*.
- **Meta-parser:** In this step, we define the targeted information that needs to be harvested for all services including: service offer template (service name, service price, service provider), and service reviews.
- **Meta-database:** Finally, we store the extracted information for each service offer as a *"Service Object"*, and lastly we establish a meta-database which has around 5294 SaaS service offers.

6   Results and Statistics

In this section, we present the results and details of the SaaS datasets and the statistics on the harvested data. The harvested data is distributed among two datasets as follows:

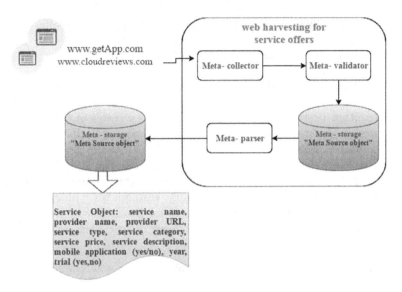

Fig. 1. Research framework

1. **SaaS offers dataset:** contains the primary information on SaaS offers harvested from the targeted Web resources. The SaaS offers data collection shows that each offer comprises the following attributes: service name, service provider name, provider URL, service rate, service description, year founded, mobile application (yes/no), starting price, service type, service category, free trial (yes/no).

   The data harvesting took place between February 2015 and August 2015, and the total number of SaaS offers harvested was 5294. Our constructed dataset illustrates that the majority of harvested offers are from getApp, which provides around 5146 service offers. Table 1 presents details on the SaaS dataset with respect to the resources used to collect the SaaS offers. Also, the results of this study, as illustrated in Table 2, indicate that the total number of unique SaaS offers is 3184, and surprisingly, we found that around 2110 are duplicated. This result may be explained by the fact that some services belong to more than one category (Table 2 shows that around 1512 service offers belong to more than one category). Also, the study found that, so far, the maximum number of service offers per category is five. In addition, Table 3 shows that there is no feedback data recorded in our constructed dataset from www.cloudreviews.com, whereas around 6343 were collected from www.getApp.com. This table shows the distribution of service reviews by service category. The data in the table indicates that the operations management application received the highest feedback from users, followed by customer management. Moreover, as can be seen from Fig. 1, only 14 % of offers provide a URL. The data collected shows that the majority of service offers published do not have a provider link, which accounts for 86 % of the constructed dataset.

2. **SaaS reviews/feedbacks dataset:** contains the collected reviews/feedback that have been made by services' users.

3. **SaaS sentiment polarity dataset:** Sentiment analysis was applied on the Saas reviews/feedback dataset to determine the tone of each SaaS post/review as being either positive, negative, or neutral. As a result of this analysis, we have another dataset, namely "the SaaS sentiment polarity dataset". The results obtained from the sentiment analysis on the SaaS reviews are shown in Table 4. More details on this analysis can be found in [15]. This dataset is a very useful for training machine learning algorithms and for further study. To conclude, all these can be accessed online through www.bluepagesdataset.com as the first publicly available datasets for SaaS offer information, SaaS reviews and feedback, and the SaaS sentiment polarity dataset.

Table 1. Summary of harvested SaaS offers per web resources

	www.getApp.com	www.cloudreviews.com
Harvested offers	5146	148
Execution time	3 minutes	1 minutes
Total harvested offers	5294	

Table 2. Summary of Unique/Duplicated harvested SaaS offers per web resource

	www.getApp.com	www.cloudreviews.com	Total
Unique offers	3038	146	3184
Duplicated offers	2108	2	2110
Total harvested offers			5294

7   SaaS Harvesting Challenges

At any point, Web portals may update service offers, therefore the Web scraper needs to be able to update or revisit Web resources to identify the changes that have taken place and update the downloaded data. Additionally, the number of SaaS services increases as well as the number of web portals, therefore it is an ongoing process to keep data up-to-date. From our experience with Web harvesting in this work, to achieve database availability, the challenges are:

– the customized Web scraper needs to monitor sites/pages for changes and updates.
– adding more repositories to our dataset requires designing and adding more code to our customized scraper.
– building a dynamic scraper that can handle the addition of more repositories and page source changes.

Table 3. Number of reviews/feedback per service type

	Service types	Reviews/Feedbacks
T1	Finance & accounting	1055
T2	Marketing	797
T3	Communications applications	389
T4	collaboration applications	1448
T5	Sales	1263
T6	Project management	1203
T7	Customer management	1655
T8	IT management	997
T9	Customer service & support	763
T10	Operations management	1822
T11	Business Intelligence & Analytics	208
T12	HR & Employee management	931

SAAS PROVIDER URL DISTRIBUTION

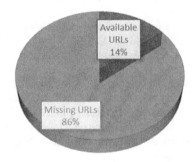

Fig. 2. SaaS provider URL distributions

Table 4. Summary of sentiment analysis

Polarity of reviews	Number of reviewers
Positive	2487
Neutral	1312
Negative	201
Total	4000

8   Conclusions

Finding and selecting relevant Software as a Service (SaaS) offers is mainly done manually by scanning through a number of suggestions from general search engines, such as Google or Bing. The dynamic discovery of SaaS service offers is necessary, especially when the number of services on the Web and the number of Web portals continues to significantly increase. Our study presented the implementation of a "Web scraper" to discover and investigate the number of SaaS offers available on the Web. We harvested SaaS offers from targeted web portals. The results provide an overview of the current status of SaaS offers and knowledge on the Web. An interesting result shows that some SaaS services are categorized according to business function, and some services belong to more than one category. For future work, our objective is to construct a large SaaS knowledge base and we will continue harvesting more Web sources as well as develop effective tools to dynamically update our datasets.

References

1. Gartner says worldwide it spending is forecast to grow 0.6 percent in 2016. http://www.gartner.com/newsroom/id/3186517. Accessed 08 Aug 2016
2. Worldwide SaaS and cloud software 2015–2019 forecast and 2014 vendor shares - 257397. https://www.idc.com/getdoc.jsp?containerId=257397. Accessed 08 Aug 2016
3. Cloud reviews — cloud hosting — managed cloud — cloud storage & apps. http://www.cloudreviews.com/. Accessed 08 Aug 2016
4. Business software reviews, SaaS & cloud applications directory – getapp. https://www.getapp.com/. Accessed 08 Aug 2016
5. Afify, Y., Moawad, I., Badr, N., Tolba, M.: A semantic-based software-as-a-service (SaaS) discovery and selection system. In: 2013 8th International Conference on Computer Engineering & Systems (ICCES), pp. 57–63. IEEE (2013)
6. Han, T., Sim, K.M.: An ontology-enhanced cloud service discovery system. In: Proceedings of the International MultiConference of Engineers and Computer Scientists, vol. 1, pp. 17–19 (2010)
7. Kang, J., Sim, K.M.: Cloudle: an agent-based cloud search engine that consults a cloud ontology. In: Proceedings of the International Conference on Cloud Computing and Virtualization, pp. 312–318. Citeseer (2010)
8. Noor, T.H., Sheng, Q.Z., Alfazi, A., Ngu, A.H., Law, J.: CSCE: a crawler engine for cloud services discovery on the world wide web. In: 2013 IEEE 20th International Conference on Web Services (ICWS), pp. 443–450. IEEE (2013)
9. Weikum, G., Theobald, M.: From information to knowledge: harvesting entities and relationships from web sources. In: Proceedings of the Twenty-Ninth ACM SIGMOD-SIGACT-SIGART Symposium on Principles of Database Systems, pp. 65–76. ACM (2010)
10. Vasudevan, M., Haleema, P., Iyengar, N.C.S.: Semantic discovery of cloud service catalog published over resource description framework. Int. J. Grid Distrib. Comput. $\mathbf{7}$(6), 211–220 (2014)
11. Kang, J., Sim, K.M.: Cloudle: a multi-criteria cloud service search engine. In: 2010 IEEE Asia-Pacific Services Computing Conference (APSCC), pp. 339–346. IEEE (2010)

12. Tahamtan, A., Beheshti, S.A., Anjomshoaa, A., Tjoa, A.M.: A cloud repository and discovery framework based on a unified business and cloud service ontology. In: 2012 IEEE Eighth World Congress on Services, pp. 203–210. IEEE (2012)
13. Kerrigan, M., Mocan, A., Tanler, M., Fensel, D.: The web service modeling toolkit - an integrated development environment for semantic web services. In: Franconi, E., Kifer, M., May, W. (eds.) ESWC 2007. LNCS, vol. 4519, pp. 789–798. Springer, Heidelberg (2007). doi:10.1007/978-3-540-72667-8_57
14. Alkalbani, A., Shenoy, A., Hussain, F.K., Hussain, O.K., Xiang, Y.: Design and implementation of the hadoop-based crawler for SaaS service discovery. In: 2015 IEEE 29th International Conference on Advanced Information Networking and Applications, pp. 785–790. IEEE (2015)
15. Alkalbani, A.M., Ghamry, A.M., Hussain, F.K., Hussain, O.K.: Sentiment analysis and classification for software as a service reviews. In: 2016 IEEE 30th International Conference on Advanced Information Networking and Applications (AINA), pp. 53–58. IEEE (2016)

Cloud Monitoring Data Challenges: A Systematic Review

Asif Qumer Gill[✉] and Sarhang Hevary

School of Software, University of Technology Sydney, Ultimo, NSW 2007, Australia
asif.gill@uts.edu.au, sarhang.hevary@alumni.uts.edu.au

Abstract. Organizations need to continuously monitor, source and process large amount of operational data for optimizing the cloud computing environment. The research problem is: what are cloud monitoring data challenges – in particular virtual CPU monitoring data? This paper adopts a Systematic Literature Review (SLR) approach to identify and report cloud monitoring data challenges. SLR approach was applied to initially identify a large set of 1861 papers. Finally, 24 of 1861 relevant papers were selected and reviewed to identify the five major challenges of cloud monitoring data: monitoring technology, virtualization technology, energy, availability and performance. The results of this review are expected to help researchers and practitioners to understand cloud computing data challenges and develop innovative techniques and strategies to deal with these challenges.

Keywords: Big data · Cloud computing · Capacity planning · Monitoring · And virtual CPU

1 Introduction

Cloud computing is a virtual data-intensive environment, which runs multiple virtual machines in large scalable clusters [11]. Cloud computing is one such new modes that supports pay-as-you-go and on-demand services (e.g. software as a service, platform as a service, infrastructure as a service) to enable business agility and flexibility [10]. Cloud computing seems to offer lucrative benefits [3], however, organizations need to actively monitor and analyze the operational data about the quality of cloud services and utilization of underlying virtual resources such as CPU, memory, storage and network [6]. This is also important to verify and identify any service performance related issues including Service Level Agreement (SLA). Monitoring is also important to track and control the expenses associated with the cloud service resource utilization [4].

There are a number of tools (e.g. AWS Cloud Watch) that claim to support the data monitoring including data acquisition and processing [e.g. 9]. However, to effectively adopt or develop specific cloud monitoring data tools, organizations need to identify and understand the fundamental challenges of cloud monitoring data. The understanding of challenges will help organizations in making informed decisions about the development and improvement of specific cloud monitoring data sourcing and processing tools for different types of cloud resources (e.g. CPU data, storage device data) at different levels (e.g. resource utilization data, health data). Cloud monitoring data is a broad topic. This

© Springer International Publishing AG 2016
A. Hirose et al. (Eds.): ICONIP 2016, Part I, LNCS 9947, pp. 72–79, 2016.
DOI: 10.1007/978-3-319-46687-3_8

paper mainly focuses on the monitoring challenges of virtual CPU utilization data within the overall context of cloud monitoring. Thus, the main research question is: what are cloud monitoring challenges – in particular virtual CPU utilization monitoring data?

This paper applied the well-known Systematic Literature Review (SLR) method [7] to systematically search, identify and synthesize the challenges of virtual CPU utilization monitoring data. This paper is organized as follows. Firstly, it discusses the research method. Secondly, it presents the research findings. Finally, it discusses the research findings and future research directions before concluding.

2   Research Method

This paper applied the SLR guidelines [7] for systematically searching, selecting, reviewing and synthesizing the cloud monitoring data challenges from relevant academic and industry publications (2011–2015). This study included the paper written in English language, which were selected from five well-known electronic databases (Table 3).

Table 1 presents the keyword or terms that were used during the first attempt to search the topic. All of the terms from the search category field "Monitoring of Virtual CPU" were also joined using the "AND" or "OR" operator to examine different combinations. In addition to this, search history features were used to combine different returned searched results to narrow down to the desired topics. Table 2 presents the paper selection criteria stages, which were applied to systematically identify the relevant papers for this study.

Table 1. Search keywords

Search category	Keywords/Phrases
Monitoring of virtual CPU (using advanced search interfaces)	Virtual Processor; Virtual CPU; Monitoring Tool; Monitoring Technique; Cloud Computing; Virtual Machine; SaaS Monitoring Technology.
Monitoring of virtual CPU (using advance command search)	(("virtual processor") OR ("virtual CPU")) AND (("monitoring tool*") OR ("monitoring technique*")) and (("cloud computing*")) (("*virtual processor*") OR ("*virtual CPU*")) AND (("*monitoring tool*") OR ("*monitoring technique*")) and (("*cloud computing*"))

Initial search of keywords and filtration (based on title) across five selected databases resulted in a large number of 1861 papers. Second filtration resulted in 97 papers (based on the review of abstract), and finally, third filtration stage (based on the exploration of paper contents) resulted in 24 relevant papers for this review (Table 3). Please note that for the first and second search filtration stages, the items such as news, eBooks and tutorials were excluded as the contents were not suitable for this academic study. Finally, only those papers were selected that satisfied the five Assessment Criteria – Final Search Filtration (Table 2).

Table 2. Paper selection criteria

Filtration stage	Method	Assessment criteria
First search filtration	Explore the title	Title = **search keyword (s)** Yes = accepted No = rejected
Second search filtration	Explore the abstract	Abstract = **CPU OR Virtual CPU OR VCPU** Yes = accepted No = rejected
Final search filtration	Explore the content	1. Address **Virtual CPU OR Monitoring Tool OR Cloud Computing** 2. Well-referenced 3. Objective is clearly defined 4. Well-presented argument and justified 5. Clearly stated findings (Yes = accepted, No = rejected)

Table 3. Search results

Database	1st Search filtration	2nd filtration	Final count
Web of Science	462	40	3
IEEE	913	30	10
Google Scholar	67	16	7
Gartner	12	6	2
Scopus	407	5	2
Total	**1861**	**97**	**24**

3   Findings

The selected 24 papers were analyzed and interpreted in order to answer the research question in hand. The detailed review of the selected papers resulted in five major challenge categories as shown in Table 4. These challenge categories are: monitoring technology, virtualization technology, energy, availability and performance. These

Table 4. Findings – data monitoring challenges categories

Challenge categories	Sources	Frequency (number of studies)	Percentage
C1. Monitoring Technology	S14, S15, S16, S17, S18, S19, S20, S21	8	**34 %**
C2. Virtualization technology	S1, S2, S3, S4, S5, S6	6	**25 %**
C3. Energy	S10, S11, S12, S13	4	**17 %**
C4. Availability	S7, S8, S9	3	**13 %**
C5. Performance	S22, S23, S24	3	**13 %**

categories were extracted after the careful review of the papers by using the analysis techniques from the well-known Grounded Theory [2], which is useful for identifying the relevant concepts and categories from a large volume of qualitative data or text [1]. Table 4 presents the identified major challenge categories and corresponding literature sources (see Appendix for selected literature sources S1–S24).

3.1 Monitoring Technology

Monitoring of a specific remote virtual CPU resource is a challenge in the complex distributed cloud environment. Monitoring technology category is heavily referenced in the literature (e.g. 34 % of the selected studies) and can help to resolve this challenge. (Table 4). This challenge category has identified three key underlying monitoring technology data challenges: (1) lack of data standardization, (2) live resource monitoring data, and (3) interoperability of data. Lack of monitoring data standards (e.g. templates, format) hinder the ability to integrate disparate monitoring tools and different types of virtual CPU data [e.g. S14]. The accurate monitoring data of a live virtual CPU resource in a cluster, which can be dynamically added, updated or removed, is challenging [e.g. S16]. Finally, third challenge is about the inability of the monitoring technology to support the interoperability of monitoring data across different cloud platforms [e.g. S18].

3.2 Virtualization Technology

Monitoring of the virtualized CPU resource data can also be impacted by the hypervisor, which is used to virtualize the physical CPU resource [5]. Interaction between the monitoring tool and hypervisor is important for collecting the utilization data. This is the second highly referenced category (25 % of the selected studies) (Table 4). This challenge category has identified two key underlying challenges: (1) dual monitoring data and (2) heterogeneous virtual environment data. Hypervisor and virtualized CPU share physical resources and the challenge is that both need to be monitored for collecting the correct utilization data. Thus, this dual monitoring challenge needs to be addressed to accurately collect the utilization data [e.g. S1]. Heterogeneous virtual environment, containing different types of hypervisors and virtual CPUs, poses the challenge of dealing with different hypervisors' APIs and monitoring tools' APIs for monitoring, collecting and processing large amount of data in different formats [e.g. S2].

3.3 Energy

This is the third referenced category (17 % of the selected studies) (Table 4). This challenge category has identified two key underlying energy related challenges: (1) energy utilization data and (2) energy efficiency. Virtualized cloud environments claim to offer low energy utilization. These challenges draws our attention to the challenge of collecting and processing large amount of monitoring data using minimal energy or power. Energy utilization data needs to be monitored and optimized for energy efficiency [e.g. S10 and S11].

3.4   Availability

It is not about the CPU resource monitoring data. The monitoring should also provide the capability to collect and provide the virtual CPU availability data [e.g. S8]. This is the fourth referenced category (13 % of the selected studies) (Table 4). This challenge category has identified two key underlying availability related challenges: (1) SLA verification data, and (2) detecting and alerting data. The availability, in the context of virtual CPU utilization, is important and draws our attention to the challenge of monitoring availability SLAs, and then processing it for detecting and alerting any related issues or breaches [e.g. S7 and S8].

3.5   Performance

Finally, this category draws our attention to the computational performance challenge of both the virtual machines and monitoring. This is the fifth referenced category (13 % of the selected studies) (Table 4). This challenge category has identified two key challenges: (1) performance identification, and (2) detecting and alerting. The monitoring should provide the capability to collect and provide the data about the performance of the virtual CPU resource and monitoring technology to help detecting and alerting any performance issues [e.g. S24]. Based on the performance results, we can dynamically adjust the utilization and control the number of virtual CPUs assigned to a physical CPU [e.g. S24]. Further, performance results can lead to the consolidation and de-consolidation of the virtual CPUs and underlying physical resources [e.g. S22 and S23].

4   Discussion

The effective utilization of cloud requires monitoring the hypervisors and the virtual environments. Monitoring data growth and velocity are increasing, and different monitoring standards, architectures, tools, and APIs are required to monitor the resource usage and capture a large amount of operational data [8]. However, the monitoring of cloud, in particular virtual CPUs hosted on a heterogeneous environment, poses several challenges. This paper addresses this research problem and systematically identify the five key challenges categories and underlying challenges.

Firstly, our findings highlighted that monitoring (34 %) and virtualization (24 %) technology were the most important challenge categories with respect to virtual CPU monitoring data collection and processing. Thus, we can classify these two as core challenge categories. Other challenge categories such as energy (17 %), availability (13 %), and performance (13 %) were classified as secondary. This is because they were not heavily reported, although, they could impact the monitoring (Table 4).

Secondly, our findings highlighted that the effective monitoring of the heterogeneous environment requires monitoring standards and frameworks for monitoring data integration and interoperability across different cloud platforms. This is important to facilitate the effective adoption of cloud.

Thirdly, our findings highlighted that the capturing and processing of the monitoring data are not enough. Monitoring capability should also support detecting any issues and alerting or taking corrective actions or adjustments. This leads to the identification of whole new area of research about smart data-driven and analytics-enabled adaptive monitoring. Monitoring of large and complex environment generates huge amount of data, which draws our attention to another area of research about BigData analytics for cloud monitoring data.

Similar to any other SLR studies, this paper has some limitations. One limitation could be the use of finite number of selected literature databases and studies. This paper included studies from well-known databases, and we have full confidence that the selected databases and studies provided us with the relevant and recent literature to address the research question in hand. One may argue about the possible bias in the selection of studies and inaccuracy of analysis. To mitigate this risk, we developed and applied relevant search string and keywords, systematic study selection criteria (Tables 1–2) and analysis techniques from well-known Grounded Theory [2]. This was done to ensure that the relevant studies were not omitted.

Despite possible limitations, this paper provided useful insights for both practitioners and researchers interested in the area of cloud monitoring data capturing and processing. For instance, practitioners may be interested in developing new tools, formats and standards for exchanging monitoring data across different cloud platforms. Researchers may be interested in developing new frameworks for BigData analytics enabled smart and adaptive monitoring.

5   Conclusion

This paper presented a SLR of virtual CPU utilization monitoring data challenges. This paper systematically searched, identified and reviewed a set of twenty-four relevant papers. The detailed review of selected papers provided us with the five major challenge categories. This study provided a knowledge-base of monitoring data challenges to practitioners and researchers who have interest in cloud computing. The findings of this paper can be further used in developing monitoring data sharing and processing standards, formats and tools to facilitate the effective cloud monitoring data management. The findings of this paper will provide necessary inputs to further research and develop the BigData analytics enabled framework for smart and adaptive cloud monitoring data.

Appendix: Selected Papers Included in This Review

[S1] Anand, Ankit, et al. "Resource usage monitoring for KVM based virtual machines." Advanced Computing and Communications (ADCOM), 2012 18th Annual International Conference on. IEEE, 2012.

[S2] Du, Jiaqing, Nipun Sehrawat, and Willy Zwaenepoel. "Performance profiling of virtual machines." ACM SIGPLAN Notices 46.7 (2011): 3-14.

[S3] P. Vijaya Vardhan Reddy, P. Vijaya Vardhan Reddy, and Dr Lakshmi Rajamani Dr. Lakshmi Rajamani. "Performance Evaluation of Hypervisors in the Private Cloud based on System Information using SIGAR Framework and for System Workloads using

Passmark." International Journal of Advanced Science and Technology 70 (2014): 17-32.

[S4] Reddy, P., and Lakshmi Rajamani. "Performance comparison of different operating systems in the private cloud with KVM hypervisor using SIGAR framework." Communication, Information & Computing Technology (ICCICT), 2015 International Conference on. IEEE, 2015.

[S5] Reddy, P. Vijaya Vardhan, and Lakshmi Rajamani. "Evaluation of Different Hypervisors Performance in the Private Cloud with SIGAR Framework." International Journal of Advanced Computer Science and Applications (IJACSA) 5.2 (2014).

[S6] Reddy, P., and Lakshmi Rajamani. "Virtualization overhead findings of four hypervisors in the CloudStack with SIGAR." Information and Communication Technologies (WICT), 2014 Fourth World Congress on. IEEE, 2014.

[S7] Houlihan, Ryan, et al. "Auditing cloud service level agreement on VM CPU speed." Communications (ICC), 2014 IEEE International Conference on. IEEE, 2014.

[S8] Huang, Qiang, et al. "Auditing CPU Performance in Public Cloud." Services (SERVICES), 2013 IEEE Ninth World Congress on. IEEE, 2013.

[S9] Alhamazani, Khalid, et al. "An overview of the commercial cloud monitoring tools: research dimensions, design issues, and state-of-the-art." Computing (2014): 1-21.

[S10] Madani, N., et al. "Power-aware Virtual Machines consolidation architecture based on CPU load scheduling." Computer Systems and Applications (AICCSA), 2014 IEEE/ACS 11th International Conference on. IEEE, 2014.

[S11] Vrbsky, Susan V., et al. "Decreasing power consumption with energy efficient data aware strategies." Future Generation Computer Systems 29.5 (2013): 1152-1163.

[S12] Katsaros, Gregory, et al. "A service framework for energy-aware monitoring and VM management in Clouds." Future Generation Computer Systems 29.8 (2013): 2077-2091.

[S13] Verma, Akshat, Puneet Ahuja, and Anindya Neogi. "pMapper: power and migration cost aware application placement in virtualized systems." Middleware 2008. Springer Berlin Heidelberg, 2008. 243-264.

[S14] Katsaros, Gregory, et al. "A Self-adaptive hierarchical monitoring mechanism for Clouds." Journal of Systems and Software 85.5 (2012): 1029-1041.

[S15] Povedano-Molina, Javier, et al. "DARGOS: A highly adaptable and scalable monitoring architecture for multi-tenant Clouds." Future Generation Computer Systems 29.8 (2013): 2041-2056.

[S16] Smit, Michael, Bradley Simmons, and Marin Litoiu. "Distributed, application-level monitoring for heterogeneous clouds using stream processing." Future Generation Computer Systems 29.8 (2013): 2103-2114.

[S17] Lee, Hyungro, et al. "Towards Understanding Cloud Usage through Resource Allocation Analysis on XSEDE."

[S18] Ranjan, Rajiv, et al. "A note on software tools and techniques for monitoring and prediction of cloud services." Software: Practice and Experience 44.7 (2014): 771-775.

[S19] Serrano, Nicolas, Gorka Gallardo, and Josune Hernantes. "Infrastructure as a Service and Cloud Technologies." IEEE Software 2 (2015): 30-36.

[S20] Manvi, Sunilkumar S., and Gopal Krishna Shyam. "Resource management for Infrastructure as a Service (IaaS) in cloud computing: A survey." Journal of Network and Computer Applications 41 (2014): 424-440.

[S21] Dhingra, Mohit, J. Lakshmi, and S. K. Nandy. "Resource usage monitoring in clouds." Proceedings of the 2012 ACM/IEEE 13th International Conference on Grid Computing. IEEE Computer Society, 2012.

[S22] Reddy, P., and Lakshmi Rajamani. "Performance comparison of different operating systems in the private cloud with KVM hypervisor using SIGAR framework." Communication, Information & Computing Technology (ICCICT), 2015 International Conference on. IEEE, 2015.

[S23] Janpan, Tanasak, Vasaka Visoottiviseth, and Ryousei Takano. "A virtual machine consolidation framework for CloudStack platforms." Information Networking (ICOIN), 2014 International Conference on. IEEE, 2014.

[S24] Miao, Tianxiang, and Haibo Chen. "FlexCore: Dynamic virtual machine scheduling using VCPU ballooning." Tsinghua Science and Technology 20.1 (2015): 7-16.

References

1. Alzoubi, Y.I., Gill, A.Q., Al-Ani, A.: Empirical studies of geographically distributed agile development communication challenges: a systematic review. Inf. Manag. **53**(1), 22–37 (2016)
2. GTI, What is Grounded Theory? (2008). http://www.groundedtheory.com/what-is-gt.aspx
3. Gill, A.Q., Bunker, D., Seltsikas, P.: Moving forward: emerging themes in financial services technologies' adoption. Commun. Assoc. Inf. Syst., 36, 205–230 (2015)
4. Gill, A.Q.: Adaptive Cloud Enterprise Architecture. World Scientific, Singapore (2015)
5. Gill, A.Q., Bunker, D., Seltsikas, P.: An empirical analysis of cloud, mobile, social and green computing: financial services it strategy and enterprise architecture. In: 2011 IEEE Ninth International Conference on Dependable, Autonomic and Secure Computing (DASC), pp. 697–704. IEEE (2011)
6. Jamail, N.S.M., Atan, R., Abdullah, R., Said, M.Y.: Development of SLA monitoring tools based on proposed DMI in cloud computing. Trans. Mach. Learn. Artif. Intell. **3**(1), 01 (2015)
7. Kitchenham, B.A., Charters, S.: Procedures for performing systematic literature reviews in software engineering. Keele University & Durham University, UK (2007)
8. Kowall, J., Fletcher, C.: Modernize your monitoring strategy by combining unified monitoring and log analytics tools, (2014). Gartner http://www.gartner.com/document/code/257830?ref=grbody&refval=2809724
9. Meng, S., Liu, L.: Enhanced monitoring-as-a-service for effective cloud management. IEEE Trans. Comput. **62**(9), 1705–1720 (2013)
10. NIST: NIST Cloud Computing Reference Architecture (2011). http://www.nist.gov/customcf/get_pdf.cfm?pub_id=909505
11. Smith, S., Gill, A. Q., Hasan, H., Ghobadi, S.: An enterprise architecture driven approach to virtualisation. In: Proceedings of PACIS 2013 (2013)

Locality-Sensitive Linear Bandit Model for Online Social Recommendation

Tong Zhao[1,2](✉) and Irwin King[1,2]

[1] Shenzhen Key Laboratory of Rich Media Big Data Analytics and Applications,
Shenzhen Research Institute, The Chinese University of Hong Kong, Shenzhen, China
[2] Department of Computer Science and Engineering,
The Chinese University of Hong Kong, Shatin, N.T., Hong Kong
{tzhao,kingg}@cse.cuhk.edu.hk

Abstract. Recommender systems provide personalized suggestions by learning users' preference based on their historical feedback. To alleviate the heavy relying on historical data, several online recommendation methods are recently proposed and have shown the effectiveness in solving data sparsity and cold start problems in recommender systems. However, existing online recommendation methods neglect the use of social connections among users, which has been proven as an effective way to improve recommendation accuracy in offline settings. In this paper, we investigate how to leverage social connections to improve online recommendation performance. In particular, we formulate the online social recommendation task as a contextual bandit problem and propose a *Locality-sensitive Linear Bandit* (LS.Lin) method to solve it. The proposed model incorporates users' local social relations into a linear contextual bandit model and is capable to deal with the dynamic changes of user preference and the network structure. We provide a theoretical analysis to the proposed *LS.Lin* method and then demonstrate its improved performance for online social recommendation in empirical studies compared with baseline methods.

Keywords: Social recommendation · Linear bandits · Online learning

1 Introduction

Recommender systems are ubiquitous in online applications ranging from e-commence websites to content recommendation services such as Yahoo! and Digg, since they can effectively help users to relieve the problem of *information overload* by filtering out irrelevant information and also provide users relevant information according to their personal preferences. Most techniques used in recommender systems are heavily relying on users' historical feedback (e.g., in the form of a rating to an item or a click behavior) in order to learn reliable models of a user's preferences. However, the user's preferences may be diverse and changed dynamically. Moreover, for new users, there are obviously no sufficient historical records for recommendation methods to learn the true preference.

© Springer International Publishing AG 2016
A. Hirose et al. (Eds.): ICONIP 2016, Part I, LNCS 9947, pp. 80–90, 2016.
DOI: 10.1007/978-3-319-46687-3_9

To alleviate the critical reliance of recommender systems on historical information, online recommendation methods are proposed and have shown the success in solving the cold start and diversified recommendation problems. Online recommendation methods aim at building models that can continue to learn and improve their performance automatically as long as users provide feedback (i.e., ratings on items or clicks on links) to them. To achieve the self-learning goal, online recommendation algorithms always need to explore some uncertain results to identify users' unknown preferences or to adapt to the dynamic changes of users' preferences. To strike a balance between uncertain exploration and conservative exploitation to achieve high quality recommendation performance, several recent studies formulate this online recommendation task as a multi-armed bandit problem [9,12,20].

The multi-armed bandit (MAB) problem [6] has been extensively studied and becomes increasingly popular in the machine learning community recently. It can be formulated as a sequential decision-making problem, where in each of the T rounds the player needs to select one arm from a set of arms, each having an unknown distribution of rewards. In each round, only the reward of the selected arm can be observed. Thus, the player has to make this decision based on its historical decisions. The goal of the player is to maximize his/her cumulative rewards during the T rounds. When the arms are represented by their feature vectors that can be observed by the player, this problem is known as the *contextual bandit* problem [1,10,20], which has been successfully adopted to recommender systems. Though existing contextual bandit algorithms have shown their contributions to online recommendation, some new challenges come alongside with the rapid development of social media. In many recent applications, users are always not independent, but have strong social connections among them. Such augmented social network information provides an important source of evidence, reflecting the affinities between users and their friends, and has shown its contribution in offline social recommendation tasks.

To leverage the knowledge from social networks, in this paper, we propose the *Locality-sensitive Linear Bandit Model* (LS.Lin) for online social recommendation. Specifically, our model assumes that each user in the social network has a set of unknown parameters and the feedback (ratings or clicks) of users comes from a linear combination of items' contexts and users' parameters. We show that our proposed method follows a robust theoretical bound as other linear bandit models and it demonstrates improved performance for online social recommendation in empirical studies, compared with other contextual bandit methods.

2   Preliminary

We first introduce some general notations for matrices and vectors. Note we shall use player and user exchangeably in the following sections. Let $G = (V, E)$ denote a social network, where V is a set of users ($|V| = M$) and $E \subset V \times V$ is a set of relationships between users. $N_u(G)$ represents the neighbors of user

u in graph G. We use $\|x\|_p$ to denote the p-norm of a vector $x \in R^d$ and $\mathbf{I}$ to denote identity matrix. For a positive definite matrix $A \in R^{d \times d}$, the weighted norm of vector x is defined as $\|x\|_A = \sqrt{x^T A x}$. Let $\lambda_{min}(A)$ denote the smallest eigenvalue of the positive definite matrix A. det(A) denotes the determinant value of matrix A. The inner product is represented as $\langle \cdot, \cdot \rangle$ and weighted inner product $x A^T y = \langle x, y \rangle_A$.

Now we introduce the contextual bandit problem and show how to involve it into recommender systems. Let T be the number of rounds and k be the number of arms. At time $t \in T$, a player u observes k available arms with their arm-specific feature vectors $\{x_{t,a(i)}\} \in \mathbb{R}^d$, where $i \in 1, 2, ..., k$. Without loss of generality, we assume $\|x_{t,a(i)}\|_2 \leq L$ in this paper, where L is a hyperparameter. The player is then asked to make a decision and select an arm a_t to play and once the arm a_t is chosen, the player receives a reward r_{t,a_t} from it. Specifically, for a particular player u, we use a_t^u to represent u's decision at time t. Note that no reward information from other arms can be observed at this time. For each arm $a(i)$, the reward $r_{t,a(i)}$ is assumed to be a linear function with $x_{t,a(i)}$ as follows,

$$r_{t,a(i)} = x_{t,a(i)}^T \theta_u^\star + \eta_t, \tag{1}$$

where $\theta_u^\star$ is the unknown parameter of player u, and η_t is a zero mean random variable. The goal of a contextual bandit problem is to design a strategic algorithm in order to maximize the user's expected cumulative reward $\mathbb{E}[\sum_{t=1}^{T} r_{t,a_t}]$ over T rounds. The difference between the player's total reward and the total reward of the optimal strategy is called the *pseudo-regret* of the algorithm and it can be written as,

$$R_T = \sum_{t=1}^{T} (r_{t,a_t^\star} - r_{t,a_t}), \tag{2}$$

where $a_t^\star$ represents the player's optimal choice at round t. Similarly, we can use $a_t^{\star u}$ to represent a particular user u's optimal choice at time t. Clearly, maximizing the learner's expected cumulative reward is equivalent to minimize the regret of the algorithm. Therefore, our goal is to design an algorithm whose regret is as small as possible.

3   Locality-Sensitive Linear Bandit Model

Although contextual bandit algorithms have been proven to be an effective online recommendation technique to solve the cold-start and data sparsity problem in recommender systems, how to leverage social network information into contextual bandit algorithms is still not well understood. Different from traditional contextual bandit based online recommendation methods, the users in online social recommendation tasks are not assumed to be independent, but form a social network by setting up links with others.

Motivated by most of offline social recommendation methods [21, 22, 24], we propose the *Locality-sensitive Linear Bandit* (LS.Lin) method based on the assumption that user u's unknown parameter should be similar with his/her local neighbors, $i \in N_u(G)$. Formally, LS.Lin still follows a linear reward formulation to link arm's contextual features with a user's parameters, in which the parameter can be estimated by using ridge regression [1,10]. Moreover, LS.Lin method also requires to minimize the following regularization term when solving the ridge regression,

$$\|\theta_u^\star - \theta_{u'}^\star\|_F^2, \tag{3}$$

where $\| \cdot \|_F$ is the Frobenuis norm and $\theta_{u'}^\star$ represents an *integrated social regularized parameter* from the neighbors of user u, $N_u(G)$

Therefore, seamlessly combining Eqs. (1) and (3) together, we can estimate user u's parameter $\widehat{\theta}_u$ by solving the following regression problem,

$$\frac{1}{2}\|b_{u,t} - X_{u,t}\widehat{\theta}_u\|_F^2 + \frac{\lambda}{4}\|\widehat{\theta}_u\|_F^2 + \frac{\lambda}{4}\|\widehat{\theta}_u - \widehat{\theta}_{u'}\|_F^2 \tag{4}$$

where λ is the weight for regularization terms, $X_{u,t}$ is a matrix whose rows are x_{1,a_1^u}, x_{2,a_2^u},...,x_{t,a_t^u} corresponding to user u's historical selected contexts and $b_{u,t}$ is the corresponding historical rewards vector. $\widehat{\theta}_u$ is the estimation of user u's true but unknown parameters $\theta_u^\star$ and $\widehat{\theta}_{u'}$ is the estimated social regularized parameter. Thus, $\widehat{\theta}_u$ can be estimated as Eq. 5.

$$\widehat{\theta}_u = (X_{u,t}X_{u,t}^T + \lambda I)^{-1}(X_{u,t}b_{u,t} + \frac{\lambda}{2}\widehat{\theta}_{u'}), \tag{5}$$

Since in real online social recommendation scenario, different users may not be involved simultaneously (users rate movies or click links at different time), it is crucial to differentiate the confidence interval of the estimated parameters for different users with different number of historical contexts and rewards, especially when constructing a similarity regularization term ($\widehat{\theta}_{u'}$). To achieve this goal, we record the number of feedback from each user u till time t as n_u, and employ a weighted combination based on softmax functions to produce a reliable social regularization term for the proposed LS.Lin method.

As depicted above, the basic idea of LS.Lin model is to maintain a confidence set for the true parameter of each user u with the help of his/her local neighbor's parameters. Specifically, for each round, the confidence set is constructed from his historical contexts x_{1,a_1^u},...,x_{t,a_t^u}, the corresponding rewards $\{r_{t,a_t^u}\}$ and his/her neighbors' estimated parameters $\widehat{\theta}_{u'}$. In the following Theorem 1, we show that by estimating parameter as Eq. 5, the true parameter $\theta_u^\star$ always falls into the confidence set with high probability. Thus, the algorithm can efficiently compute an upper confidence bound of rewards for each arm, $\{\hat{r}_{t,a(1)}, ..., \hat{r}_{t,a(k)}\}$. Finally, the arm with highest upper confidence bound will be given to the oracle as the recommendation in this round. The algorithm of proposed model is shown as Algorithm 1 and the time complexity over T rounds is $O(T(d^3 + kd))$, where k is the number of arms and d is the number of features of each arm. In Algorithm 1, at each time t, we first use a *softmax* function to build up a weighted vector for u's neighbors according to their historical playing times and then use it

---

Input: $\lambda, \alpha_1, \alpha_2, ..., \alpha_T$
Initialization:
for *each user u* **do**
| $A_u^0 \leftarrow \lambda \mathbf{I}^{d \times d}, \mathbf{b_u} \leftarrow \mathbf{0}^d$
end
Simulation:
for *round* $t \leftarrow 1, ..., T$ **do**
  for *each user u* **do**
    for $v \in N_u(G)$ **do**
    | $p_v \leftarrow \frac{\exp(n_v)}{\sum_{v' \in N_u(G)} \exp(n_{v'})}$
    end
    $\widehat{\theta}_{u'} \leftarrow \sum_{v \in N_u(G)} p_v \widehat{\theta}_v$
    $\widehat{\theta}_u \leftarrow A_u^{-1}(b_u + \frac{\lambda}{2}\widehat{\theta}_{u'})$
    for $i \in 1, ..., k$ **do**
    | $\hat{r}_{t,a(i)} \leftarrow x_{t,a(i)}^T \widehat{\theta}_u + \alpha_t \sqrt{x_{t,a(i)}^T A_u^{-1} x_{t,a(i)}}$
    end
    Choose the arm $a_t^u \leftarrow \arg\max_{a(i)} \hat{r}_{t,a(i)}$
    Observe rewards r_{t,a_t^u}
    $A_u^t \leftarrow A_u^{t-1} + x_{t,a_t^u} x_{t,a_t^u}^T$
    $b_u \leftarrow b_u + r_{t,a_t^u} x_{t,a_t^u}$
    $n_u \leftarrow n_u + 1$
  end
end

Algorithm 1. LS.Lin simulation algorithm.

to construct the social regularization term $\widehat{\theta}_{u'}$. In this way, we can offer more confidence on u's neighbors with more historical records to build a reliable social regularization term. Moreover, the softmax-based combination allows LS.Lin to handle the change of social networks by only changing components involved in the computation for $\widehat{\theta}_{u'}$. After u finishes playing and providing feedback to the system, only u's parameters, $X_{u,t}$, $b_{u,t}$ and n_u, will be updated.

Now, we show the theoretical analysis of the proposed LS.Lin algorithm. The main theorem is stated as follows.

Theorem 1 (Cumulative regret analysis of Algorithm 1). *Without loss of generality, assume that* $\|\theta_u^*\|_2 \leq 1$, $\|x_{t,a(i)}\|_2 \leq L$ *and* $r_{t,a(i)} \in [0,1]$. *Given* $\delta \geq 0$ *and set* $\alpha_t = \sqrt{d\log(\frac{1+\frac{tL^2}{\lambda}}{\delta})} + \frac{3}{2}\lambda^{\frac{1}{2}}$, *with probability at least* $1 - \delta$, *the cumulative regret of LS.Lin satisfies*

$$R_T \leq 4M\sqrt{dT\log(1 + \frac{TL^2}{\lambda d})}(\sqrt{d\log(\frac{1+\frac{TL^2}{\lambda}}{\delta})} + \frac{3}{2}\lambda^{\frac{1}{2}}),$$

for any $T \geq 0$.

To prove Theorem 1, we first state a modified concentration result derived in [1]. This result shows that the true parameter θ_u^* lies within an ellipsoid centered at $\widehat{\theta}_u$ for all rounds $t \geq 0$ with high probability, as Theorem 2 shows.

Theorem 2. *Assume for each user u, $r_{t,a_t^u} = x_{t,a_t^u}^T \theta_u^* + \eta_t$, $\|\theta_u^*\|_2 \leq 1$, $\|x_{t,a_t^u}\|_2 \leq L$ and $r_{t,a_t^u} \in [0,1]$. Define $A_u^t = \lambda I + \sum_t x_{t,a_t^u} x_{t,a_t^u}^T$. Then with probability at least $1 - \delta$, for all round $t \geq 0$,*

$$\|\widehat{\theta}_u - \theta_u^*\|_{A_a^t} \leq \sqrt{d \log(\frac{1 + \frac{tL^2}{\lambda}}{\delta})} + \frac{3}{2}\lambda^{\frac{1}{2}}. \tag{6}$$

Proof of Theorem 2.

$$\begin{aligned}
\widehat{\theta}_u &= (X_{u,t}X_{u,t}^T + \lambda I)^{-1}(X_{u,t}b_{u,t} + \frac{\lambda}{2}\widehat{\theta}_{u'}) \\
&= (X_{u,t}X_{u,t}^T + \lambda I)^{-1}(X_{u,t}(X_{u,t}^T\theta_u^* + \eta_t) + \frac{\lambda}{2}\widehat{\theta}_{u'}). \\
&= \theta_u^* - \lambda(X_{u,t}X_{u,t}^T + \lambda I)^{-1}\theta_{u^*} + (X_{u,t}X_{u,t}^T + \lambda I)^{-1}X_{u,t}\eta_t \\
&\quad + \frac{\lambda}{2}(X_{u,t}X_{u,t}^T + \lambda I)^{-1}\widehat{\theta}_{u'}.
\end{aligned} \tag{7}$$

Thus,

$$\begin{aligned}
x^T\widehat{\theta}_u - x^T\theta_u^* &= -\lambda x^T(X_{u,t}X_{u,t}^T + \lambda I)^{-1}\theta_{u^*} \\
&\quad + x^T(X_{u,t}X_{u,t}^T + \lambda I)^{-1}X_{u,t}\eta_t + \frac{\lambda}{2}x^T(X_{u,t}X_{u,t}^T + \lambda I)^{-1}\widehat{\theta}_{u'} \\
&= \langle x, X_{u,t}^T\eta_t\rangle_{A_u^t{}^{-1}} - \lambda\langle x, \theta_u^*\rangle_{A_u^t{}^{-1}} + \frac{\lambda}{2}\langle x, \widehat{\theta}_{u'}\rangle_{A_u^t{}^{-1}},
\end{aligned} \tag{8}$$

where $A_u^t = X_{u,t}X_{u,t}^T + \lambda I$. Now by following Lemma 9 (Self-normalized bound for vector-valued martingales) from [1], we can easily complete the proof.   □

To show Theorem 1, we still need to provide some lemmas besides Theorem 2.

Lemma 1. Let $A_u^t = \lambda I + \sum_t^T x_{t,a_t^u}x_{t,a_t^u}^T$, assume $\|x_{t,a_t^u}\|_2 \leq L$, then $det(A_u^t) \leq (\lambda + TL^2/d)^d$.

Proof: Let $\alpha_1, ..., \alpha_d$ be the eigenvalues of A_u^t. Then $det(A_u^t) = \prod_i \alpha_i$, $trace(A_u^t) = \sum_i \alpha_i$. According to inequality of arithmetic and geometric means, we can obtain

$$det(A_u^t) \leq (\frac{trace(A_u^t)}{d})^d,$$

since

$$trace(A_u^t) = trace(\lambda I + \sum_t^T x_{t,a_t^u}x_{t,a_t^u}^T) = \lambda d + \sum_t^T \|x_{t,a_t^u}\|_2^2 \leq \lambda d + TL^2,$$

with $\|x_{t,a_t^u}\|_2 \leq L$, we have $det(A_u^t) \leq (\lambda + TL^2/d)^d$.   □

Lemma 2. *Let* $\{x_t^u\}$ *be a sequence*, $A_u^t = \lambda\mathbf{I} + \sum_{s=1}^{t} x_{s,a_s^u} x_{s,a_s^u}^T$, *then if* $\|x_{t,a_t^u}\|_2 \le L$ *and* $\lambda_{min}(\lambda\mathbf{I}) \ge \max(L^2, 1)$,

$$\sum_{s=1}^{t} \|x_{s,a_s^u}\|_{A_u^{s-1-1}}^2 \le 2\log\frac{det(A_u^t)}{det(\lambda\mathbf{I})} \le 2d\log(1 + \frac{tL^2}{\lambda d}).$$

This lemma directly follows Lemma 11 from [1]. Now we could show the proof of Theorem 1.

Proof of Theorem 1. We first show the regret of LS.Lin in each user and each round can be bounded as follows,

$$
\begin{aligned}
regret_t^u &= x_{t,a_t^{*u}}^T \theta_u^* - x_{t,a_t^u}^T \theta_u^* \\
&\le x_{t,a_t^u}^T \tilde\theta_u - x_{t,a_t^u}^T \theta_u^* = x_{t,a_t^u}(\tilde\theta_u - \widehat\theta_u) + x_{t,a_t^u}^T(\widehat\theta_u - \theta_u^*) \\
&\le \|x_{t,a_t^u}\|_{A_u^t-1}(\|\tilde\theta_u - \widehat\theta_{t-1}\|_{A_u^t} + \|\widehat\theta_{t-1} - \theta_u^*\|_{A_u^t}) \\
&\le 2\alpha_t\|x_{t,a_t^u}\|_{A_u^{t-1}},
\end{aligned}
\tag{9}
$$

where $\alpha_t = \sqrt{d\log(\frac{1+\frac{tL^2}{\lambda}}{\delta})} + \frac{3}{2}\lambda^{\frac{1}{2}}$, $\tilde\theta_u$ is the optimal parameter for x_{t,a^u} following the condition $\|\tilde\theta_u\|_2 \le 1$ and the first inequality follows the fact that $x_{t,a_t^u}^T \tilde\theta_u$ is the optimal reward in round t. The last inequality follows Theorem 2. Then the cumulative regret bound of LS.Lin can be obtained as follows,

$$R_T = \sum_{u=1}^{M}\sum_{t}^{T} regret_t^u \le M\sqrt{T\sum_{t}^{T} regret_t^2}$$

$$\le M\sqrt{T\sum_{t}^{T}(2\alpha_t\|x_{t,a_t^u}\|_{A_a^t-1})^2} \le M\sqrt{4\alpha_T^2 T\sum_{t}^{T}(\|x_{t,a_t^u}\|_{A_a^t-1})^2}$$

$$\le 4M\sqrt{dT\log(1 + \frac{TL^2}{\lambda d})}(\sqrt{d\log(\frac{1+\frac{TL^2}{\lambda}}{\delta})} + \frac{3}{2}\lambda^{\frac{1}{2}}).$$

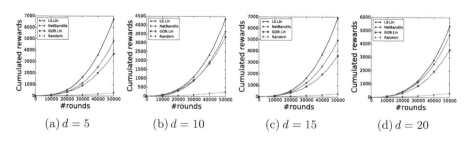

(a) $d = 5$ (b) $d = 10$ (c) $d = 15$ (d) $d = 20$

Fig. 1. Cumulative rewards analysis with different feature space (Last. fm).

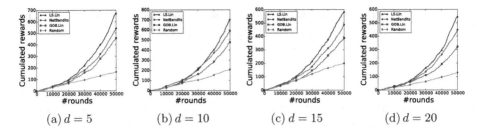

(a) $d = 5$ (b) $d = 10$ (c) $d = 15$ (d) $d = 20$

Fig. 2. Cumulative rewards analysis with different feature space (Delicious).

4 Experimental Results

4.1 Experimental Setup

We conduct experiments on two real world datasets, Last.fm and Delicious, which are both publicly available [8]. Last.fm dataset contains 1,892 users and 17,632 items with 11,946 tags, 12,717 social relations and 92,834 feedbacks. Delicious dataset contains 1,867 users and 69,226 items with 53,388 tags, 7,688 social relations and 104,799 feedbacks. Both datasets contain users' feedback to items. For example, in Last.fm data, users can choose artists and in Delicious dataset, users can bookmark URLs. Here we view artists and URLs as arms and use users' binary feedback as payoff information on each arm. For each arm, we follow the operation of [9,12]: we represent arms by the TF-IDF value of their corresponding tags and then employ PCA to obtain the principle components as final context features. We choose different numbers of principle components and report the experimental results in the following section.

In order to demonstrate the benefits of our approach, we compare the proposed LS.Lin method with three baseline methods: (1) **Random**, (2) **NetBandits** [12] and (3) **GOB.Lin** [9]. Random selection is the simplest strategy that randomly selects an arm for a user in each round. The other two methods are introduced in above sections and since both NetBandits and GOB.Lin have proven their better performance than other contextual bandit algorithms, we would not include other methods for comparison.

4.2 Performance Comparisons

Figures 1 and 2 show the experimental results of average cumulative rewards on two datasets. We find that the proposed model always performs better than baseline methods on the two datasets. The reasons we consider are: (i) although NetBandits [12] is built on an interesting observation that even when a user is randomly selected for promotion, other users close to the selected user in the network will be influenced. Their algorithm, which uses one user's selected context to update all other close users' historical contexts, is unsuitable; (ii) Since GOB.Lin allows one users context to update all users parameters in the social network, the context may mislead the updating of unrelated users through the whole network propagation. Different from these baseline methods, LS.Lin only use users' local social relations to construct a reliable social regularization term and thus avoids the uncertain impact from context propagation.

5   Related Work

Offline Social Recommendation. Integrating social network information into recommender systems has become a popular research direction in recent years. There are also several methods [17,22,27,30,31] that focus on incorporating social network knowledge into recommender systems. Fang et al. [11] decomposed the social trust relations between users into multiple aspects to improve rating estimation accuracy. Guo et al. [15] proposed *TrustSVD* by extending SVD++ [18] to incorporate social trust relations. Ma et al. [21] proposed a SoRec model to collaboratively factorize both rating matrix and social network matrix. Hu et al. [16] combined SoRec and topic matrix factorization to improve recommendation performance. However, none of these methods consider the online setting for social recommendation, where historical records are limited and the structure of social network may change over time.

Multi-armed Bandit. Multi-armed bandit [3,4,6] has attracted great research interests in recent years, ranging applications from online advertising [25,28], routing [5], web search and ranking [2,26], game playing and optimization [13], recommendation [19,29], etc. The contextual bandit problem, which allows arms to be observed with their features, arises as a natural extension of traditional bandit problems and has shown success in online recommendation [1,10,20]. Although much effort has been done in contextual bandit problems, only a few recent papers investigate social networked bandit problems. [7] considered a non-contextual bandit problem in social networks. [9] proposed a GOB.Lin method and [12] proposed a disjoint linear bandit algorithm. [14] investigated the online community bandit problems and [23] further studied this problem with a dynamic community formation setting.

6   Conclusion and Future Work

In this paper, we study the online social recommendation problem and propose a *Locality-sensitive Linear Bandit Model* to solve it. We provide a theoretical analysis to it and conduct experiments to evaluate the performance of our method on two real world datasets. The results demonstrate the effectiveness of the proposed model compared with baselines. For future work, we are interested in investigating how to use bandit algorithms to model complex user behaviors in social networks, such as collaboration or competition.

Acknowledgments. The work described in this paper was partially supported by the Research Grants Council of the Hong Kong Special Administrative Region, China (No. CUHK 14208815 of the General Research Fund), and 2015 Microsoft Research Asia Collaborative Research Program (Project No. FY16-RES-THEME-005).

References

1. Abbasi-Yadkori, Y., Pal, D., Szepesvari, C.: Improved algorithms for linear stochastic bandits. In: NIPS (2011)
2. Agarwal, D., Chen, B.-C., Elango, P.: Explore/exploit schemes for web content optimization. In: ICDM (2009)
3. Audibert, J.-Y., Munos, R., Szepesvari, C.: Exploration-exploitation tradeoff using variance estimates in multi-armed bandits. Theoret. Comput. Sci. **410**(19), 1876–1902 (2009)
4. Auer, P., Cesa-Bianchi, N., Fischer, P.: Finite-time analysis of the multiarmed bandit problem. Mach. Learn. **47**, 235–256 (2002)
5. Awerbuch, B., Kleinberg, R.: Online linear optimization and adaptive routing. J. Comput. Syst. Sci. **74**(1), 97–114 (2008)
6. Bubeck, S., Cesa-Bianchi, N.: Regret analysis of stochastic and nonstochastic multi-armed bandit problems. Found. Trends Mach. Learn. **5**, 1–122 (2012)
7. Buccapatnam, S., Eryilmaz, A., Shroff, N.B.: Multi-armed bandits in the presence of side observations in social networks. OSU Tech. rep. (2013)
8. Cantador, I., Brusilovsky, P., Kuflik, T.: 2nd workshop on information heterogeneity and fusion in recommender systems (hetrec 2011). In: Recsys (2011)
9. Cesa-Bianchi, N., Gentile, C., Zappella, G.: A gang of bandits. In: NIPS (2013)
10. Chu, W., Li, L., Reyzin, L., Schapire, R.E.: Contextual bandits with linear payoff functions. In: AISTAS (2011)
11. Fang, H., Bao, Y., Zhang, J.: Leveraging decomposed trust in probabilistic matrix factorization for effective recommendation. In: AAAI (2014)
12. Fang, M., Tao, D.: Networked bandits with disjoint linear payoffs. In: KDD (2014)
13. Gai, Y., Krishnamachari, B., Jain, R.: Combinatorial network optimization with unknown variables: multi-armed bandits with linear rewards and individual observations. TON **20**(5), 1466–1478 (2012)
14. Gentile, C., Li, S., Zappella, G.: Online clustering of bandits. In: ICML (2014)
15. Guo, G., Zhang, J., Yorke-Smith, N.: Trustsvd: collaborative filtering with both the explicit and implicit influence of user trust and of item ratings. In: AAAI (2015)
16. Hu, G.-N., Dai, X.-Y., Song, Y., Huang, S.-J., Chen, J.-J.: A synthetic approach for recommendation: combining ratings, social relations, and reviews. In: IJCAI (2015)
17. Hu, L., Sun, A., Liu, Y.: Your neighbors affect your ratings: on geographical neighborhood influence to rating prediction. In: SIGIR (2014)
18. Koren, Y.: Factorization meets the neighborhood: a multifaceted collaborative filtering model. In: KDD (2008)
19. Lacerda, A., Santos, R.L., Veloso, A., Ziviani, N.: Improving daily deals recommendation using explore-then-exploit strategies. Inf. Retr. **18**(2), 95–122 (2015)
20. Li, L., Chu, W., Langford, J., Schapire, R.E.: A contextual-bandit approach to personalized news article recommendation. In: WWW (2010)
21. Ma, H., Yang, H., Lyu, M.R., King, I.: Sorec: social recommendation using probabilistic matrix factorization. In: CIKM
22. Ma, H., Zhou, D., Liu, C., Lyu, M.R., King, I.: Recommender systems with social regularization. In: WSDM (2011)
23. Nguyen, T.T., Lauw, H.W.: Dynamic clustering of contextual multi-armed bandits. In: CIKM (2014)
24. Noel, J., Sanner, S., Tran, K., Christen, P., Xie, L., Bonilla, E.V., Abbasnejad, E., Penna, N.D.: New objective functions for social collaborative filtering. In: WWW (2012)

25. Pandey, S., Olston, C.: Handling advertisements of unknown quality in search advertising. In: NIPS (2006)
26. Radlinski, F., Kleinberg, R., Joachims, T.: Learning diverse rankings with multi-armed bandits. In: ICML (2008)
27. Shen, Y., Jin, R.: Learning personal + social latent factor model for social recommendation. In: Proceedings of SIGKDD, pp. 1303–1311 (2012)
28. Slivkins, A.: Multi-armed bandits on implicit metric spaces. In: NIPS (2011)
29. Tang, L., Jiang, Y., Li, L., Zeng, C., Li, T.: Personalized recommendation via parameter-free contextual bandits (2015)
30. Zhao, T., Li, C., Li, M., Ding, Q., Li, L.: Social recommendation incorporating topic mining and social trust analysis. In: CIKM (2013)
31. Zhao, T., McAuley, J.J., King, I.: Leveraging social connections to improve personalized ranking for collaborative filtering. In: CIKM (2014)

An Online-Updating Approach on Task Recommendation in Crowdsourcing Systems

Man-Ching Yuen[✉], Irwin King, and Kwong-Sak Leung

Department of Computer Science and Engineering,
The Chinese University of Hong Kong, Sha Tin, Hong Kong
{mcyuen,king,ksleung}@cse.cuhk.edu.hk

Abstract. In crowdsourcing systems, task recommendation can help workers to find their right tasks faster as well as help requesters to receive good quality output quicker. A number of previous works adopted active learning for task recommendation in crowdsourcing systems to achieve certain accuracy with a very low cost. However, the model updating methods in previous works are not suitable for real-world applications. In our paper, we propose a generic online-updating method for learning a factor analysis model, ActivePMF on TaskRec (Probabilistic Matrix Factorization with Active Learning on Task Recommendation Framework), for crowdsourcing systems. The larger the profile of a worker (or task) is, the less important is retraining its profile on each new work done. In case of the worker (or task) having large profile, our algorithm only retrains the whole feature vector of the worker (or task) and keeps all other entries in the matrix fixed. Besides, our algorithm runs batch update to further improve the performance. Experiment results show that our online-updating approach is accurate in approximating to a full retrain while the average runtime of model update for each work done is reduced by more than 90 % (from a few minutes to several seconds).

1 Introduction

Crowdsourcing aims to outsource a task to people on the Internet to reduce the production cost [1]. In recent years, crowdsourcing systems attract much attentions at present [8]. In a crowdsourcing system, a requester has to verify the quality of every answer submitted by workers, and it is very time-consuming. Alternatively, requesters highly rely on redundancy of answers provided by multiple workers with varying expertise, but massive redundancy is very expensive and time-consuming. The available worker history makes it possible to mine workers preference on tasks and to provide favorite recommendations. Task recommendation can help requesters to receive good quality output quicker as well as help workers to find their right tasks faster. Probabilistic Matrix Factorization (PMF) [7] is the state-of-the-art approach for recommendation systems. A factorization model has to be trained and learned before the model can be applied for prediction. In real-world applications, the performance of a factorization model is highly affected by how the model is updated, and thus dynamic

© Springer International Publishing AG 2016
A. Hirose et al. (Eds.): ICONIP 2016, Part I, LNCS 9947, pp. 91–101, 2016.
DOI: 10.1007/978-3-319-46687-3_10

updating a model is very important [6]. When updating a worker's profile, the profile will not change much if the worker having large profile; while the profile will have great change if the worker having small profile. Therefore, it does not make sense to retrain the model from scratch whenever a worker of large profile completes a task, because the performance improvement by retraining the model in the case is tiny but the cost of retraining model is high. Moreover, when a large number of workers are working in the crowdsourcing system at the same period of time, the computational complexity is very high if the model is retrained after each worker completes a task. Batch update can reduce both user waiting time and computational complexity.

To overcome the weakness mentioned above, this paper proposes a task recommendation framework for quality assurance for crowdsourcing systems. Our contribution are: (1) This paper proposes a generic online-updating method for learning a factor analysis model, ActivePMF on TaskRec (Probabilistic Matrix Factorization with Active Learning on Task Recommendation Framework). Our approach considers the varying expertise of workers for different tasks in real crowdsourcing scenarios. The most informative task and the most skillful worker are selected to learn the factor analysis model. (2) Our proposed online-updating methods are generic and applicable for all PMF models. (3) Complexity analysis shows that our model is efficient and is scalable to large datasets. (4) Experimental results show that the prediction of online-updating ActivePMF on TaskRec model approximates to that of a full retrain of ActivePMF on TaskRec model while the running time of online-updating algorithm is significantly lower than that of a full retrain of the model (reduced by more than 90 %).

The rest of this paper is organized as follows. Section 2 presents the related works. Section 3 presents our task recommendation framework for quality assurance in crowdsourcing systems. Section 4 presents our proposed Online-Updating Probabilistic Matrix Factorization with Active Learning algorithm. Section 5 describes our experiments. Section 6 concludes our paper.

2   Related Work

2.1   Task Recommendation in Crowdsourcing Systems

Recently, many research works [2,3,9–11] proposed recommendation systems based on a Probabilistic Matrix Factorization (PMF) model to improve the output quality in crowdsourcing systems. Jung and Lease [3] proposed to use a PMF model to infer unobserved labels to reduce the bias of the existing crowdsourced labels, thus improve the quality of labels. Later, Jung [2] proposed to use a PMF model to improve the quality of crowdsourcing tasks. Experimental results proved that the strength of PMF over Singular Value Decomposition (SVD) and baseline methods. However, it does not consider a huge number of tasks on crowdsourcing systems in reality. Yuen et al. [9,11] considered various task categories in real scenarios in crowdsourcing systems and proposed a PMF model for task recommendation in crowdsourcing systems. They proved that considering task categories in PMF can improve the performance. However, it does not consider to reduce the labeling cost by applying active learning approach. Later, Yuen et al. [10] proposed an active learning model for task recommendation systems, but new workers have to wait for a long time before having a list of preferred tasks recommended due to lack of worker performance history for all new workers. Moreover, some previous works proposed various ways to improve the performance of recommendation systems for real-world scenarios [4, 6]. However, no previous work considers dynamic-updating for reducing the user waiting time in task recommendation in crowdsourcing systems.

2.2 Our Motivation

Our motivation is the observation of the need to improve system performance in terms of model update time per completed task in task recommendation for real-world scenarios [4]. We propose an online-updating method for learning a factor analysis model, ActivePMF on TaskRec, to recommend tasks in crowdsourcing systems. The prediction of our online-updating approach approximates that of a full retrain of the model, whilst the running time of our online-updating approach is significantly lower than that of a full retrain of the model.

3 Task Recommendation Framework

Our task recommendation framework (TaskRec) is based on matrix factorization method, to perform factor analysis to learn the worker latent feature, the task latent feature and the task category latent feature. For quality assurance, ActivePMF is used to select the most informative task to be learned and select the best worker to query from. We define the problem of task recommendation in crowdsourcing systems as follows:

Definition 1 *Task recommendation problem: Given a worker w_i, a set of tasks $VS = \{v_j\}_{j=1}^n$ and a set of ratings $R = \{r_{ij}\}$ associated between worker w_i and task v_j, rank the ratings in R and select the top few tasks in VS for task recommendation to worker w_i.*

To facilitate our discussions, Table 1 defines basic terms and notations used throughout this paper.

3.1 Probabilistic Matrix Factorization on Task Recommendation Framework

Our model consists of three parts. First, we connect workers' task preferring information with workers' category preferring information through the shared worker latent feature space. Second, we connect workers' task preferring information with tasks' category grouping information through the shared task latent feature space. Third, we connect workers' category preferring information with tasks' category grouping information through the shared category latent feature space. The graphical model of the TaskRec framework is represented in Fig. 1.

Table 1. Basic notations throughout this paper

Notation	Description
$WS= \{w_i\}_{i=1}^m$	WS is the set of workers, w_i is the i-th worker, m is the total number of workers
$VS= \{v_j\}_{j=1}^n$	VS is the set of tasks, v_j is the j-th task, n is the total number of tasks
$CS= \{c_k\}_{k=1}^o$	CS is the set of task categories, c_k is the k-th task category, o is the total number of task categories
$l \in \mathbb{R}$	l is the number of dimensions of latent feature space
$W \in \mathbb{R}^{l \times m}$	W is the worker latent feature matrix
$V \in \mathbb{R}^{l \times n}$	V is the task latent feature matrix
$C \in \mathbb{R}^{l \times o}$	C is the task category latent feature matrix
$R = \{r_{ij}\}, R \in \mathbb{R}^{m \times n}$	R is the worker-task preferring matrix, r_{ij} is the extent of the favor of task v_j for worker w_i
$U = \{u_{ik}\}, U \in \mathbb{R}^{m \times o}$	U is the worker-category preferring matrix, u_{ik} is the extent of worker w_i's preference for task category c_k
$D = \{d_{jk}\}, D \in \mathbb{R}^{n \times o}$	D is the task-category grouping matrix, d_{jk} indicates the task category c_k that task v_j belongs to
$N(x\|\mu,\sigma^2)$	Probability density function of Gaussian distribution (mean μ, variance σ^2)
I_{ij}^R	I_{ij}^R is the indicator function that is equal to 1 if the entry r_{ij} is observed and equal to 0 otherwise

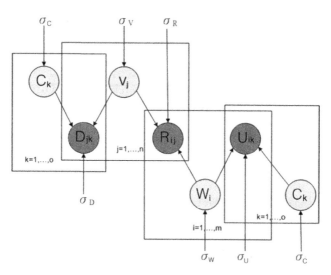

Fig. 1. Graphical Model for TaskRec

Table 2. Transformation of workers' behaviors into values

Worker behavior		Value
Worker's work done is accepted by requester	$\longrightarrow$	5
Worker's work done is rejected by requester	$\longrightarrow$	4
Worker completes a task and submits the work done	$\longrightarrow$	3
Worker selects a task to work on but not complete it	$\longrightarrow$	2
Worker browses the detailed information of a task	$\longrightarrow$	1
Worker does not browse the detailed information of a task	$\longrightarrow$	0

By using a worker-task preferring matrix, we can measure the extend the worker prefer to work the task and provide output that accepted by requesters. Unlike traditional recommendation systems, workers do not have to give ratings to tasks to indicate the extent of their favor of each task. To have ratings on tasks, we transform workers' behaviors into values as shown in Table 2.

According to the graphical model of the TaskRec framework described in Fig. 1, the local minimum can be found by performing the gradient descent on W_i, V_j and C_k. The detailed derivation can be found in [11].

4   Online-Update on Active Learning for Concurrent Selection of Task and Worker

In this section, we present our Online-Updating Probabilistic Matrix Factorization with Active Learning, which is presented in Algorithm 1.

4.1   Random New Task Selection for Reliable Worker

To learn the most accurate classifier with the least number of workdone, we first query all new tasks, as given in Eq. (1), and select the most reliable worker in the task category to query from, as given in Eq. (2).

$$v^* = \{v_j | \exists v_j \in VS; I_{ij} = 0, \forall w_i \in WS\}. \tag{1}$$

$$w^* = \arg \max_{w_i \in WS} u_{ik} \quad where \quad d_{jk} = 1, v_j = v^*. \tag{2}$$

In Algorithm 1, Step 6 represents the process of new task selection for the most reliable worker in the category.

Algorithm 1 Online-Updating on ActivePMF

Input:

    Partially observed worker-task matrix, R;

    Threshold, $Threshold$;

    Batch size, $BatchSize$;

    Active-sampling heuristic, h (use uncertainty-sampling using the Maximum Difference between predicted rate and observed rate: $\arg\max_{v_j \in VS} \sum_{i=1}^{m} \frac{1}{\sum_{i=1}^{m} I_{ij}} \left| I_{ij}^{R} \left(g \left(W_i^T V_j \right) - r_{ij} \right) \right|$);

Output:

    Full worker-task matrix R_{full} valued within the interval $[0, 1]$ predicting unobserved entries of R;

Initialize:

    $R_{tmp} = R$; /* currently observed data */

    $NumQueries = 0$; /*num of queries done by workers*/

1: $\bar{R}_{full} = \text{PMF}(R_{tmp})$; /* compute full matrix $\bar{R}_{full}$ */

2: Un = set of unobserved entries of R_{tmp};

3: $NewT$ = Select all new tasks;

4: $NewW$ = Select all new workers;

5: Set = ActiveSelect(h, $\bar{R}_{full}$, Un); /* select the most uncertain unobserved instances (Maximum Difference between predicted rate and observed rate) from Un-New using h and current predictions $\bar{R}_{full}$ */

6: **if** $|NewT| > 0$ **then**

    Select a new task v^* from New using Eq. (1);

    Select the most reliable worker w^* for task v^* using Eq. (2);

    Request worker w^* to work on task v^*;

    Add the rate to R_{tmp};

    $NumQueries = NumQueries + 1$;

7: **else**

8:     **if** $|NewW| > 0$ **then**

    Select the most uncertain task v^{**} from Set using Eq. (3);

    Select a new worker w^{**} for task v^{**} using Eq. (4);

    Request worker w^{**} to work on task v^{**};

    Add the rate to R_{tmp};

    $NumQueries = NumQueries + 1$;

9:     **else**

    Select the most uncertain task v^{***} from Set using Eq. (5);

    Select the most reliable worker w^{***} for task v^{***} using Eq. (6);

    Request worker w^{***} to work on task v^{***};

    Add the rate to R_{tmp};

    $NumQueries = NumQueries + 1$;

10:     **end if**

11: **end if**

12: **if** (Worker Profile $>= Threshold$) and (Task Profile $>= Threshold$) **then**

    Update the feature vectors of the selected worker $w_{m'}$;

    Update the feature vectors of the selected task $v_{n'}$;

13: **else**

14:     **if** (Worker Profile $>= Threshold$) **then**

    Update the feature vectors of the selected worker $w_{m'}$;

15:     **else**

16:         **if** (Task Profile $>= Threshold$) **then**

        Update the feature vectors of the selected task $v_{n'}$;

17:         **else**

        Update the feature vectors of all workers and all tasks;

18:         **end if**

19:     **end if**

20: **end if**

21: **if** (No new incoming work done) **then**

22:     **return** $\bar{R}_{full}$;

23: **end if**

24: **if** ($NumQueries$ mod $BatchSize = 0$) **then**

    $NumQueries = 0$;

    Go to Step 1;

25: **else**

    Go to Step 2;

26: **end if**

4.2 Uncertainty Sampling for Task Selection for Randomly Selected New Worker

The algorithm assumes a particular active learning heuristic specified as an input, and we adopt uncertainty-sampling [5] using the Maximum Difference between predicted rate and observed rate as in Eq. (3) to choose the most uncertain task, that requires minimization of uncertainty. To let new workers having a list of preferred tasks recommended but not having to work on a large amount of tasks beforehand, we randomly select a new worker (if any) to query from, as given in Eq. (4).

$$v^{**} = \arg \max_{v_j \in VS} \sum_{i=1}^{m} \frac{1}{\sum_{i=1}^{m} I_{ij}} \left| I_{ij}^{R} \left(g \left(W_i^T V_j \right) - r_{ij} \right) \right|. \tag{3}$$

$$w^{**} = \{ w_i | \exists w_i \in WS; I_{ij} = 0, \forall v_j \in VS \}, \tag{4}$$

where the logistic function $g(x)$ bounds the range of $W_i^T V_j$ within the range $[0, 1]$.

In Algorithm 1, Step 8 represents the process of most uncertain task selection for new worker.

4.3 Uncertainty Sampling for Task Selection for Reliable Worker

The algorithm assumes a particular active learning heuristic specified as an input, and we adopt uncertainty-sampling [5] using the Maximum Difference between predicted rate and observed rate as in Eq. (5) to choose the most uncertain task, that requires minimization of uncertainty. To select the most reliable worker for the most uncertain task, we select the worker with the maximum worker-category preferring score where the category that the task belongs to as in Eq. (6).

$$v^{***} = \arg \max_{v_j \in VS} \sum_{i=1}^{m} \frac{1}{\sum_{i=1}^{m} I_{ij}} \left| I_{ij}^{R} \left(g \left(W_i^T V_j \right) - r_{ij} \right) \right|. \tag{5}$$

$$w^{***} = \arg \max_{w_i \in WS} u_{ik} \quad where \quad d_{jk} = 1, v_j = v^{***}. \tag{6}$$

where the logistic function $g(x)$ bounds the range of $W_i^T V_j$ within the range $[0, 1]$.

In Algorithm 1, Step 9 represents the most uncertain task selection for the most reliable worker.

After annotation, the selected task is removed from the unlabeled data set. Then, the selected task and its rate are added to the set of labeled dataset.

Next, the algorithm has to update the model.

4.4   Partial Update

The impact on retraining the whole learning model decreases as the profile size of the worker (or task) increases. Especially when work done by new workers or work done on task having small profile, updating the feature matrix is crucial. For a new worker, each work done by him will result in much change in his task preference in his worker profile; while for a worker that has already completed a lot of tasks, each work done by him will not change much in his worker profile. Updating feature vectors for a worker (or task) having smaller profile results in a much better model. As a result, for a worker (or task) having large profile, we observe that the model learned from retraining the feature vector of the worker (or task) only is approximate to that learned from a full retrain.

In Algorithm 1, Step 12, 14 and 16 represent the process of a partial update. For a worker (or task) having profile size larger than the threshold, the algorithm retrains the feature vector of the worker (or task) and keep all other entries in the matrix unchanged; otherwise, the algorithm retrains the feature vectors of all workers and all tasks.

4.5   Batch Update

The time for retraining a learning model is proportional to the computational complexity of the model and the amount of information stored in the model; while the amount of information depends on the number of workers, the number of tasks and the number of work done. Retraining a large learning model takes a long time. For a large real-world crowdsourcing system, it is inefficient if the whole model is retrained from scratch once a worker completes a task.

In Algorithm 1, Step 24 and 25 represent the process of a batch update. When the number of work done is smaller than the batch size, the learning model is not retrained. On the other hand, when the number of work done is larger than the batch size, the algorithm retrains the learning model.

4.6   Complexity Analysis

To compute the complexity of our Online-Updating ActivePMF, we consider both the computation of the gradient descent methods and the computation of selecting the most uncertain task for the most reliable worker. The main computation of the gradient descent methods is evaluating objective function E and corresponding gradients on variables. Because of the sparsity of matrices R, U, and D, the complexity of evaluating the objective function is $\mathcal{O}\left(n_R l + n_U l + n_D l\right)$, where n_R, n_U and n_D are the number of non-zero entries in matrices R, U, and D respectively, and l is the number of dimensions of latent feature space. For the computation of selecting the most uncertain task for the most reliable worker, the complexity of selecting the most uncertain task in Eq. (5) is $\mathcal{O}\left(n_R l\right)$, the complexity of selecting the most reliable worker in Eq. (6) is $\mathcal{O}\left(m\right)$, and thus the total complexity of assigning a task to a worker is $\mathcal{O}\left(m + n_R l\right)$. As a result, the total complexity for one iteration is $\mathcal{O}\left(m + n_R l + n_U l + n_D l\right)$. It means that the complexity is linear with respect to the number of workers and the number of observations in the three sparse matrices. The complexity analysis shows that Online-Updating ActivePMF can scale to very large datasets.

5   Experimental Analysis

In this section, our experiments are intended to address the following two research questions:

1. How is the partial update on Online-Updating ActivePMF approach compared with the full-retrain of ActivePMF approach?
2. How is the batch update on Online-Updating ActivePMF approach compared with the full-retrain of ActivePMF approach?

5.1   Description of Dataset

Our dataset is retrieved from the recent NAACL 2010 workshop on crowdsourcing, which has made publicly available all the data collected as part of the workshop[1]. The statistics about our dataset: No of workers is 1,592; No of different tasks is 6,639; No of categories is 43; Total HITs from all tasks is 19,815; No of ratings is 19,815. We categorize the dataset by both languages and keywords of tasks given by MTurk [9].

5.2   Evaluation Metrics

To compare the prediction quality of our Online-Updating ActivePMF method with the full retrain of ActivePMF approach, we use the Mean Absolute Error (MAE) and the Root Mean Squared Error (RMSE) as the comparison metrics.

5.3   Performance Comparison

We compare our approach with the full retrain of ActivePMF approach, where Probabilistic Matrix Factorization (PMF) [7] is the state-of-the-art approach for recommendation systems.

   In the comparison, we randomly select 20 % of ratings from the dataset as training data, randomly choose 60 % of ratings from the dataset as active dataset, and leave the remaining 20 % as prediction performance testing. For the value transformation, we have 10,411 approved tasks (value transformed to 5), 9,399 submitted tasks (value transformed to 3) and only 5 rejected tasks (value transformed to 4). Most rejected tasks are removed in our dataset. We set $\theta_W = \theta_V = \theta_C = 0.00004$, set $\theta_U = 0.0001$ and $\theta_D = 0.01$, set the number of latent features $k = 20$. The MAE results, the RMSE results and the runtimes of model update are reported in Table 3.

---

[1] NAACL 2010 workshop: http://sites.google.com/site/amtworkshop2010/data-1.

Table 3. Comparison on a full-retrain with online-updating approach on ActivePMF model learning (Feature k = 20; No of work done = 11,000)

ActivePMF model	MAE	RMSE	Runtime (min)
Full Retrain	**0.0156**	**0.0845**	3.839
Online-Updating (P = 0.001; Batch = 1)	**0.0156**	**0.0845**	3.374
Online-Updating (P = 0.001; Batch = 10)	0.0191	0.0914	0.675
Online-Updating (P = 0.001; Batch = 50)	0.0313	0.1353	0.142
Online-Updating (P = 0.001; Batch = 100)	0.0515	0.2103	0.137
Online-Updating (P = 0.001; Batch = 150)	0.0768	0.2977	0.049
Online-Updating (P = 0.001; Batch = 200)	0.0513	0.2033	0.038
Online-Updating (P = 0.001; Batch = 500)	0.1022	0.3445	**0.017**

By using the partial update method, the prediction quality of ActivePMF with partial updates (threshold $t = 0.001$) approximates to that of ActivePMF with full retrain, but the average runtime per workdone of ActivePMF with partial updates (threshold $t = 0.001$) is greatly reduced by 12 % compared with that of ActivePMF with full retrain.

By using the batch update method, the average runtime on model update per work done can be further reduced. As batch size increases, both the MAE results and the RMSE results also increase, while the average runtime on model update per work done decreases significantly. Batch size 10 is the best choice among all the choices in Table 3.

6   Conclusion

In this paper, we have proposed a generic online-updating method for learning a factor analysis model, ActivePMF on TaskRec, for crowdsourcing systems. Our algorithm improves the runtime of model update significantly but the prediction accuracy still approximates to that of full retrain. Experiment results show that the average runtime of model update for each workdone is reduced by more than 90 % (decreases from a few minutes to several seconds).

Acknowledgment. This research was in part supported by grants from the National Grand Fundamental Research 973 Program of China (No. 2014CB340405), the Research Grants Council of the Hong Kong Special Administrative Region, China (Project No. CUHK 14203314), and Microsoft Research Asia Regional Seed Fund in Big Data Research (Grant No. FY13-RES-SPONSOR-036).

References

1. Howe, J.: The rise of crowdsourcing. Wired **14**(6), 1–4 (2006)
2. Jung, H.J.: Quality assurance in crowdsourcing via matrix factorization based task routing. In: International Conference on World Wide Web (2014)
3. Jung, H.J., Lease, M.: Improving quality of crowdsourced labels via probabilistic matrix factorization. In: Human Computation Workshop at the 26th AAAI (2012)
4. Karimi, R., Freudenthaler, C., Nanopoulos, A., Schmidt-Thieme, L.: Active learning for aspect model in recommender systems. In: Proceedings of IEEE CIDM 2011, pp. 162–167 (2011)
5. Lewis, D.D., Gale, W.A.: A sequential algorithm for training text classifiers. In: Croft, B.W., van Rijsbergen, C.J. (eds.) SIGIR 1994, pp. 3–12. Springer, London (1994)
6. Rendle, S., Schmidt-Thieme, L.: Online-updating regularized kernel matrix factorization models for large-scale recommender systems. In: Proceedings of RecSys 2008, pp. 251–258. ACM, New York (2008)
7. Salakhutdinov, R., Mnih, A.: Probabilistic matrix factorization. In: Proceedings of the Twenty-First Annual Conference on Neural Information Processing Systems, NIPS 2007, Curran Associates Inc. (2007)
8. Yuen, M.-C., King, I., Leung, K.-S.: A survey of crowdsourcing systems. In: Social-Com 2011, pp. 766–773. IEEE Computer Society (2011)
9. Yuen, M.-C., King, I., Leung, K.-S.: TaskRec: probabilistic matrix factorization in task recommendation in crowdsourcing systems. In: Huang, T., Zeng, Z., Li, C., Leung, C.S. (eds.) ICONIP 2012. LNCS, vol. 7664, pp. 516–525. Springer, Heidelberg (2012). doi:10.1007/978-3-642-34481-7_63
10. Yuen, M.-C., King, I., Leung, K.-S.: Probabilistic matrix factorization with active learning for quality assurance in crowdsourcing systems. In: Proceedings of the IADIS International Conference WWW/Internet 2015, Ireland (2015)
11. Yuen, M.-C., King, I., Leung, K.-S.: TaskRec: a task recommendation framework in crowdsourcing systems. Neural Process. Lett. **41**(2), 223–238 (2015)

Neural Data Analysis

Rhinal-Hippocampal Information Flow Reverses Between Memory Encoding and Retrieval

Juergen Fell[1(✉)], Tobias Wagner[1], Bernhard P. Staresina[1,2], Charan Ranganath[3], Christian E. Elger[1], and Nikolai Axmacher[4,5]

[1] Department of Epileptology, University of Bonn, Bonn, Germany
juergen.fell@ukb.uni-bonn.de
[2] Department of Psychology, University of Birmingham, Birmingham, UK
[3] Center for Neuroscience and Department of Psychology, University of California, Davis, USA
[4] Department of Psychology, Ruhr-University Bochum, Bochum, Germany
[5] German Center for Neurodegenerative Diseases, Bonn, Germany

Abstract. The medial temporal lobe is crucial for the encoding and retrieval of episodic long-term memories. It is widely assumed that memory encoding is associated with information transfer from sensory regions via the rhinal cortex into the hippocampus. Retrieval of information should then be associated with transfer in the reverse direction. However, experimental evidence for this mechanism is still lacking. Here, we show in human intracranial EEG data during two independent recognition memory paradigms that rhinal-hippocampal information flow significantly changes its directionality from encoding to retrieval. Using a novel phase-based method to analyze directional coupling of oscillations, coupling values were more positive (i.e., from rhinal cortex to the hippocampus) during encoding as compared to retrieval. These effects were observed in the delta (1–3 Hz) range where rhinal-hippocampal post-stimulus phase synchronization increased most robustly across both experiments.

Keywords: Directional coupling · Long-term memory · Intracranial EEG · Medial temporal lobe · Hippocampus · Rhinal cortex

1 Introduction

Processes within the human medial temporal lobe are crucial for episodic long-term memory [1–3]. It is assumed that encoding of new events depends on information flow from sensory cortices via peri- and entorhinal cortex (here together referred to as rhinal cortex, RC) to the hippocampus (HC), which supports the rapid formation of novel memory traces. Retrieval of these events after a short time period should be associated with information flow in the reverse direction [4, 5]. To our knowledge, no previous study has tested this prediction, possibly due to methodological challenges: First, recording neural activity from medial temporal brain structures is difficult in humans and ideally requires intracranial EEG electrodes, which are only implanted in specific populations of pharmacorefractory epilepsy patients. Second, EEG activity is characterized by oscillatory networks [6, 7], but investigating directional interactions between oscillators is a methodologically complex issue [8].

© Springer International Publishing AG 2016
A. Hirose et al. (Eds.): ICONIP 2016, Part I, LNCS 9947, pp. 105–114, 2016.
DOI: 10.1007/978-3-319-46687-3_11

Here, we report results from two intracranial EEG experiments in which time-resolved analyses of directional coupling were performed in order to characterize the flow of information between RC and HC during memory encoding and retrieval. EEG was recorded via medial temporal depth electrodes in epilepsy patients undergoing presurgical evaluation. In both experiments, a large number of trial-unique pictures either with complex landscapes and houses (experiment 1; n = 11 patients), or with cut-out depictions of faces and houses (experiment 2; n = 7 patients), was presented for encoding and had to be recognized again among a smaller number of lures during retrieval (Fig. 1). To investigate directional information flow, we first identified the frequency band which in both experiments and during both, encoding and retrieval, consistently showed task-related changes in RC-HC phase synchronization (i.e., increases in phase synchronization after a stimulus is presented as compared to baseline). Then, we estimated directional RC-HC coupling within this frequency band with a phase-based method particularly well suited to capture oscillatory EEG dynamics [9, 10]. Finally, directional coupling values for the encoding and the retrieval phase were compared with non-parametric label permutation tests [11]. Because RC-HC

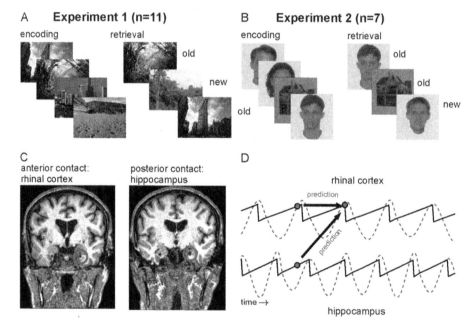

Fig. 1. Analysis of directional coupling during memory encoding and retrieval. (A) In the first experiment (n = 11), recognition memory was tested for complex landscapes and buildings. **(B)** In the second experiment (n = 7), participants encoded pictures of faces and houses, and memory was afterwards inquired in a recognition test. **(C)** Post-implantation MRI from one patient showing one intracranial EEG electrode contact in the rhinal cortex (left) and in the hippocampus (right). **(D)** Directional coupling between rhinal cortex and hippocampus was analyzed based on phase dynamics of delta (1–3 Hz) oscillations in the two regions. In principle, the influence of activity in one region on activity in the other region was quantified by the improvement of prediction of activity in region A at time t by knowledge of activity in region B at time t-τ.

information transfer is thought to depend on the output of early rhinal memory operations [12], we furthermore evaluated the temporal dependency between directional coupling effects and the rhinal N400 component.

2   Methods

2.1   Participants and Experimental Design

Both experiments were performed by patients with pharmaco-refractory temporal lobe epilepsy undergoing invasive presurgical investigation. The first experiment (involving memory encoding and recognition of landscapes and buildings) was conducted by 11 patients (39.4 ± 10.1 years), the second experiment (involving encoding and recognition of houses and faces) by 7 patients (35.3 ± 13.2 years). Previous analyses of these experiments have been published before (first experiment: [13]; second experiment: [14]), and details about the paradigms can be found in these papers.

In brief, in the first experiment (Fig. 1A), we presented complex pictures of landscapes and buildings in four separate learning sessions, which took place on two different days (two on each day). In each learning session, 80 pictures were presented (presentation time 1200 ms, inter-trial interval 1800 ± 200 ms), and subjects had to indicate whether they were presented a landscape (half of the items) or a building. On one day, the first learning session was followed by an afternoon "nap" of around 60 min duration, while subjects did not take a nap on the other day. The second learning session followed 15 min after the nap (or around 90 min after the first learning session on the day without nap) and contained 80 new pictures on each day. During the subsequent retrieval phase, which followed the second learning session after an interval of 15 min, subjects were shown the 160 items presented before on that day, randomly intermixed with 80 new items, and were asked to indicate via button-press whether they remembered each item or not (presentation time 1200 ms, inter-trial interval 2000 ms ± 200 ms).

The second experiment (Fig. 1B) consisted of several (between 2 and 8, depending on the participant's availability) consecutive blocks, each of which contained a familiarization phase, an encoding phase, and a retrieval phase. During the familiarization phase, four stimuli which afterwards served as "repeats" (see below) were presented four times each in random order. These data were not analyzed here. During the encoding phase, 112 pictures from three different classes were presented (presentation time 2500 ms, inter-trial interval 1500 ms), and subjects had to rate their pleasantness by either pressing the left mouse button or the right mouse button. Items from class one (80/112 items, "expected") consisted of trial-unique pictures of a house or a face, presented on a green or a red background. In each block, all expected items showed either faces or houses, and had either red or a green background, and this changed throughout the blocks. Items from class two (16/112 items, "unexpected") showed trial-unique pictures from the other category and with the other background color. Finally, items from class three (16/112 items, "repeats") consisted of the four pictures shown during the familiarization phase, which had the same category and background color as the expected items (but were not trial-unique). During retrieval, we presented 72 test items (presentation time 5000 ms, inter-trial interval 1500 ms). Thirty-two of them were old items shown on "expected" trials, 16 were new items from the same category and background color as in the expected trials, 16 were old items shown on "unexpected" trials, and 8 were novel items from the same category and background color as the unexpected trials. On each trial, subjects indicated on a four-point scale whether they believed an item had been presented before ("sure old", "unsure old", "unsure new", "sure new"). Notably, in our previous paper [14], we compared EEG activity during presentation of expected and unexpected items in the encoding phase; in the current analysis, we focused on "expected" items during both encoding and retrieval.

2.2   Recordings

The location of electrode contacts was ascertained by MRI in each patient (see Fig. 1C for a typical example). Electrodes (AD-Tech, Racine, WI, USA) had 10 cylindrical platinum-iridium contacts and a diameter of 1.3 mm. All recordings were performed using a Schwarzer recording system (Schwarzer GmbH, Munich, Germany). The EEG data were referenced to linked mastoids, recorded at a sampling rate of 1000 Hz, and band-pass filtered (0.01 Hz [6 dB/octave] to 300 Hz [12 dB/octave]). EEG trials were visually inspected for artifacts (e.g., epileptiform spikes), and trials with artifacts were excluded from further analysis. Group statistical analyses were performed by analyzing data from one rhinal and one hippocampal contact in each patient. All recordings were taken from the nonepileptic hemisphere (i.e., contralateral to the epileptogenic focus), to minimize the possibility of artifact contamination. In each patient, we selected one rhinal and one hippocampal contact based on the following criteria: (1) anatomical localization in post-implantation MRI scans; (2) low contamination by electrical and epileptiform artifacts; (3) high overall t-value within the analyzed epochs (averaged across all conditions and then compared to baseline), indicating relatively consistent responses across trials. Recordings were performed at the Department of Epileptology, University of Bonn, Germany. The studies were approved by the local ethics committee, and all patients gave written informed consent.

2.3   Analysis of Phase Synchronization

Phase synchronization was quantified based on calculating circular variance [15]. To avoid edge effects, data were demeaned before phase estimation and the borders of the time windows were cut off afterwards (keeping the central 1200 out of 4096 data points). For the analysis of synchronization strength, we used wavelet transformation (Morlet wavelet) to derive phases for 23 different center frequencies (logarithmically scaled). The synchronization values were averaged across a baseline interval from -200 ms to 0 ms, as well as across a poststimulus interval from 0 to 1000 ms.

2.4   Analysis of Directional Coupling

The time-resolved measure used here is an extended variant of the directional coupling measure proposed by Rosenblum and Pikovsky [9], which is based on the concept of phase synchronization. This time-resolved directional coupling measure has been described in detail previously ([10]; see Fig. 1D). Phase values were estimated using a combination of a band-pass filter (1st order forward/backward Butterworth filter) and a subsequent Hilbert transformation. It was shown that this transformation is equivalent to the usage of the wavelet transformation applied above [16]. In the following, the phases of two systems (1,2) – in our case: EEG recordings from rhinal cortex and hippocampus – are denoted as $\Phi_{1,2}(t_j^r)$. With $j = 1,\dots,n$ as a time index related to an arbitrary time point (e.g. an external cue) of an ensemble of realizations $r = 1,\dots,m$. We modeled the slope of the phase of one system as a function of the phases of both systems. For example, the phase dynamics of system 1 were quantified by the phase slope of system 1 between two time points with a time delay τ,

$$\Delta\phi_1(t_j^r) = \phi_1(t_{j+\tau}^r) - \phi_1(t_j^r)$$

These phase dynamics were modeled by two-dimensional Fourier series

$$\Delta\phi_1 \approx \sum_{k,l} a_j^{k,l} \exp(i[k\phi_1(t_j^r) + l\phi_2(t_j^r)])$$

with k, l denoting the order of the Fourier terms, by approximating the Fourier coefficients $a_j^{k,l}$ in a least-square sense over the m realizations. In this model, the phase dynamic of system 1 depends on previous phase states of system 1 and system 2. Following Rosenblum and Pikovsky [9], we used fixed combinations of orders k, l and a fixed value of τ, corresponding to the average length of one oscillatory cycle. The influence from system 2 onto system 1 was calculated via

$$c_j^2(1|2) = 1/(a_j^{0,0})^2 \sum_{k,l} (a_j^{k,l})^2$$

With $l \neq 0$, as described previously [17]. With the analogously calculated influence of system 1 onto system 2, the time-resolved directional interaction follows as

$$D_j(1,2) = c_j(2|1) - c_j(1|2)$$

Therefore $D_j > 0$ reflects a predominant influence of system 1 onto system 2, $D_j < 0$ a predominant influence of system 2 onto 1, while $D_j \approx 0$ corresponds to a bidirectional or no influence between the systems.

The calculation of directional coupling requires large numbers of trials, which need to be equal across conditions [10]. Therefore, we cumulated all encoding trials from the two subsequent days in experiment 1, and compared these to an equal number of randomly drawn retrieval trials. In experiment 2, the minority of trials which differed in terms of background color and picture category ("unexpected items") were excluded from further analysis, and again an equal number of encoding and retrieval trials was selected. All statistical results were corrected on the cluster level for multiple comparisons using a nonparametric permutation-based approach [11]. This procedure effectively corrects the alpha level for multiple comparisons on an assumption-free basis regarding the sampling distribution under the null hypothesis. We first calculated for each time point whether coupling directions differed significantly (using a paired-samples t-test with an uncorrected threshold of p < .05). Then, we added the t-values for contiguous significant time points, resulting in one sum t-value for each cluster. Next, we shuffled the data across trials within each condition (encoding and retrieval) and for each patient, thereby randomly re-assigning rhinal cortex activity from trial i to hippocampal activity from trial j, and again extracted clusters as in the empirical data. For each permutation, only the largest surrogate cluster was taken into account. Finally, corrected p values were obtained as the rank of each empirical data cluster within the sorted distribution of surrogate data.

2.5   Results

First, we investigated the strength of stimulus-related phase synchronization in different frequency bands during encoding and retrieval. We calculated a four-way ANOVA with "band" (delta: 1–3 Hz, theta: 4–8 Hz, alpha: 9–12 Hz, beta: 13–25 Hz, gamma: 26–45 Hz), "encoding vs. retrieval" and "poststimulus vs. baseline" as repeated measures and "study" as between-subject factor. We found an interaction of "poststimulus vs. baseline" with "band" ($F_{4,64}$ = 4.428; p = 0.003), indicating that both experiments were associated with significantly different task-related changes of phase synchronization in the different frequency bands. Importantly, there were no main effects of, or interactions with, the factors "study" or "encoding vs. retrieval", indicating similar stimulus-related effects across encoding and retrieval and in both studies.

Given the "poststimulus vs. baseline" × "band" interaction, we next calculated separate three-way ANOVAs in all bands with the factors "encoding vs. retrieval" and "poststimulus vs. baseline" as repeated measures and "study" as between-subject factor. In the delta frequency range, there was a significant main effect of "poststimulus vs. baseline" ($F_{1,16}$ = 7.941; p = 0.012), reflecting a significant enhancement of rhinal-hippocampal delta synchronization during memory processing in both paradigms. For the beta band, there were significant interactions for "poststimulus vs. baseline" × "encoding vs. retrieval" ($F_{1,16}$ = 6.959; p = 0.018) and "poststimulus vs. baseline" × "encoding vs. retrieval" × "study" ($F_{1,16}$ = 4.627; p = 0.047). In the beta range, we thus compared phase synchronization during the experiment as compared to the baseline phase separately during encoding and retrieval using T-tests. However, none of these tests became significant. There were no main effects of, or interactions with, the factor "poststimulus vs. baseline" in the theta, alpha, or gamma frequency range. As poststimulus (task-related) increases in RC-HC phase synchronization were exclusively evident in the delta (1–3 Hz) frequency range across the two experiments, further analyses of directional coupling focused on rhinal and hippocampal phase dynamics in this frequency band.

Next, we compared the direction of information flow during encoding and retrieval. As described in detail in the Methods section, we used a surrogate-based non-parametric statistical approach to search for significant clusters of differences between these two experimental conditions [11]. In the first experiment, we found that the direction of information flow significantly differed between encoding and retrieval from 244 to 247 ms and from 251 to 295 ms after stimulus presentation, with an increased RC→HC coupling during encoding as compared to retrieval ($p_{corr} < .05$; Fig. 2A). Similar results were obtained for the second experiment: Here, the direction of information flow reversed between 113 and 161 ms, again with a significantly increased RC→HC coupling during encoding as compared to retrieval (Fig. 2B).

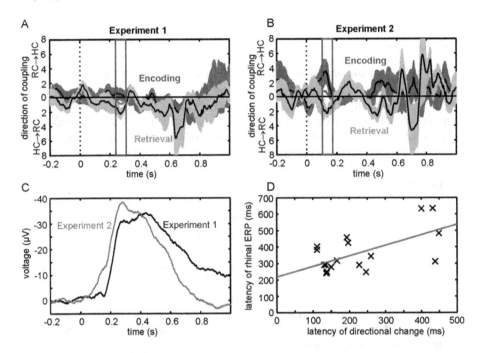

Fig. 2. Direction of rhinal-hippocampal coupling changes between memory encoding and retrieval. (A, B) In two separate recognition memory experiments, we found that during memory encoding, directional coupling values from RC to HC are more positive than during retrieval (average values across patients and standard errors of the mean are shown; difference values are depicted by red lines where significant). These effects occurred slightly earlier in the second as compared to the first paradigm. **(C)** These latency differences were paralleled by a numerically earlier peak of the average encoding-related rhinal event-related potential (N400 component) in the second as compared to the first experiment. **(D)** There was a significant inter-individual correlation between the peak latencies of the rhinal N400 components during encoding and the latencies of directional coupling effects ($R = 0.62$; $p = 0.006$).

Finally, we investigated the functional relevance of these effects in greater detail. Hippocampal memory processes are thought to depend on the output of rhinal memory operations evaluating, for instance, stimulus novelty [18]. Since early rhinal memory operations are considered to be reflected by the N400 component [12], we tested whether latencies of the rhinal N400 component predicted inter-individual differences in the latencies of directional RC-HC coupling effects (Fig. 2C). Corresponding to the earlier directional coupling effect in experiment 2 compared to experiment 1, the average N400 peak during encoding occurred numerically earlier in experiment 2 (see Fig. 2C; individual peak latencies did not significantly differ in both experiments). More importantly, directional coupling latencies across both experiments correlated significantly with the peak latencies of the N400 component during encoding (R = 0.62; p = 0.006; Fig. 2D), and a similar trend was evident during retrieval (R = 0.41; p = 0.09). Hereby, maximal directional coupling differences occurred at times corresponding to the pre-peak slope of the rhinal N400 component.

2.6 Discussion

Our data show that the direction of information flow between RC and HC changes significantly between memory encoding as compared to retrieval, with a stronger RC→HC coupling during encoding and stronger HC→RC coupling during retrieval. The timing of directional coupling differences fits well with single-cell and intracranial EEG data indicating that earliest hippocampal responses to visual stimuli occur around 200 ms [14, 19], i.e. in close temporal proximity to the observed increase of RC→HC information transfer during encoding.

An electrophysiological marker of early rhinal memory operations, which are thought to precede and provide a necessary basis for hippocampal processes, is the rhinal N400 component [12]. This component has been found to be related, for instance, to semantic preprocessing and evaluation of stimulus novelty [18, 20, 21]. Thus, we hypothesized that rhinal N400 peak latencies and the latencies of directional coupling effects may be interindividually correlated. Indeed, we observed such an interrelation. Generally, the peak of an event-related potential rather corresponds to the endpoint than to the initiation of a neural process. Together with the finding that maximal directional coupling differences occurred at times corresponding to the pre-peak slope of the rhinal N400 component, this suggests that RC-HC information transfer crucially depends on the output of early rhinal memory operations.

Furthermore, our results are in line with findings showing that phase synchronization of oscillations across RC and HC is highly relevant for memory operations [7, 22]. During both encoding and retrieval, we consistently detected prominent stimulus-related synchronization enhancements in the delta frequency range. Based on this initial finding, we demonstrated a reversal of directional RC-HC coupling of delta oscillations during encoding compared to retrieval. Our results are in accordance with recent intracranial EEG data indicating a prominent role of human medial temporal delta oscillations for memory encoding and retrieval [23–27]. In this sense, our findings moreover support the idea that rodent medial temporal theta oscillations, which are an essential element in models of hippocampal encoding and retrieval operations [4, 28], may be functionally most closely paralleled by human delta oscillations [24–27, 29].

As a methodological remark, the applied method for estimating directional coupling, although particularly well suited to capture oscillatory dynamics, requires large amounts of data to provide robust estimates [9, 10]. Therefore, we used as many trials as possible and were unable to analyze relevant subconditions, for instance, encoding trials with subsequently remembered versus forgotten items or retrieval trials with correct versus incorrect responses. Still, the interindividual variance of the directional coupling values appears to be quite large, in particular, during the second 500 ms after stimulus presentation (Fig. 2A and B). Hence, application of the described method to more extended encoding and retrieval data is needed to validate our findings. Furthermore, an improved method for the quantification of directional oscillatory coupling being less reliant on large amounts of data would be desirable.

To summarize, our data show that the direction of oscillatory coupling within the human medial temporal lobe reverses between memory encoding and retrieval. Further studies, including analysis of larger EEG data sets and recordings of action potentials in animal experiments, will be required to validate our findings and to elucidate the cellular mechanisms underlying this effect.

References

1. Scoville, W.B., Milner, B.: Loss of recent memory after bilateral hippocampal lesions. J. Neurol. Neurosurg. Psychiatry **20**, 11–21 (1957)
2. Squire, L.R., Stark, C.E., Clark, R.E.: The medial temporal lobe. Ann. Rev. Neurosci. **27**, 279–306 (2004)
3. Eichenbaum, H., Yonelinas, A.P., Ranganath, C.: The medial temporal lobe and recognition memory. Ann. Rev. Neurosci. **30**, 123–152 (2007)
4. Hasselmo, M.E.: What is the function of hippocampal theta rhythm? Linking behavioral data to phasic properties of field potential and unit recording data. Hippocampus **15**, 936–949 (2005)
5. Carr, M.F., Frank, L.M.: A single microcircuit with multiple functions: state dependent information processing in the hippocampus. Curr. Opin. Neurobiol. **22**, 704–708 (2012)
6. Buzsáki, G., Draguhn, A.: Neuronal oscillations in cortical networks. Science **304**, 1926–1929 (2004)
7. Fell, J., Axmacher, N.: The role of phase synchronization in memory processes. Nat. Rev. Neurosci. **12**, 105–118 (2011)
8. Osterhage, H., Mormann, F., Wagner, T., Lehnertz, K.: Detecting directional coupling in the human epileptic brain: limitations and potential pitfalls. Phys. Rev. E **77**, 011914 (2008)
9. Rosenblum, M.G., Pikovsky, A.S.: Detecting direction of coupling in interacting oscillators. Phys. Rev. E **64**, 045202 (2001)
10. Wagner, T., Fell, J., Lehnertz, K.: Detection of transient directional couplings based on phase synchronization. New J. Phys. **12**, 053031 (2010)
11. Maris, E., Oostenveld, R.: Nonparametric statistical testing of EEG- and MEG-data. J. Neurosci. Methods **164**, 177–190 (2007)
12. Fernández, G., Effern, A., Grunwald, T., Pezer, N., Lehnertz, K., Dümpelmann, M., Van Roost, D., Elger, C.E.: Real-time tracking of memory formation in the human rhinal cortex and hippocampus. Science **285**, 1582–1585 (1999)
13. Axmacher, N., Haupt, S., Fernández, G., Elger, C.E., Fell, J.: The role of sleep in declarative memory consolidation - direct evidence by intracranial EEG. Cereb. Cortex **18**, 500–507 (2008)
14. Axmacher, N., Cohen, M.X., Fell, J., Haupt, S., Dümpelmann, M., Elger, C.E., Schlaepfer, T.E., Lenartz, D., Sturm, V., Ranganath, C.: Intracranial EEG correlates of expectancy and memory formation in the human hippocampus and nucleus accumbens. Neuron **65**, 541–549 (2010)
15. Lachaux, J.P., Rodriguez, E., Martinerie, J., Varela, F.J.: Measuring phase synchrony in brain signals. Hum. Brain Mapp. **8**, 194–208 (1999)
16. Bruns, A.: Fourier-, Hilbert- and wavelet-based signal analysis: are they really different approaches? J. Neurosci. Methods **137**, 321–332 (2004)

17. Kralemann, B., Cimponeriu, L., Rosenblum, M.G., Pikovsky, A.S., Mrowka, R.: Phase dynamics of coupled oscillators reconstructed from data. Phys. Rev. E **77**, 066205 (2008)
18. Staresina, B.P., Fell, J., Do Lam, A.T., Axmacher, N., Henson, R.N.: Memory signals are temporally dissociated in and across human hippocampus and perirhinal cortex. Nat. Neurosci. **15**, 1167–1173 (2012)
19. Mormann, F., Kornblith, S., Quiroga, R.Q., Kraskov, A., Cerf, M., Fried, I., Koch, C.: Latency and selectivity of single neurons indicate hierarchical processing in the human medial temporal lobe. J. Neurosci. **28**, 8865–8872 (2008)
20. Nobre, A.C., McCarthy, G.: Language-related field potentials in the anterior-medial temporal lobe: II. Effects of word type and semantic priming. J. Neurosci. **15**, 1090–1098 (1995)
21. Grunwald, T., Beck, H., Lehnertz, K., Blümcke, I., Pezer, N., Kurthen, M., Fernández, G., Van Roost, D., Heinze, H.J., Kutas, M., Elger, C.E.: Evidence relating human verbal memory to hippocampal N-methyl-D-asparate receptors. Proc. Natl. Acad. Sci. U.S.A. **96**, 12085–12089 (1999)
22. Jutras, M.J., Buffalo, E.A.: Synchronous neural activity and memory formation. Curr. Opin. Neurobiol. **20**, 150–155 (2010)
23. Fell, J., Ludowig, E., Rosburg, T., Axmacher, N., Elger, C.E.: Phase-locking within human mediotemporal lobe predicts memory formation. Neuroimage **43**, 410–419 (2008)
24. Lega, B.C., Jacobs, J., Kahana, M.: Human hippocampal theta oscillations and the formation of episodic memories. Hippocampus **22**, 748–761 (2011)
25. Clemens, Z., Borbély, C., Weiss, B., Eröss, L., Szücs, A., Kelemen, A., Fabó, D., Rásonyi, G., Janszky, J., Halász, P.: Increased mesiotemporal delta activity characterizes virtual navigation in humans. Neurosci. Res. **76**, 67–75 (2013)
26. Watrous, A.J., Tandon, N., Conner, C.R., Pieters, T., Ekstrom, A.D.: Frequency-specific network connectivity increases underlie accurate spatiotemporal memory retrieval. Nat. Neurosci. **16**, 349–356 (2013)
27. Watrous, A.J., Lee, D.J., Izadi, A., Gurkoff, G.G., Shahlaie, K., Ekstrom, A.D.: A comparative study of human and rat hippocampal low-frequency oscillations during spatial navigation. Hippocampus **23**, 656–661 (2013)
28. Hanslmayr, S., Staudigl, T.: How brain oscillations form memories - a processing based perspective on oscillatory subsequent memory effects. Neuroimage **85**, 648–655 (2014)
29. Buzsáki, G., Logothetis, N., Singer, W.: Scaling brain size, keeping timing, evolutionary preservation of brain rhythms. Neuron **80**, 751–764 (2013)

Inferred Duality of Synaptic Connectivity in Local Cortical Circuit with Receptive Field Correlation

Kohei Watanabe[1], Jun-nosuke Teramae[1,2($\boxtimes$)], and Naoki Wakamiya[1,2]

[1] Graduate School of Information Science and Technology,
Osaka University, Suita, Japan
teramae@ist.osaka-u.ac.jp
[2] Center for Information and Neural Networks, Osaka University, Suita, Japan

Abstract. Synaptic connections in local cortical circuit are highly heterogeneous and nonrandom. A few strong synaptic connections often form "cluster" that is a tightly connected group of several neurons. Global structure of the clusters, however, has not been clarified yet. It is unclear whether clusters distribute independently and isolated in cortical network, or these clusters are a part of large-scale of global network structure. Here, we develop a network model based on recent experimental data of V1. In addition to reproducing previous result of highly skewed EPSPs, the model also allows us to study mutual relationship and global feature of clusters. We find that the network consists with two largely different sub-networks; a small-world network consists only of a few strong EPSPs and a random network consists of dense weak EPSPs. In other words, local cortical circuit shows a duality, and previously reported clusters are results of local observation of the global small-world network.

1 Introduction

Local cortical circuit is highly nonrandom [4,10,11]. Distribution of synaptic strength, i.e. amplitude distribution of excitatory postsynaptic neurons (EPSPs) is largely skewed to the right hand side, which is well described by long-tail distributions, typically the lognormal distribution [2,8]. The lognormal distribution of synaptic strength means that whereas almost all connections are weak, a few synaptic connections are extremely strong comparing with the weak typical value of amplitude of EPSPs. Actually, while typical amplitude of EPSP is less than $1\,\mathrm{mV}$, a few connections have about $10\,\mathrm{mV}$ EPSPs, which is dozens of times larger than the typical strength.

A few strong EPSPs distribute neither randomly nor homogeneously in local cortical circuit. Physiological experiments using multi-electrode array revealed that strong synaptic connections often form "cluster" structure [11]. For pairs of excitatory neurons, probability of bidirectional connection is higher than naively expected from probability of unidirectional connections and the former probability increases as strengths of EPSPs of these connection increases [11]. Moreover,

A. Hirose et al. (Eds.): ICONIP 2016, Part I, LNCS 9947, pp. 115–122, 2016.
DOI: 10.1007/978-3-319-46687-3_12

for triplet of excitatory neurons, distribution of cluster structure that characterized patterns of connections among these three neurons, called "network motif",
clearly shows that connected network motives, or clusters, are formed with significantly high probability than random network [11].

Synaptic connectivity including skewed distribution of EPSPs and cluster
structure in the network actually affects spiking dynamics of neurons in local
circuit. Spontaneous ongoing activity, for example, that is measured actually in
vivo is realized more robustly in networks with the skewed EPSP distribution
than these with normally distributed EPSPs [5,7,12,13]. Moreover, it is shown
by theoretical studies that existence of cluster structures in network often enrich
variety of the spontaneous firing state [1,6,9], which can responsible for various
computational function in the brain like as working memory for example.

However, global nature of the cluster structure has not been clarified yet.
Because previous studies of cluster structure treated groups of neurons consist
of up to three neurons, it is still unknown whether these clusters distribute
independently and are isolated in the network or they are the tips of the icebergs
of a global structure of the local circuit.

In this paper, in order to solve the problem, we develop a network model of
neurons based on the data obtained from the latest physiology experiment of
primary visual cortex (V1) that reveals positive relationship between synaptic
connectivity between pairs of neurons and correlation of receptive fields of them
[3]. In addition to reproduce previous results of cluster structure, we show that
we can approach global relationship among them owing to mediation of receptive
fields of neurons at edges of each synaptic connection. As a result, we show that
previously reported clusters are a just observed parts of global "small world"
feature of the network. We also show that the network consists two sub-networks
with different network topology; the small-world network consists of a few strong
EPSPs and apparently random network consists of major and dense weak EPSPs.

2   Network Model

We develop our network model based on recently reported data of neurons in V1
that shows positive relationship between synaptic connectivity between neurons
and correlation between receptive fields of them. In the paper, the authors find
that both connection probability and mean amplitude of synaptic connections, if
it exists, increases almost monotonically as correlation of receptive fields of them
increases. In other words, neurons are connected more tightly if their receptive
fields are more similar.

We first prepare 10000 of excitatory model neurons and assign them receptive
fields virtually using the 2-dim Gabor function,

$$G(x', y') = A \exp(-\frac{x'^2}{2\sigma_x^2} - \frac{y'^2}{2\sigma_y^2}) \cos(2\pi f x' + \phi) \qquad (1)$$

$$x' = (x - c_x)\cos\theta - (y - c_y)\sin\theta \qquad (2)$$

$$y' = (x - c_x)\sin\theta + (y - c_y)\cos\theta \tag{3}$$

where A is amplitude, (c_x, c_y) is center of the Gaussian, σ_x and σ_y are standard deviations of the Gaussian perpendicular, θ is orientation, f is frequency, and ϕ is phase. Parameters of the Gabor function for each excitatory neuron are chosen uniformly and randomly from a ranges of parameters that are adjusted as they reproduce seemingly natural shape of receptive fields and reported distribution of correlation coefficient between receptive fields. Correlation coefficient between receptive fields is defined as the simple pixel-wise inner product of 2-dim images of receptive fields. Actual range of parameters of the Gabor filter are $A = 1.0$, $0.35 \leq c_x, c_y \leq 0.65$, $0 \leq \theta, \phi \leq 2\pi$ and $1 \leq s \leq 4, \sigma_x = \sigma_y = 0.20/s, f = 1.25s$. Figure 1 shows obtained distribution of correlation coefficient of pairs of neuron models. As similar to experimental report, the distribution has a sharp peak around zero correlation coefficient. Moreover, we have confirmed that decay curve of the distribution is well fitted by the exponential function and decay rate of the function agree well to the experimental result (Fig. 2). Figure 3 shows examples of assigned receptive fields.

Based on the assigned receptive fields, we construct synaptic connections for all pairs of neurons in the network. Because experimentally reported connection probability between neurons [3] is well fitted by $p(c) = 0.55c^2 + 0.22c + 0.064$, where c is correlation coefficient between receptive fields of these neurons, we measure correlation coefficient of assigned receptive fields of a pair of neurons and put each unidirectional connection between them independently with the connection probability $p(c)$. Then, we assign amplitude of EPSP of each connection, if exists, as explained below.

Experimental report provided the average amplitude of EPSPs as a function of correlation coefficient between receptive fields between presynaptic and post-synaptic neurons. Detailed values of these amplitude, however, are not explicitly given. In order to overcome the limitation, we assume that distributions of amplitude of EPSPs conditioned by the mean are described as the exponential function $p(a|m) = 1/m\exp(-a/m)$, where m is the mean amplitude. Thus, the whole distribution of EPSPs (Fig. 4) is given as;

$$P_a(a) = \int dc\, P_c(c) \int dm\, \delta(m - f_m(c))p(a|m), \tag{4}$$

where $P_c(c)$ is the distribution function of correlation coefficient between receptive fields given in Fig. 1 and $f_m(c)$ is the mean of EPSP as a function of the correlation coefficient, which we approximate as $f_m(c) = 0.28\exp(2.5c)$ based on the experimental report, Extended data Fig. 2b of [3]. Thus, final procedure to chose amplitude of EPSP of each synaptic connection is as follows; we first measure correlation coefficient c between receptive fields of presynaptic and post-synaptic neurons and derive the mean of the EPSP as $m = f_m(c)$, then randomly choose an actual value of EPSP from the exponential distribution $p(a|m)$. If the random sampling gives biologically implausible value of EPSP that is larger than 20 mV, we simply discard the value and resample a value from the same distribution.

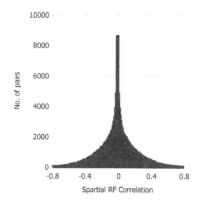

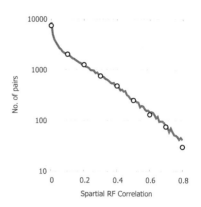

Fig. 1. Obtained distribution of correlation coefficient of pairs of neuron models. As similar to experimental report, the distribution has a sharp peak around zero correlation coefficient.

Fig. 2. Decay curve of the distribution (blue line) is well fitted by the exponential function and decay rate of the function agree well to the experimental result (red points). (Color figure online)

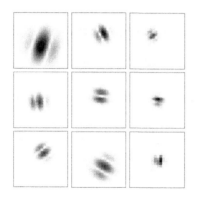

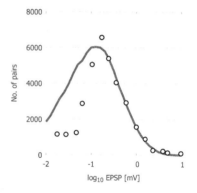

Fig. 3. Examples of assigned receptive fields.

Fig. 4. Obtained distribution of EPSPs of neuron models (blue line) is similar to experimental report (red points). (Color figure online)

3   Results

3.1   Amplitude Distribution of EPSP

In order to verify the proposed model, we numerically obtain amplitude distribution of EPSPs of the network model (Fig. 4). As shown in the figure, the obtained distribution agrees well with previously reported experimental data. As with experimental observation, distribution of our model shows a bell-shaped curve when it is plotted in a semi-logarithmic plot. The average amplitude of

EPSP of the model is about $0.25\,\text{mV}$ while a few connections have about $10\,\text{mV}$ strengths. These results imply that our model precisely reproduces highly heterogeneous nature of the synaptic strength, i.e. coexistence of sparse strong and dense weak connections.

Note that our model gives the lognormal EPSP distribution even though we assumed that conditional EPSP distributions are exponential functions, thus short-tailed distributions. This is because superposition of exponential functions with different means, Eq. (4), can give similar almost long-tailed function like the lognormal function.

3.2   Global Network Statistics Imply Duality of Cortical Circuit

Based on previous literature [11] and above results that show highly heterogeneous nature of synaptic connections in the model, it is naturally expected that topology of the network including distribution of cluster structures has different tendency depending on strength of synaptic connections or amplitude of EPSPs. In order to study the possibility and characterize network topology as a function of amplitude of EPSPs, we divide connections or edges of the model network into 20 group by dividing amplitude of EPSPs equally in logarithmic scale. Then, we construct 20 sub-networks from the model network consisting of edges belonging each sub-group (Fig. 5). Each sub-network, therefore, consists only of approximately same strength connections. Note that while numbers of nodes are almost the same over these sub-networks, numbers of edges and therefore connection rate are very different over sub-networks because of heterogeneous amplitude distribution of EPSPs.

For these sub-networks, we measure network statistics, the cluster coefficient (CC) and the mean shortest path length (MPL). These statistics, however, strongly depend on connection rate of networks. In order to remove influence of connection rate from these statistics, we artificially construct uniform random networks with the same number of nodes and edges of each sub-network. We evaluate CC and MPL of each sub-network as relative values of them from corresponding uniform random network with same numbers of nodes and edges.

Figure 6 shows results of these network statistics. Thick line and dotted line are CCs and MPLs normalized by random uniform networks respectively. As we expected, normalized CC rapidly increases from about $1\,\text{mV}$, which agrees well with the previous literatures and means synapses with a few strong EPSPs form cluster with higher probability than that expected from random network. The normalized MPLs, however, are kept to almost unity for all sub-networks even though the relative CCs are rapidly increasing for sub-networks with strong EPSPs. If clusters with a few strong EPSPs are, as indicated from previous literature, isolated in the network, the normalized MPL must take very large value. The result, therefore, implies that, despite previous view of distributed clusters, clusters of a few EPSPs are connected each other and may form a large component i.e. a global structure.

In order to study details of above result of connected clusters, we decompose each sub-networks into connected components and measure the number

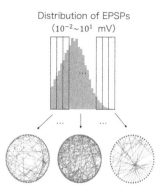

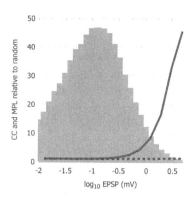

Fig. 5. We divide connections or edges of the model network into 20 group by dividing amplitude of EPSPs equally in logarithmic scale and construct 20 sub-networks from the model network consisting of edges belonging each sub-group.

Fig. 6. Thick line and dotted line are CCs and MPLs normalized by random uniform networks respectively. Whereas the normalized MPLs are close to unity for all sub-networks, normalized CC rapidly increases from about 1 mV.

of connected components and size of the maximum components, so called the giant components. Figure 7 shows these results. The size of the maximum components rapidly decreases for sub-networks with extremely large EPSPs, $\log(\text{EPSP}) \geq 0.8$, since the number of edges of these networks are extremely small. However, the size is kept to almost the same number with the network size itself up to the sub-network with $\log(\text{EPSP}) < 0.8$ whose relative CC is enough high, as shown in Fig. 6. Moreover, the numbers of components remain

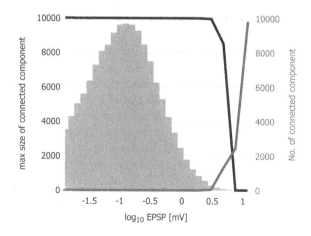

Fig. 7. The number of connected component (red) and size of the maximum connected component, or the giant component (blue) of each sub-network. (Color figure online)

small for these sub-networks with log(EPSP) < 0.8. These results means that, excepting extremely large EPSPs, sub-networks actually consists with one or a few giant components whose size is compatible the size of these network themselves even though CCs of a sub-network is high.

4 Discussion

We have revealed using the model network that cortical network consists on two networks with significantly different features, a small-world network of sparse strong connections and a random network of dense weak connections. Previously reported clusters are not independently distributed in the network. Rather, they are parts of the small-world structure, i.e. cliques of the small-world network.

Topology of the network significantly affects spiking dynamics of neurons in the network. Therefore, it is an important future task to elucidate what kind of spiking dynamics of population of neurons appears on the network with the duality. Also, it is an important to consider functional roles of the duality in higher-order function of cortex, learning rule or synaptic plasticity that gives the duality.

Acknowledgement. This work was partially supported by the Ministry of Internal Affairs and Communications with a contract entitled "R&D for fundamental technology for energy-saving network control compatible to changing communication status" in FY2015 and Kakenhi 25430028 and JP16H01719.

References

1. Brunel, N.: Dynamics of sparsely connected networks of excitatory and inhibitory spiking neurons. J. Comput. Neurosci. **8**(3), 183–208 (2000)
2. Buzsáki, G., Mizuseki, K.: The log-dynamic brain: how skewed distributions affect network operations. Nat. Rev. Neurosci. **15**(4), 264–278 (2014)
3. Cossell, L., Iacaruso, M.F., Muir, D.R., Houlton, R., Sader, E.N., Ko, H., Hofer, S.B., Mrsic-Flogel, T.D.: Functional organization of excitatory synaptic strength in primary visual cortex. Nature **518**, 399–403 (2015)
4. Ecker, A.S., Berens, P., Keliris, G.A., Bethge, M., Logothetis, N.K., Tolias, A.S.: Decorrelated neuronal firing in cortical microcircuits. Science **327**(5965), 584–587 (2010)
5. Ikegaya, Y., Sasaki, T., Ishikawa, D., Honma, N., Tao, K., Takahashi, N., Minamisawa, G., Ujita, S., Matsuki, N.: Interpyramid spike transmission stabilizes the sparseness of recurrent network activity. Cereb. Cortex **23**(2), 293–304 (2013)
6. Klinshov, V.V., Teramae, J., Nekorkin, V.I., Fukai, T.: Dense neuron clustering explains connectivity statistics in cortical microcircuits. PloS One **9**(4), e94292 (2014)
7. Kriener, B., Enger, H., Tetzlaff, T., Plesser, H.E., Gewaltig, M.O., Einevoll, G.T.: Dynamics of self-sustained asynchronous-irregular activity in random networks of spiking neurons with strong synapses. Front. Comput. Neurosci. **8**, 136 (2014)
8. Lefort, S., Tomm, C., Sarria, J.C.F., Petersen, C.C.: The excitatory neuronal network of the c2 barrel column in mouse primary somatosensory cortex. Neuron **61**(2), 301–316 (2009)

9. Litwin-Kumar, A., Doiron, B.: Slow dynamics and high variability in balanced cortical networks with clustered connections. Nat. Neurosci. **15**(11), 1498–1505 (2012)
10. Renart, A., De La Rocha, J., Bartho, P., Hollender, L., Parga, N., Reyes, A., Harris, K.D.: The asynchronous state in cortical circuits. Science **327**(5965), 587–590 (2010)
11. Song, S., Sjöström, P.J., Reigl, M., Nelson, S., Chklovskii, D.B.: Highly nonrandom features of synaptic connectivity in local cortical circuits. PLoS Biol. **3**(3), e68 (2005)
12. Teramae, J., Fukai, T.: Computational implications of lognormally distributed synaptic weights. Proc. IEEE **102**(4), 500–512 (2014)
13. Teramae, J., Tsubo, Y., Fukai, T.: Optimal spike-based communication in excitable networks with strong-sparse and weak-dense links. Sci. Rep. **2** (2012)

Identifying Gifted Thinking Activities Through EEG Microstate Topology Analysis

Li Zhang[1,2], Mingna Cao[1], and Bo Shi[1(✉)]

[1] Department of Medical Imaging, Bengbu Medical College, Bengbu 233030, Anhui, China
{li_zhang,mingna_mit,shibo}@bbmc.edu.cn
[2] Research Center for Learning Science, Southeast University, Nanjing 210096, Jiangsu, China

Abstract. EEG microstate of the brain has been suggested to reflect functional significance of cognitive activity. In this paper, from math-gifted and non-gifted adolescents' EEG during a reasoning task, four classes of microstate configuration were extracted based on clustering analysis approach. Computations of multiple parameters were down for each class of EEG microstate. Between-groups statistical and discriminating analyses for these parameters discovered significant functional differences between math-gifted and non-gifted subjects in momentary microstates, involving mean duration and occurrence of EEG electric field configuration. Additionally, the topological differences between the two groups vary across classes and reflect functional disassociation of cognitive processing of the reasoning task. Our study suggests that the microstate classes can be used as the effective EEG features for identifying mental operations by individuals with typical cognitive ability differences.

Keywords: EEG microstate topology · Math-gifted adolescents · Cluster-based analysis · Microstate class

1 Introduction

Electroencephalography (EEG) microstates are defined as the transient, patterned, and quasi-stable topologies with short periods (80–120 ms). During the duration of a microstate, the global topology remains a fixed electric field configuration, but strength might vary and polarity invert [1]. In previous neuroscience studies, EEG microstate sequence of the human brain has been found to be associated with disease, mental disorder, modalities of thinking activity etc., and EEG microstate parameters were suggested to index functional significance of cognitive activity of the brain [2–5].

Math-gifted adolescents have shown significant differences in cognitive performance and functional brain activity, as compared with non-gifted individuals. The previous empirical studies have discovered that, during reasoning, mental imagery, or creative thinking processes, math-gifted individuals primarily displayed superior central executive function of the prefrontal cortex, enhanced fronto-parietal brain network, and greater involvement of the right hemisphere in information processing [6].

© Springer International Publishing AG 2016
A. Hirose et al. (Eds.): ICONIP 2016, Part I, LNCS 9947, pp. 123–130, 2016.
DOI: 10.1007/978-3-319-46687-3_13

In this study, topological microstates of brain activities of math-gifted and non-gifted subjects were extracted over the time process of EEG recording. According to well-established standard approach to microstate analysis, similar spatial configurations were clustered into four typical classes of topological maps. Relevant parameters of each class were calculated for each group respectively, and between-groups comparisons were conducted to further discover the functional correlations with math-gifted brain. Based on the findings from EEG microstate variance across humans with different reasoning ability levels, the relevant cognitive psychological mechanism of math-gifted adolescents was analyzed and discussed.

2   Materials and Methods

2.1   Participants

Eight gifted adolescents (five males and three females) aged 16.5 ± 0.7 (mean $\pm$ SD) with high intelligence level and specific aptitude in mathematics were recruited in this experiment. The control group was composed of seven normal adolescents (five males and two females) aged 16.3 ± 0.8 (mean $\pm$ SD), who had average-level performance in mathematical and intelligence tests. Exclusion criteria of subjects included left hand-edness, neurological illness, and history of brain injury. All subjects were given informed consent and the study was approved by the Academic Committee of the Research Center for Learning Science, Southeast University, China.

2.2   Experimental Task

A deductive reasoning task was performed by each subject. Deductive reasoning is the process that draws a conclusion from given premises, which is regarded as an essential element of human thinking and cognitive ability. In this study, the deduction task with categorical arguments is composed of three basic letter items, such as "S", "M", and "P", which can constitute a three-stage (major premise, minor premise, and conclusion) reasoning model [7]. Figure 1 shows the valid, invalid and baseline samples of three-stage reasoning process.

	Major premise	Minor premise	Conclusion
Valid sample	No A is B	All X are B	No A is X
Invalid sample	Some D are E	All K are D	Some K are E
Baseline sample	Some C are C	All D are D	No E is E

Fig. 1.   Some samples of valid, invalid and baseline trials of deductive reasoning with categorical arguments.

During experiment process, the stimuli of the three-stage reasoning task were presented along the timeline, as shown in Fig. 2. The letter items of each trial were randomly selected from 26 letters of the English alphabet. All the subjects were asked to judge whether the conclusion was correct or wrong.

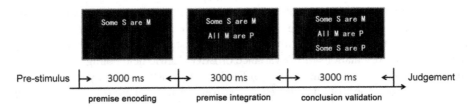

Fig. 2. Experiment protocol: Timeline of stimuli presentation of three-stage deductive reasoning task.

For each subject, the time length of this experimental task was about 30 min. Valid, invalid, and baseline trials of the reasoning task were presented randomly in the E-Prime 2.0 procedure. Before the formal experiment, each subject conducted a practice session composed of five trials.

2.3   EEG Recording and Preprocessing

EEG signals were recorded by Neuroscan international 10–20 system, which includes 60 data electrodes covering frontal, parietal, temporal, occipital regions, 2 reference electrodes located at the bilateral mastoids, and 4 surface electrodes monitoring ocular movements and eye blinks.

The EEG signals were band-pass filtered between 1 Hz and 30 Hz. The EEG trials were extracted by using a time window of 3000 ms, which covers the second stage of deductive reasoning, *i.e.*, premise integration, since it is viewed as the period of actual reasoning [7]. Through further baseline-correction and artifacts rejection, 192 effective trials were retained for math-gifted subjects and 176 trials for non-gifted subjects. Additionally, the signals in pre-stimulus periods were used as the eyes-open epochs of resting state. The independent component analysis (ICA) was used to clear visible artifacts, *e.g.*, the components of possible ocular and muscle movements.

2.4   Extraction and Analysis of EEG Microstate Classes

The extraction and analysis of microstates were conducted by using the Cartool EEG analysis software (http://www.fbmlab.com/cartool-software/) and the EEGLAB Toolbox [8]. For each trial, the EEG time points of global field power (GFP) peaks with maximal potential strength were collected first, which are considered as optimal representations of stable electric field configuration. The GFP at time point t is the empirical standard deviation across all the EEG signals,

$$GFP(t) = \sqrt{\frac{1}{n} \sum_{i=1}^{n} (v_i(t) - \overline{v(t)})^2} \tag{1}$$

where $v(t) = (v_1(t), \ldots, v_n(t))$, which is the vector of potentials of signals at time t. Here, n is the number of EEG electrodes, and $\overline{v(t)} = \dfrac{1}{n}\sum\limits_{i=1}^{n} v_i(t)$.

After collecting the EEG signals at the peaks of GFP, these microstates were analytically clustered into mean classes of EEG maps, *i.e.*, mean microstate topologies, through k-means clustering algorithm [9]. The microstate classes were considered as the maximal representations of the variance of EEG electric field configuration. After that, four typical classes of microstates of EEG signals was produced by the clustering analysis method. The EEG data for each trial were then assigned to these microstate topographies.

According to the definitions from previous studies [4, 5], we computed three microstate parameters for each class, including "mean duration", "occurrence" and "coverage". "Mean duration" refers to the averaged time length lasted for a given microstate topology, which can physiologically index the temporal stability of non-overlapping patterns of synchronous activation of the brain regions related to the momentary EEG measurements. "Occurrence" represents the mean number of distinct microstates of a specified class emerged in 1 s time window. The parameter can quantify how often each class of brain state is activated, so it is considered as an index of the relative utilization of different cognitive resources [5]. "Coverage" is the percentage of time points of a given microstate class covering a task course [4].

2.5   Statistical Analysis on Microstate Measures

The differences in microstate measures derived from single-trial samples were examined by the analysis of variance (ANOVA) in the Matlab Statistics Toolbox.

To assess the task effect on EEG microstates, the differences in mean duration, occurrence, and coverage between resting state and task period were tested by using the one-way ANOVA, with time period (rest/task) serving as the within-subjects factor.

Group differences of the microstate measures in reasoning task were then assessed by one-way ANOVAs, with each measure of each microstate serving as between-subjects factor. To reveal the difference of each class in topological distribution, EEG data with assigned class were statistically tested between the two groups for each pair of channels. The Bonferroni Corrections were conducted in the multiple comparisons, with significance level set to 0.05.

To further validate the effectiveness of the microstate parameters in identifying math-gifted and non-gifted brain states, microstate duration, occurrence, and coverage of the four classes were combined to construct a 12-dimension initial feature vector $d_1, d_2, \ldots, d_{12}$. The single-trial samples from the two groups were then labeled with "math-gifted" and "non-gifted". Linear discrimination analysis (LDA), support vector machine (SVM) and Naive Bayes classifiers with 10-fold cross validation were respectively adopted to perform a between-groups discrimination.

3   Results and Discussions

3.1   Electric Configurations of EEG Microstate Topologies

As with the normative microstate maps that have been revealed by most studies [1–5], the four mean classes of microstates during the reasoning task are produced for the two groups in our study. As shown in Fig. 3, class A reflects frontal to parietal activation, class B shows mostly frontal and medial to slightly less occipital activity, class C covers right-frontal to left-posterior activity, and class D involves left-frontal to right-posterior activity.

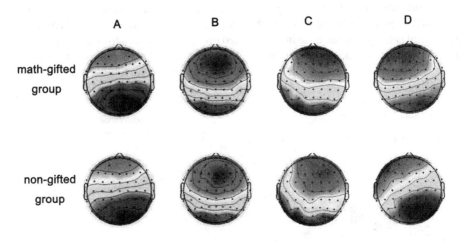

Fig. 3. EEG microstate topologies of four classes (From left to right: Classes A, B, C and D) retrieved from k-means clustering algorithm for math-gifted and non-gifted group respectively.

3.2   Task Effect on Microstate Measures

The ANOVA tests reveal significant task effect on the measures of the four microstate classes, as illustrated in Table 1.

Table 1. The changes of microstate measures from resting state to reasoning task. p value indicates significance level of ANOVA, in which * represents $p < 0.05$, and ** denotes $p < 0.01$.

Microstate class	Mean duration				Occurrence				Coverage			
	A	B	C	D	A	B	C	D	A	B	C	D
p value	–	*	**	*	–	*	**	*	–	*	*	*
Resting → reasoning	–	↗	↘	↘	–	↘	↗	↗	–	↗	↗	↗

In the reasoning process, mean duration of microstate class B was significantly increased as compared with the baseline resting state, but the measures of classes C and D were decreased. The results indicate higher temporal stability of brain topology of microstate class B. Additionally, from resting state to reasoning process, occurrence of classes C and D was increased significantly, which means more frequent utilization of the two classes of microstates than the other brain resources. Significant increases of coverage were found in microstate classes B, C and D, which denotes more total usage time in the task course relative to resting state. There is no significant task-related change in microstate class A.

3.3   Between-Groups Differences in Microstate Measures

The results with significant difference between the math-gifted and control groups are illustrated in Fig. 4. While performing the reasoning task, the math-gifted adolescents show shorter duration of microstate classes A, B and D than the control subjects, whereas microstate class C has lasted longer in the math-gifted group. The four classes of micro-states display higher occurrence in math-gifted group as compared to control group. The shorter duration and more frequent usage of microstate A, B and D support the opinion that math-gifted adolescents have higher flexibility of brain topology in cognitive processing [12].

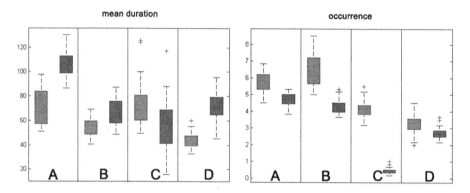

Fig. 4. Statistical boxplots of ANOVA tests between the math-gifted and control groups ($p < 0.05$). Left: mean duration; right: occurrence. In each plot, from left to right: microstate measures of classes A, B, C, and D. The red boxplots represent the math-gifted group and the blue boxplots indicate the control group. (Color figure online)

However, in microstate class C, there are longer mean duration and higher occur-rence in math-gifted group. It should be noted that, microstate class C refers to the brain activity ranging from right-frontal to left-posterior brain regions. These regions are highly involved in spatial information processing, reasoning and creative thinking, which have been suggested as the important indications of precocious mathematical ability of gifted adolescents [10, 11]. Higher temporal stability (duration) and more active brain topology (occurrence) of microstate class C might reflect an optimized state of fronto-parietal network of math-gifted brain during reasoning process, which has been suggested by Zhang et al.' study [12].

Topological representation of between-groups difference in each microstate class are illustrated by Fig. 5. In microstate class A, the math-gifted group has higher activity in parietal regions (Fig. 5A), which could be associated with spatial relationship processing of the reasoning task. The difference in microstate class B is discovered in occipital regions (Fig. 5B). The brain area is considered to be connected to visual processing of cognitive materials.

A   **B**   **C**   **D**

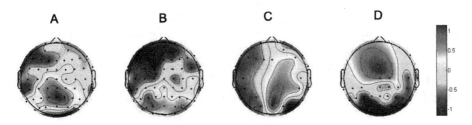

Fig. 5. Brain mapping of significant differences between the math-gifted and control groups. From left to right: microstate classes A, B, C, and D. The value of each channel is derived from a multiplication between EEG data difference and log p value of ANOVA test between the two groups.

Microstate class C reveals higher task-related activity of the math-gifted brain in right frontal to posterior regions, which reflects the dominance of right hemisphere involvement in information processing (Fig. 5C). Additionally, the math-gifted group shows stronger brain activity in left-frontal to medial brain regions in microstate class D (Fig. 5D). Previous studies have discovered that, three-stage reasoning task without concrete content basically activates the left-lateral fronto-parietal brain network and also requires highly imaginative situation in the right hemisphere [7]. Since the enhanced reliance on the right hemisphere function has been suggested as the important neural characteristic of math-gifted brain [10], microstate class C might be the most key momentary brain state that can be connected to specific aptitude in mathematics. Moreover, the group difference in microstate class D could be viewed as a reflection of neurocognitive differences in basic reasoning abilities, such as executive controlling function of anterior brain regions, functional interaction in fronto-parietal network etc., which are more relevant to higher level of general intelligence of math-gifted adolescents.

Furthermore, the accuracy for classifying math-gifted and non-gifted EEG data is 63.5–66.3 % (Table 2), suggesting the discriminant validity of microstate features in identifying the mental operations of individuals with typical cognitive ability differences.

Table 2. Classification accuracy of LDA, SVM, and Naive Bayes in identifying "math-gifted" and "non-gifted" EEG samples.

Classifier	LDA	SVM	Naive Bayes
Accuracy	65.5 %	66.3 %	63.5 %

4   Conclusions

By extracting the cognitive microstates of math-gifted and non-gifted adolescents during a reasoning task, our study discovers the significant association of four microstate classes with the math-gifted brain in mean duration and occurrence. Moreover, the topological differences between the two groups varied across classes, suggesting the functional disassociation of different microstate classes for reflecting individual differences in cognitive processing of reasoning problems. Specifically, microstate class C with topological difference in the right hemisphere shows more characteristics related to specific ability in mathematics. Functional network research on different microstate classes of math-gifted brain, especially on class C, is worthy to be systematically explored further.

Acknowledgements. This work was supported in part by the National Basic Research Program of China under Grant 2015CB351704, and by the Natural Science Foundation of Anhui Province Ministry of Education under Grant KJ2016A470.

References

1. Van de Ville, D., Britz, J., Michel, C.M.: EEG microstate sequences in healthy humans at rest reveal scale-free dynamics. Proc. Natl. Acad. Sci. U. S. A. **107**, 18179–18184 (2010)
2. Gärtner, M., Brodbeck, V., Laufs, H.: A stochastic model for EEG microstate sequence analysis. NeuroImage **104**, 199–208 (2015)
3. Khanna, A., Pascual-Leone, A., Michel, C.M.: Microstates in resting-state EEG: current status and future directions. Neurosci. Biobehav. Rev. **49**, 105–113 (2015)
4. Milz, P., Faber, P.L., Lehmann, D.: The functional significance of EEG microstates—associations with modalities of thinking. NeuroImage **125**, 643–656 (2016)
5. Nishida, K., Morishima, Y., Yoshimura, M.: EEG microstates associated with salience and frontoparietal networks in frontotemporal Dementia, Schizophrenia and Alzheimer's Disease. Clin. Neurophysiol. **124**, 1106–1114 (2013)
6. Zhang, L., Gan, J.Q., Wang, H.: Neurocognitive mechanisms of mathematical giftedness: a literature review. Appl. Neuropsychol. Child. 1–16 (2016)
7. Prado, J., Chadha, A., Booth, J.R.: The brain network for deductive reasoning: a quantitative meta-analysis of 28 neuroimaging studies. J. Cogn. Neurosci. **23**, 3483–3497 (2011)
8. Delorme, A., Makeig, S.: EEGLAB: an open source toolbox for analysis of single-trial EEG dynamics including independent component analysis. J. Neurosci. Methods **134**, 9–21 (2004)
9. Koenig, T., Prichep, L., Lehmann, D.: Millisecond by millisecond, year by year: normative EEG microstates and developmental stages. Neuroimage **16**, 41–48 (2002)
10. Prescott, J., Gavrilescu, M., Cunnington, R.: Enhanced brain connectivity in math-gifted adolescents: an fMRI study using mental rotation. Cogn. Neurosci. **1**, 277–288 (2010)
11. Desco, M., Navas-Sanchez, F.J., Sanchez-Gonzalez, J.: Mathematically gifted adolescents use more extensive and more bilateral areas of the fronto-parietal network than controls during executive functioning and fluid reasoning tasks. Neuroimage **57**, 281–292 (2011)
12. Zhang, L., Gan, J.Q., Wang, H.: Optimized gamma synchronization enhances functional binding of fronto-parietal cortices in mathematically gifted adolescents during deductive reasoning. Front. Hum. Neurosci. **8**, 430 (2014)

Representation of Local Figure-Ground by a Group of V4 Cells

M. Hasuike[1(✉)], Y. Yamane[2], H. Tamura[2], and K. Sakai[1]

[1] Department of Computer Science, University of Tsukuba, 1-1-1 Tennodai,
Tsukuba, Ibaraki 305-8753, Japan
{hasuike,sakai}@cvs.cs.tsukuba.ac.jp
[2] Graduate School of Frontier Biosciences, Osaka University, 1-4 Yamadaoka,
Suita-shi, Osaka 560-0871, Japan
{yukako,tamura}@bpe.es.osaka-u.ac.jp,
http://www.cvs.cs.tsukuba.ac.jp

Abstract. Figure-ground (FG) segregation is a crucial function of the inter-
mediate-level vision. Physiological studies on monkey V2 have reported border-
ownership (BO) selective cells that signal the direction of figure along a local
border. However, local borders in natural images are often complicated and they
often do not provide a clue for FG segregation. In the present study, we hypothesize that a population of V4 cells represents FG by means of surface rather than
border. We investigated this hypothesis by the computational analysis of neural
signals from multiple cells in monkey V4. Specifically, we applied Support
Vector Machine as an ideal integrator to the cellular responses, and examined
whether the responses carry information capable of determining correct local FG.
Our results showed that the responses from several tens of cells are capable of
determining correct local FG in a variety of natural image patches while single-
cell responses hardly determine FG, suggesting a population coding of local FG
by a small number of cells in V4.

Keywords: Visual cortex · V4 · Figure ground segregation · Natural image ·
Electrophysiology · Population coding · Support Vector Machine

1 Introduction

Figure-ground segregation is a crucial step towards object recognition. Physiological
studies have reported that a number of cells in monkey V2 are selective to Border
Ownership (BO) that shows the direction of figure along the border [e.g., 1, 2]. Recent
computational studies have reported that model BO-selective cells based on surround
modulation can discriminate BO in natural contours with 66 % correct [3, 4]. Our
computational study further showed that an integration of the model BO-selective cells
increases the correct rate to 85 % with 10 model cells. [5]. BO may be considered as a
contour-based clue for the determination of FG. We consider that the next step is the
construction of a surface-based representation of FG. Physiological studies on monkeys
have reported that V4 cells appear to be sensitive to the organization of FG [6, 7]. Our
physiological study reported that a number of cells in V4 were selective to FG [8],

© Springer International Publishing AG 2016
A. Hirose et al. (Eds.): ICONIP 2016, Part I, LNCS 9947, pp. 131–137, 2016.
DOI: 10.1007/978-3-319-46687-3_14

although their performance on FG discrimination with natural contours was barely above chance. Our computational study reported that an integration of the responses of 50 FG-selective cells in V4 were capable of determining FG in natural contour patch with 66 % correct [8]. In the present study, we further investigate the representation of FG in V4 with the consideration of difference responses. Physiological studies often use difference responses, such as $R_{BO-right} - R_{BO-left}$ as an index for the representation of BO. In the present study, we focus on such difference responses and investigate the surface-based population representation of FG in monkey V4.

Our hypothesis is that V4 cells code FG by surface-based representation from difference responses. To test the surface-based representation, we presented pairs of stimuli that were slightly translated. The translation enabled the recording from the responses to figure and ground, respectively, of stimulus patches where figure and ground parts, not the border, were projected on to the classical receptive fields of the cells. We examined whether the difference responses ($R_{figure} - R_{ground}$) of single cells were capable of determining correct FG. We further examined whether an integration of the responses of multiple cells increases the discrimination rate of FG. Because the way to integrate multiple responses is controversial, we used Support Vector Machine (SVM) as an ideal integrator. If the machine is capable of discriminating correct FG from the cellular responses, it indicates that the cellular responses included sufficient information for judging FG. An analysis of single cell responses showed that about 20 % of cells exhibited significance to FG while the discrimination rates for natural image patches were very low, slightly above the chance, for the most of cells. The integration of the paired responses from 40 cells increased the discrimination rate to 66 %. The integration of the unpaired responses also showed a similar discrimination rate. These results suggest that a group of a small number of V4 cells provides a surface-based representation of FG without the necessity of difference responses.

2 Methods

We recorded multiple single unit activities from V4 of two analgesized and immobilized macaque monkeys (Macaca fuscata) using 32 channel array electrodes. Eight hundred forty patches that were sampled from natural images with the constrains of diversity [11] were presented for ten times. The experiments were approved by the ethics committee of the institute and performed in accordance with The Code of Ethics of the World Medical Association (Declaration of Helsinki). Stimuli presented to the monkey were comprised of four set of small patches. The first set consisted of single squares and two overlapping squares, which were similar to those used in previous studies on BO [1, 2], With a variation in contrast, translation and rotation, a total of 64 square stimuli were presented, as examples are shown in Fig. 1(Left). The second set consisted of natural contours with one side filled with black and the other side white. We used Human Marked Contours (HMC) available in Berkeley Segmentation Dataset [9, 10] that were drawn by 10 human participants. We chose pseudo-randomly 105 subregions (69×69 pixel) from the HMC with the constrain of diversity, as similar to our previous experiment [11]. Figure 1(Right). shows a few examples of filled patches. Note that purely-random choice of image patches end up with a set of similar contours such as straight lines because the distribution of contour characteristics are highly non-uniform in natural scenes. To assure the diversity of stimuli (not to chose similar contours), we controlled the distribution of the degree of convexity, closure and parallel of stimuli (uniformly chosen from each range of these characteristics) [11]. With a variation in contrast and mirror image, a total of 420 filled patches were presented. The third set consisted of the

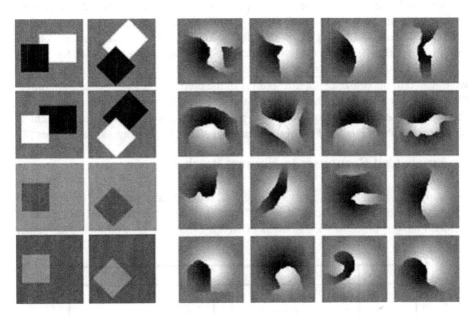

Fig. 1. Examples of square stimuli (left) and filled patches whose borders represent natural contours (right).

	Original		Mirror	
	Fill-Black	Fill-White	Fill-Black	Fill-White
Left				
Right				

Fig. 2. Examples of patches. A patch filled with black (fill-black, left) was translated to the right (fill-black, right). We prepared contrast-reversed and mirrored patches. The red dots indicates the CRF center. (Color figure online)

patches of the second set, which were slightly translated orthogonally with respect to the border near the CRF. Figure 2 shows the example of patches. This set aimed to obtain responses to Figure and Ground in each patch. We analyzed the difference in responses to Figure and Ground ($R_{figure} - R_{ground}$). With a variation in translation, a total of 840 filled patches were presented. Subtracting the response to "ground" patch from that to "figure" patch, we obtained differential response that may extract FG signal. The patches in the fourth set were the original natural-image patches of the second set (a total of 210). To obscure the boundary of stimulus and background, we attenuated contrast towards the periphery with a Gaussian function. The stimuli were presented for 200 ms with a blank interval of 200 ms in a random order with the repetition of 10. The patches were scaled to maximize cellular responses in the preparation phase of the experiment (2.5°–20°). The CRF of single cells was estimated from the responses to grating stimuli that were presented at 25 locations. We performed a psychophysical experiment to obtain the veridical label of FG for the stimulus patches taken from natural images [8].

3   Single Cell Responses to Natural Patches

We analyzed whether the responses of single cells in monkey V4 depend on FG organization of stimuli with respect to their CRF center. The electrophysiological recording included the responses to 2430 neurons of which 1725 cells showed their CRF center

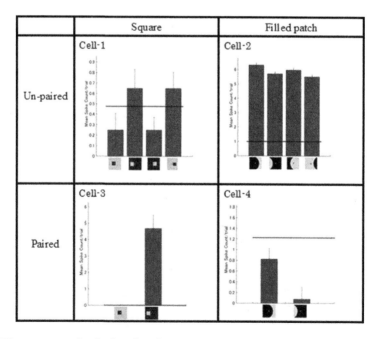

Fig. 3. The responses of a single cell to the square (left) and filled (right) stimuli in the unpaired (top) and paired (bottom) response analyses. Error bars indicate SE. The mean responses per trial (200 ms) are shown. Icons indicate categories, i.e. figure-ground and contrast. The red dots in icons illustrate the location of the CRF center. (Color figure online)

located within stimulus patches. We did not take visual response and spike counts into account for the analysis. To examine whether FG and contrast are significant factors for the responses, we performed a two-way ANOVA with the two factors at the CRF center. Twenty percent (351/1725) of cells showed the significance to FG ($p < 0.05$), 30 % to contrast, and 12 % to the interaction. Figure 3(Top) shows example responses of a single cell that showed significance to FG. For paired responses, 12 % of cells showed the significance to FG, with an example shown in Fig. 3(Bottom). We also determined the correct rate for the discrimination of FG. The mean discrimination rate among all FG-selective cells for all stimulus patches was 51 % with 0.6 % Standard Error (SE), indicating that single cells barely discriminate FG. These results indicate that 10–20 % of V4 cells exhibit the selectivity to local FG, however, it does not directly mean that these single cells are capable of representing FG.

4   Discrimination of FG from a Group of Cellular Responses

The determination of FG in natural image patches was difficult for single cells, leading to the idea that multiple cells work simultaneously as a population for the veridical determination of FG. To examine this hypothesis, we applied SVM to the responses of cells so as to maximize the correct rate of FG discrimination. We used the responses of 139 V1 cells and 1586 V4 cells for the analysis. Specifically, we used SVM to

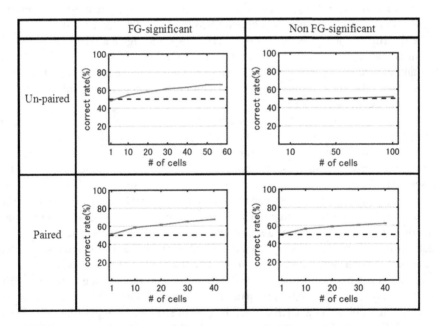

Fig. 4. The discrimination rates of the SVM models as a function of the number of cell for the integration. The results for the un-paired and paired response analyses are shown on the top and bottom rows, respectively. The results for FG significant and non-FG-significant cells are shown in the left and right columns, respectively. The error bars indicate SE.

discriminate FG at the center of CRF from the cellular responses to filled-contour stimuli (set 3). We repeated the learning 200 times with five-fold cross-validation. We implemented this procedure by using LIBSVM software package [12] with Gaussian kernels. Optimal parameters were determined by Grid Search Method. The learning was successful with more than 99 % correct. The mean correct rates are shown in Fig. 4 as a function of the number of cells. With unpaired responses, the correct rates of FG significant cells increase as the number of cells increases. With 50 cells, the mean correct rate reaches 66 % with 50 cells (SE = 0.1 %). The mean correct rates of non-FG significant cells were in the range of chance for the entire range of the number of cells examined (up to 100). These results indicate that FG significant cells, in fact, carry the information necessary for the determination of FG. With paired responses, the mean correct rate increases as the number of cells increases regardless of FG significance though the magnitude is significantly different ($p < 0.05$). The mean correct rate for FG significant cells reaches 67 % with 40 cells (SE = 0.2 %). It appears that non-FG significant cells also carry information relevant to FG determination in some extent, specifically when cellular responses are paired. In real-life situations, such paired responses are not accessible. Further investigation on the functional role of paired responses is expected.

5   Conclusion and Discussions

Our results showed that a population of V4 cells is capable of discriminating FG from a variety of natural image patches. Since how the visual system integrates cellular responses has not been revealed, we applied SVM for the integration with an expectation that the discrimination rate is maximized. In the un-paired signal analysis, the result of the computational experiment showed the mean correct rate of 66 % with 50 cells for filled natural patches, similarly in the paired signal analysis, the mean correct rate was 67 % with 40 cells. These results suggest that a group of V4 cells are capable of discriminating local FG in natural image.

In the paired signal analysis where we subtracted a signal evoked by a ground region from that by a figure region for each patch, the mean correct rate of non-FG-significant cells increased as the number of cells for the integration increased. This result indicates that single cells are not capable of discriminating FG but a group of the cells is capable of discriminating FG. This phenomenon may be caused by the extraction of FG information through the difference between two paired images in which only FG differ. SVM might be capable of discriminating FG in the paired signal analysis because of its astonishing discrimination power. Because cortical cells may not be accessible to the paired signals in real situations, FG significant cells that are labeled based on single cell responses may contribute substantially to the FG discrimination as a population.

We paired patches so that the contrast of figure regions remains the same (e.g., Fill-Black Left and Fill-Black Right were paired). However, when we focus on the contrast of a region that fall onto the CRF center, the pair of stimulus changes (e.g., Fill-Black Left and Fill-White Right should be paired). In the present study, we assumed that neurons represent the contrast of FG, not that of a region projected onto the CRF. It may be important to analyze further the effects of contrast.

Spatial extent of the receptive field of V4 cells is typically several to ten degree in central vision, suggesting that FG selective cells may represent local FG, not global FG. A psychophysical study [9] reported that human subjects show 70 % correct in local FG determination from small patches with respect to the global inspection of the scene. In the present study, the integration of the responses from 40–60 V4 neurons achieved 67 % correct in local FG discrimination, suggesting that a group of cells may underlie the perception of local FG. The global figure-ground segregation of an entire scene would require further mechanisms [e.g., 13].

Acknowledgment. This work was supported by grant-in-aids from JSPS and MEXT of Japan (KAKENHI 26280047, 25135704 (Shitsukan)), and from RIEC of Tohoku University (H25-A09).

References

1. Zhou, H., Friedman, H.S., von der Heydt, R.: Coding of border ownership in monkey visual cortex. J. Neurosci. **20**, 6594–6611 (2000)
2. Zhang, N.R., von der Heydt, R.: Analysis of the context integration mechanisms underlying figure-ground organization in the visual cortex. J. Neurosci. **30**, 6482–6496 (2010)
3. Sakai, K., Nishimura, H.: Surrounding suppression and facilitation in the determination of border ownership. J. Cogn. Neurosci. **18**, 562–579 (2006)
4. Sakai, K., Nishimura, H., Shimizu, R., Kondo, K.: Consistent and robust determination of border-ownership based on asymmetric surrounding contrast. Neural Netw. **33**, 257–274 (2012)
5. Nakata, Y., Sakai, K.: Structures of surround modulation for the border-ownership selectivity of V2 cells. In: Huang, T., Zeng, Z., Li, C., Leung, C.S. (eds.) ICONIP 2012, Part I. LNCS, vol. 7663, pp. 383–391. Springer, Heidelberg (2012)
6. Pasupathy, A., Connor, C.E.: Population coding of shape in area V4. Nat. Neurosci. **5**(12), 1332–1338 (2002)
7. Poort, J., Raudies, F., Wanning, A., Lamme, V.A.F., Neumann, H., Roelfsema, P.R.: The role of attention in figure-ground segregation in areas V1 and V4 of the visual cortex. Neuron **75**, 143–156 (2012)
8. Hasuike, M., Ueno, S., Minowa, D., Yamane, Y., Tamura, H., Sakai, K.: Figure-ground segregation by a population of V4 cells. In: Arik, S., et al. (eds.) ICONIP 2015. LNCS, vol. 9490, pp. 617–622. Springer, Heidelberg (2015). doi:10.1007/978-3-319-26535-3_70
9. Fowlkes, C.C., Martin, D.R., Malik, J.: Local figure-ground cues are valid for natural images. J. Vis. **7**(8), 2, 1–9 (2007)
10. Berkeley Segmentation Dataset; Figure-Ground Assignments in Natural Images. http://www.eecs.berkeley.edu/Research/Projects/CS/vision/grouping/fg/
11. Sakai, K., Matsuoka, S., Kurematsu, K., Hatori, Y.: Perceptual representation and effectiveness of local figure-ground cues in natural contours. Front. Psychol. (2015). doi: 10.3389/fpsyg.2015.01685
12. Chang, C.C., Lin, C.J.: LIBSVM - a library for support vector machines. http://www.csie.ntu.edu.tw/~cjlin/libsvm/
13. Jones, H.E., Andolina, I.M., Shipp, S.D., Adams, D.L., Cudeiro, J., Salt, T.E., Sillito, A.M.: Figure-ground modulation in awake primate thalamus. PNAS **112**(22), 7085–7090 (2015)

Dynamic MEMD Associated with Approximate Entropy in Patients' Consciousness Evaluation

Gaochao Cui[1,2(✉)], Qibin Zhao[1,2], Toshihisa Tanaka[3],
Jianting Cao[1,2], and Andrzej Cichocki[2]

[1] Saitama Institute of Technology, Fusaiji 1690, Fukaya, Saitama 3690293, Japan
e4001hbx@sit.ac.jp
[2] Brain Science Institute, RIKEN, Hirosawa 2-1, Wakoshi, Saitama3510198, Japan
[3] Tokyo University of Agriculture and Technology,
2-24-16 Nakacho, Koganei-shi, Tokyo 184-8588, Japan

Abstract. Electroencephalography (EEG) based preliminary examination has been widely used in diagnosis of brain diseases. Based on previous studies, clinical brain death determination also can be actualized by analyzing EEG signal of patients. Dynamic Multivariate empirical mode decomposition (D-MEMD) and approximate entropy (ApEn) are two kinds of methods to analyze brain activity status of the patients in different perspectives for brain death determination. In our previous studies, D-MEMD and ApEn methods were always used severally and it cannot analyzing the patients' brain activity entirely. In this paper, we present a combine analysis method based on D-MEMD and ApEn methods to determine patients' brain activity level. Moreover, We will analysis three different status EEG data of subjects in normal awake, comatose patients and brain death. The analyzed results illustrate the effectiveness and reliability of the proposed methods.

Keywords: Electroencephalography (EEG) · Multivariate empirical mode decomposition (MEMD) · Dynamic-MEMD · Approximate entropy (ApEn)

1 Introduction

Electroencephalography (EEG) is a recording of voltage fluctuations resulting from ionic current flows within the neurons of the brain and refers to the recording of the brain's spontaneous electrical activity over a short period of time. The healthy subjects in normal awake have high brain activity because of their healthy brain cells. On the contrary, the patient in brain death state has no normal brain cells, so that their brain activity is extremely low. The concept of brain death was proposed in 1960's. It is defined as the irreversible and complete loss of all brain activity including involuntary activity necessary to sustain life due to total necrosis of cerebral neurons following loss of blood flow and oxygenation [1]. According to this definition, the Japanese established the criterion to diagnose the brain death can be shown as Fig. 1.

© Springer International Publishing AG 2016
A. Hirose et al. (Eds.): ICONIP 2016, Part I, LNCS 9947, pp. 138–146, 2016.
DOI: 10.1007/978-3-319-46687-3_15

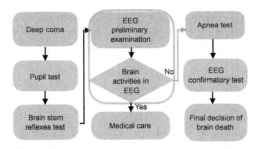

Fig. 1. The process of diagnosing brain death in clinical.

For supporting the diagnosis of brain death, we have proposed an EEG preliminary examination method as a reliable yet safety and rapid way for the determination of brain death [2]. That is, after items (1)–(3) have been verified, and an EEG preliminary examination along with real-time recorded data analysis method is applied to detect the brain wave activity at the bedside of patients. To extract informative features from noisy EEG signals and evaluate their significance, approximate entropy (ApEn) measure and Multivariate empirical mode decomposition (MEMD) were proposed for the EEG analysis in our previous study. ApEn is to extract informative features from noisy EEG signals and evaluate their statistical significance, several complexity measures are developed for the quantitative EEG analysis [3]. MEMD, which is extended of empirical mode decomposition (EMD), have been used for EEG to evaluate the brain activity through calculate power spectrum within the frequency band multiplied by recorded EEG time. In the previous study, we used MEMD or ApEn method to analysis patients' data to evaluate EEG activity. We have proposed MEMD to calculate the energy of EEG of randomly chosen interval of one second [1]. However, by using MEMD, it is difficult to observe EEG energy variation of subjects. The most important things for clinical medicine application, considering the safety of the patient, it has low reliability and certain risk to evaluate patients' status if we use only one method.

In this paper, we present a combine analysis method based on dynamic ApEn and D-MEMD to analysis the real-life recorded EEG signal. By using D-MEMD and dynamic ApEn methods, we can not only denoise the original EEG data but also calculate the EEG energy of subjects with the time series. In addition to this, we observe EEG energy variation of subjects to increase the reliability and show three examples of healthy subject in normal awake, comatose patient and brain death. The analyzed results illustrate the effectiveness and performance of the proposed method in calculation of EEG energy for evaluating consciousness level.

2   Methods of Data Analysis

2.1   EMD and MEMD Algorithms

The original signal was decomposed into a finite set of amplitude- and/or frequency- modulated components by EMD method [6,8]. These components termed intrinsic mode functions (IMFs), which represent its inherent oscillatory modes [7]. More specifically, for a real-valued signal $x(t)$, the standard EMD finds a set of K IMFs $\{c_k(t)\}_{k=1}^{K}$, and a monotonic residue signal $r(t)$, so that

$$x(t) = \sum_{k=1}^{K} c_k(t) + r(t). \tag{1}$$

IMFs $c_k(t)$ are defined so as to have symmetric upper and lower envelopes, with the number of zero crossings and the number of extrema differing at most by one. The process to obtain the IMFs is called sifting algorithm. Moreover, the first complex extension of EMD was proposed in [4]. An extension of EMD to analyze complex/bivariate data which operates fully in the complex domain was first proposed in [5], termed rotation-invariant EMD (RI-EMD). In MEMD, we choose a suitable set of direction vectors in n-dimensional spaces by using: (i) uniform angular coordinates and (ii) low-discrepancy pointsets. The multivariate extension of EMD suitable for operating on general nonlinear and non-stationary n-variate time series is summarized in the following.

1. Choose a suitable pointset for sampling on an $(n-1)$ sphere.
2. Calculate a projection, denoted by $\{p^{\theta_k}(t)\}_{t=1}^{T}$, of the input signal $\{\mathbf{v}(t)\}_{t=1}^{T}$ along the direction vector $\mathbf{x}^{\theta_k}$, for all k (the whole set of direction vectors), and giving $\{p^{\theta_k}(t)\}_{k=1}^{K}$ as the set of projections.
3. Find the time instants $\{t_i^{\theta_k}\}$ corresponding to the maxima of the set of projected signals $\{p^{\theta_k}(t)\}_{k=1}^{K}$.
4. Interpolate $[t_i^{\theta_k}, \mathbf{v}(t_i^{\theta_k})]$ to obtain multivariate envelope curves $\{\mathbf{e}^{\theta_k}(t)\}_{k=1}^{K}$.
5. For a set of K direction vectors, the mean $\mathbf{m}(t)$ of the envelope curves is calculated as

$$\mathbf{m}(t) = \frac{1}{K} \sum_{k=1}^{K} \mathbf{e}^{\theta_k}(t). \tag{2}$$

6. Extract the 'detail' $d(t)$ using $d(t) = x(t) - m(t)$. If the 'detail' $d(t)$ fulfills the stoppage criterion for a multivariate IMF, apply the above procedure to $x(t) - d(t)$, otherwise apply it to $d(t)$.

The stoppage criterion for multivariate IMFs is similar to the standard one in EMD, which requires IMFs to be designed in such a way that the number of extrema and the zero crossings differ at most by one for S consecutive iterations of the sifting algorithm. The optimal empirical value of S has been observed to be in the range of 2–3 [4].

2.2 Dynamic MEMD Algorithm

The D-MEMD is an adaptive algorithm of the MEMD. We have defined the EEG energy using the power spectrum within the frequency band multiplied by recorded EEG time [2]. To observe EEG energy variation of subjects, we extend MEMD in the temporal domain along time-coordinate of EEG signal.

Supposing a multivariate EEG data series $\mathbf{v}(t)$ consisting of N segments (epochs) $\{\mathbf{v}_n(t)\}_{n=1}^N$, the MEMD can be carried out through each segment.

The Dynamic MEMD is defined as the MEMD applied to all segments such that

$$\mathbf{v}(t) = [\mathbf{v}_1(t), \ldots, \mathbf{v}_N(t)]$$

$$= \left[\sum_{k=1}^{K_1} c_{k,1}(t) + r_1(t), \ldots, \sum_{k=1}^{K_N} c_{k,n}(t) + r_N(t) \right] \tag{3}$$

where $\mathbf{r}_N(t)$ are residue signals and $\{\mathbf{c}_{k,n}(t)\}_{k=1}^{K_n}$ are IMF components with K_n ($n = 1, \ldots, N$) being the number of IMFs for the segmented nth signal $\mathbf{v}_n(t)$.

Consequently, in our experiment, we remove the residue signal $\mathbf{r}_N(t)$ and Q IMFs from $\{\mathbf{c}_{k,n}(t)\}_{k=1}^{K_n}$ which is not expected, and combine the $(N - Q)$ IMFs to be the denoised signal. We have defined the EEG energy using the power spectrum within the frequency band multiplied by recorded EEG time. Thus we change the denoised signal from time domain to frequency domain by Fast Fourier Transformation and integrate it to compute the EEG energy.

2.3 Approximate Entropy

Given a time series $\{x(n)\}$, ($n = 1, \cdots, N$), to compute the ApEn$(\mathbf{x}(n), m, r)$ (m: length of the series of vectors, r: tolerance parameter) of the sequence, the series of vectors of length m, $\mathbf{v}(k) = [x(k), x(k+1), \cdots, x(k+m-1)]$ is firstly constructed from the signal samples $\{x(n)\}$. Let $D(i, j)$ denote the distance between two vectors $\mathbf{v}(i)$ and $\mathbf{v}(j)$ ($i, j \leq N - m + 1$), which is defined as the maximum difference in the scalar components of $\mathbf{v}(i)$ and $\mathbf{v}(j)$, or

$$D(i, j) = \max_{l=1,\cdots,m} |v_l(i) - v_l(j)|. \tag{4}$$

Then, we further compute the $N^{m,r}(i)$, which represents the total number of vectors $\mathbf{v}(j)$ whose distance with respect to the generic vector $\mathbf{v}(i)$ is less than r, or $D(i, j) \leq r$. Now define $C^{m,r}(i)$, the probability to find a vector that differs from $\mathbf{v}(i)$ less than the distance r. And $\phi^{m,r}$, the natural logarithmic average over all the vectors of the $C^{m,r}(i)$ probability as

$$C^{m,r}(i) = \frac{N^{m,r}(i)}{N - m + 1}, \tag{5}$$

$$\phi^{m,r} = \frac{\sum_{i=1}^{N-m+1} \log C^{m,r}(i)}{N - m + 1}. \tag{6}$$

For $m + 1$, repeat above steps and compute $\phi^{m+1,r}$. ApEn statistic is given by

$$ApEn(\mathbf{x}(n), m, r) = \phi^{m,r} - \phi^{m+1,r}. \tag{7}$$

The typical values $m = 2$ and r between 10% and 25% of the standard deviation of the time series $\{x(n)\}$ are often used in practice [9].

Furthermore, base on the algorithm for computing ApEn of one sequence, we extend it in the temporal domain along timecoordinate of EEG signal. Supposing an EEG data series $\mathbf{S}_N$ consists of N sequence intervals $\{x_i(n)\}$, the ApEn measure is carried out through each interval. We define the dynamic ApEn measure of given EEG signal as

$$\mathbf{ApEn}(\mathbf{S}_N, m, r) = [ApEn(\mathbf{x}_1(n), m, r), ..., ApEn(\mathbf{x}_N(n), m, r)] \qquad (8)$$

Consequently, in our experiment, the $\mathbf{ApEn}(\mathbf{S}_N$, m, r) statistic measures the variation the of complexity of a EEG data series $\mathbf{S}_N$. The occurrence of irregular pattern of one interval is excepted to be followed by the next in brain-death EEG.

3   Experiments and Results

3.1   EEG Signal Recording

A portable EEG system (NEUROSCAN ESI) was used to record the healthy subjects' brain signal in normal awake. Six exploring electrodes (Fp1, Fp2, F3, F4, F7 and F8) as well as GND were placed on the forehead, and two electrodes (A1, A2) as the reference were placed on the earlobes based on the standardized 10–20 system. The sampling rate of EEG was $1000\,\mathrm{Hz}$ and the resistances of the electrodes were set to less than $10\,\mathrm{k\Omega}$. With the same setting of healthy subjects EEG recording experiments, the EEG data was directly recorded at the bedside of the patients in the intensive care unit (ICU) from a hospital of Shanghai, China.

3.2   The Result for Healthy Subject, Comatose Patient and Brain Death Using Dynamic MEMD

Firstly, let us show dynamic EEG energy of healthy subject, comatose patient and brain death by using D-MEMD. The example for healthy subject's EEG examination was performed in August 2013. By applying D-MEMD algorithms described in Part 2.2, we obtain EEG energy variation of healthy subject (Fig. 3(a)) in 60 seconds. In Fig. 3(a), EEG energy of each channel are between 1.43×10^4 and 8.65×10^4. The comatose case is concerned with a male patient. By the same way of healthy subject to analysis the EEG data of this patient by D-MEMD, we obtain the EEG energy variation of comatose patient in 60 s (Fig. 3(b)). This patient's EEG energy of each channel is between 1.05×10^4 to 4.2×10^4 (Fig. 2(b)) that reflects a high intensity of brain activity. With the same analysis for brain death, we still analyzed 60 seconds EEG data by using D-MEMD as an example. Figure 2(c) shows each channel's EEG energy. This patient's maximum value of 6 channels' EEG energy is only 7.03×10^3, the value is extremely low.

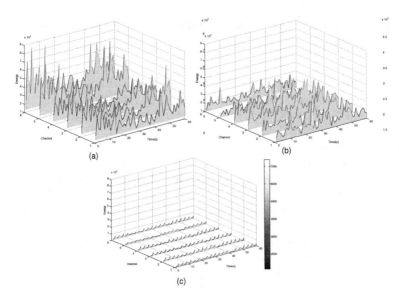

Fig. 2. Results for a dynamic EEG energy analysis used D-MEMD.

From the analysis result by using D-MEMD, we can see that the brain activity of health subject and comatose patient were more obvious than the brain death one. It means that, the brain cell of comatose patient has smaller extent of damage than the brain death patient.

3.3 The Result for Healthy Subject, Comatose Patient and Brain Death Using Dynamic ApEn

First, we use dynamic ApEn to analysis the patients' EEG signal. In our previous study, when the patients in Quasi-Brain-Death state, ApEn value will be approximate to 1, or greater than 1. However the patients' brain activity in the coma state produces ApEn of a low number but not approximate to 0. The result can be seen in Fig. 3(a), the average results of each channel are from 0.165 to 0.291. This result indicates that patient still having spontaneous brain activity. Then, we use the same method to analysis second patient's EEG. It can be seen from the Fig. 3(b), comparing with the first patient, ApEn measure distribution of each channel is mostly over 0.9, and the average results of each channel are from 0.707 to 1.1, and gives us a much higher ApEn value of approximate to 1. From this result above, we suspect the patient was in the quasi-brain-death state.

Last, we also analyzed a health subject and the result were shown in Fig. 3(c). From this result, we can see that the average value of each channel is from 0.079 to 0.222.

3.4 The Comparison for Healthy Subject, Comatose Patient and Brain Death

Furthermore, we analyzed 5 healthy subjects, 5 comatose patients and 5 brain deaths' EEG data by using D-MEMD and Dynamic ApEn methods. The EEG energy of all subjects is shown in Fig. 4. The EEG energy of healthy subject is between 2.52×10^4 and 3.22×10^5, the EEG energy of comatose is between 1.65×10^4 and 3.05×10^5, the EEG energy of brain death is under 1.00×10^4. From these results, the EEG energy of healthy subject and comatose patients is significantly higher than brain death. And we find the brain activity of comatose patients whose EEG energy is close to the brain deaths' are not high. We speculate that they are brain damage. But another part of comatose patients' EEG energy is close to, even more than the healthy subject's. These patients still have high brain activity.

In Fig. 4, we also use ApEn method to analysis the same EEG data. In Fig. 4, we calculate the average ApEn value of 6 channels of each subjects. From the result in Fig. 4, we can see that the average ApEn value of each patients in brain death state were from 0.82 to 1.09. And the average ApEn value of each patients

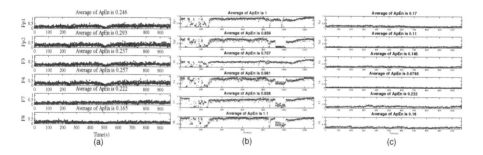

Fig. 3. Results for a dynamic EEG analysis used dynamic ApEn.

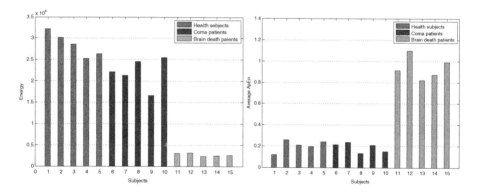

Fig. 4. The EEG energy and ApEn value of 5 healthy subject, 5 comatose patients and 5 brain deaths.

in coma state were from 0.13 to 0.23. The ApEn value of health subjects were from 0.12 to 0.26.

4 Conclusion

In this paper, we focus on a novel data analysis method based on D-MEMD and ApEn to analysis EEG recorded from the healthy subjects, comatose patients and brain deaths and observe the state changes of patients' consciousness. By using D-MEMD and ApEn, we can not only denoised the original EEG data but also calculate the EEG energy of subjects with the time series. Two methods were used to analysis same patients' EEG from different perspectives improving the reliability of the analysis results. In addition to this, we recorded EEG energy variation of subjects and compared them. The result is that EEG energy of healthy subjects is extremely high and show a high brain activity. EEG energy of brain death is extremely low and demonstrate that brain death has no brain activity. In comatose patients, a part of patients' EEG energy is close to the brain deaths'. We speculate that they are brain damage. Another part of comatose patients' EEG energy is close to, even more than the healthy subjects'. They are no-brain-damage and still have high brain activity. The analyzed results illustrate the effectiveness and performance of the proposed method in calculation of EEG energy for evaluating consciousness level and increase the reliability.

References

1. Yin, Y., Zhu, H., Tanaka, T., Cao, J.: Analyzing the EEG energy of healthy human, comatose patient and brain death using multivariate empirical mode decomposition algorithm. In: Proceedings of the 2012 IEEE International Conference on Signal Processing, vol. 1, pp. 148–151. IEEE Press (2012)
2. Yin, Y., Cao, J., Shi, Q., Mandic, D., Tanaka, T., Wang, R.: Analyzing the EEG energy of quasi brain death using MEMD. In: Proceedings of the Asia-Pacific Signal and Information Processing Association Annual Summit and Conference (CD-ROM) (2011)
3. Rehman, N., Mandic, D.: Multivariate empirical mode decomposition. Proc. R. Soc. A **466**(2117), 1291–1302 (2010)
4. Tanaka, T., Mandic, D.: Complex empirical mode decomposition. IEEE Signal Process. Lett. **14**(2), 101–104 (2006)
5. Altaf, M., Gautama, T., Tanaka, T., Mandic, D.: Rotation invariant complex empirical mode decomposition. In: Proceedings of the IEEE International Conference on Acoustics, Speech, Signal Processing (ICASSP 2007), Honolulu, HI, pp. 1009–1012 (2007)
6. Rehman, N., Mandic, D.: Empirical mode decomposition for trivariate signals. IEEE Trans. Signal Process. **58**(3), 1059–1068 (2010)
7. Huang, N., Wu, M., Long, S., Shen, S., Qu, W., Gloersen, P., Fan, K.: A confidence limit for the empirical mode decomposition and Hilbert spectral analysis. Proc. R. Soc. Lond. A **459**, 2317–2345 (2003)

8. Huang, N., Shen, Z., Long, S., Wu, M., Shih, H., Zheng, Q., Yen, N., Tung, C., Liu, H.: The empirical mode decomposition and the hilbert spectrum for nonlinear and non-stationary time series analysis. Proc. R. Soc. Lond. A **454**, 903–995 (1998)
9. Pincus, S.M.: Approximate entropy (ApEn) as a measure of system complexity. Proc. Natl. Acad. Sci. **88**, 110–117 (1991)

Robotics and Control

Neural Dynamic Programming for Event-Based Nonlinear Adaptive Robust Stabilization

Ding Wang[1(✉)], Hongwen Ma[1], Derong Liu[2], and Huidong Wang[3]

[1] The State Key Laboratory of Management and Control for Complex Systems,
Institute of Automation, Chinese Academy of Sciences, Beijing 100190, China
{ding.wang,mahongwen2012}@ia.ac.cn
[2] School of Automation and Electrical Engineering,
University of Science and Technology Beijing, Beijing 100083, China
derong@ustb.edu.cn
[3] School of Management Science and Engineering,
Shandong University of Finance and Economics, Jinan 250014, China
huidong.wang@ia.ac.cn

Abstract. In this paper, we develop an event-based adaptive robust stabilization method for continuous-time nonlinear systems with uncertain terms via a self-learning technique called neural dynamic programming. Through system transformation, it is proven that the robustness of the uncertain system can be achieved by designing an event-triggered optimal controller with respect to the nominal system under a suitable triggering condition. Then, the idea of neural dynamic programming is adopted to perform the main controller design task by building and training a critic network. Finally, the effectiveness of the present adaptive robust control strategy is illustrated via a simulation example.

Keywords: Adaptive dynamic programming · Adaptive robust stabilization · Event-based control · Neural dynamic programming · Neural network

1   Introduction

When dealing with the control design of uncertain systems, the robustness and the related optimality issues have been studied by many researchers [1–3]. This brings in a great research interest to the optimal regulation design. The adaptive or approximate dynamic programming (ADP) method was originally proposed by Werbos [4] as an effective avenue to conquer the phenomenon of "curse of dimensionality" arising in optimal control problems. It is implemented by solving the Hamilton-Jacobi-Bellman (HJB) equation through the function approximation structures, usually referring neural networks. Hence, it also can be called neural dynamic programming [5]. The novel ideas of ADP have been utilized for designing feedback controller with optimality [6–11].

Unlike the time-triggered control methods, in the event-triggered control mechanism, the sampling instant for updating the feedback controller is determined by a certain triggering condition, rather than relying on a fixed sampling

© Springer International Publishing AG 2016
A. Hirose et al. (Eds.): ICONIP 2016, Part I, LNCS 9947, pp. 149–157, 2016.
DOI: 10.1007/978-3-319-46687-3_16

interval. This always results in a significant reduction on computation and communication resources. Recently, the combination of event-triggering mechanism and ADP method has achieved considerable attention [12,13]. Note that under the new mechanism, the ADP-based controller is only updated when an event is triggered, and hence, the computational burden of learning and updating can be greatly saved. However, it is apparent to find that the dynamical uncertainties of the controlled plant are not always considered in the existing work of ADP-based event-triggered feedback design, which motivates our research. Consequently, in this paper, we investigate the event-based adaptive robust stabilization for continuous-time nonlinear systems using neural dynamic programming.

2   Problem Description

In this work, we consider a class of input-affine continuous-time nonlinear systems

$$\dot{x}(t) = f(x(t)) + g(x(t))u(t) + Z(x(t)), \tag{1}$$

where $Z(x(t))$ is the uncertain dynamics satisfying

$$Z(x) = G(x)d(\varphi(x)), d^\mathsf{T}(\varphi(x))d(\varphi(x)) \leq h^\mathsf{T}(\varphi(x))h(\varphi(x)). \tag{2}$$

In system (1), $x(t) \in \mathbb{R}^n$ is the state vector and $u(t) \in \mathbb{R}^m$ is the control vector, $f(\cdot)$ and $g(\cdot)$ are differentiable in their arguments with $f(0) = 0$. In (2), the terms $G(\cdot) \in \mathbb{R}^{n \times r}$ with $\|G(x)\| \leq G_{\max}$ and $\varphi(\cdot)$ with $\varphi(0) = 0$ are fixed functions reflecting the structure of uncertainty, $d(\cdot) \in \mathbb{R}^r$ is an uncertain function satisfying $d(0) = 0$, and $h(\cdot) \in \mathbb{R}^r$ is a known function satisfying $h(0) = 0$.

We study how to stabilize system (1) adaptively. Note the nominal system of (1) is

$$\dot{x}(t) = f(x(t)) + g(x(t))u(t). \tag{3}$$

Like other literature, we let $x(0) = x_0$ be the initial state vector and assume that $f + gu$ is Lipschitz continuous on a set Ω in $\mathbb{R}^n$ containing the origin and that the system (3) is controllable. The following lemma presents the achievement of robustness of (1).

Lemma 1 (cf. [10]). *Assume that there exist a continuously differentiable cost function $V(x)$ which satisfies $V(x) > 0$ for all $x \neq 0$ and $V(0) = 0$, a bounded function $\Gamma(x)$ satisfying $\Gamma(x) \geq 0$, and a feedback control function $u(x)$, such that*

$$(\nabla V(x))^\mathsf{T} Z(x) \leq \Gamma(x); \tag{4a}$$

$$U(x, u) + (\nabla V(x))^\mathsf{T}(f(x) + g(x)u) + \Gamma(x) = 0, \tag{4b}$$

where $\nabla(\cdot) \triangleq \partial(\cdot)/\partial x$ is employed to denote the gradient operator, $U(x, u) = Q(x) + u^\mathsf{T} Ru$, and $Q(x) = x^\mathsf{T} Qx$ with $Q = Q^\mathsf{T} \geq 0$ and $R = R^\mathsf{T} > 0$. Under the *action of the control input $u(x)$, there exists a neighborhood of the original state, such that system (1) is locally asymptotically stable. In addition, define the cost function of system (3) as*

$$J(x_0, u) = \int_0^\infty \{U(x(\tau), u(x(\tau))) + \Gamma(x(\tau))\} d\tau, \tag{5}$$

which ensures that $J(x_0, u) = V(x_0)$ holds.

According to Lemma 1, the cost function $V(x)$, the bounded function $\Gamma(x)$, and feedback control $u(x)$ satisfying (4a) and (4b) can guarantee the robust stabilization of system (1). It is important to note that the optimal cost function and optimal control law of system (3) can provide specific forms of the cost function and feedback control. Hence, we should make great effort to cope with the adaptive optimal control problem of system (3) with $V(x_0)$ considered as the cost function. The designed optimal feedback control must be admissible. For system (3), it can be observed that (4b) is an infinitesimal version of the cost equation $V(x)$ and is the nonlinear Lyapunov equation.

In the following, for consistency, we generally take $J(x)$ to denote the cost function, instead of $V(x)$. In light of the classical optimal control theory, we define the Hamiltonian of transformed problem as $H(x, u, \nabla J(x)) = U(x, u) + (\nabla J(x))^\mathsf{T}(f + gu) + \Gamma(x)$. Let Ω be a compact subset of $\mathbb{R}^n$ and $\Psi(\Omega)$ be the set of admissible controls on Ω. The optimal cost function of system (3) is defined as $J^*(x_0) = \min_{u \in \Psi(\Omega)} J(x_0, u)$. Note that the optimal cost $J^*(x)$ satisfies the continuous-time HJB equation of the form $0 = \min_{u \in \Psi(\Omega)} H(x, u, \nabla J^*(x))$. Hence, the optimal control of system (3) is

$$u^*(x) = -\frac{1}{2} R^{-1} g^\mathsf{T}(x) \nabla J^*(x). \tag{6}$$

From [10], for any continuously differentiable function $V(x)$, if we define $\Gamma(x) = h^\mathsf{T}(\varphi(x)) h(\varphi(x)) + \frac{1}{4}(\nabla V(x))^\mathsf{T} G(x) G^\mathsf{T}(x) \nabla V(x)$, then $(\nabla V(x))^\mathsf{T} Z(x) \leq \Gamma(x)$. Using the optimal control $u^*(x)$ and this specific form of $\Gamma(x)$, the HJB equation becomes

$$0 = U(x, u^*) + (\nabla J^*)^\mathsf{T}(f + gu^*) + h^\mathsf{T}(\varphi(x)) h(\varphi(x)) + \frac{1}{4}(\nabla J^*)^\mathsf{T} GG^\mathsf{T} \nabla J^* \tag{7}$$

with $J^*(0) = 0$. Based on (6), (7) can also be written as

$$0 = Q(x) + (\nabla J^*(x))^\mathsf{T} f(x) + h^\mathsf{T}(\varphi(x)) h(\varphi(x))$$
$$- \frac{1}{4}(\nabla J^*(x))^\mathsf{T} g(x) R^{-1} g^\mathsf{T}(x) \nabla J^*(x) + \frac{1}{4}(\nabla J^*)^\mathsf{T} G(x) G^\mathsf{T}(x) \nabla J^* \tag{8}$$

with $J^*(0) = 0$. Since it is always difficult to solve the nonlinear optimal control analytically, kinds of ADP-based methods combining the idea of reinforcement learning have been proposed to get the approximate solution [6]. However, nearly all of the existing methods are implemented predicated on the time-driven formulation, which generally speaking, is time-consuming. Thus, in the following, we turn to the event-driven adaptive robust controller design using neural dynamic programming.

3 Event-Based Robust Control via Neural Dynamic Programming

In this section, the adaptive robust stabilization method under event-based framework is established, followed by the event-triggered ADP implementation via neural network.

3.1 Adaptive Robust Stabilization with Event-Based Formulation

Let $\mathbb{N} = \{0, 1, 2, \dots\}$ denote the set of all non-negative integers. In order to employ the event-triggering method to control the sampled-data system, we first define a monotonically increasing time sequence of triggering instants as $\{s_j\}_{j=0}^{\infty}$ with $s_0 = 0$ such that $s_j < s_{j+1}$, $j \in \mathbb{N}$. Assume that the nominal system (3) is sampled at the triggering instants s_j, which results in the sampled state vector $x(s_j) \triangleq \hat{x}_j$ for all $t \in [s_j, s_{j+1})$, $j \in \mathbb{N}$. Then, the event-triggered controller $\mu(\hat{x}_j)$ is updated based on the sampled state $\hat{x}_j$ rather than the current state $x(t)$. That is to say, the controller $\mu(\hat{x}_j)$ is executed at every triggering instants. By using a zero-order hold, the corresponding control sequence $\{\mu(\hat{x}_j)\}_{j=0}^{\infty}$ becomes a continuous-time input signal $\mu(\hat{x}_j, t)$, i.e.,

$$\mu(\hat{x}_j, t) = u(\hat{x}_j) = u(x(s_j)), \forall t \in [s_j, s_{j+1}), j \in \mathbb{N}. \tag{9}$$

For simplicity, we use the notation $\mu(\hat{x}_j)$ instead of $\mu(\hat{x}_j, t)$. Define the event-triggered error between the sampled state and current state as

$$e_j(t) = \hat{x}_j - x(t), \forall t \in [s_j, s_{j+1}), j \in \mathbb{N}. \tag{10}$$

Thus, the event-triggered optimal control problem can be described. For the corresponding sampled-data system

$$\dot{x}(t) = f(x) + g(x)\mu(x(t) + e_j(t)). \tag{11}$$

Based on (6) and (9), the event-triggered optimal controller can be given by

$$\mu^*(\hat{x}_j) = -\frac{1}{2}R^{-1}g^{\mathsf{T}}(\hat{x}_j)\nabla J^*(\hat{x}_j), \forall t \in [s_j, s_{j+1}), \tag{12}$$

where $\nabla J^*(\hat{x}_j) = (\partial J^*(x)/\partial x)|_{x=\hat{x}_j}$. Under the framework of event-triggering mechanism, the HJB Eq. (8) can be rewritten as

$$H(x, \mu^*(\hat{x}_j), \nabla J^*(x))$$

$$= Q(x) + (\nabla J^*(x))^{\mathsf{T}} f(x) + h^{\mathsf{T}}(\varphi(x)) h(\varphi(x)) - \frac{1}{2}(\nabla J^*(x))^{\mathsf{T}} g(x) R^{-1} g^{\mathsf{T}}(\hat{x}_j) \nabla J^*(\hat{x}_j)$$

$$+ \frac{1}{4}(\nabla J^*(\hat{x}_j))^{\mathsf{T}} g(\hat{x}_j) R^{-1} g^{\mathsf{T}}(\hat{x}_j) \nabla J^*(\hat{x}_j) + \frac{1}{4}(\nabla J^*(x))^{\mathsf{T}} GG^{\mathsf{T}} \nabla J^*(x) \tag{13}$$

with $J^*(0) = 0$. Note that unlike the time-triggered HJB Eq. (8), the new formula (13) is in fact the event-triggered HJB equation.

In event-triggered control, a triggering condition should be designed to determine the event-triggering instants and guarantee the stability of the closed-loop system.

Assumption 1 (cf. [13]). *The control law $\mu(x)$ is Lipschitz continuous with respect to the event-triggered error, i.e., $\|\mu(x(t)) - \mu(\hat{x}_j)\| \leq \mathcal{L}\|e_j(t)\|$, where $\mathcal{L}$ is a positive real constant and $\mu(x) = u(x)$.*

Theorem 1. *For the nominal system (3), suppose that $J^*(x)$ is the solution of the HJB equation and $\mu^*(\hat{x}_j)$ is the event-triggered optimal controller. For all $t \in [s_j, s_{j+1}), j \in \mathbb{N}$, if the triggering condition is defined as*

$$\|e_j(t)\|^2 \leq \frac{(1 - \eta_1^2)\lambda_{\min}(Q)\|x\|^2 + \|r\mu^*(\hat{x}_j)\|^2}{\mathcal{L}^2\|r\|^2} \triangleq \|e_T\|^2, \tag{14}$$

where e_T denotes the threshold, $\lambda_{\min}(Q)$ is the minimal eigenvalue of Q, and $\eta_1 \in (0,1)$ is a designed sample frequency parameter, then the closed-loop form of system (1) with event-triggered controller (12) is asymptotically stable.

    Proof. Choose $L_1(t) = J^*(x)$ as the Lyapunov function. The derivative of $L_1(t)$ is

$$\dot{L}_1(t) = (\nabla J^*(x))^{\mathsf{T}} (f(x) + g(x)\mu^*(\hat{x}_j) + Z(x)). \tag{15}$$

From (6), we have $g^{\mathsf{T}}(x)\nabla J^*(x) = -2Ru^*(x)$. The derivative (15) can be rewritten as

$$\dot{L}_1(t) \leq (\nabla J^*)^{\mathsf{T}}f + (\nabla J^*)^{\mathsf{T}}g(x)\mu^*(\hat{x}_j) + \Gamma \leq -Q(x) + u^{*\mathsf{T}}(x)Ru^*(x) - 2u^{*\mathsf{T}}(x)R\mu^*(\hat{x}_j). \tag{16}$$

Adding and subtracting $\mu^{*\mathsf{T}}R\mu^*$, letting $R = r^{\mathsf{T}}r$, and using Assumption 1, it yields

$$\dot{L}_1(t) \leq -\eta_1^2\lambda_{\min}(Q)\|x\|^2 + (\eta_1^2 - 1)\lambda_{\min}(Q)\|x\|^2 + \mathcal{L}^2\|r\|^2\|e_j(t)\|^2 - \|r\mu^*(\hat{x}_j)\|^2. \tag{17}$$

If the triggering condition (14) is satisfied, then it follows from (17) that $\dot{L}_1(t) \leq -\eta_1^2\lambda_{\min}(Q)\|x\|^2 < 0$ for any $x(t) \neq 0, t \in [s_j, s_{j+1})$. This proves that the triggering condition (14) can guarantee the asymptotic stability of the uncertain system (1). ∎

3.2   Neural Dynamic Programming and Implementation Process

According to the universal approximation property, $J(x)$ can be reconstructed by a neural network on a compact set Ω as $J(x) = \omega_c^{\mathsf{T}}\sigma_c(x) + \varepsilon_c(x)$, where $\omega_c \in \mathbb{R}^l$ is the ideal weight vector, $\sigma_c(x) \in \mathbb{R}^l$ is the activation function, l is the number of neurons in the hidden layer, and $\varepsilon_c(x)$ is the approximation error of the neural network. Then, $\nabla J(x) = (\nabla\sigma_c(x))^{\mathsf{T}}\omega_c + \nabla\varepsilon_c(x)$. Under the framework of ADP, since the ideal weight vector is unknown, a critic neural network can be built in terms of the estimated weight elements as $\hat{J}(x) = \hat{\omega}_c^{\mathsf{T}}\sigma_c(x)$ to approximate the cost function. Then, we have $\nabla\hat{J}(x) = (\nabla\sigma_c(x))^{\mathsf{T}}\hat{\omega}_c$. According to (12), we describe the event-triggered optimal control law as $\mu(\hat{x}_j) = -\frac{1}{2}R^{-1}g^{\mathsf{T}}(\hat{x}_j)((\nabla\sigma_c(\hat{x}_j))^{\mathsf{T}}\omega_c + \nabla\varepsilon_c(\hat{x}_j))$. In addition, the event-triggered approximate optimal control law can be formulated as

$$\hat{\mu}(\hat{x}_j) = -\frac{1}{2}R^{-1}g^{\mathsf{T}}(\hat{x}_j)(\nabla\sigma_c(\hat{x}_j))^{\mathsf{T}}\hat{\omega}_c. \tag{18}$$

As for the Hamiltonian, when taking the neural network into account, it becomes

$$H\big(x,\mu(\hat{x}_j),\omega_c\big) = U(x,\mu(\hat{x}_j))+h^{\mathsf{T}}(\varphi(x))h(\varphi(x))+\omega_c^{\mathsf{T}}\nabla\sigma_c(x)\big(f(x)+g(x)\mu(\hat{x}_j)\big)$$
$$+\frac{1}{4}\omega_c^{\mathsf{T}}\nabla\sigma_c(x)G(x)G^{\mathsf{T}}(x)(\nabla\sigma_c(x))^{\mathsf{T}}\omega_c \triangleq e_{cH}, \tag{19}$$

where $e_{cH} = -(\nabla\varepsilon_c(x))^{\mathsf{T}}\big(f(x)+g(x)\mu(\hat{x}_j)\big) - \frac{1}{2}\omega_c^{\mathsf{T}}\nabla\sigma_c(x)G(x)G^{\mathsf{T}}(x)\nabla\varepsilon_c(x) - \frac{1}{4}(\nabla\varepsilon_c(x))^{\mathsf{T}}G(x)G^{\mathsf{T}}(x)\nabla\varepsilon_c(x)$ represents the residual error due to the neural network approximation. Using (18), the approximate Hamiltonian can be obtained by

$$\hat{H}\big(x,\mu(\hat{x}_j),\hat{\omega}_c\big) = U(x,\mu(\hat{x}_j))+h^{\mathsf{T}}(\varphi(x))h(\varphi(x))+\hat{\omega}_c^{\mathsf{T}}\nabla\sigma_c(x)\big(f(x)+g(x)\mu(\hat{x}_j)\big)$$
$$+\frac{1}{4}\hat{\omega}_c^{\mathsf{T}}\nabla\sigma_c(x)G(x)G^{\mathsf{T}}(x)(\nabla\sigma_c(x))^{\mathsf{T}}\hat{\omega}_c \triangleq e_c. \tag{20}$$

By letting the error of estimating the critic network weight be $\tilde{\omega}_c = \omega_c - \hat{\omega}_c$ and combining (19) with (20), we find that

$$e_c = -\tilde{\omega}_c^{\mathsf{T}}\nabla\sigma_c(x)\big(f(x)+g(x)\mu(\hat{x}_j)\big) + \frac{1}{4}\tilde{\omega}_c^{\mathsf{T}}\nabla\sigma_c(x)G(x)G^{\mathsf{T}}(x)(\nabla\sigma_c(x))^{\mathsf{T}}\tilde{\omega}_c$$
$$-\frac{1}{2}\omega_c^{\mathsf{T}}\nabla\sigma_c(x)G(x)G^{\mathsf{T}}(x)(\nabla\sigma_c(x))^{\mathsf{T}}\tilde{\omega}_c + e_{cH}, \tag{21}$$

which shows the relationship between the terms e_c and e_{cH}.

For training the critic network, it is desired to design $\hat{\omega}_c$ to minimize the objective function $E_c = 0.5e_c^{\mathsf{T}}e_c$. Note that the approximated control law (18) is often used for conducting the learning stage because of the unavailability of the optimal control law $\mu(\hat{x}_j)$. At present, we employ the standard steepest descent algorithm to tune the weight vector as $\dot{\hat{\omega}}_c = -\alpha_c(\partial E_c/\partial\hat{\omega}_c)$, which, based on (20), is in fact

$$\dot{\hat{\omega}}_c = -\alpha_c\bigg(U(x,\hat{\mu}(\hat{x}_j))+h^{\mathsf{T}}(\varphi(x))h(\varphi(x))+\phi^{\mathsf{T}}\hat{\omega}_c+\frac{1}{4}\hat{\omega}_c^{\mathsf{T}}\nabla\sigma_c(x)G(x)G^{\mathsf{T}}(x)(\nabla\sigma_c(x))^{\mathsf{T}}\hat{\omega}_c\bigg)$$
$$\times\bigg(\phi+\frac{1}{2}\nabla\sigma_c(x)G(x)G^{\mathsf{T}}(x)(\nabla\sigma_c(x))^{\mathsf{T}}\hat{\omega}_c\bigg), \tag{22}$$

where $\phi = \nabla\sigma_c(x)(f(x)+g(x)\hat{\mu}(\hat{x}_j))$ and $\alpha_c > 0$ is the designed learning rate of the critic network. Then, recalling $\dot{\tilde{\omega}}_c = -\dot{\hat{\omega}}_c$ and (21), we can further derive that the error dynamical equation of approximating the cost function by the critic network is

$$\dot{\tilde{\omega}}_c = -\alpha_c\bigg(\phi^{\mathsf{T}}\tilde{\omega}_c-\frac{1}{4}\tilde{\omega}_c^{\mathsf{T}}\nabla\sigma_c(x)G(x)G^{\mathsf{T}}(x)(\nabla\sigma_c(x))^{\mathsf{T}}\tilde{\omega}_c+\frac{1}{2}\omega_c^{\mathsf{T}}\nabla\sigma_c(x)G(x)G^{\mathsf{T}}(x)(\nabla\sigma_c(x))^{\mathsf{T}}\tilde{\omega}_c-e_{cH}\bigg)$$
$$\times\bigg(\phi+\frac{1}{2}\nabla\sigma_c(x)G(x)G^{\mathsf{T}}(x)(\nabla\sigma_c(x))^{\mathsf{T}}\omega_c-\frac{1}{2}\nabla\sigma_c(x)G(x)G^{\mathsf{T}}(x)(\nabla\sigma_c(x))^{\mathsf{T}}\tilde{\omega}_c\bigg). \tag{23}$$

Actually, we observe that the closed-loop sampled-data system is an impulsive dynamical system with flow dynamics for all $t \in [s_j, s_{j+1})$ and jump dynamics for all $t = s_{j+1}, j \in \mathbb{N}$. When defining an augmented state vector as $z = [x^\mathsf{T}, \hat{x}_j^\mathsf{T}, \tilde{\omega}_c]^\mathsf{T}$ and basing on (10), (11), and (23), the dynamics of impulsive system can be described by

$$
\begin{cases}
\dot{z} = \begin{bmatrix} [(f(x) + g(x)\hat{\mu}(\hat{x}_j))^\mathsf{T}, 0]^\mathsf{T} \\ -\alpha_c \left(\phi^\mathsf{T} \tilde{\omega}_c - \frac{1}{4} \tilde{\omega}_c^\mathsf{T} B \tilde{\omega}_c + \frac{1}{2} \omega_c^\mathsf{T} B \tilde{\omega}_c - e_{cH} \right) \left(\phi + \frac{1}{2} B \omega_c - \frac{1}{2} B \tilde{\omega}_c \right) \end{bmatrix}, & t \in [s_j, s_{j+1}); \\
z(t) = z(t^-) + \left[0, (x - \hat{x}_j)^\mathsf{T}, 0 \right]^\mathsf{T}, & t = s_{j+1},
\end{cases}
$$

where $B \triangleq \nabla \sigma_c(x) G(x) G^\mathsf{T}(x) (\nabla \sigma_c(x))^\mathsf{T}$, $z(t^-) = \lim_{\varrho \to 0} z(t - \varrho)$, and the term 0 represents a null vector with appropriate dimension.

At last, we turn to the stability issue of the closed-loop system. Before proceeding, the following assumption is needed, as often used in ADP literature like [8,11,12].

Assumption 2. *The dynamics $g(x)$ is Lipschitz continuous such that $\|g(x) - g(\hat{x}_j)\| \leq A\|e_j(t)\|$, where A is a positive constant. It is also upper bounded such that $\|g(x)\| \leq g_{\max}$, where $g_{\max}$ is a positive constant. Besides, the derivative of the activation function, i.e., $\nabla \sigma_c(x)$ is Lipschitz continuous such that $\|\nabla \sigma_c(x) - \nabla \sigma_c(\hat{x}_j)\| \leq B\|e_j(t)\|$, where B is a positive constant. The derivative term $\nabla \sigma_c(x)$ is also upper bounded such that $\|\nabla \sigma_c(x)\| \leq \nabla \sigma_{c\max}$, where $\nabla \sigma_{c\max}$ is a positive constant.*

Here, we present the closed-loop stability result. Suppose that Assumption 2 holds. The tuning law for the critic network is given by (22). Then, the closed-loop system (11) is asymptotically stable if the adaptive triggering condition

$$
\|e_j(t)\|^2 \leq \frac{(1 - \eta_2^2)\lambda_{\min}(Q)\|x\|^2 + \|r\hat{\mu}(\hat{x}_j)\|^2}{2\ell^2 \|\hat{\omega}_c\|^2 \|R^{-1}\|} \triangleq \|\hat{e}_T\|^2 \tag{24}
$$

is satisfied, where $\eta_2 \in (0, 1)$ is the parameter to be designed as the sample frequency and $\ell^2 = A^2 \nabla \sigma_{c\max}^2 + B^2 g_{\max}^2$. The proof is omitted here due to the space constraint.

4   Simulation

We consider an input-affine continuous-time nonlinear system with an uncertain term

$$
\dot{x} = \begin{bmatrix} -x_1 - 2x_2 \\ x_1 - x_2 - \cos x_1 \sin x_2^2 \end{bmatrix} + \begin{bmatrix} 1 \\ -1 \end{bmatrix} u(x) + \begin{bmatrix} p_1 x_1 \sin x_2^2 \\ 0 \end{bmatrix}, \tag{25}
$$

where $x = [x_1, x_2]^\mathsf{T} \in \mathbb{R}^2$ and $u(x) \in \mathbb{R}$ are the state and control vectors, while $Z(x) = [p_1 x_1 \sin x_2^2, 0]^\mathsf{T}$ (with $p_1 \in [-2, 2]$) represents the uncertainty. Letting $\varphi(x) = x$ and considering the uncertain structure, we can select $G(x) = [1, 0]^\mathsf{T}$, $d(\varphi(x)) = p_1 x_1 \sin x_2^2$, and $h(\varphi(x)) = 2x_1 \sin x_2^2$. Let $Q(x) = 2x^\mathsf{T} x$ and $R = I$ (I denotes an identity matrix with suitable dimension). The activation

156   D. Wang et al.

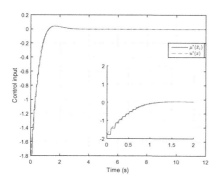

Fig. 1. Event and time based controls.

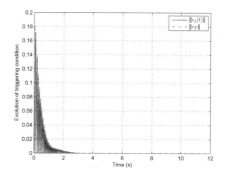

Fig. 2. Evolution of the triggering condition.

function is selected as $\sigma_c(x) = [x_1^2, x_1 x_2, x_2^2]^\mathsf{T}$. Denote the weight vector as $\hat{\omega}_c = [\hat{\omega}_{c1}, \hat{\omega}_{c2}, \hat{\omega}_{c3}]^\mathsf{T}$. We set the learning rate as $\alpha_c = 0.1$ and let the initial state of system (25) be $x_0 = [1, -1]^\mathsf{T}$.

In the simulation process, we add a probing noise to guarantee the persistency of excitation condition. We experimentally choose $\eta_2 = 0.6$ and $\ell = 12$. The sampling time is chosen as 0.1 s. We observe that the weight vector of the critic network converges to $[0.8013, -0.2200, 0.7583]^\mathsf{T}$. We also find that the event-triggered controller only needs 1640 samples of the state while the time-triggered controller uses 3500 samples. Next, we choose $p = -2$ to evaluate the robust control performance. Let $\mathcal{L} = 12$ and $\eta_1 = 0.5$. The sampling time is chosen as 0.02s for the uncertain system (25). The Fig. 1 compares the performance of control inputs obtained under the event-triggered and the time-triggered frameworks, where the latter is approached by the former gradually. The Fig. 2 displays the evolution of triggering condition during the robust control implementation. The above results verify the excellent control performance.

5   Conclusion

A novel event-based ADP formulation is developed to design the adaptive robust control for a class of uncertain nonlinear systems under a suitable triggering condition. An artificial neural network is constructed for implementing the ADP technique and establishing the event-driven approximate optimal control law with simulation study.

Acknowledgment. This work was supported in part by the National Natural Science Foundation of China under Grants 61233001, 61273140, 61304086, 61402260, 61533017, and U1501251, in part by Beijing Natural Science Foundation under Grant 4162065, in part by Shandong Province Higher Educational Science and Technology Program (J13LN42), in part by The Excellent Young and Middle-Aged Scientist Award Foundation of Shandong Province (BS2013DX043), and in part by the Early Career Development Award of SKLMCCS.

References

1. Haddad, W.M., Chellaboina, V.: Nonlinear Dynamical Systems and Control: A Lyapunov-Based Approach. Princeton University Press, Princeton (2008)
2. Lin, F.: Robust Control Design: An Optimal Control Approach. Wiley, UK (2007)
3. Wang, D., Liu, D., Li, H.: Policy iteration algorithm for online design of robust control for a class of continuous-time nonlinear systems. IEEE Trans. Autom. Sci. Eng. **11**(2), 627–632 (2014)
4. Werbos, P.J.: Beyond regression: New tools for prediction and analysis in the behavioural sciences. Ph.D. dissertation, Harvard University, Cambridge, MA (1974)
5. Si, J., Wang, Y.T.: On-line learning control by association and reinforcement. IEEE Trans. Neural Netw. **12**(2), 264–276 (2001)
6. Lewis, F.L., Vrabie, D., Vamvoudakis, K.G.: Reinforcement learning and feedback control: using natural decision methods to design optimal adaptive controllers. IEEE Control Syst. Mag. **32**(6), 76–105 (2012)
7. Luo, B., Wu, H.N., Huang, T., Liu, D.: Data-based approximate policy iteration for affine nonlinear continuous-time optimal control design. Automatica **50**(12), 3281–3290 (2014)
8. Yang, X., Liu, D., Ma, H., Xu, Y.: Online approximate solution of HJI equation for unknown constrained-input nonlinear continuous-time systems. Inf. Sci. **328**, 435–454 (2016)
9. Mu, C., Ni, Z., Sun, C., He, H.: Air-breathing hypersonic vehicle tracking control based on adaptive dynamic programming. IEEE Trans. Neural Netw. Learn. Syst. (in press). doi:10.1109/TNNLS.2016.2516948
10. Wang, D., Liu, D., Li, H., Ma, H., Li, C.: A neural-network-based online optimal control approach for nonlinear robust decentralized stabilization. Soft Comput. **20**(2), 707–716 (2016)
11. Zhao, D., Zhang, Q., Wang, D., Zhu, Y.: Experience replay for optimal control of nonzero-sum game systems with unknown dynamics. IEEE Trans. Cybern. **46**(3), 854–865 (2016)
12. Sahoo, A., Xu, H., Jagannathan, S.: Neural network-based event-triggered state feedback control of nonlinear continuous-time systems. IEEE Trans. Neural Netw. Learn. Syst. **27**(3), 497–509 (2016)
13. Vamvoudakis, K.G.: Event-triggered optimal adaptive control algorithm for continuous-time nonlinear systems. IEEE/CAA J. Automatica Sin. **1**(3), 282–293 (2014)

Entropy Maximization of Occupancy Grid Map for Selecting Good Registration of SLAM Algorithms

Daishiro Akiyama, Kazuya Matsuo, and Shuichi Kurogi$^{(\boxtimes)}$

Kyushu Institute of Technology, Tobata, Kitakyushu, Fukuoka 804-8550, Japan
akiyama@kurolab.cntl.kyutech.ac.jp, {matsuo,kuro}@cntl.kyutech.ac.jp
http://kurolab.cntl.kyutech.ac.jp/

Abstract. This paper analyzes entropy of occupancy grid map (OGM) for evaluating registration performance of SLAM (simultaneous localization and mapping) algorithms. So far, there are a number of SLAM algorithms having been proposed, but we do not have general measure to evaluate the registration performance of point clouds obtained by LRF (laser range finder) for SLAM algorithms. This paper analyzes to show that good registration seems corresponding to large overlap of point clouds in OGM as well as large entropy, large uncertainty and low information of OGM. This analysis indicates a method of entropy maximization of OGM for selecting good registration of SLAM algorithms. By means of executing numerical experiments, we show the validity and the effectiveness of the entropy of OGM to evaluate the registration performance.

Keywords: Entropy maximization of occupancy grid map · Evaluation of registration performance · ICP-SLAM · Registration of point clouds of LRF range data

1 Introduction

This paper analyzes entropy of occupancy grid map (OGM) for evaluating the registration performance of SLAM (simultaneous localization and mapping) algorithms. Recently, we have presented a method to improve the accuracy of localization and mapping obtained by ICP-SLAM (iterative closest point - SLAM) algorithm [1]. The method has utilized CAN2 (competitive associative net) [2] for learning piecewise linear approximation of nonlinear functions to deal with point clouds of range data obtained by LRF (laser range finder) characterized as involving lack of data called black spots, quantization error owing to range and angular resolution (e.g. 10 mm and 0.25°, respectively), data density different from region to region. Furthermore, the method has utilized LOOCV (leave-one-out cross-validation) to reduce the propagation error owing to iterative pairwise registration of ICP-SLAM. We have shown that the effectiveness of the method in reducing LOOCV-MSE (mean square error) of point clouds

© Springer International Publishing AG 2016
A. Hirose et al. (Eds.): ICONIP 2016, Part I, LNCS 9947, pp. 158–167, 2016.
DOI: 10.1007/978-3-319-46687-3_17

approximated by the CAN2. Since LOOCV-MSE is a measure of iterative and relative error between approximated point clouds registered at the current step and point cloud expected to perform better registration, this measure cannot be used to compare with other registrations even for the registrations obtained by the same method using CAN2s with different number of units. As a result, we could not have optimized the number of units of the CAN2 for obtaining better registration. Here, note that, as shown in a research review of range image registration [3], the performance of a number of algorithms is evaluated by means of registration error for synthetic data, but no general measure to evaluate registration performance for non-synthetic data has been mentioned.

In the field of medical image registration [4,5], maximization of mutual information (MI) is widely used for registering multimodality images such as CT and MRI, where the analysis of MI as well as Shannon entropy is described from the point of view of measure of information and registration. Since OGM represents the posterior probability of existence of objects, the entropy can be naturally introduced and has a possibility to be used as a measure of information and registration. It seems important to analyze good registration of range data from the point of view of information and entropy because good registration is not always obtained owing to the amount of information of given data involving various noise and resolution different from region to region. Incidentally, from the perspective of the amount of information, entropy reduction of OGM is utilized as information gain in robot exploration [6,7]. In this paper, we analyze entropy of OGM from the perspective of good registration of point clouds, and show that good registration corresponds to large overlap of registered point clouds in OGM as well as large entropy, large uncertainty and low information of OGM.

In the next section, we show a formulation of OGM and entropy of OGM, and describe the relationship between good registration of point clouds of range data and entropy of OGM, and then the validity and the effectiveness of the method is examined by means of numerical experiments in Sect. 3.

2   Large Entropy of OGM for Good Registration Performance

2.1   Measurement Data

For a better readability, we first describe specifications of our equipments. We have used a pioneer 3-AT mobile robot from MobileRobots Inc. equipped with a Hokuyo UTM-30LX LRF mounted on the top as shown in Fig. 1(a). The specification of UTM-30LX is as follows; the distance scanning range is 0.1 to 30 m and the angular scanning range is $\phi_{max} = 270°$, while the accuracy is ± 30 mm for the distance from 0.1 to 10 m and ± 50 mm from 10 to 30 m, and the angular resolution is $\Delta_\phi = 0.25°$. We use an application of ICP-SLAM with icp-classic option provided by MRPT (Mobile Robot Programming Toolkit) [8]. It is an offline application which provides an estimated trajectory of the robot poses (positions and orientations) and an estimated 2D point map after a running of

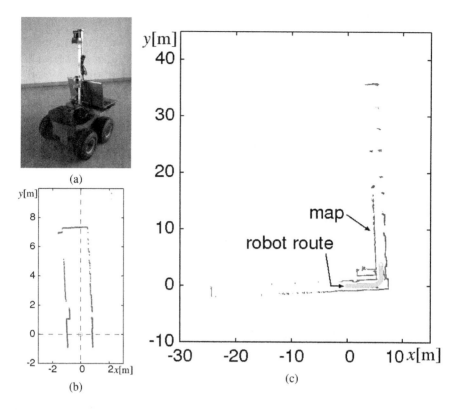

Fig. 1. (a) Mobile robot (P3-AT) equipped with LRF (Hokuyo UTM-30LX) mounted on the top to scan the environment horizontally and a note PC on the robot used in the experiment, (b) a point cloud obtained by a single 2D scan of the LRF, and (c) a map of registered range data and a robot route estimated by the ICP-SLAM application.

the robot, where the point map is represented as registered point clouds of LRF range data as shown in Fig. 1(c).

Now, let $z_t = (D_t, \widehat{\boldsymbol{x}}_t^{(0)})$ denote the measurement data at a discrete time $t \in I^{\text{time}} = \{1, 2, \cdots \}$. Here, $D_t = \{r_t^{(i)} \mid i \in I^{\text{scan}}\}$ denotes a point cloud of range (distance) data obtained by the LRF, where $r_t^{(i)}$ for $I^{\text{scan}} = \{0, 1, 2, \cdots, i_{\max} = \phi_{\max}/\Delta_\phi\}$ is the ith scanned range data. The notation $\widehat{\boldsymbol{x}}_t^{(0)} = (\widehat{x}_t^{(0)}, \widehat{y}_t^{(0)}, \widehat{\theta}_t^{(0)})^T$ denotes the pose of the robot estimated by SLAM method. Then, the scanned range data $r_t^{(i)}$ corresponds to the position $\boldsymbol{x}_t^{(i)} = r_t^{(i)} \boldsymbol{R}(i\Delta_\phi - \phi_r + \widehat{\theta}_t^{(0)}) \boldsymbol{e}_x + (\widehat{x}_t^{(0)}, \widehat{y}_t^{(0)})^T$ on the global map, where $\boldsymbol{e}_x = (1, 0)^T$, $\boldsymbol{R}(\theta) = \begin{pmatrix} \cos\theta & -\sin\theta \\ \sin\theta & \cos\theta \end{pmatrix}$ denotes the rotation matrix, and ϕ_r the orientation of the robot on the LRF coordinate system.

2.2 OGM

Let $p(m \mid z_{1:t})$ be the posterior probability of OGM, denoted by m, given the measurements $z_{1:t} = z_1 z_2 \cdots z_t$ from time 1 to t. We divide the map m into $N^{\mathrm{gm}} \times N^{\mathrm{gm}}$ grid cells $m^{[i]}$ whose centers are given by $\boldsymbol{x}^{[i]} = (x^{[i]}, y^{[i]})^T = \left((k + 0.5)\Delta^{\mathrm{gm}} + x_{\min}^{\mathrm{gm}}, (l + 0.5)\Delta^{\mathrm{gm}} + y_{\min}^{\mathrm{gm}} \right)^T$ for $i = lN^{\mathrm{gm}} + k$ and $k, l = 0, 1, 2, \cdots, N^{\mathrm{gm}} - 1$. Here, Δ^{gm} represents the grid size, and the absolute value of $x_{\min}^{\mathrm{gm}} = y_{\min}^{\mathrm{gm}}$ (=-40 m in the experiments shown below) is set as large as all registered data $\boldsymbol{x}_t^{(i)}$ and their all rotations on the global map m are in the square area bounded by the four points $(\pm x_{\min}^{\mathrm{gm}}, \pm y_{\min}^{\mathrm{gm}})$, where the necessity of rotations is shown below. The notation $p(m^{[i]})$ represents the probability that $m^{[i]}$ is occupied by some obstacle, and $p(\overline{m^{[i]}}) = 1 - p(m^{[i]})$ the probability that $m^{[i]}$ is not occupied. For simplicity as shown in [7], we assume $p(m \mid z_{1:t}) = \prod_i p(m^{[i]} \mid z_{1:t})$. From Bayesian inference (BI) using conditional independence assumption (CIA), known as naive BI, we derive

$$p(m^{[i]} \mid z_{1:t}) = \frac{1}{Z_t} p(z_t \mid m^{[i]}, z_{1:t-1}) \, p(m^{[i]} \mid z_{1:t-1}) \quad (\because \text{BI}) \tag{1}$$

$$\simeq \frac{1}{Z_t} p(z_t \mid m^{[i]}) \, p(m^{[i]} \mid z_{1:t-1}) \quad (\because \text{CIA}) \tag{2}$$

$$= \frac{1}{Z_t} \frac{p(z_t) \, p(m^{[i]} \mid z_t)}{p(m^{[i]})} \, p(m^{[i]} \mid z_{1:t-1}) \quad (\because \text{BI}) \tag{3}$$

where Z_t represents the normalization coefficient for holding $p(m^{[i]} \mid z_{1:t}) + p(\overline{m^{[i]}} \mid z_{1:t}) = 1$, and CIA assuming $p(z_t \mid m^{[i]}, z_{1:t-1}) \simeq p(z_t \mid m^{[i]})$ is shown effective in many applications of naive BI [9]. Similarly, we have

$$p(\overline{m^{[i]}} \mid z_{1:t}) = \frac{1}{Z_t} \frac{p(z_t) p(\overline{m^{[i]}} \mid z_t)}{p(\overline{m^{[i]}})} p(\overline{m^{[i]}} \mid z_{1:t-1}). \tag{4}$$

From (3) and (4) and log-odds (logarithm of odds) given by

$$l_{1:t}^{[i]} = \log \left(p(m^{[i]} \mid z_{1:t}) \Big/ p(\overline{m^{[i]}} \mid z_{1:t}) \right), \tag{5}$$

we have

$$l_{1:t}^{[i]} = l_{1:t-1}^{[i]} + l_t^{[i]} - l_0. \tag{6}$$

Here, $l_0 = \log p(m^{[i]})/p(\overline{m^{[i]}})$ represents the log-odds of the prior of occupancy, and $l_t^{[i]}$ is obtained by the inverse sensor model (slightly modified from [7] for computational efficiency), or

$$l_t^{[i]} = \log \frac{p(m^{[i]} \mid z_t)}{p(\overline{m^{[i]}} \mid z_t)} = \begin{cases} l_{\mathrm{occ}} & \text{if } \tilde{i} \in I^{\mathrm{scan}} \text{ and } |\tilde{r}^{[i]} - r_t^{(\tilde{i})}| < \alpha/2, \\ l_{\mathrm{free}} & \text{if } \tilde{i} \in I^{\mathrm{scan}} \text{ and } \tilde{r}^{[i]} \le r_t^{(\tilde{i})} - \alpha/2, \\ l_0 & \text{otherwise}, \end{cases} \tag{7}$$

1:	**Algorithm occupancy_grid_mapping**$\left(\{l^{[i]}_{1:t-1}\}, z_t \right)$
2:	for all $t \in I^{\text{time}}$ and $k \in I^{\text{scan}}$ do
3:	$\quad i_0 = \text{round}((\widehat{\boldsymbol{x}}^{(0)}_t - \boldsymbol{x}^{\text{gm}}_{\text{min}})/\Delta^{\text{gm}})$
4:	$\quad$ obtain four grid positions $\boldsymbol{i}(\pm 1/2, \pm \alpha/2)$ by means of the function given by
5:	$\quad\quad \boldsymbol{i}(a,b) = \text{round}((\boldsymbol{R}((k+a)\Delta_\phi - \phi_r + \widehat{\theta}^{(0)}_t)(r^{(k)}_t + b, 0)^T + \widehat{\boldsymbol{x}}^{(0)}_t - \boldsymbol{x}^{\text{gm}}_{\text{min}})/\Delta^{\text{gm}})$
6:	$\quad$ for all $\boldsymbol{i} = (i_x, i_y)$ in the smallest rectangle involving $\boldsymbol{i}_0$ and $\boldsymbol{i}(\pm 1/2, \pm \alpha/2)$ do
7:	$\quad\quad l^{[i]}_{1:t} = l^{[i]}_{t-1} + l^{[i]}_t - l_0 \quad$ for $\quad i = i_y N^{\text{gm}} + i_x$
8:	$\quad$ endfor
9:	endfor
10:	return$\{l^{[i]}_{1:t}\}$

Fig. 2. OGM algorithm modified from [7] for computational efficiency. The function round($\cdot$) denotes round half up function, or round(x) = floor($x + 0.5$) for a real number x, and round($\boldsymbol{x}$) = (round(x), round(y)) for a vector $\boldsymbol{x} = (x, y)$.

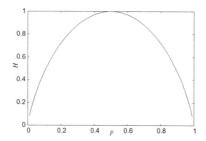

Fig. 3. Shannon's entropy $H = -p \log_2 p - (1 - p) \log_2 (1 - p)$ of the probability p of binary random variable. We encounter infinity calculation for $-\log_2 p$ with $p \to 0$ and $-\log_2 (1 - p)$ with $p \to 1$ in practice although H converge to 0 theoretically.

where α indicates the thickness of obstacle (we set $\alpha = 100\,\text{mm}$ in the experiments shown below), $\tilde{i} = \text{round}(\tilde{\phi}^{[i]}/\Delta_\phi)$ for $\tilde{\phi}^{[i]} = \text{atan2}(y^{[i]} - \widehat{y}^{(0)}_t, x^{[i]} - \widehat{x}^{(0)}_t) - \widehat{\theta}^{(0)}_t + \phi_r$ and $\tilde{r}^{[i]} = \|\boldsymbol{x}^{[i]} - \widehat{\boldsymbol{x}}^{(0)}_t\|$. The log-odds $l_{\text{occ}}(> l_0)$ and $l_{\text{free}}(< l_0)$ indicate the cell being occupied or not, respectively. We use the probability $p_0 = 0.5$, $p_{\text{occ}} = 0.9$ and $p_{\text{free}} = 0.1$ which correspond to $l_0 = \log p_0/(1 - p_0) = 0$, $l_{\text{occ}} = \log(p_{\text{occ}}/p_{\text{free}}) \simeq 2.2$ and $l_{\text{free}} = \log(p_{\text{free}}/p_{\text{occ}}) \simeq -2.2$ in the experiments shown below. In order to obtain $l^{[i]}_{1:t}$ from $l^{[i]}_{1:t-1}$ for all $i \in I^{\text{scan}}$ and $t \in I^{\text{time}}$, we use the algorithm shown in Fig. 2 which is modified from [7] for computational efficiency.

2.3   Shannon's Entropy of OGM and Relationship with Good Registration

From (5), we have the probability $p(m^{[i]} \mid z_{1:t}) = 1/(1 + \exp(-l^{[i]}_{1:t}))$. Here, in order to avoid infinity calculation practically (not theoretically) in the following step (see Fig. 3), we truncate $p(m^{[i]} \mid z_{1:t})$ so as to be in between $\tilde{p}_{\text{occ}} = 0.99$ and $\tilde{p}_{\text{free}} = 0.01$. Then, Shannon's entropy for binary random variable $m^{[i]}$ is obtained by

$$H(m^{[i]}) = p(m^{[i]} \mid z_{1:t}) \log(p(m^{[i]} \mid z_{1:t})) + p(\overline{m^{[i]}} \mid z_{1:t}) \log(p(\overline{m^{[i]}} \mid z_{1:t})), \quad (8)$$

and we use the entropy of the map m given by

$$H(m) = \left\langle H(m^{[i]}) \right\rangle_{i \in I^{\mathrm{gm}}} = \frac{1}{|I^{\mathrm{gm}}|} \sum_{i \in I^{\mathrm{gm}}} H(m^{[i]}) \quad (9)$$

where $\langle \cdot \rangle$ denotes the mean and the subscript indicates the range of the mean. The notation $|I^{\mathrm{gm}}| = (N^{\mathrm{gm}})^2$ denotes the number of elements in $I^{\mathrm{gm}} = \{0, 1, 2, \cdots, (N^{\mathrm{gm}})^2 - 1\}$. Note that the above $H(m)$ is a normalized version of the entropy shown in [6,7].

Let us examine how $H(m)$ works for evaluating registration performance. From (6) and (7) with $l_{\mathrm{occ}} = -l_{\mathrm{free}}$ and $l_0 = 0$, we have $l^{[i]}_{1:t} = \sum_{k=1}^{t} l^{[i]}_k = (n^{[i]}_{\mathrm{occ}} - n^{[i]}_{\mathrm{free}}) l_{\mathrm{occ}}$, where $n^{[i]}_{\mathrm{occ}}$ and $n^{[i]}_{\mathrm{free}}$ indicate the number of overlaps of the ith cell occupied and unoccupied, respectively, by the registered point $x^{(j)}_t$ (via (7)) for all $t \in I^{\mathrm{time}}$ and $j \in I^{\mathrm{scan}}$. Then, we have $p(m^{[i]} \mid z_{1:t}) = \left(1 + (1/p_{\mathrm{occ}} - 1)^{n^{[i]}_{\mathrm{occ}} - n^{[i]}_{\mathrm{free}}}\right)^{-1}$. Now, let $n^{[i]}_{\mathrm{overlap}} = |n^{[i]}_{\mathrm{occ}} - n^{[i]}_{\mathrm{free}}|$, and let us call it the amount of overlap. Then, $H(m^{[i]}) = 1$ for $n^{[i]}_{\mathrm{overlap}} = 0$, and $H(m^{[i]}) < 0.35$ for $n^{[i]}_{\mathrm{overlap}} \geq 1$ and $p_{\mathrm{occ}} = 0.9$. Here, when good registration is achieved, the grid cell $m^{[i]}$ corresponding to obstacle and free area is expected to have large amount of overlap $n^{[i]}_{\mathrm{overlap}} \simeq n^{[i]}_{\mathrm{occ}}$ and $n^{[i]}_{\mathrm{overlap}} \simeq n^{[i]}_{\mathrm{free}}$, respectively. Then, the number of non-overlap cells with $n^{[i]}_{\mathrm{occ}} = 0$, which have the largest entropy $H(m^{[i]}) = 1$, is expected large because occupied and unoccupied areas assigned by $x^{(j)}_t$ via (7) are constant for any registration (i.e., translation and rotation) and the number of registered points $x^{(j)}_t$ for all $t \in I^{\mathrm{time}}$ and $j \in I^{\mathrm{scan}}$ is constant. As a result, good registration is expected corresponding to large entropy $H(m) = \left\langle H(m^{[i]}) \right\rangle_i$.

Incidentally, this property of large entropy seems to contradict the properties of entropy of medical images such that low entropy corresponds to good registration as well as small uncertainty, low information, and small dispersion of probability [4,5]. This superficial contradiction arises from the difference of the definition of probability. Namely, the (joint) probability distribution of medical images represents uncertainty of transformation between two images and good registration has small entropy as well as small uncertainty. On the other hand, the probability distribution of OGM represents uncertainty of grid occupancy, and the relationship between large entropy and good registration has not been clarified, so far.

However, from the above analysis of overlap of registered points and entropy of OGM, we may say that good registration corresponds to large overlap of registered points and large entropy of OGM. From the perspective of information theory, large entropy corresponds to large uncertainty and small information. Furthermore, the best registration may correspond to the principle of maximum entropy which states that "you should select that distribution which leaves you

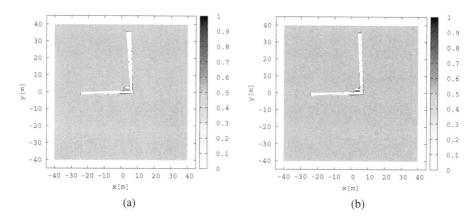

Fig. 4. OGM registered by (a) ICP-SLAM and (b) CAN2+LOOCV whose mean entropy are $\langle H(m)\rangle_{\theta_j} = 0.981605$ and 0.983487, respectively.

the largest remaining uncertainty (i.e., the maximum entropy) consistent with your constraints" [10]. This principle may be useful when we embed additional constraint, such that two walls should intersect at right angle, then the best registration will be obtained by maximizing the entropy under this constraint. Although we do not examine this property in this paper, it is one of interesting future research studies.

One of the problems to use large entropy of OGM for good registration is that the value is not independent from the grid size and the origin and orientation of OGM. The grid size should be tuned appropreately because information loss of measured data will occur for very large grid size, and inefficient computational cost will be exhasted for very small grid size. We obtain the mean entropy $\langle H(m)\rangle_{\theta_j\in\Theta^M}$ for rotated points $\boldsymbol{x}_t^{(i)} = \boldsymbol{R}(\theta_j)\left(r_t^{(i)}\boldsymbol{R}(i\Delta_\phi - \phi_r + \widehat{\theta}_t^{(0)})\boldsymbol{e}_x + \widehat{\boldsymbol{x}}_t^{(0)}\right)$ for $\theta_j \in \Theta^M = \{2\pi j/M \mid j = 0, 1, 2, \cdots, M-1\}$. The influence of some shift of the origin to the value of entropy may be reduced by the above processing of rotation mean because a change of rotation is represented by a chnge of shift as $\boldsymbol{R}(\theta + \Delta\theta)\boldsymbol{x}_t^{(i)} = \boldsymbol{R}(\theta)\boldsymbol{x}_t^{(i)} + \Delta\boldsymbol{x}_t^{(i)}$.

3   Experimental Results

We have operated the robot to run around a corner of a corridor in our department building as shown in Fig. 1(c), and we have 45 pairs of robot poses and LRF point clouds for the sampling rate 1 s. We have executed ICP-SLAM followed by our method [1], which we denote CAN2+LOOCV, to improve the performance of registration obtained by ICP-SLAM. For CAN2+LOOCV, we have used CAN2s with different number of units $N = 70, 80, 90, 100, 110$ and terminated the iterations at $t_{\text{iterate}} = 1000$ for reducing performance measure called LOOCV-MSE. And then, we have selected the registration for $N = 90$ and $t_{\text{iterate}} = 356$ which has achieved the maximum mean entropy $\langle H(m) \rangle_{\theta_j \in \Theta^M}$ for $M = 9$ rotations and the grid size $\Delta^{\mathrm{gm}} = 100\,\mathrm{mm}$ (see below for details). We show the OGM registered by ICP-SLAM and the selected CAN2+LOOCV in Fig. 4. We can see that non-parallel walls are observed at the vertical corridor (white region) in (a) and they are corrected in (b). Furthermore, (b) has achieved larger mean entropy $\langle H(m) \rangle_{\theta_j}$ than (a), which indicates that the OGM in (b) of CAN2+LOOCV has achieved better registration than the OGM in (a) of ICP-SLAM.

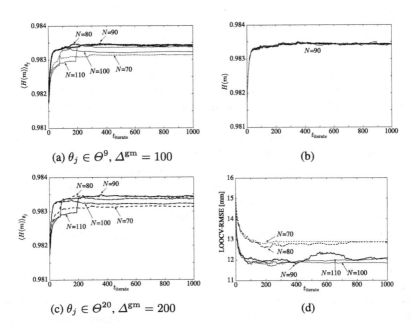

Fig. 5. Registration performance measure vs. the iteration to improve registration performance. The measures are (a) the mean entropy $\langle H(m) \rangle_{\theta_j}$ for the mean of $M = 9$ rotations of $\theta_j \in \Theta^M$ and the grid size $\Delta^{\mathrm{gm}} = 100\,\mathrm{mm}$, (b) superimposed entropy $H(m)$ for $M = 9$ rotations and $N = 90$, (c) the mean entropy $\langle H(m) \rangle_{\theta_j}$ for $M = 20$ rotations and $\Delta^{\mathrm{gm}} = 200\,\mathrm{mm}$, (d) LOOCV-RMSE. The largest mean entropy $\langle H(m) \rangle_{\theta_j}$ is 0.983487 at $t_{\mathrm{iterate}} = 356$ in (a), and 0.983504 at $t_{\mathrm{iterate}} = 357$ in (c).

In Fig. 5, performance measure of registration vs. the iteration of improving registration performance is shown. From (a), we can see that the largest $\langle H(m) \rangle_{\theta_j}$ for the mean of $M = 9$ rotations of $\theta_j \in \Theta^M$ is achieved by $N = 90$ and $t_{\mathrm{iterate}} = 356$, while the difference and the variance of constituent $H(m)$ for $N = 90$ is shown in (b). From these results, we would like to say that the registrations achieved by the largest mean entropy for $N = 80$ and 90 through i_{iterate} between 200 and 400 are not so different, while they are better than the registrations for $N = 70$, 100 and 110. Here, this statement may underestimate the results for $N = 110$ and 100, namely, $\langle H(m) \rangle_{\theta_j} = 0.983431$ and 0.983333 for

$N = 110$ and 100 at $i_{\text{iterate}} = 235$ and 152, respectively, are smaller than those for $N = 90$ and 80 although we do not have checked the significance. These results seem reasonable from the point of view that the number of piecewise linear regions from 80 to 90 (or 110) may be appropreate to approximate each of 45 LRF point clouds, and the LOOCV method to reduce the registration error has appropreate range of the number of iterations. By means of comparing (a) and (c) for $(\Delta^{\text{gm}}, M) = (100, 9)$ and $(200, 20)$, respectively, we may say that almost the same largest mean entropy is obtained robustly for the grid size Δ^{gm} from 100 to 200 mm and the number of rotations M from 9 to 20. For larger $\Delta^{\text{gm}} = 300$, 400 and 500 mm, the largest $\langle H(m) \rangle_{\theta_j}$ is achieved by $N = 90$, 80 and 80, respectively, and i_{iterate} being slightly different from those for $\Delta^{\text{gm}} = 200$ mm. These results suggest that largest entropy of OGM is robustly useful for a wide range of grid sizes in selecting good registration and good model parameters such as N and i_{iterate}. This is possible when the information loss of OGM by the increase of grid size is almost uniform for all registrations and all model parameters.

From (d), we can see that LOOCV-RMSE decreases with the increase of iterations with fluctuation until about 200 iterations, which is supposed to reflect the improvement of registration performance. However, after 200 iterations, it fluctuates and we cannot decide when good registration is achieved. Furthermore, as we have mentioned in Sect. 1, LOOCV-MSE is a measure of iterative and relative error and cannot be used to evaluate the performance of registrations for different N, which is understandable by means of comparing (a), (c) and (d).

4   Conclusion

We have analysed entropy of OGM from the point of view of good registration and overlap of registered point clouds in OGM. Then, we have shown that good registration seems corresponding to large overlap as well as large entropy, large uncertainty and low information of OGM. We have executed numerical experiments and shown the validity and the effectiveness of the entropy to evaluate the performance of registration. The principle of maximum entropy suggests that we examine and develop a method to maximize the entropy under some constraints, such as two walls should intersect at right angle (see Sect. 2.3), which is for our future research studies.

References

1. Kurogi, S., Yamashita, Y., Yoshikawa, H., Hirayama, K.: Accuracy improvement of localization and mapping of ICP-SLAM via competitive associative nets and leave-one-out cross-validation. In: Loo, C.K., Yap, K.S., Wong, K.W., Teoh, A., Huang, K. (eds.) ICONIP 2014. LNCS, vol. 8835, pp. 160–169. Springer, Heidelberg (2014). doi:10.1007/978-3-319-12640-1_20
2. Kurogi, S., Koya, H., Nagashima, R., Wakeyama, D., Nishida, T.: Range image registration using plane extraction by the CAN2. In: Proceedings of CIRA2009, pp. 346–550 (2009)

3. Salvi, J., Matabosh, C., Foli, D., Forest, J.: A review of recent range image registration methods. Image Vis. Comput. **25**, 578–596 (2007)
4. Pluim, J.P.W., Maintz, J.B.A., Viergever, M.A.: Mutual information based registration of medical images: a survey. IEEE Trans. Med. Imaging **22**(8), 986–1004 (2003)
5. Chanda, B., Majumder, D.D.: Digital Image Processing and Analysis. Prentice Hall of India Pvt. Ltd., New Delhi (2008)
6. Stachniss, C., Grisetti, G., Burgard, W.: Information gain-based exploration using Rao-Blackwellized particle filters
7. Thrun, S., Burgard, W., Fox, D.: Probabilistic Robotics. MIT Press, Cambridge (2005)
8. http://www.mrpt.org/
9. Chen, Y., Medioni, G.: Object modeling by registration of multiple range images. In: IEEE International Conference on Robotics and Automation, pp. 2724–2729 (1991)
10. Lloyd, S., Penfield Jr., P.: Maximum entropy. In: 6.050J/2.110J Information and Entropy. MIT, Cambridge (2003)

Analysis of an Intention-Response Model Inspired by Brain Nervous System for Cognitive Robot

Jae-Min Yu and Sung-Bae Cho[✉]

Department of Computer Science, Yonsei University, Seoul, Republic of Korea
{yjam,sbcho}@yonsei.ac.kr

Abstract. A service robot requires natural and interactive interaction with users without explicit commands. It is still one of the difficult problems to generate robust reactions for the robot in the real environment with unreliable sensor data to satisfy user's requests. This paper presents an intention-response model based on mirror neuron and theory of mind, and analyzes the performance for a humanoid to show the usefulness. The model utilizes the modules of behavior selection networks to realize prompt response and goal-oriented characteristics of the mirror neuron, and performs reactions according to an action plan based on theory of mind. To cope with conflicting goals, behaviors of the sub-goal unit are generated using a hierarchical task network. Experiments with various scenarios reveal that appropriate reactions are generated according to external stimuli.

Keywords: Intention-response · Hybrid architecture · Behavior selection network · Planning · Hierarchical task network

1 Introduction

To facilitate interaction with user and agents such as conversational agents, train-booking agents, intelligent agents for smartphone, and robots, we need to make them to understand user's intention and respond to user's actions or command reflecting the intention. In this regard, a new interface for the intention-response is demanded in various research fields [1]. Because it requires techniques for recognizing the user intention from various sensory information and responding to it, we proposed a model to imitate cognitive process of the human brain inspired by mirror neuron and theory of mind [2]. As many researchers have used simple methods like rule-based systems that lack the rationale from the brain science point of view, the interfaces have a difficulty to process reactions flexibly like humans [3]. It is also difficult to represent relations between the intention and the action. We presented an intention-response

This work was supported by the Industrial strategic technology development program, 10044828, Development of augmenting multisensory technology for enhancing significant effect on service industry funded by the Ministry of Trade, industry & Energy (MI, Korea).

© Springer International Publishing AG 2016
A. Hirose et al. (Eds.): ICONIP 2016, Part I, LNCS 9947, pp. 168–176, 2016.
DOI: 10.1007/978-3-319-46687-3_18

model based on brain nervous system composed of the mirror neuron system and the theory of mind system.

In this paper, we fine tune the model and analyze the performance by categorizing user intentions into simple ones for sub-goals and complex ones for task goals. The model for simple intentions based on mirror neuron system uses modules of behavior selection networks (BSNs) to respond to low-level intentions. The BSN modules mimicked the mirror neuron have difficulty to generate a behavioral sequence to handle complex intentions with conflicting goals. To cope with this problem, the model for complex intentions based on theory of mind system employs planning by a hierarchical task network to control the modules. The planner can configure a sequence of behaviors automatically using the conditions defined independently of the modules. The model has the advantage of having a well-defined structure which makes it more scalable. To verify the feasibility of the model, we implemented it on a humanoid robot, NAO, and attempted to analyze the performance with various scenarios in real situations.

2   Related Works

2.1   Mirror Neuron and Theory of Mind

In the literature of human brain research, the two systems called the mirror neuron and theory of mind are well known for understanding other's state through observation; we can understand other's intentions through the systems. The mirror neuron system consists of anterior intraparietal sulcus (aIPS), premotor cortex (PMC), and superior temporal sulcus (STS) [4]. The aIPS relates to goal-oriented actions, the PMC identifies goals or actions based on the previous memory, and the STS parses motions into a meaningful sequence [6]. Simple actions composed of sub-goals activate these areas [7]. The theory of mind system consists of temporo-parietal junction (TPJ) and medial prefrontal cortex (mPFC) [5]. The TPJ is crucial for the representation of goals and intentions, and the mPFC plays a role in reflective reasoning of actions and judgments, including goals and intentions [8].

2.2   Hybrid Control System

The hybrid control architecture to generate robot behavior is categorized into the two types: reactive and deliberative control [9]. Reactive control allows the robot to select an appropriate action instantaneously in the given environment because it uses local information obtained from sensory information. On the other hand, deliberative control manages the plan about the global environment to achieve high-level goals.

Some of the relevant studies are shown in Table 1. Min proposed the goal-oriented BSN system to generate behaviors of the delivery service robot [10]. The system used BSN and priority-based sequence plan which can be changed by the user's input and the robot generates behaviors over the sequence. Yun developed a humanoid control system using MBSN and predefined sequence [11]. A BSN in the modules was selected in a fixed order by the planner. This system connected the emotion and reactivity of the robot in multiple layers, and adopted the planning facility with probability. Quintero proposed a control system of autonomous robots using automated planning technique and 2-layer actions [13]. The hybrid control system cannot only solve the problems of each method, but also maximize the advantages of each method. However, since previous studies have not sufficiently considered the scalability problem and unstable environment, there are limitations on achieving long-term goals through solving complex problems in the real world.

Table 1. Previous studies on hybrid control system

Authors	Methods	Domain
Min et al. [10]	BSN and priority-based sequence	Mobile service robot
Yun et al. [11]	MBSN and predefined sequence	Humanoid service robot
Christopher et al. [12]	Control parameter modulations and probability-based sequence	Mobile robot navigation
Quintero et al. [13]	2-layer action	Autonomous mobile robot

3 System Architecture

3.1 Overview

This paper is based on an intention-response model for a humanoid NAO that was originally proposed at our previous study [14]. First, the intention is divided into simple and complex intentions. The simple intention is explained by the mirror neuron system which can understand mere actions instinctively and mimic the user's behaviors. The complex intention is described by the theory of mind that can recognize abstract state of the user and infer user's inner mind without direct observation. The entire process of the model is shown in Fig. 1. Simple intention-response model using the modularized BSN (MBSN) is activated by understanding the low-level intention including a sub-goal. On the other hand, complex intention-response using the planning-driven BSN is recruited by understanding the high-level intention in terms of the task goal.

Definition 1. Intention-Response:
$I = \{SI, CI\}$ is a set of the user's intentions, where $SI = \{si_1, si_2, ..., si_n\}$ is a set of simple intentions based on the mirror neuron system, and $CI = \{ci_1, ci_2, ..., ci_n\}$ is a set of complex intentions based on the theory of mind. Let $\mathscr{S}$ be the state, and t be the time. $G = \{SG, TG\}$ is a set of goals, $SG = \{sg_1, sg_2, ..., sg_n \mid \forall sg_i \leftarrow \exists si_m\}$ is a set of the sub-goals, and $TG = \{tg_1, tg_2, ..., tg_n \mid tg_i = \{sg_a, sg_b, ..., sg_d\}, \forall tg_n \leftarrow \exists ci_i\}\}$ is a set of the task goal. The response $\mathscr{R}$ according to the user's intention can be defined as follows:

$$S(t) \xrightarrow{\mathscr{R}} S(t + \alpha) \cong \exists (sg_i \vee tg_i)$$

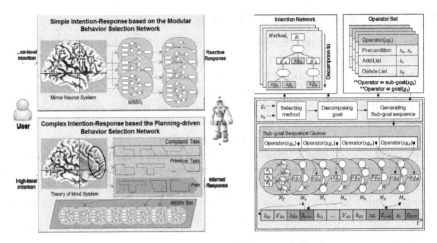

Fig. 1. Overview of the intention-response process (left) and the process of complex intention-response (right)

3.2 Simple Intention-Response Modelling

For responding to simple intention, the model is based on the BSN which generates suitable reactions rapidly and copes with the uncertain environment flexibly. However, the BSN has problems such as conflicting goals and slow reaction time. To solve these problems, Tyrrell proposed modular BSN, each of which is designed for only one goal [15]. The module allows the scalability and reusability easier than the BSN.

Definition 2. Maes' BSN $B = \{E_B, N_B, G_B\}$:
$E_B = \{e_{B1}, e_{B2}, ..., e_{Bn}\}$ is a set of the environments, $N_B = \{n_{B1}, n_{B2}, ..., n_{Bn}\}$ is a set of behavioral nodes, and $G_B = \{g_{B1}, g_{B2}, ..., g_{Bn}\}$ is a set of goals. Therefore, Tyrrell' MBSN M can be defined by $M = \{EM = \{em_i\} \mid \forall em_i \in E_B, NM = \{nm\} \mid \forall nm \in N_B, G_M \cong g_{Bi} \}$. The model has the relevance to simple intention-response based on the mirror neuron system. The mirror neuron system is activated by external stimulus rapidly and is used to understand actions including a sub-goal. Similarly, the model consists of the stimulus, behavior nodes, and a sub-goal.

Definition 3. Simple intention-response:
$SR = \{S, N, SG\}$ is a set of simple intention-response pairs. Let $U = \{u_1, u_2, ..., u_n\}$ be a set of user's actions. $S = \{U, E_M\}$ is a set of stimuli, and $N = \{n_1, n_2, ..., n_n\}$ is a set of behaviors for the intention-response. The module is selected by the recognized intention. After that, the model generates the response for user's intention.

$$S(t) \xrightarrow[\mathcal{R}:\, S \Rightarrow N \supseteq \{n_a, n_b, ..., n_d\}]{} \exists sg_i \cong si_i$$

3.3   Complex Intention-Response Modelling

Complex intention-response method for responding to high-level intentions requires the ability that can solve complex problems by achieving long-term goals. The BSN can generate the response reactively to achieve a sub-goal containing simple intention. However, since it cannot make sequence to solve the complex problem, it is not suitable for responding to complex intention. To work out this problem, we present planning by hierarchical task network. When the goal is set, sequence is made by actions which are added into a queue to achieve the goal.

If an agent is faced with unexpected situations, it would fail to realize its sub-goals, and then it cannot perform the next plan. To solve this problem, we modify the module so that a failure of sub-goals can be managed, and the plan can be revised. If a node is abnormally repeated or the number of total steps is unusually overflowed, that is considered a failure about achieving the goal.

Definition 4. Complex intention-response:
$CR = \{S, N, TG\}$ is a set of complex intention-response. Let $U = \{u_1, u_2, ..., u_n\}$ be a set of user's actions. $S = \{U\}$ is a set of stimuli, and $N = \{n_1, n_2, ..., n_n\}$ is a set of behaviors for the intention-response. The method based on theory of mind can cope with user's abstract intentions deliberatively through the process of decomposing the complex intention into simple intentions using the planning by hierarchical task network.

4   Experiments

In this section, we apply the intention-response model to NAO, a humanoid robot platform, and show the usefulness of the model. We show that it can overcome previously mentioned problems of requiring a reactive control system while performing behaviors corresponding to user's intention. The NAO is a useful robot platform to implement human cognitive structure, but this type of robots needs more sophisticated control process on many joints that might cause errors. Therefore, our experiments were performed on both the Webot simulation and a real world environments as shown in Fig. 2.

4.1   Analysis of Simple Intention-Response

We design a BSN to confirm imitation and achieving a sub-goal in the simple intention-response. Figure 3 shows a BSN designed for the activity moving a box to the left or the right. For example, 'stand by' is one of the stimuli, and 'attention' is contained in behaviors. Lastly, 'move(object)' is the sub-goal for responding to the simple intention. The observed behaviors trigger spreading activation energy between the nodes through links, and one node is finally selected.

As a result of the experiment, the model makes proper behaviors according to the intention of the observed objects as shown in Fig. 3(right). We can confirm that behaviors which should be performed at that time are activated appropriately.

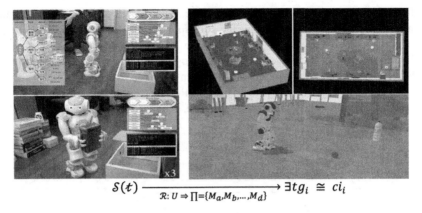

$$S(t) \xrightarrow[\mathcal{R}:\, U \Rightarrow \prod = \{M_a, M_b, \ldots, M_d\}]{} \exists tg_i \cong ci_i$$

Fig. 2. Real world (left) and virtual simulation (right) environments

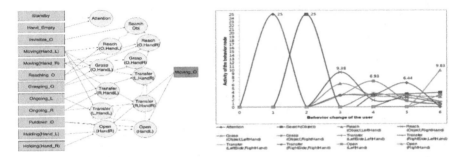

Fig. 3. The BSN module for moving a box (left) and response process (right)

4.2   Analysis of Complex Intention-Response

Scenario: A user requests to NAO that the OBJECT be delivered from PLACE _A to PLACE_B, when the NAO's current position is PLACE _A.

The experiment was based on the scenario to verify appropriate reactions to the changing environment. Accordingly, we aim to confirm the suitable responses to accomplish a target goal. As a result of the experiment, BSN planning queue is constructed automatically. When user's intention to clean a room is input, four intermediate purposes are generated (detect the object, pick up the object, detect the place, and putdown the object). Through the observation of environment around, the node having the highest activation energy is chosen at every time step. The node chosen is performed right away. That is, the value of activation can be zero or not.

Figure 4 shows the activation nodes of agent robot and internal processing of command. After moving to the object, NAO picked up the object, and the module which can put down object is performed in succession. At the same time, pickup module was deleted in planning queue of right top. When all modules in planning queue is performed completely, intention-response process is terminated. Whenever user tries conversation to NAO, this system analyzes the intention. If the intention is complicated, system creates a planning queue which performs a task in regular sequence. Left side of the figure shows the corresponding flow of scenario and internal components. X-axis of the figure means the flow of time. When a complex intention like "Please throw away garbage to trash" is recognized, the model generates a queue of behavior sequence of responding to the intention. For instance, GrabA to detect garbage, Putdown to throw away garbage, and Return to user. In this figure, we can confirm the generation and elimination of queue.

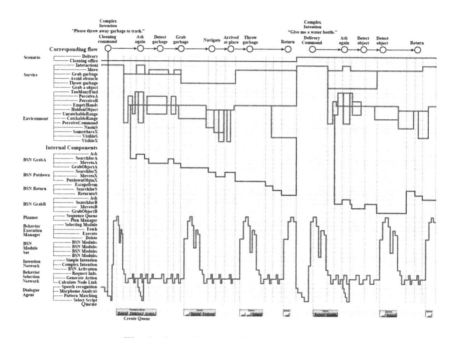

Fig. 4. Internal states of a response process

Figure 5 shows the result after running the experiments 15 times. This result shows the response rate, success rate and average execution time of service process. Detection requires the longest time, and this condition leads to high success rate accordingly. In movement, direction of walk or the number of walk can be changed on experiment environment. That is the reason why success rate is low despite of high response rate. We can confirm that if intermediate process is not performed appropriately, final goal is influenced by intermediate process. Average execution time of action is one third of detection. The reason of this difference can be explained by understanding environment of BSN. Because all environment conditions of behavior nodes are satisfied, BSN can perform the final goal directly without detection or movement.

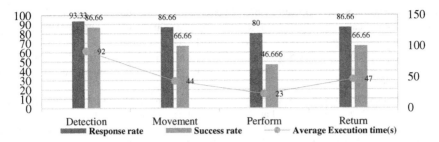

Fig. 5. Experiment results according to iterations

5   Concluding Remarks

This paper evaluated the service process of agent robot which is based on intention-response model. This model can select appropriate response according to user's various requests and generate corresponding behaviors. The experiment was based on the scenario to verify appropriate reactions to the changing environment. As a result, we can confirm the usefulness of this model. Also, processes of internal system of agent robot were presented to explain this model. Through the series of iterating experiments, we got quantitative results which can be evaluated on various aspects. In the future, we need to validate the accuracy and effectiveness comparing to other response methods, and conduct additional iterative experiments to show the reliability.

References

1. Liu, J., Wong, C.K., Hui, K.K.: An adaptive user interface based on personalized learning. IEEE Intell. Syst. **18**(2), 52–57 (2003)
2. Kuniyoshi, Y., Yorozu, Y., Ohmura, Y., Terada, K., Otani, T., Nagakubo, A., Yamamoto, T.: From humanoid embodiment to theory of mind. In: Iida, F., Pfeifer, R., Steels, L., Kuniyoshi, Y. (eds.) Embodied Artificial Intelligence. LNCS (LNAI), vol. 3139, pp. 202–218. Springer, Heidelberg (2004)
3. Nicolescu, R.K.M., Nicolescu, A.T.M., Bebis, C.K.G.: Understanding human intentions via hidden markov models in autonomous mobile robots, pp. 367–374. ACM/IEEE Hum. Robot, Interaction (2008)
4. Duijnhoven, D.V.: The role of the mirror neuron system in action understanding and empathy, Bachelor thesis Cognitive Neuroscience (2010)
5. Amodio, D.M., Frith, C.D.: Meeting of minds: the medial frontal cortex and social cognition. Nat. Rev. Neurosci. **7**(4), 268–277 (2006)
6. Cross, E.S., Hamilton, A.F.D.C., Grafton, S.T.: Building a motor simulation de novo: observation of dance by dancers. Neuroimage **31**(3), 1257–1267 (2006)
7. Hamilton, A.F.D.C., Grafton, S.T.: Goal representation in human anterior intraparietal sulcus. Neuroscience **26**(4), 1133–1137 (2006)
8. Saxe, R., Powell, L.J.: It's the thought that counts specific brain regions for one component of theory of mind. Psychol. Sci. **17**(8), 692–699 (2006)
9. Murphy, R.R.: Introduction to AI Robotics. The MIT Press, Cambridge (2000)

10. Min, H.-J., Cho, S.-B.: Generating optimal behavior of mobile robot using behavior network with planning capability. In: IEEE International Symposium on Computational Intelligence in Robotics and Automation, vol. 1, pp. 186–191 (2003)
11. Yun, S.-J., Lee, M.-C., Cho, S.-B.: P300 BCI based planning behavior selection network for humanoid robot control. In: Proceedings 9th International Conference on Natural Computation, pp. 354–358 (2013)
12. Lee-Johnso, C.P., Carnegie, D.A.: Mobile robot navigation modulated by artificial emotions. IEEE Trans. Syst. Man Cybern. Part B Cybern. **40**(2), 468–480 (2010)
13. Quintero, E.A., García-Olaya, Á., Borrajo, D., Fernández, F.: Control of autonomous mobile robots with automated planning. J. Phys. Agents **5**(1), 3–13 (2011)
14. Chae, Y.-J., Cho, S.-B.: An intention-response model based on mirror neuron and theory of mind. In: 10th International Conference on Natural Computation, pp. 380–385 (2014)
15. Tyrrell, T.: An evaluation of Maes's bottom-up mechanism for behavior selection. Adapt. Behav. **2**(4), 307–348 (1994)

Dynamic Surface Sliding Mode Algorithm Based on Approximation for Three-Dimensional Trajectory Tracking Control of an AUV

Kai Zhang[1], Tieshan Li[1(✉)], Yuqi Wang[1], and Zifu Li[2]

[1] Dalian Maritime University, Navigation College, Dalian, China
{2277493720,394562998}@qq.com, tieshanli@126.com
[2] Jimei University, Navigation College, Xiamen, China
lzfxmjmul019@163.com

Abstract. In this paper, a novel dynamic surface sliding mode control method is proposed for three-dimensional trajectory tracking control of autonomous underwater vehicle (AUV) in the presence of model errors. To enhance the robustness, the sliding mode control approach is modified by employing dynamic surface control (DSC). The radial basis function neural network (RBFNN) approximation technique is used for approximating model errors, furthermore the norm of the ideal weighting vector in neural network system is considered as the estimation parameter, such that only one parameter is adjusted. The proposed controller guarantees uniform ultimate boundedness (UUB) of all the signals in the closed-loop system via Lyapunov stability analysis, while the tracking errors converge to a small neighborhood of the desired trajectory. Finally, simulation studies are given to illustrate the performance of the proposed algorithm.

Keywords: Autonomous underwater vehicle(AUV) · Three-dimensional trajectory tracking · Dynamic surface control(DSC) · Sliding mode control (SMC) · Radial basis function neural network(RBFNN)

1 Introduction

Nowadays the ocean space is an important competition field of military and economic powers in the world, and many countries are vigorously developing deep sea exploration technology [1]. It is well known that an underwater vehicle is the favored solution to be deployed in many undersea applications especially in the military field

This work is supported in part by the National Natural Science Foundation of China (Nos. 51179019, 61374114), the Fundamental Research Program for Key Laboratory of the Education Department of Liaoning Province (LZ2015006), the Fundamental Research Funds for the Central Universities under Grant 3132016313, the Hong Kong Research Grants Council under Project no.: CityU113212, Fujian Provincial Department of education Projection (JAJ09148), and The Pan Jinlong project of Jimei University (ZC2012019).

© Springer International Publishing AG 2016
A. Hirose et al. (Eds.): ICONIP 2016, Part I, LNCS 9947, pp. 177–184, 2016.
DOI: 10.1007/978-3-319-46687-3_19

and the oil and gas industry [2]. Therefore the precise trajectory tracking of AUV is required, so far various control strategies have been proposed for tracking of AUV. Due to the simple algorithm, the traditional sliding mode control (SMC) has been successfully applied to dynamic positioning and motion control of AUV [3], nevertheless, the main disadvantage of the control algorithm was chattering effect. To enhance the algorithm, it is a commonly used method that SMC was combined with other control algorithm, such as fuzzy control [4, 5], dynamic surface control [6] and genetic algorithm [7]. Recently, a simple adaptive neural tracking controller was proposed for an uncertain AUV system in [8]. In the algorithm, only one parameter needs to be estimated online regardless of the number of the NN nodes, so it is convenient to implement in applications. However, the accuracy of filter is not satisfied, which has effect on the robustness of controller [9].

According to the above observations, this paper designs a dynamic surface sliding mode control algorithm based on radial basis function neural network(RBFNN) for trajectory tracking control of an AUV in the presence of model errors. The controller is designed by combination of sliding mode control, dynamic surface control and RBFNN approximation technique. A simulation is carried out for an AUV, the simulation results show that the designed controller achieves a good performance for three-dimensional trajectory tracking control.

2 Problem Formulation and Preliminaries

The mathematical model for underwater vehicle motion of six degrees of freedom can be obtained in the general form [8].

$$M(\eta)\ddot{\eta} + C(v,\eta)\dot{\eta} + G(\eta) + D(v,\eta)\dot{\eta} + \tau_d + \Delta(v,\eta) = \tau \tag{1}$$

where, $\eta = \begin{bmatrix} x & y & z & \phi & \theta & \psi \end{bmatrix}^T$ is the position and Euler angles vector with respect to earth-fixed coordinate system:$v = \begin{bmatrix} u & v & w & p & q & r \end{bmatrix}^T$ is the velocity vector with respect to body-fixed coordinate system; $M(\eta)$ is the inertia matrix, $C(v,\eta)$ is the so-called centripetal-Coriolis matrix. Both $M(\eta)$ and $C(v,\eta)$ are related to rigid-body dynamics and added mass forces and moments.$g(\eta)$ is the gravitational or restoring forces vector, $D(v,\eta)$ denotes the general damping coefficients matrix derived from potential damping, skin friction, and damping due to vortex shedding etc. τ_d is an immeasurable environmental disturbance vector due to waves or cable traction etc. τ denotes the external forces vector provided by rudders, thrusters or etc. $\Delta(v,\eta)$ denotes modeling errors or system perturbation. In some cases, the modeling errors do affect the control performance of underwater vehicle, especially in low speed maneuvering.

The control objective of this paper is to track the desired trajectory η_d of AUV in the presence of the model error, i.e. $\lim_{t\to\infty}\|\eta - \eta_d\| < \delta, \quad \delta > 0$.

Assumption 1: The desired trajectory η_d is a sufficiently smooth function of t and η_d, $\dot{\eta}_d$, $\ddot{\eta}_d$ are bounded, that is, there exists a positive constant B_0 satisfy $\eta_d^2 + \dot{\eta}_d^2 + \ddot{\eta}_d^2 \leq B_0$.

Assumption 2: $\Delta(v, \eta)$ is bounded, that is, there exists a positive unknown constant $\Delta * (v, \eta)$ such that $|\Delta_i(v, \eta)| \le \Delta * (v, \eta), 1 \le i \le 6$.

Notation 1: $\lambda_{\max}(A)$ and $\lambda_{\min}(A)$ denote the largest and smallest eigenvalues of a square matrix A, respectively. $\|\cdot\|$ stands for Frobenius norm of matrices and Euclidean norm of vectors, i.e., given a matrix B and a vector Q, the Frobenius norm and Euclidean norm are given by $\|B\|^2 = tr(B^T B) = \sum_{i,j} b_{ij}^2$ and $\|Q\|^2 = \sum_i q_i^2$.

Notation 2: In this paper, radial basis function NNs are employed to approximate unknown nonlinear functions. Given any real continuous function $f : \Omega \to R$, then there exists a basis function vector $S : R^m \to R^l$ and ideal weight vector $W^* \in R^l$ such that $f = W^{*T} S(Z) + \varepsilon$, where $Z \in \Omega \subset R^m$ is the input vector, $\varepsilon \in R$ is the NN approximation error satisfying $|\varepsilon| \le \bar{\varepsilon}$ with $\bar{\varepsilon} > 0$, and W^* is an optimal weight vector of W and is defined as

$$W^* = \arg \min_{W \in R^l} \left\{ \sup_{Z \in \Omega} |f(x) - W^T S(Z)| \right\}$$

where the weight vector $W = [w_1, \cdots, w_l]^T \in R^l$, the NN node number $l > 1$, and $S(Z) = [s_1(Z), s_2(Z), \ldots, s_l(Z)]^T$ with

$$s_i(Z) = \frac{1}{\sqrt{2\pi}\eta_i} \exp\left(-\frac{(Z - \mu_i)^T (Z - \mu_i)}{2\eta_i^2} \right), \quad i = 1, \cdots, l$$

where $\mu_i = [\mu_{i1}, \mu_{i2}, \cdots, \mu_{im}]^T$ is the center of the receptive field and η_i is the width of the Gaussian function.

3 Controller Design and Stability Analysis

The detailed design procedure is described in the following steps. It mainly includes 2 steps.

Step 1. Choosing $x_1 = \eta$; $x_2 = \dot{\eta}$, then (1) can be rewritten as follows.

$$\dot{x}_1 = x_2$$
$$\dot{x}_2 = -M^{-1}(\eta)C(v, \eta)x_2 - M^{-1}(\eta)D(v, \eta)x_2 - M^{-1}(\eta)G(\eta) - M^{-1}(\eta)\tau_d - M^{-1}(\eta)\Delta(v, \eta) + M^{-1}(\eta)\tau$$

Define the error surface

$$z_1 = x_1 - \eta_d \tag{2}$$

$$\dot{z}_1 = x_2 - \dot{\eta}_d \tag{3}$$

Choose the Lyapunov function candidate

$$V_1 = \frac{1}{2} z_1^T z_1 \tag{4}$$

$$\dot{V}_1 = z_1^T \dot{z}_1 \tag{5}$$

Choose virtual control law for the first subsystem

$$a_1^0 = -c_1 z_1 + \dot{\eta}_d \tag{6}$$

where a_1 is a new state variable and can be obtained by introducing a first-order filter with a time constant e as follows.

$$e\dot{a}_1 + a_1 = a_1^0 \quad a_1(0) = a_1^0(0) \tag{7}$$

Define the filter error

$$h_1 = a_1 - a_1^0 \tag{8}$$

Define the second error surface

$$z_2 = x_2 - a_1 \tag{9}$$

$$\dot{z}_2 = \dot{x}_2 - \dot{a}_1 \tag{10}$$

Substituting (9) and (8) into (3) gives

$$\dot{z}_1 = z_2 + h_1 + a_1^0 - \dot{\eta}_d \tag{11}$$

Step2. Consider the position tracking, virtual control and filter error, choose the following Lyapunov function candidate.

$$V_2 = V_1 + \frac{1}{2} z_2^T z_2 + \frac{1}{2} h_1^T h_1 \tag{12}$$

Due to $\dot{z}_2 = \dot{x}_2 - \dot{a}_1, \dot{h}_1 = \frac{a_1^0 - a_1}{e} - \dot{a}_1^0 = \frac{-h_1}{e} + c_1 \dot{z}_1 - \ddot{\eta}_d$, so

$$
\dot{V}_2 = z_1^T(z_2 + h_1 + a_1^0 - \dot{\eta}_d) + z_2^T(-M^{-1}(\eta)C(v,\eta)x_2 - M^{-1}(\eta)D(v,\eta)x_2 - M^{-1}(\eta)G(\eta)
$$
$$
- M^{-1}(\eta)\tau_d - M^{-1}(\eta)\Delta(v,\eta) + M^{-1}(\eta)\tau - \dot{a}_1) + h_1^T(\frac{-h_1}{e} + B_2) \tag{13}
$$

where $B_2 = c_1 \dot{z}_1 - \ddot{\eta}_d, B_2(i)$ is a continuous function and has a maximum value $M_2(i)$ i.e. $|B_2(i)| \le M_2(i)$, please refer to [10] for details. ($1 \le i \le 6$, it will be used throughout this paper).

Combining the definition of sliding mode control, define the sliding surface is $s = z_2$, now, if Δ is known, we can design the dynamic surface sliding mode control.

$$\tau^* = M(\eta)[M^{-1}(\eta)C(v,\eta)x_2 + M^{-1}(\eta)D(v,\eta)x_2 + M^{-1}(\eta)G(\eta) + M^{-1}(\eta)\tau_d$$
$$+ M^{-1}(\eta)\Delta(v,\eta) + \dot{a}_1 - \zeta.\mathrm{sgn}(z_2) - c_2 z_2] \tag{14}$$

where $c_2 > 0, \zeta > 0$, and ζ is a constant.

The RBFNN is used to approximate the $-M^{-1}(\eta)\Delta(v,\eta)$, we have

$$-M^{-1}(\eta)\Delta(v,\eta) = W_2^{*T}S_2(Z) + \varepsilon_2 \tag{15}$$

$$z_2^T W_2^{*T} S_2(Z) + z_2^T \varepsilon_2 \leq \frac{\lambda_2^T \|z_2\|^2 \|S_2(Z)\|^2}{2b_2^2} + \frac{b_2^2}{2} + \frac{\|z_2\|^2}{2} + \frac{\|\bar{\varepsilon}_2\|^2}{2} \tag{16}$$

Remark 1: $\lambda_2^T = \|W_2^*\|^2$ is the norm of the ideal weighting vector in a neural network. Since W_2^* is unknown, λ_2^T will be replaced by its estimation value in the following design procedure. Throughout this paper, let $\lambda_2 - \hat{\lambda}_2 = \tilde{\lambda}_2$.

Choose the Lyapunov function candidate

$$V_3 = V_2 + \frac{1}{2}\tilde{\lambda}_2^T \Gamma_2^{-1} \tilde{\lambda}_2 \tag{17}$$

where $\tilde{\lambda}_2 = \lambda_2 - \hat{\lambda}_2, \Gamma_2 = \Gamma_2^T > 0$ is a gain constant matrix, so

$$\dot{V}_3 = \dot{V}_2 - \tilde{\lambda}_2^T \Gamma_2^{-1} \dot{\hat{\lambda}}_2 \tag{18}$$

Choose the final controller

$$\tau = C(v,\eta)x_2 + D(v,\eta)x_2 + G(\eta) + \tau_d - \frac{M(\eta)z_2 \hat{\lambda}_2^T \|S_2(Z)\|^2}{2b_2^2} + M(\eta)\dot{a}_1$$
$$- \zeta.M(\eta)\mathrm{sgn}(z_2) - c_2 M(\eta)z_2 \tag{19}$$

Substituting (6) (13) (16) and (19) into (18), choose

$$c_1 \geq 1 + a_0, c_2 \geq 1 + a_0, \frac{1}{e} \geq \frac{1}{2} + \frac{M_2^2}{2} + a_0$$

Then (18) can be written as follows.

$$\dot{V}_3 \leq -a_0\|z_1\|^2 - \zeta\mathrm{sgn}(z_2)z_2 - a_0\|z_2\|^2 + \frac{b_2^2}{2} + \frac{\tilde{\lambda}_2^T \|z_2\|^2 \|S_2(Z)\|^2}{2b_2^2}$$
$$+ \frac{\|\bar{\varepsilon}_2\|^2}{2} - \frac{\|M_2\|^2 \|h_1\|^2}{2} - a_0\|h_1\|^2 + \frac{\|B_2\|^2 \|h_1\|^2}{2} + \frac{1}{2} - \tilde{\lambda}_2^T \Gamma_2^{-1} \dot{\hat{\lambda}}_2 \tag{20}$$

Because of $|B_2| \leq M_2$, so

$$
\begin{aligned}
\dot{V}_3 \leq & -a_0\|z_1\|^2 - a_0\|z_2\|^2 - a_0\|h_1\|^2 - \zeta\mathrm{sgn}(z_2)z_2 \\
& - \tilde{\lambda}_2^T \Gamma_2^{-1} \dot{\hat{\lambda}}_2 + \frac{\tilde{\lambda}_2^T \|z_2\|^2 \|S_2(Z)\|^2}{2b_2^2} + \frac{\|\bar{\varepsilon}_2\|^2}{2} + \frac{b_2^2}{2} + \frac{1}{2}
\end{aligned} \tag{21}
$$

Choose adaptive law

$$
\dot{\hat{\lambda}}_2 = \Gamma_2 \left(\frac{\|z_2\|^2 \|S_2(Z)\|^2}{2b_2^2} - \sigma_2\left(\hat{\lambda}_2 - \lambda_2^0\right) \right) \tag{22}
$$

Noting the follow.

$$
\sigma_2 \tilde{\lambda}_2^T \left(\hat{\lambda}_2 - \lambda_2^0\right) \leq -\frac{\sigma_2 \tilde{\lambda}_2^T \tilde{\lambda}_2}{2} + \frac{\sigma_2\left(\lambda_2 - \lambda_2^0\right)^2}{2} \leq -a_0 \tilde{\lambda}_2^T \Gamma_2^{-1} \tilde{\lambda}_2 + \frac{\sigma_2\left(\lambda_2 - \lambda_2^0\right)^2}{2} \tag{23}
$$

where, $\frac{\sigma_2}{2\lambda_{\max}\left(\Gamma_2^{-1}\right)} \geq a_0, \lambda_2^0$ are initial values of $\lambda_2, \sigma_2 > 0, \lambda_2^0$ and σ_2 are design constants.

Substituting (22) and (23) into (21)

$$
\dot{V}_3 \leq -2a_0 V_3 + D - \zeta\mathrm{sgn}(z_2)z_2 \leq -2a_0 V_3 + D \tag{24}
$$

where $D = \frac{1}{2} + \frac{b_2^2}{2} + \frac{\|\bar{\varepsilon}_2\|^2}{2} + \frac{\sigma_2\left(\lambda_2 - \lambda_2^0\right)^2}{2}$.

From (24), one has

$$
V_3(t) \leq \frac{D}{2a_0} + \left(V_3(t_0) - \frac{D}{2a_0}\right)e^{-(t-t_0)} \tag{25}
$$

It follows that, for any $\mu_1 > (D/a_0)^{1/2}$, there exists a constant $T > 0$ such that $\|z_1(t)\| \leq \mu_1$ for all $t \geq t_0 + T$, and the tracking errors can be made small, since $(D/a_0)^{1/2}$ can arbitrarily be made small if the design parameters are appropriately chosen.

4   Simulation Result

In this section, an example is given to show the efficiency of the proposed controller. We will use the nonlinear model of Naval Postgraduate School AUV II [11].

The reference trajectory is

$$
\eta_d = [\,3\sin(t)\quad 3\sin(2t)\quad 25\sin(4t)\quad 12\sin(0.25t)\quad 12\sin(0.5t)\quad 25\sin(2t)\,]^T
$$

Modeling error is assumed $\Delta(v, \eta) = 6\sin\xi$ ($\xi = z_1 + \dot{z}_1$). The disturbance is

$$\tau_d = [2+2\sin(t) \quad 4+2\cos(t) \quad 1+2\sin(2t) \quad 1+2\cos(t) \quad 3+2\sin(t) \quad 2+2\cos(t)]^T$$

The designed parameters of the above controller are given as $c_1 = 12, c_2 = 15$, $e = 0.1, \Gamma_2 = diag\{0.5\}, \sigma_2 = 0.5, \lambda_2^0 = 0$. The number of the NN nodes are chosen as $l = 25$, the centers of basis function are evenly distributed in $[-1,1] \times [-1,1]$ with the width $\eta_1 = 5$.

The simulation result is shown in Fig. 1, it can be observed that the tracking performance is satisfactory under the disturbance and model errors.

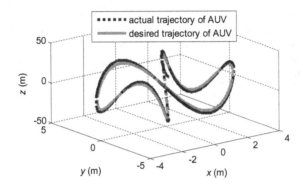

Fig. 1. Trajectory of AUV

5 Conclusion

In this paper, employing dynamic surface technique and sliding mode technique, a dynamic surface sliding mode control based on neural network (NN) method is proposed for trajectory tracking control of AUV in the presence of model errors. This algorithm enhances the robustness of this system with less computational burden. Simulation result shows the performance of the proposed algorithm.

References

1. Qi, D., Feng, J.F., Yang, J.: Longitudinal motion control of AUV based on fuzzy sliding mode method. J. Control Sci. Eng. Article ID 7428361, 7 pages (2016)
2. Isnail, Z.H., Mokhar, M.B.M., Putranti, V.W.E., Dunnigan, M.W.: A robust dynamic region-based control scheme for an autonomous underwater vehicle. Ocean Eng. **111**, 155–165 (2016)
3. Peng, Y.F.: Robust intelligent sliding model control using recurrent cerebellar model articulation controller for uncertain nonlinear chaotic systems. Chaos, Solitons Fractals **39**(1), 150–167 (2009)

4. Lakhekar, G.V., Saundarma, V.D.L.: Novel Adaptive fuzzy sliding mode controller for depth control of an underwater vehicles. In: The 2013 IEEE International Conference on Fuzzy Systems (FUZZ), Hyderabad, India, pp. 1–7 (2013)

5. Guo, J., Chiu, F.-C., Huang, C.-C.: Design of a sliding mode fuzzy controller for the guidance and control of an autonomous underwater vehicle. Ocean Eng. 30(16), 2137–2155 (2003)

6. Chen, W., Wei, Y.H., Zeng, J.H., Jia, X.Q., Wang, Z.P.: Adaptive control of AUV's pitch angle based on dynamic surface sliding mode. Fire Control Command Control 41(6), 73–76 (2016)

7. Lin, F.J., Chou, W.D.: An induction motor servo drive using sliding-mode controller with genetic algorithm. Electr. Power Syst. Res. 64(2), 93–108 (2003)

8. Miao, B.B., Li, T.S., Luo, W.L.: A DSC and MLP based robust adaptive NN tracking control for underwater vehicle. Neurocomputing 111, 184–189 (2013)

9. Zou, J., Li, L.: Research on dynamic surface sliding mode intelligent control algorithm based on approximation. Comput. Technol. Dev. 25(6), 6–11 (2015)

10. Wang, D., Huang, J.: Neural network-based adaptive dynamic surface control for a class of uncertain nonlinear systems in strict-feedback form. IEEE Trans. Neural Networks 16(1), 195–202 (2005)

11. Fossen, T.I.: Guidance and control of ocean vehicles. Wiley, New York (1998)

Bio-Inspired/Energy-Efficient Information Processing: Theory, Systems, Devices

Exploiting Heterogeneous Units for Reservoir Computing with Simple Architecture

Gouhei Tanaka[1]([✉]), Ryosho Nakane[1], Toshiyuki Yamane[2], Daiju Nakano[2],
Seiji Takeda[2], Shigeru Nakagawa[2], and Akira Hirose[1]

[1] Graduate School of Engineering, The University of Tokyo, Tokyo 113-8656, Japan
`gouhei@sat.t.u-tokyo.ac.jp, nakane@cryst.t.u-tokyo.ac.jp,`
`ahirose@ee.t.u-toyo.ac.jp`
[2] IBM Research - Tokyo, Kawasaki, Kanagawa 212-0032, Japan
`{tyamane,dnakano,seijitkd,snakagw}@jp.ibm.com`

Abstract. Reservoir computing is a computational framework suited for sequential data processing, consisting of a reservoir part and a read-out part. Not only theoretical and numerical studies on reservoir computing but also its implementation with physical devices have attracted much attention. In most studies, the reservoir part is constructed with identical units. However, a variability of physical units is inevitable, particularly when implemented with nano/micro devices. Here we numerically examine the effect of variability of reservoir units on computational performance. We show that the heterogeneity in reservoir units can be beneficial in reducing the prediction error in the reservoir computing system with a simple cycle reservoir.

Keywords: Reservoir computing · Sequential data processing · Simple cycle reservoir · Heterogeneous neurons · Energy efficiency

1 Introduction

Recurrent neural networks are capable of producing high-dimensional complex dynamics due to feedback connections, which has often been utilized for information processing of sequential data [1]. The training methods for recurrent neural networks have been proposed, including the backpropagation through time algorithm, the real-time recurrent learning, and the extended Kalman filter method [2]. These algorithms try to adapt all the connection weights by minimizing the total error between the network output sequence and the desired output sequence. Since they have relatively high time complexity, their practical applications with large-scale networks are still not realized. Reservoir computing is one of the potent frameworks that can overcome the problem of the training cost in recurrent neural networks for energy efficient computing [3,4]. The reservoir computing framework was established by combining the concepts from the echo state network (ESN) [5–7] and the liquid state machine [8].

© Springer International Publishing AG 2016
A. Hirose et al. (Eds.): ICONIP 2016, Part I, LNCS 9947, pp. 187–194, 2016.
DOI: 10.1007/978-3-319-46687-3_20

The reservoir computing system consists of the reservoir part and the read-out part. The reservoir part is used for mapping the input sequence to a high-dimensional spatiotemporal pattern. The readout part is used for adjusting the output connection weights so that the spatiotemporal pattern generated by the reservoir is appropriately mapped to the desired output sequence. Since not all the weights but only the output weights are adapted, the reservoir computing can save the learning time compared with the conventional recurrent neural networks. Moreover, the fixed reservoir can be implemented with nonlinear physical systems and devices, including optoelectronics [9], memristors [10], and wave phenomena [11,12].

In the standard ESN [5], the reservoir is given as a randomly connected recurrent neural network. The performance of reservoir computing relies on the number of neurons and the weight matrix in the reservoir, which govern the length of the history of input sequence that can be embedded into its spatiotemporal dynamics. For constructing a good mapping from an input sequential data to an output one, the reservoir is required to satisfy the echo state property [5] which indicates the property that the influence of the input stream is gradually attenuated with time. This means that the mapping represented by the reservoir should be neither expanding nor highly contracting. Hence, the spectral radius of the weight matrix is often set to be less than and close to unity, corresponding to the edge of chaos [2]. However, the random reservoir topology is not mandatory. A deterministically designed reservoir with simple ring architecture is comparable to the standard random reservoir in their computational performance [13]. The simple cycle reservoir enables theoretical analyses of reservoir computing properties such as memory capacity. In addition, it is favorable for hardware implementation because only local connections and uniform weights are needed.

In this study, we incorporate variability into the neuron units in the simple cycle reservoir, motivated by two aspects. One is that the simple cycle reservoir with identical units seems to be too uniform to produce rich nonlinear dynamics. The unit variability is expected to diversify the dynamics of individual units. The other is that the variability of the reservoir units are inevitable when they are implemented with physical devices, particularly with nano/micro devices. We examine how heterogeneity of the reservoir units impacts on the reservoir dynamics and its computational capability. We show that the variability in the reservoir units can improve the performance of the simple cycle reservoir.

2   Methods

2.1   Model

The reservoir in the standard ESN consists of neuron units which interact with each other through weighted random connections as illustrated in Fig. 1(a). The numbers of input units, internal units, and output units are denoted by L, N, M, respectively. Then, the states of the input, internal, and output units are represented by the column vectors $\mathbf{u}(t) = (u_1(t), u_2(t), \ldots, u_L(t))^T$,

$\mathbf{x}(t) = (x_1(t), x_2(t), \ldots, x_N(t))^T$, and $\mathbf{y}(t) = (y_1(t), y_2(t), \ldots, y_M(t))^T$, respectively. The input connectivity, the reservoir connectivity, the feedback connectivity, and the output connectivity are represented by $W^{\text{in}} \in \mathbb{R}^{L \times N}$, $W \in \mathbb{R}^{N \times N}$, an $W^{\text{fb}} \in \mathbb{R}^{M \times N}$, and $W^{\text{out}} \in \mathbb{R}^{N \times M}$, respectively.

The states of the ith internal unit $(i = 1, \ldots, N)$ is updated as follows:

$$\mathbf{x}_i(t+1) = f_i \left((W_i^{\text{in}})^T \cdot \mathbf{u}(t+1) + (W_i)^T \cdot \mathbf{x}(t) + (W_i^{\text{fb}})^T \cdot \mathbf{y}(t) \right), \qquad (1)$$

where f_i stands for the activation function of the ith neuron in the reservoir and W_i^{in}, W_i, and W_i^{fb} are the ith row of the input, reservoir, and feedback weight matrices, respectively. The states of the jth output unit $(j = 1, \ldots, M)$ is given by

$$\mathbf{y}_j(t) = f^{\text{out}} \left((W_j^{\text{out}})^T \mathbf{x}(t) \right), \qquad (2)$$

where f^{out} represents the activation function of the output neurons and W_j^{out} is the jth row of the output weight matrix. Here we use $f^{\text{out}}(x) = \tanh(x)$.

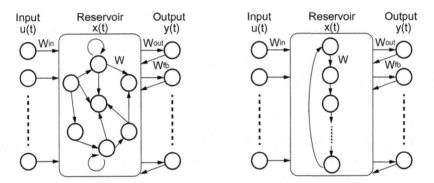

Fig. 1. Structure of the reservoir. (a) Random topology. The standard architecture in the ESN [5]. (b) Ring topology called the simple cycle reservoir [13].

When an input sequential data $\mathbf{u}(t)$ is given, the output sequence is generated by Eqs. (1)–(2). The characteristic of the reservoir computing is that the weights in the input and reservoir connections are not adapted but only the output connection weights W^{out} are determined by a learning rule. The output matrix W^{out} is obtained to minimize the error between the network output sequence $\mathbf{y}(t)$ and a desired output sequence $\mathbf{d}(t)$, given by

$$E = \langle ||\mathbf{y}(t) - \mathbf{d}(t)||^2 \rangle, \qquad (3)$$

where $\langle \cdot \rangle$ denotes an average over a time period. The minimization of E can be achieved using regression methods. Here we employ the pseudoinverse computation [2].

2.2   Reservoir Structure

We use the simple cycle reservoir as shown in Fig. 1(b), where the connectivity of the reservoir nodes has ring topology and the connection weights are uniform [13]. It is represented as the weight matrix $W = (w_{i,j})$ where $w_{i+1,i} = r$, $w_{1,N} = r$, and all the other entries are zero. For the standard reservoir, the necessary condition for the echo state property is given by $\rho(W) < 1$ where $\rho(W)$ is the spectral radius of W and the sufficient condition is given by $\bar{\sigma}(W) < 1$ where $\bar{\sigma}(W)$ is the largest singular value of W [5]. For the simple cycle reservoir [13], $\rho(W) = \bar{\sigma}(W) = r$. The simple cycle reservoir is comparable to the standard reservoir in the performance of time series predictions and its memory capacity can be theoretically derived [13].

2.3   Heterogeneous Units

In most studies on reservoir computing, the units of the reservoir have been assumed to be identical. The hyperbolic tangent function is normally used as the nonlinear activation function of the units in ESNs. In the standard reservoir, the diversity of the dynamics of the reservoir units are brought about by the random weight matrix. However, in the simple cycle reservoir with identical units, the dynamics generated by the individual units become uniform. The total system can be essentially reduced to a lower-dimensional system. This is unbenefited for producing high-dimensional spatiotemporal dynamics. Thus, we introduce the variability in the activation function of the reservoir units, which are represented as follows:

$$f_i(x) = \tanh(\beta_i x), \tag{4}$$

where the parameter β_i, corresponding to the slope of the function at the origin, controls the nonlinearity of the function. Although there are many ways to introduce variability in β_i, for simplicity we assume that β_i is randomly generated from the uniform distribution in the range $[1 - v, 1]$, where v ($0 \leq v \leq 1$) is the control parameter representing the degree of variability.

2.4   Simulation Setting

Initially, we give the internal state $\mathbf{x}(0)$ and the output weight matrix W^{out} randomly. The weights of input connections have the same absolute value p but the signs are randomly assigned. After a washout period with length T_{init}, the sample sequential data with length T_{trn} are used for training the output weights and subsequently the sequential data with length T_{test} are used for testing the generalization ability of the reservoir. The computational performance is evaluated using the normalized mean squared error (NMSE) defined as follows:

$$NMSE = \frac{\langle ||\mathbf{y}(t) - \mathbf{d}(t)||^2 \rangle}{\langle \mathbf{d}(t)^2 \rangle}. \tag{5}$$

We use the following benchmark tasks on sequential data processing, which have been widely used to test the performance of reservoir computing.

(1) *Mackey-Glass equation* [14]:

$$\frac{dy(t)}{dt} = \frac{ay(t-\tau)}{1+y(t-\tau)^{10}} - by(t),\qquad(6)$$

where $a = 0.2$, $b = 0.1$, and $\tau = 30$. The dataset was generated by numerically solving this equation with time step $\Delta t = 1$ [15]. The task is to predict the value of $y(t+1)$ from the past values up to time t.

(2) *Laser dataset*: The Santa Fe Laser dataset is a crosscut through periodic to chaotic intensity pulsations of a real laser [16]. The task is the same as that in the previous one. The simple cycle reservoir has been applied to this task [13].

(3) *NARMA 10th-order system*: The nonlinear auto-regressive moving average (NARMA) system of order 10 is described as follows:

$$y(t+1) = 0.3y(t) + 0.05y(t)\sum_{i=0}^{9} y(t-i) + 1.5u(t-9)u(t) + 0.1,\qquad(7)$$

where $u(t)$ is the input sequence which is randomly sampled from the uniform distribution in the range $[0, 0.5]$. This task is widely used in the literature of recurrent neural networks and reservoir computing [7,13].

(4) *NARMA 20th-order system*: The NARMA system of order 20 is described as follows [13]:

$$y(t+1) = \tanh\left(0.3y(t) + 0.05y(t)\sum_{i=0}^{19} y(t-i) + 1.5u(t-19)u(t) + 0.01\right),\qquad(8)$$

where $u(t)$ is generated as in the previous task. This task is more difficult compared with the NARMA 10th-order system due to the dependence of the current state on the longer history of inputs.

3   Results

In the following numerical experiments, the input and output data were scaled and shifted appropriately for each dataset. For the dataset generated by the Mackey-Glass equation, we set $N = 2$, $T_{\text{init}} = 500$, $T_{\text{trn}} = 1000$, $T_{\text{test}} = 1000$, $p = 0.87$, and $r = 1$. The result of the test performance is shown in Fig. 2(a). The plot for the variability parameter $v = 0$ corresponds to the result for the simple cycle reservoir with the identical units [13]. As the variability parameter v is increased, the NMSE is gradually decreased. Namely, the variability of the units can improve the computational performance.

For the Laser dataset, we set $N = 100$, $T_{\text{init}} = 500$, $T_{\text{trn}} = 2000$, $T_{\text{test}} = 3000$, $p = 0.87$, and $r = 0.7$. The result is shown in Fig. 2(b). As the variabililty increases, the prediction error decreases and reaches the bottom at around 0.7. The error slightly increases for further increase in v, but it is much lower than the case without variability.

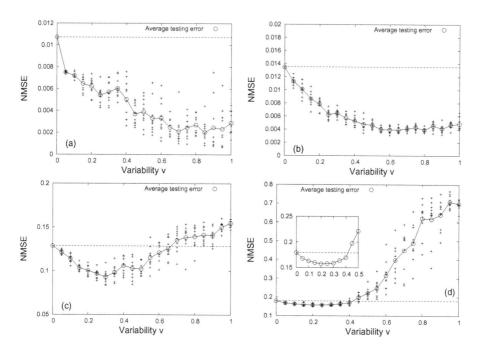

Fig. 2. The performance of the simple cycle reservoir with heterogeneous neurons. The NMSE for the test set is plotted against the variability parameter v. The crosses represent the results of 10 trials for each parameter value. The open circle indicates the average of the 10 trials. (a) Mackey-Glass equation. (b) Santa Fe Laser dataset. (c) NARMA 10th-order system. (d) NARMA 20th-order system.

For the NARMA 10th-order system, we set $N = 100$, $T_{\text{init}} = 200$, $T_{\text{trn}} = 1000$, $T_{\text{test}} = 1000$, $p = 0.87$, and $r = 0.86$. Figure 2(c) shows the result, where the variability can yield a better result if v is less than around 0.5 but for a larger value of v the result is worse than the case without variability.

For the NARMA 20th order system, we set $N = 100$, $T_{\text{init}} = 500$, $T_{\text{trn}} = 1000$, $T_{\text{test}} = 1000$, $p = 0.87$, and $r = 0.95$. The result for the NARMA 20th-order system is similar to that for the NARMA 10-th order system as shown in Fig. 2(d). Although a large value of v significantly increases the prediction error, there exists a range of v in which the variability has a positive effect (the inset).

To clarify the conditions that the performance is improved by the unit variability, we indicated the parameter regions (black) for good computational performance in Fig. 3. The performance increases with the variability v for the range of r in Figs. 3(a), (b), whereas there is a optimal range of v in Figs. 3(c), (d). There is a correlation between the values of r and v, suggesting that the effective spectral radius is determined not only by r but by β_i. It remains to explicitly give the formula for the spectral radius in the simple cycle reservoir with heterogeneous units.

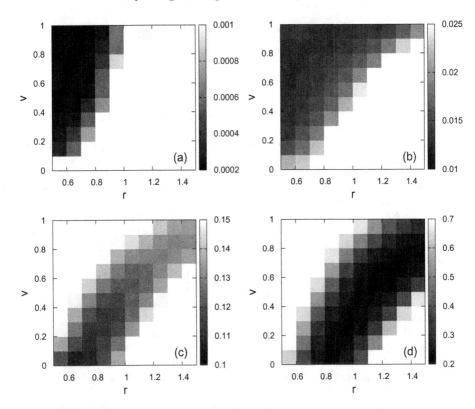

Fig. 3. The parameter region for good prediction performance on the (r, v)-plane. The color bar indicates the NMSE for the test set. (a) The Mackey-Glass equation. (b) The Santa Fe Laser dataset. (c) The NARMA 10th-order system. (d) The NARMA 20th-order system.

4   Conclusions

We have proposed to exploit heterogeneity in the reservoir units for improving the computational performance in the reservoir computing with the simple cycle architecture. We have introduced variability in the slope parameter in the hyperbolic tangent activation functions of the reservoir units. Numerical experiments have shown that both the unit variability and the connection weight govern the performance on the benchmark tasks for sequential information processing.

Our result is beneficial for hardware implementation of reservoir computing because of the simple reservoir structure and the unavoidable unit variability when implemented with nano/micro devices. For verification of the effectiveness of our method, we need further numerical experiments using other datasets. It is significant to clarify the conditions under which the unit variability works well. The mathematical mechanism of the positive role of the unit variability still remains to be investigated.

Acknowledgments. This work was partially supported by JSPS KAKENHI Grant Number 16K00326 (GT).

References

1. Haykin, S.: Neural Networks. A Comprehensive Foundation, 2nd edn. Prentice Hall, Englewood Cliffs (1998)
2. Jaeger, H.: Tutorial on training recurrent neural networks, covering BPPT, RTRL, EKF and the "echo state network" approach. GMD-Forschungszentrum Informationstechnik (2002)
3. Schrauwen, B., Verstraeten, D., Van Campenhout, J.: An overview of reservoir computing: theory, applications and implementations. In: Proceedings of the 15th European Symposium on Artificial Neural Networks, pp. 471–482 (2007)
4. LukošEvičIus, M., Jaeger, H.: Reservoir computing approaches to recurrent neural network training. Comput. Sci. Rev. **3**(3), 127–149 (2009)
5. Jaeger, H.: The "echo state" approach to analysing and training recurrent neural networks-with an erratum note. German National Research Center for Information Technology GMD Technical Report **148**, Bonn, Germany 34 (2001)
6. Jaeger, H.: Short term memory in echo state networks. GMD-Forschungszentrum Informationstechnik (2001)
7. Jaeger, H.: Adaptive nonlinear system identification with echo state networks. In: Advances in neural information processing systems, pp. 593–600 (2002)
8. Maass, W., Natschläger, T., Markram, H.: Real-time computing without stable states: a new framework for neural computation based on perturbations. Neural Comput. **14**(11), 2531–2560 (2002)
9. Paquot, Y., Duport, F., Smerieri, A., Dambre, J., Schrauwen, B., Haelterman, M., Massar, S.: Optoelectronic reservoir computing. Scientific Reports **2**, 287 (2012)
10. Kulkarni, M.S., Teuscher, C.: Memristor-based reservoir computing. In: 2012 IEEE/ACM International Symposium on Nanoscale Architectures (NANOARCH), pp. 226–232 (2012)
11. Katayama, Y., Yamane, T., Nakano, D., Nakane, R., Tanaka, G.: Wave-based neuromorphic computing framework for brain-like energy efficiency and integration. IEEE Trans. Nanotechnology (Accepted)
12. Yamane, T., Katayama, Y., Nakane, R., Tanaka, G., Nakano, D.: Wave-based reservoir computing by synchronization of coupled oscillators. In: Arik, S., Huang, T., Lai, W.K., Liu, Q. (eds.) Neural Information Processing, pp. 198–205. Springer, Switzerland (2015)
13. Rodan, A., Tiňo, P.: Minimum complexity echo state network. IEEE Trans. Neural Netw. **22**(1), 131–144 (2011)
14. Mackey, M.C., Glass, L.: Oscillation and chaos in physiological control systems. Science **197**(4300), 287–289 (1977)
15. Wyffels, F., Schrauwen, B., Verstraeten, D., Stroobandt, D.: Band-pass reservoir computing. In: IEEE International Joint Conference on Neural Networks (IEEE World Congress on Computational Intelligence), pp. 3204–3209. IEEE (2008)
16. Weigend, A., Gershenfeld, N.: Time series prediction: forecasting the future and understanding the past. In: Proceedings of a NATO Advanced Research Workshop on Comparative Time Series Analysis, held in Santa Fe, New Mexico (1994)

Graceful Degradation Under Noise on Brain Inspired Robot Controllers

Ricardo de Azambuja[1,2]($\boxtimes$), Frederico B. Klein[1], Martin F. Stoelen[1], Samantha V. Adams[1], and Angelo Cangelosi[1]

[1] School of Computing, Electronics and Mathematics, Plymouth University, Plymouth, UK
{ricardo.deazambuja,frederico.klein,martin.stoelen,samantha.adams, a.cangelosi}@plymouth.ac.uk
[2] CAPES Foundation, Ministry of Education of Brazil, Brasilia 70040-020, Brazil
http://www.plymouth.ac.uk, http://www.capes.gov.br

Abstract. How can we build robot controllers that are able to work under harsh conditions, but without experiencing catastrophic failures? As seen on the recent Fukushima's nuclear disaster, standard robots break down when exposed to high radiation environments. Here we present the results from two arrangements of Spiking Neural Networks, based on the Liquid State Machine (LSM) framework, that were able to gracefully degrade under the effects of a noisy current injected directly into each simulated neuron. These noisy currents could be seen, in a simplified way, as the consequences of exposition to non-destructive radiation. The results show that not only can the systems withstand noise, but one of the configurations, the Modular Parallel LSM, actually improved its results, in a certain range, when the noise levels were increased. Also, the robot controllers implemented in this work are suitable to run on a modern, power efficient neuromorphic hardware such as SpiNNaker.

Keywords: SNN · Liquid state machines · Robot control · Noise · Graceful degradation · Robustness

1 Introduction

Five years have passed since Fukushima's nuclear disaster and current robot technology is still not ready for such a big challenge. The high level of radiation in areas close to the reactors was lethal for human beings and the robots sent to the site have severely suffered from it, hence making clear the need for more research. Modern computers, and therefore robot controllers, are designed around digital circuits and, despite several advances in manufacturing processes, design and simulation, they are still not immune to it. Digital systems also suffer from non-destructive radiation, since it can generate Single-Event Upsets (SEU) or "soft-errors". A SEU is an alteration in a logic state as a result of an energetic particle entering the microelectronic device [11]. In addition to man-made radiation sources, space and terrestrial environments are also subjected to cosmic rays and naturally available radioactive isotopes.

© Springer International Publishing AG 2016
A. Hirose et al. (Eds.): ICONIP 2016, Part I, LNCS 9947, pp. 195–204, 2016.
DOI: 10.1007/978-3-319-46687-3_21

There is evolutionary pressure for natural information processing systems to be fault tolerant. If damaged or malfunctioning neuronal cells were to change drastically the overall behaviour of the organism, it would restrict chances of survival. According to [16], graceful degradation is defined as "graded, probabilistic deficits, with some sparing of function, and with performance strongly influenced by the frequency or familiarity of the stimulus and/or its degree of consistency with other items". As such, we can identify graceful degradation in a number of neural systems. For example, in motor control neurologic disorders, Essential Tremor (characterised by periodic 4–12 Hz low amplitude movements) is among the most common (prevalence $\approx 4.0\,\%$ among aged 40 years or older), while motor disorder with choreoathetotic and ballistic movements i.e. "excessive, spontaneous, irregularly timed, non-repetitive, randomly distributed and abrupt in character" has a prevalence $\approx 0.01\,\%$ [2,9,22]. As another example, in memory encoding and consolidation, engram cells distribute learned information so that no individual neuron is responsible for a particular information, but their collective activation. Recent advances are being made in elucidating the biology behind this [18,19].

Efficiency is another characteristic seen all around nature designs. The best example is the human brain, since cortex and cerebellum together spend on average around 15W [6]. On the other hand, the Human Brain Project expects to simulate the whole brain, in the cellular level, using an exascale computer or 60 MW [12].

In an attempt to start developing solutions for the current problems robotic systems encounter when exposed to an environment with a high level of radiation, we propose in this work the use of biologically inspired robot controllers [1] for a more nature-like graceful degradation, instead of a catastrophic failure, when exposed to it. Modular and Monolithic designs of a special type of feedback enhanced parallel Liquid State Machines (LSM) [14,15] are exposed to different noise levels, in a simulated environment, and the results analysed with a robotic task as the benchmark. White Gaussian noise is injected directly into the neuron model, which could be seen as an example of the result from the non-destructive effects of radiation. Additionally, LSM are modelled based on Spiking Neural Networks (SNN), therefore power efficiency could be easily acchieved implementing the SNN in a neuromorphic hardware such as SpiNNaker [5], BrainScaleS [21] or Silicon Neurons (SiN) [7] which could also improve the reliability even further.

2   Materials and Methods

The investigation presented here was based on an earlier work[1], where a new humanoid robot control framework using parallel, diverse and noisy groups of biologically inspired LSM was introduced [1]. This robot controller was able to reproduce trajectories (shapes) previously learned from a teacher, but the effects of varying noise levels were not studied.

---

[1] Source code available at github.com/ricardodeazambuja/IJCNN2016.

In this new work, eleven different noise levels (100 trials each), starting from the standard one defined in [8] and going up to 100 % above that (see Sect. 2.2), were employed to verify the noise effect on two different parallel LSM configurations: Modular and Monolithic (see Sect. 2.1). The final analysis was done through the robot's resultant movement performing the benchmark task of drawing a square shape on a table (see Sect. 2.3).

2.1 Modular and Monolithic Parallel LSM

The idea of breaking an LSM into multiple liquids (or simplified models of cortical columns) in parallel to increase the computational power was initialy presented in [15], but only in [1] was an external feedback loop, as suggested in [8,14], explored for this particular situation. Also the parallel system presented in [15] had an external output layer (*readout*) shared among all neurons contrasting with the one presented in [1] where each liquid was trained individually and had its own *readout* resulting in a system with improved learning capabilities. Those two approaches are called here the *Monolithic Parallel LSM* (Fig. 1b) and *Modular Parallel LSM* (Fig. 1a), respectively. To facilitate comparisons, the same random seeds from [1], therefore the same *liquids*, were employed here, but the readout layers were trained again as the Monolithic approach has not been tested before.

2.2 Neuron Model and Noise Levels

The neuron model applied in this work, the Leaky Integrate and Fire (LIF) partially represented by the Eq. 1 (for more details see [1]), has its membrane reset voltage (V_{reset}) drawn from a uniform distribution ([13.8 mV, 14.49 mV]) when the neural network is created and generates a spike when it reaches 15 mV ($V_{threshold}$). On the algorithmic level, the membrane voltage is always clamped

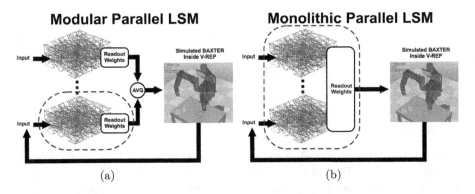

Fig. 1. The Modular approach (a) uses individual readout layers for each liquid. The Monolithic approach (b) has only one readout layer shared among all its neurons. Both systems reuse the same five LSM (liquids) from [1], but with retrained readouts.

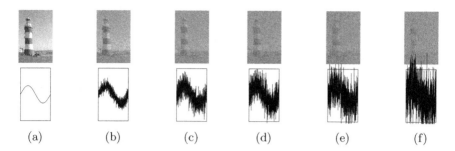

Fig. 2. As an easy way to visualize the noise effects, the photographs (top row) had added to their greyscale values (0 to 255) noise proportional to how A_{noise} affects the membrane voltage, varying it from 0.0 (a) to 1.0 (f). On the bottom row, noise is applied to a sinusoid whilst keeping the same scale.

between $-15\,\text{mV}$ and $+15\,\text{mV}$, although its rest potential is $0\,\text{mV}$ and it is set back to the reset value (V_{reset}) after every spike. Consequently, most of the time, the neuron membrane will fluctuate between V_{reset} and $V_{threshold}$ or, in the worst scenario, with $\Delta V \approx 1.2\,\text{mV}$.

The simulation of a faulty system through the injection of noise (see Sect. 1) is accomplished using the i_{noise} variable from Eq. 1. Its value is drawn from a Gaussian distribution ($\mu = 0$ and $\sigma = 1nA$) multiplied accordingly to what we call here *noise level* (A_{noise}). Having a noise level of 100 %, 110 %, 120 %, ..., 200 % means the multiplier value goes from 1.0 up to 2.0. The parameters were defined according to what was presented in [8,15], hence $c_m = 30nF$ and $\tau_m = 30\,\text{ms}$. This yields, ignoring other noise sources, a Signal-to-noise ratio (SNR) of approximately $\left(\frac{\Delta V/mV}{A_{noise}}\right)^2$. Thus the system has its SNR varied from 1.44 to 0.36 (see Fig. 2).

$$\frac{dv(t)}{dt} = \frac{i_e(t) + i_i(t) + i_{offset} + i_{noise}(t)}{c_m} + \frac{v_{rest} - v(t)}{\tau_m} \tag{1a}$$

$$\frac{di_e(t)}{dt} = -\frac{i_e(t)}{\tau_{syn_e}} \tag{1b}$$

$$\frac{di_i(t)}{dt} = -\frac{i_i(t)}{\tau_{syn_i}} \tag{1c}$$

2.3 Benchmark Task

The benchmark test consisted of the simultaneous control of four joints (Fig. 3) of a simulated BAXTER robot in order to draw a square shape on top of a table (for more details see [1]). All analyses are done on the robot's taskspace (Cartesian space) instead of joint space. Although being a two dimentional shape drawn on a surface, the system follows a human-inspired movement [4] and, for that reason, must keep in control a total of four dimensions: X, Y, Z and time.

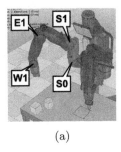
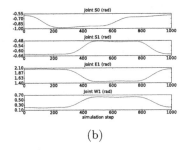

(a)   (b)

Fig. 3. Simulated BAXTER robot inside V-REP [17], with joint names indicated on its right arm, drawing the square on top of a table (a). Joint curves necessary to command the robot to generate the square shape (b).

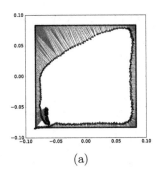
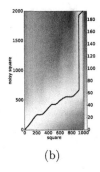

(a)   (b)

Fig. 4. An example of how the DTW fits a very distorted square (green triangles) against a perfect one (blue squares) (a). The path that minimises the accumulated distance (b). The DTW path cost for the distorted square is 42.12. (Color figure online)

Cost Calculation. Using the Dynamic Time Warping (DTW) [20] the total distance defined by the path formed with the minimum values of the accumulated distance (Fig. 4, right-hand side) can be easily applied to compare the quality between different shapes. If the shapes are exactly the same, that distance is minimal and forms a straight diagonal line. The use of this same algorithm in a robotic task was already presented in [1].

3 Results and Discussion

Eleven distinct levels of noise were tested here for both, Modular and Monolithic, approaches (Sect. 2.1) with A_{noise} varying from 1.0 to 2.0 (Sect. 2.2). These experiments resulted in a total of 2,200 simulations, where each one consisted of 3,000 spiking neurons (five 600 neurons liquids in parallel). After every run, the joint values produced were loaded into the simulated Baxter robot inside V-REP to verify the final movement executed for the benchmark task and the results processed by the DTW algorithm (Sect. 2.3).

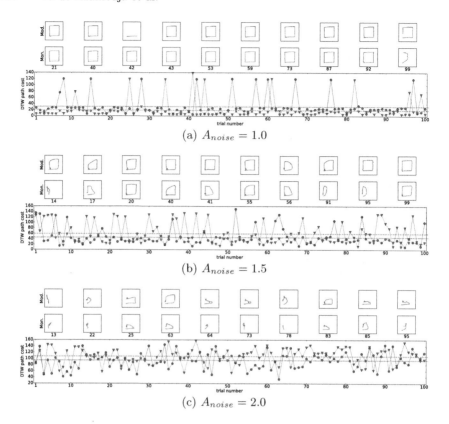

(a) $A_{noise} = 1.0$

(b) $A_{noise} = 1.5$

(c) $A_{noise} = 2.0$

Fig. 5. Each plot shows the DTW path cost (bottom) for all trials and some of the shape outcomes (top) comparing the Modular (blue, circles) and the Monolithic approaches (red, triangles). The shapes (top) were selected based on the sorted cost values of both configurations to show a more comprehensive set of examples. Average values plotted as horizontal dashed lines (bottom). (Color figure online)

The DTW path cost values generated from three different noise intensities (A_{noise} equal to 1.0, 1.5 and 2.0) are presented in Fig. 5 with all one hundred trials (bottom) and ten examples of the final shapes generated (top). Clearly as the noise is increased, the square shapes become strongly degraded, but the Modular approach still can produce some rectangular forms even with $A_{noise} = 2.0$ or a noise level twice that injected during the readout training phase (see Fig. 2 for a visual hint about noise levels). However, when using the standard noise level ($A_{noise} = 1.0$), the Monolithic approach had a better performance with an average cost value about 39 % smaller than the Modular one. This type of system, sometimes, get stuck into a value and needs noise to be able to proceed, but the DTW algorithm penalises it as the trajectory it sees, although with a nice quality, was not completed. Therefore the difference between Modular and Monolithic approaches, with $A_{noise} = 1.0$, could be explained by the limited number of simulated steps (2,000 steps).

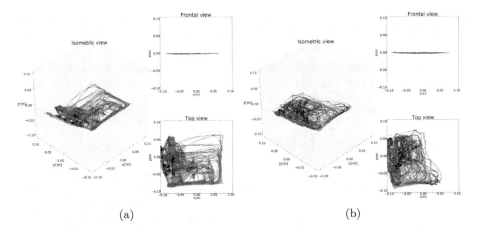

(a)   (b)

Fig. 6. Modular (a) and Monolithic (b) approaches with $A_{noise} = 2.0$.

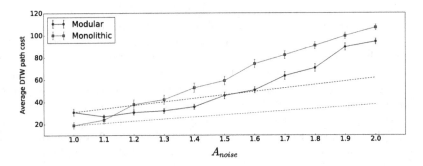

Fig. 7. Average DTW path cost and its standard error for all trials (hundred in total for each A_{noise} level). The growth, considering the first value incremented by 10 %, 20 % . . . 100 %, is shown as a dashed line.

In Fig. 6, all hundred trials with $A_{noise} = 2.0$ were plotted together on 3D Cartesian space (same scale for all views) to make it easier to evaluate them, as mean values do not work well if there are time delays among trials. Despite the fact that a strong effect on the 2D square shape is clear, the Z axis (or the height control) is barely affected (top right).

The main question raised at the introduction was about the behaviour of this kind of system when affected by different noise levels and if it would have a nature like graceful degradation. To analyse that, the DTW path cost average and standard error values were calculated and are presented in Fig. 7. The same figure also presents what would be the evolution of the cost considering the initial values incremented in steps of 10 %.

Both approaches presented here, Modular and Monolithic parallel LSM, had what is considered a graceful degradation, as with the increase of the noise the systems did not catastrophically fail, but the DTW path cost grew in a well behaved manner. Comparing both LSM configurations, the Modular approach had an almost constant behaviour up to A_{noise}=1.4 when it started growing linearly with nearly the same slope as the Monolithic approach. Therefore, the Modular system (between the A_{noise} range of 1.0 to 1.4) was able to withstand the noise better than a simple linear growth as showed by the dashed blue line (Fig. 7) whilst the Monolithic configuration always increased its DTW path cost with the increase of noise.

4   Conclusions and Future Work

The robot controllers presented here were able to withstand, or at least gracefully degrade, when exposed to different noise levels - modelled here as white Gaussian noise based currents injected into the neuron model. These noisy currents could be seen, in a simplified way, as the consequences of exposition to non-destructive radiation. It is important to develop systems that are able to be implemented using new technologies, such as neuromorphic hardware, as they seem to be one of the possible ways to bypass the declining applicability of Moore's law [23] without having to expend huge amounts of energy [12]. Also, one of the strategies to decrease energy consumption, in a quadratic way, is the reduction of the voltage supplied to the digital circuits (near-threshold voltage [10]). However, this naturally leads to a decrease in the noise immunity as the voltage margin until a transistor changes its state is reduced. Another consequence of voltage reduction is within the speed a transistor changes its state. Still, neural systems are well known to be parallel, but relatively slow systems when compared to modern digital circuits. Even if MEMS-based logic gates [3] evolve up to the point of a final product, a digital system does not degrade gracefully in normal conditions and always needs extra gates to implement error correction.

The Modular design presented here opens up the possibility for a hot-swap hardware implementation, fitting SpiNNaker very well as it is able to turn on and off chips if necessary, and also decreasing the time and memory spent during learning. Also, having smaller readout layers, the time spent during learning is smaller than when using the Monolithic setup.

The Monolithic approach uses one big readout layer while the Modular one has smaller individual output layers and a node producing the average among them. In a future work, this simple average junction could be replaced by an extra on-line learning layer with weights connecting the analogue readout outputs directly to the neuron membrane, opening the possibility to vary the amount of trust the system has to each individual LSM without the need of changing the readout weights, thus saving energy and simplifying the design.

Additionally, to extend what was presented here, other parameters could be checked to verify their influence on the robustness. One good example, easily implemented, is the number of parallel liquids and the number of neurons used with each one.

In some trials, the systems got stuck in the middle of a well defined trajectory producing high DTW path cost values (see Fig. 5a, trials 42 and 99). Our experience, after several experiments have been done using this type of system, together with the results presented in the Fig. 7, suggests a certain minimum background noise is actually necessary for this kind of system. This idea of a "good" noise is not new [13] and will be left as another avenue for future works.

All the source code necessary to generate the results presented here will be available at http://github.com/ricardodeazambuja/ICONIP2016.

Acknowledgment. This work was in part supported by the CAPES Foundation, Ministry of Education of Brazil (scholarship BEX 1084/13-5), CNPq Brazil (scholarship 232590/2014-1) and UK EPSRC project BABEL (EP/J004561/1 and EP/J00457X/1).

References

1. de Azambuja, R., Cangelosi, A., Adams, S.: Diverse, noisy and parallel: a new spiking neural network approach for humanoid robot control. In: 2016 International Joint Conference on Neural Networks (IJCNN). p. In Press (In Press)
2. Benito-Len, J., Louis, E.D.: Essential tremor: emerging views of a common disorder. Nat. Clin.l Pract. Neurol. **2**(12), 666–678 (2006)
3. Chowdhury, F.K., Choe, D., Jevremovic, T., Tabib-Azar, M.: Design of MEMS based XOR and AND gates for rad-hard and very low power LSI mechanical processors. In: 2011 IEEE Sensors, pp. 762–765, October 2011
4. Flash, T., Hogan, N.: The coordination of arm movements: an experimentally confirmed mathematical model. J. Neurosci. **5**(7), 1688–1703 (1985)
5. Furber, S.B., Lester, D.R., Plana, L.A., Garside, J.D., Painkras, E., Temple, S., Brown, A.D.: Overview of the SpiNNaker system architecture. IEEE Trans. Comput. **62**(12), 2454–2467 (2013)
6. Herculano-Houzel, S.: Scaling of brain metabolism with a fixed energy budget per neuron: implications for neuronal activity, plasticity and evolution. PLoS One **6**(3), e17514 (2011)
7. Indiveri, G., Linares-Barranco, B., Hamilton, T.J., van Schaik, A., Etienne-Cummings, R., Delbruck, T., Liu, S.C., Dudek, P., Hfliger, P., Renaud, S., Schemmel, J., Cauwenberghs, G., Arthur, J., Hynna, K., Folowosele, F., Saïghi, S., Serrano-Gotarredona, T., Wijekoon, J., Wang, Y., Boahen, K.: Neuromorphic silicon neuron circuits. Neuromorphic Eng. **5**, 73 (2011)
8. Joshi, P., Maass, W.: Movement generation with circuits of spiking neurons. Neural Comput. **17**(8), 1715–1738 (2005)
9. Kandil, M.R., Tohamy, S.A., Abdel Fattah, M., Ahmed, H.N., Farwiez, H.M.: Prevalence of chorea, dystonia and athetosis in assiut, egypt: a clinical and epidemiological study. Neuroepidemiology **13**(5), 202–210 (1994)
10. Kaul, H., Anders, M., Hsu, S., Agarwal, A., Krishnamurthy, R., Borkar, S.: Near-threshold voltage (NTV) design: opportunities and challenges. In: Proceedings of the 49th Annual Design Automation Conference, pp. 1153–1158. ACM (2012)
11. Kerns, S.E., Shafer, B.D., van Vonno, N., Barber, F.E.: The design of radiation-hardened ICs for space: a compendium of approaches. Proc. IEEE **76**(11), 1470–1509 (1988)
12. Kogge, P., Bergman, K., Borkar, S., Campbell, D., Carlson, W., Dally, W., Denneau, M., Franzon, P., Harrod, W., Hill, K., Hiller, J., Karp, S., Keckler, S., Klein, D., Lucas, R., Richards, M., Scarpelli, A., Scott, S., Snavely, A., Sterling, T., Williams, R.S., Yelick, K.: ExaScale computing study: technology challenges in achieving exascale systems. Technical report University of Notre Dame, September 2008

13. Maass, W.: Noise as a resource for computation and learning in networks of spiking neurons. Proc. IEEE **102**(5), 860–880 (2014)
14. Maass, W., Joshi, P., Sontag, E.D.: Computational aspects of feedback in neural circuits. PLoS Comput. Biol. **3**(1), e165 (2007)
15. Maass, W., Natschlger, T., Markram, H.: Real-time computing without stable states: a new framework for neural computation based on perturbations. Neural Comput. **14**(11), 2531–2560 (2002)
16. Rogers, T.T., McClelland, J.L.: Parallel distributed processing at 25: further explorations in the microstructure of cognition. Cogn. Sci. **38**(6), 1024–1077 (2014)
17. Rohmer, E., Singh, S.P., Freese, M.: V-REP: A versatile and scalable robot simulation framework. In: 2013 IEEE/RSJ International Conference on Intelligent Robots and Systems (IROS), pp. 1321–1326. IEEE (2013)
18. Roy, D.S., Arons, A., Mitchell, T.I., Pignatelli, M., Ryan, T.J., Tonegawa, S.: Memory retrieval by activating engram cells in mouse models of early Alzheimers disease. Nature **531**(7595), 508–512 (2016)
19. Ryan, T.J., Roy, D.S., Pignatelli, M., Arons, A., Tonegawa, S.: Engram cells retain memory under retrograde amnesia. Science **348**(6238), 1007–1013 (2015)
20. Sakoe, H., Chiba, S.: Dynamic programming algorithm optimization for spoken word recognition. IEEE Trans. Acoust. Speech Signal Process. **26**(1), 43–49 (1978)
21. Schemmel, J., Brderle, D., Grbl, A., Hock, M., Meier, K., Millner, S.: A wafer-scale neuromorphic hardware system for large-scale neural modeling. In: Proceedings of the 2010 IEEE International Symposium on Circuits and Systems (ISCAS 2010), pp. 1947–1950 (2010)
22. Vertrees, S.M., Berman, S.A.: Chorea in adults: background, pathophysiology, epidemiology. http://emedicine.medscape.com/article/1149854-overview. Accessed: 19 Apr 2016
23. Waldrop, M.M.: More than moore. Nature **530**(7589), 114 (2016)

Dynamics of Reservoir Computing at the Edge of Stability

Toshiyuki Yamane[1]([✉]), Seiji Takeda[1], Daiju Nakano[1], Gouhei Tanaka[2], Ryosho Nakane[2], Shigeru Nakagawa[1], and Akira Hirose[2]

[1] IBM Research - Tokyo, Kawasaki, Kanagawa 212-0032, Japan
{tyamane,seijitkd,dnakano,snakagw}@jp.ibm.com
[2] Graduate School of Engineering, The University of Tokyo, Tokyo 113-8656, Japan
gouhei@sat.t.u-tokyo.ac.jp, nakane@cryst.t.u-tokyo.ac.jp,
ahirose@ee.t.u-tokyo.ac.jp

Abstract. We investigate reservoir computing systems whose dynamics are at critical bifurcation points based on center manifold theorem. We take echo state networks as an example and show that the center manifold defines mapping of the input dynamics to higher dimensional space. We also show that the mapping by center manifolds can contribute to recognition of attractors of input dynamics. The implications for realization of reservoir computing as real physical systems are also discussed.

Keywords: Reservoir computing · Echo state network · Bifurcation phenomena · Center manifold theory · Physical reservoir

1 Introduction

Reservoir computing (RC) is an emerging special class of neural networks with a variety of engineering applications such as time series prediction, system identification, signal generation [3]. In the narrowest sense, RC is a special architecture for recurrent neural networks with two functional components. One is a fixed recurrent neural network, called reservoir, which is a (non-linear) mapping of input data to a high dimensional space. The other is adaptive filters, called readout, which extract desired results from the reservoir output. The remarkable feature of RC is that the internal connection matrix in a reservoir is initialized randomly and left unchanged. Only the readout part are trained by simple linear adaptive filters so that difference of the output of the readout and desired results is minimized. This means that RC requires relatively fewer parameters and less learning cost than traditional neural network algorithms. In addition, RC systems can be implemented not only as software but also as (nonlinear) physical systems. In fact, some interesting physical implementations have been reported so far, for example, photonic systems [6]. Embedding computation into physical dynamics is an attractive research direction since it can potentially lead to significant power reduction or high throughput if we carefully choose the physics.

© Springer International Publishing AG 2016
A. Hirose et al. (Eds.): ICONIP 2016, Part I, LNCS 9947, pp. 205–212, 2016.
DOI: 10.1007/978-3-319-46687-3_22

However, RC is often considered to be somewhat black box because of its random nature of construction. In addition, it is not still clear how the insights obtained by the software simulations should be translated to real physical systems due to lack of unified design principle for physical reservoir. The purpose of this paper is to bridge the gap between mathematical models of RC and their physical implementations. More specifically, we choose echo state networks (ESN) as an example of RC and study their dynamics when the spectral radius of internal connection matrix is close to 1 because a lot of empirical results suggest that the performance is often optimized in this case [1]. From the viewpoint of dynamical system theory, spectral radius close to one means that the dynamics lies near the critical bifurcation point, that is, the edge of stability. Therefore, we study the dynamics based on the center manifold theory [2] which can describe the dynamics at the critical bifurcation point. We discuss how slow neurons with absolute eigenvalue one and fast neurons with eigenvalues smaller than one can contribute for reservoirs to solve classification task. As for the physical implementations, we take laser systems with external feedback as an example of physical system and discuss how concepts of ESNs can be translated to such laser systems.

2   A Short Review of Echo State Networks

The ESN is a variant of RC which operates with rate coding neurons and in discrete time [3]. The most basic equation for updating reservoir state is

$$\mathbf{x}(n+1) = \tanh(W_{res}\mathbf{x}(n) + W_{in}\mathbf{u}(n)), \tag{1}$$

where $\mathbf{x}(n)$ is the N-dimensional reservoir state, W_{res} is the $(N \times N)$ reservoir internal weight matrix, W_{in} is the $N \times K$ input weight matrix, $\mathbf{u}(n)$ is the K-dimensional input signal. Note that the tanh activation function is applied to each component of a vector. Both the matrix W_{res} and W_{in} are initialized randomly and fixed throughout its operation.

To solve practical tasks, the reservoirs need to satisfy the echo state property [1]. Informally, the echo state property means that the reservoir dynamics should not be self-excitatory but be driven only by input signals. Furthermore, the reservoirs should be state forgetting in the sense that the impact of the reservoir state in the far past vanishes with time. However, it is generally difficult to check the echo state property directly for practical applications. Instead, the spectral radius (SR) of W_{res}, the maximum of the absolute value of the eigenvalues of W_{res}, is often set to be less than unity [1]. Although this spectral radius condition is not equivalent to echo state property, it offers a simple empirical alternative. In addition, the spectral radius is often a crucial key tuning parameter of ESNs. For example, the spectral radius controls the speed of fading memory in the reservoir. Spectral radius closer to one implies that reservoirs retain longer-lasting memory of the input signal. In many practical applications, the spectral radius is set close to one to optimize performance since real world applications often requires relatively long term memory of the input signal.

From the viewpoint of dynamical system theory, the spectral radius is one of bifurcation parameters of the reservoir dynamics. In our work, we restrict the input signals to be generated by a deterministic dynamical systems as $\mathbf{u}(n+1) = G(\mathbf{u}(n)), n = 0, 1, 2 \ldots$ to analyze the reservoir dynamics by bifurcation theory of dynamical systems.

3   The Center Manifold Theorem

Suppose the N dimensional dynamics $X_{n+1} = F(X_n)$ has a fix point at the origin: $F(0) = 0$ and the Jacobian matrix at the origin $DF(0)$ has eigen values N_c eigen values on the unit circles, N_s eigen values within the unit circle and N_u eigen values outside the unit circle on the complex plane $(N = N_c + N_s + N_u)$. This means that the dynamics is operating at the critical bifurcation point and is described by combination of slow variables with eigen values on the unit circle and fast variables within or outside the unit circle. Using the eigen basis, we can rewrite the dynamics $X(n + 1) = F(X(n))$ into two dynamics $X(n) = (\mathbf{x}(n), \mathbf{y}(n))$ for slow variables $\mathbf{x} \in \mathbb{R}^{N_c}$ and fast variables $\mathbf{y} \in \mathbb{R}^{N_s+N_u}$ as

$$\mathbf{x}(n + 1) = A\mathbf{x}(n) + \mathbf{f}(\mathbf{x}(n), \mathbf{y}(n)), \quad \mathbf{y}(n + 1) = B\mathbf{y}(n) + \mathbf{g}(\mathbf{x}(n), \mathbf{y}(n)) \quad (2)$$

where, A is a $N_c \times N_c$ matrix whose eigen values lie on the unit circle and B is a $(N_s + N_u) \times (N_s + N_u)$ matrix whose eigen values lie within or outside the unit circle on the complex plane. The functions $\mathbf{f}$ and $\mathbf{g}$ are sufficiently smooth nonlinear terms higher than second order and we also assume that $\mathbf{f}(0,0) = 0, D\mathbf{f}(0,0) = 0, \mathbf{g}(0,0) = 0, D\mathbf{g}(0,0) = 0$. Then, there exist a N_c dimensional (local) invariant manifold W^c, called center manifold, represented by a graph $\mathbf{y} = h(\mathbf{x})$ as

$$W^c = \{(\mathbf{x}, \mathbf{y}) : \mathbf{y} = h(\mathbf{x})\}, h : \mathbb{R}^{N_c} \to \mathbb{R}^{N_s+N_u}, \|\mathbf{x}\| < \delta, h(0) = Dh(0) = 0, \quad (3)$$

and $\mathbf{y}(n) = h(\mathbf{x}(n)) + O(e^{-\gamma n})$ holds for some constant $\gamma > 0$. The graph of the center manifold $\mathbf{y} = h(\mathbf{x})$ can be determined by the following equation

$$h(A\mathbf{x} + \mathbf{f}(\mathbf{x}, h(\mathbf{x}))) - Bh(\mathbf{x}) - \mathbf{g}(\mathbf{x}, h(\mathbf{x})) = 0. \quad (4)$$

The implication of the center manifold theorem is that the fast variables $\mathbf{y}$ are "slaves" which quickly follow the slow "master" variables $\mathbf{x}$ and the overall dynamics is finally governed by only those of $\mathbf{x}$ along the center manifold.

4   Echo State Networks at the Edge of Stability

In this section, we apply the center manifold theorem to the analysis of dynamics of ESNs. We assume that the input dynamics $\mathbf{u}(n + 1) = G(\mathbf{u}(n))$ has a fixed point at origin: $0 = G(0)$. The Jacobian matrix of entire system of the ESN and input dynamics at the origin is $\begin{pmatrix} W_{res} & W_{in} \\ 0 & DG(0) \end{pmatrix}$. Therefore, the eigen values of W_{res} and $DG(0)$ determines the dynamics around the origin.

4.1 Dynamics Without External Input

The Jacobian matrix of an ESN at the origin is W_{res} when no input is applied. As an example, we take the following two-dimensional ESN without input.

$$\mathbf{x}(n+1) = \tanh(W_{res}\mathbf{x}(n)), \quad W_{res} = \begin{pmatrix} 0.9398 & 0.6526 \\ 0.0575 & 0.3770 \end{pmatrix}. \tag{5}$$

The eigen values of W_{res} are 1.0 and 0.3168. To see the effect of SR to the ESN dynamics, we rescale the W_{res} by multiplying the gain $g > 0$ to W_{res} to control the spectral radius.

Figure 1(a) shows the the dynamics for SR = 0.2. Both neurons moves toward the origin very quickly.

Figure 1(b) shows the case SR = 1.0. The notable feature is that the coexistence of two neurons with quite different time scales. One is the slow neuron with eigen value 1.0 (x-axis) moving slowly along the center manifold toward the origin, which corresponds to long term memory. The other is the fast neuron with smaller eigenvalue 0.3168 (y-axis) attracted quickly to the center manifold and following the slow neuron. To obtain an approximate form of center manifold, we assume the graph $y = h(x)$ as infinite power series as $h(x) = a_2x^2 + a_3x^3 + a_4x^4 + \dots$. Plugging this power series into the Eq. (4) and comparing the coefficients from lower order to higher, one can determine the coefficients recursively. Then, all even order terms are found to be zero and taking up to the 7th order term, we obtain an approximate form of center manifold for (5) as $y = h(x) = 0.0582x^3 + 0.0513x^5 + 0.00649x^7$.

Figure 1(c) shows the behavior of ESN with SR above unity, 1.5. In this case, SR causes the pitchfork bifurcation to the reservoir. The origin loses its stability and two new stable fixed points appear on both sides of the origin. Note that echo state property does not hold any more in this case since to which fixed points the reservor state converges depends on the initial state of the reservoir.

4.2 Dynamics with External Input

We assume that the input dynamics is of hyperbolic type. Then the dynamics can be described locally as $\mathbf{u}(n) = (u_1(0)\lambda_1^n, \dots, u_K(0)\lambda_K^n), |\lambda_i| \neq 1$. Figure 2 shows the reservoir dynamics when input dynamics of hyperbolic type is applied. In the presence of input dynamics, the center manifold $\mathbf{u} = h(\mathbf{x})$ is not unique but dependent on input $\mathbf{u}$, that is, each trajectory corresponds to a different center manifold.

The center manifold defined by the graph $\mathbf{y} = h(\mathbf{x})$ naturally induces the mapping between the tangent space of input and reservoir dynamics. If the dimension of input dynamics K is larger than N_c, the mapping cannot be a one-to-one embedding and the information of input dynamics is inevitably lost. This means that not only spectral radius but also the dimension of center manifold N_c matters in order to capture the input dynamics correctly.

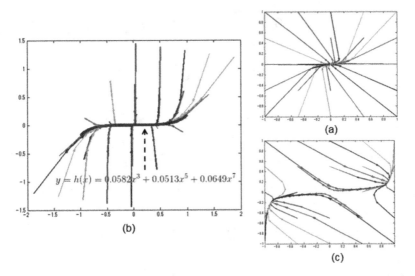

Fig. 1. (a) SR (spectral radius) = 0.2. (b) SR = 1.0. The bold line is the approximate graph of the center manifold up to 7th order. (c) SR = 1.5.

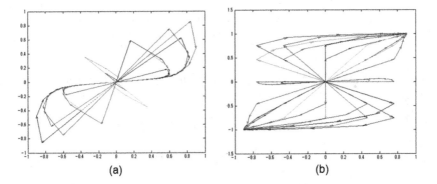

Fig. 2. Dynamics with external input. (a) $1 > \lambda_1, \lambda_2 > 0$, (b) $\lambda_1 > 1, 1 > \lambda_2 > 0$. The trajectories start with different initial points of $\mathbf{u}$.

4.3   An Example of Attractor Recognition

In this subsection, we show that how the center manifold can contribute to the recognition of input dynamics. It is already known that the RC shares a remarkable feature with kernel methods in machine learning. In fact, both reservoirs and kernels map the input data into higher dimensional space so that the input data becomes more linearly separable for classification tasks. In the reservoir of ESNs, the slow neurons can keep track of the behavior of input dynamics because they have long fading memory of the input. On the other hand, the fast neurons quickly follow slow neurons and fail to keep up with input dynamics due to fast fading memory. Then, the dynamics of $\mathbf{x}(n)$ can be viewed as

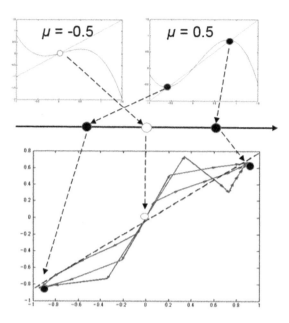

Fig. 3. Recognition of attractors for pitchfork bifurcation. Two fixed points are not separable in one dimensional space, but they become linearly separable as indicated by the dashed line after mapped to two dimensinal space by the reservor.

a "copy" (or feature vector) of input dynamics $\mathbf{u}(n)$ and the graph of center manifold $h : \mathbf{x} \mapsto (\mathbf{x}, \mathbf{y}) = (\mathbf{x}, h(\mathbf{x}))$ can be viewed as extension of the copied input dynamics.

Let us take a simple example for the input dynamics described below.

$$u(n+1) = (1+\mu)u(n) + 0.4u(n)^2 - u(n)^3. \tag{6}$$

The dynamics (6) exhibits pitchfork bifurcation; it has a unique stable fixed point at origin if $\mu < 0$ and the origin becomes unstable and two new stable fixed points appear around the origin. As can be seen from Fig. 3, the two attractors indicated by the black circles and the white circle cannot be linearly separable. However, if we use the ESN (5) with input (6), the reservoir maps the input (6) to two dimensional space by center manifold and makes the two attractors linearly separable. The problem of recognition of the shape of attractors needs long term memory of input dynamics and slow neurons play a significant role.

5   From Echo State Networks to Physical Reservoir Systems

One of the advantages of RC is that the reservoir needs no costly learning and can make use of various physical dynamics. What is necessary to obtain better performance is tuning of whole dynamics rather than learning of each intercon-nection weight. In the previous sections, we have observed that the reservoir neurons with eigen values close to one play a crucial role for the performance of ESNs. From this viewpoint, one can make an analogy between ESNs and physi-cal dynamics which seemingly have no relations to neural networks. That is, we can associate (critical) eigen modes in nonlinear physical dynamics with (slow) neurons of ESNs.

Let us take an example from the simplified Lang-Kobayashi equations describing semiconductor laser field with external feedback in long delay limit [5]:

$$\frac{dE(t)}{dt} = (1 + i\alpha)(p - |E(t)|^2)E(t) + \eta e^{-i\Omega}E(t - 1). \tag{7}$$

α is the line width enhancement factor, p the excess pump rate above the solitary laser threshold, $\eta \geq 0$ is the feedback strength, and Ω the free-running laser optical frequency. First, assuming the amplitude $|E|$ is small enough, we ignore the cubic nonlinear term. Then, the Eq. (7) is reduced to the following linear delay-differential equation:

$$\frac{dE(t)}{dt} = \eta e^{-i\Omega}E(t - 1). \tag{8}$$

Following the standard procedure to solve linear differential equations, we assume a special solution as $E(t) = Ce^{\lambda t}$. Plugging this solution into the (8), we have the characteristic equation $\lambda - \eta e^{-\lambda - i\Omega} = 0$. This equation is transcendental and can have infinitely many complex solutions $\lambda = \gamma + i\omega$ which are called eigen modes. The existence of infinitely many eigen modes means that delay-differential systems are inherently infinite dimensional. In order for the stationary solution $E(t) = 0$ to be stable, the real part γ of all eigen values have to be negative. In the critical case where $\gamma = 0$, we have $\Omega + (-1)^k\eta = (k + 1/2)\pi, k = 0, 1$ and $\omega = (-1)^k\eta$. As we increase the feedback strength η, the eigen modes become unstable one by one.

Turning back to the Eq. (7), one can see that the cubic nonlinearity prevents the amplitudes of unstablized eigen modes from growing infinitely and keeps their oscillation stable with finite amplitude. The transition from steady state $E = 0$ to stable oscillations is known as Hopf bifurcation and the resulting stable oscillations are called limit cycles. As the feedback strength η increases, the dynamics of (7) generate limit cycles by successive Hopf bifurcations. Poincaré map reduces the limit cycles to asymptotically stable fixed points and they are neutrally stable along the orbit, that is, dynamics along the orbit neither grow nor shrink and are stable in the directions across the orbit. Therefore, limit cycles can be used as alternatives of slow neurons with absolute eigen value 1 in ESNs [7]. However, when the feedback strength η becomes larger, the dynamics exhibits complicated bifurcations such as period doubling and collision of periodic branches and finally reaches chaotic region, which cannot be used as reservoirs.

The story described here is not limited to laser system but is universal for many non-linear non-equilibrium dynamics [4]. Most nonlinear physical dynamics have control parameters and undergo bifurcation phenomena as the control parameters changes. At the bifurcation point, one of the degree of freedom of the dynamics changes its stability from stable state to unstable one (or vice versa). Such control parameters, for example, the feedback strength η of laser systems, are analogous to spectral radius of ESNs. At the critical bifurcation points, some of the degree of freedom of the dynamics have eigen values with absolute value one (discrete time case) or zero real part (continuous time case) and can be asymptotically or neutrally stable, which can be used as a reservoir. Linear systems like (8) are decoupled into independent eigen modes. On the other hand, the non-linearity of dynamics can cause mode couping by interaction among eigen modes. In the case of Hopf bifurcation generating limit cycles, they can serve as coupled oscillator reservoir systems [7]. It is generally difficult to determine the detail of interactions among these eigen modes of real physical systems, but fortunately we do not necessarily need to know it since reservoirs are originally constructed randomly.

6   Conclusion

We have investigated the dynamics of reservoir computing, especially ESNs, at critical state, i.e., edge of stability. In such cases, ESNs are composed of fast vanishing neurons and slowly moving neurons which dominate the overall dynamics. Using center manifold theory, we have shown that the combination of slow and fast variables defines mapping of input time series to center subspace and extension to higher dimensional space. Extending this view of reservoirs to physical systems, we have shown that critical eigen modes of nonlinear physical systems can be candidates for reservoirs into which computation can be embedded. We have shown that dynamics with time-delay have rich behavior enough to work as reservoir, taking an example of the laser systems with external feedback.

References

1. Caluwaerts, K., et al.: The spectral radius remains a valid indicator of the echo state property for large reservoirs. In: IJCNN pp. 1–6 (2013)
2. Guckenheimer, J., Holmes, P.: Nonlinear Oscillations, Dynamical Systems, and Bifurcations of Vector Fields. Springer, Heidelberg (1983)
3. Lukoševičius, M., Jaeger, H.: Reservoir computing approaches to recurrent neural network training. Comput. Sci. Rev. **3**(3), 127–149 (2009)
4. Mori, H., Kuramoto, Y.: Dissipative Structures and Chaos. Springer, Heidelberg (2011)
5. Pieroux, D., Mandel, P.: Bifurcation diagram of a complex delay-differential equation with cubic nonlinearity. Phys. Rev. E **67**, 056213 (2003)
6. Vandoorne, K., et al.: Toward optical signal processing using photonics reservoir computing. Opt. Express **16**(15), 11182–11192 (2008)
7. Yamane, T., Katayama, Y., Nakane, R., Tanaka, G., Nakano, D.: Wave-based reservoir computing by synchronization of coupled oscillators. In: Arik, S., Huang, T., Lai, W.K., Liu, Q. (eds.) ICONIP 2015. LNCS, vol. 9491, pp. 198–205. Springer, Heidelberg (2015). doi:10.1007/978-3-319-26555-1_23

Hybrid Gravitational Search Algorithm with Swarm Intelligence for Object Tracking

Henry Wing Fung Yeung[1(✉)], Guang Liu[1], Yuk Ying Chung[1], Eric Liu[1], and Wei-Chang Yeh[2]

[1] School of Information Technologies, University of Sydney, Sydney, NSW 2006, Australia
`henrywfyeung@gmail.com`
[2] Department of Industrial Engineering and Engineering Management,
National Tsing Hua University, P.O. Box 24-60 Hsinchu 300, Taiwan, Republic of China

Abstract. This paper proposes a new approach to object tracking using the Hybrid Gravitational Search Algorithm (HGSA). HGSA introduces the Gravitational Search Algorithm (GSA) to the field of object tracking by incorporating Particle Swarm Optimization (PSO) using a novel weight function that elegantly combines GSA's gravitational update component with the cognitive and social components of PSO. The hybridized algorithm acquires PSO's exploitation of past information and fast convergence property while retaining GSA's capability in fully utilizing all current information. The proposed framework is compared against standard natural phenomena based algorithms and Particle Filter. Experiment results show that HGSA largely reduces convergence to local optimum and significantly out-performed the standard PSO algorithm, the standard GSA and Particle Filter in terms of tracking accuracy and stability under occlusion and non-linear movement in a large search space.

Keywords: Object tracking · Gravitational Search Algorithm · Particle Swarm Optimization

1 Introduction

Object tracking, defined as the process of locating a target object across a sequence of video frames, has received much attention in recent decades. Algorithms such as Kalman Filter and Particle Filter have been widely applied to accommodate the need for a fast and accurate object tracker that matches the human vision system.

Kalman Filter [1] is a state-space model which recursively compute the optimal state with lowest possible variance. This approach is highly efficient but is limited by its ability in recognizing deformed or occluded objects. Particle Filter [2] does not have such shortcomings. However, it has relatively high computation cost and suffers from low performance in the case of motion blur and disappearance of target object.

Promising results have been achieved using algorithms developed based on natural phenomena. Two prominent examples are Genetic Algorithm and Particle Swarm Optimization (PSO) [3–6]. The Genetic Algorithm was developed based on the process of evolution. It proved useful in object tracking but suffered from parameter tuning

© Springer International Publishing AG 2016
A. Hirose et al. (Eds.): ICONIP 2016, Part I, LNCS 9947, pp. 213–221, 2016.
DOI: 10.1007/978-3-319-46687-3_23

problems and low convergence speed. PSO, on the contrary, has proven to be outstanding both in terms of speed and accuracy. However, it suffers from pre-mature convergence to a local optimum and imbalance particle diversity.

The Gravitational Search Algorithm (GSA) was first introduced in Rashedi et al. (2009) as a new optimization algorithm based on the Newtonian gravity and the laws of motion [7]. Since then, variations of GSA have been widely applied in solving a diverse range of problems, including parameter tuning and function optimization [8–11]. However, it has not yet been applied to object tracking. GSA has an advantage over PSO since it utilizes all current information provided by the system of agents instead of only the local best of a particle and the global best in the swarm of PSO. This gives GSA better exploration capability and prevents pre-mature convergence to a local optimum. In contrast to PSO, GSA fails to retain the past results during the update process and has a slower convergence speed. A better algorithm can be achieved by incorporating the memory feature and fast convergence property of PSO into GSA while retaining the exploration power of GSA.

This paper proposes a new approach to object tracking using a Hybrid Gravitational Search Algorithm (HGSA) by optimally combining GSA with PSO with a novel weight function. Section 2 gives a detailed outline on the application of the standard GSA in object tracking. Section 3 describes the standard PSO algorithm. Section 4 provides the theoretical foundation for the proposed HGSA. Section 5 presents the experiment results using HGSA, the standard GSA, the standard PSO and Particle Filter algorithm. Section 6 concludes the paper.

2   GSA in the Framework of Object Tracking

GSA was originally developed from the natural phenomenon of the gravitational forces for the purpose of optimizing multi-dimensional mathematical functions. Modification is made below to incorporate GSA into the framework of object tracking.

Prior to the start of tracking, a HSV histogram representation of the target object is extracted as the feature model from the first frame and is stored as a reference for comparison. A system of I agents given in Eq. (1), each represents a potential candidate of the tracking solution, is then initialized in a given search space.

$$X = \left(x_1, \ldots, x_I\right) \tag{1}$$

Each agent has two dimensions describing its x and y coordinates which is the center of a rectangle that has similar size to the target object. During the update process, a HSV histogram representation of the said rectangle is extracted and compared to the target object using the Bhattacharyya distance, represented in the Eq. (2):

$$fit_i\left(H_i, H_T\right) = 1 - \sqrt{1 - \frac{1}{\sqrt{\bar{H}_i \bar{H}_T K^2}} \sum_{k=1}^{K} \sqrt{H_i(k) H_T(k)}} \tag{2}$$

where H_i and H_T denote the histogram representations of the ith agent where $i \in \{1, \dots, I\}$ and the target respectively. $\bar{H}$ refers to the mean of the histogram and K denotes the total number of histogram bins. The Bhattacharyya Distance is chosen for its robustness in handling deformation and rotation of target objects. The fitness values of the agents are calculated by subtracting the Bhattacharyya distances from 1 and are then used as input to Eq. (3) in calculating the inertial mass of each agent.

$$m_i(t) = \frac{fit_i(t) - worst(t)}{best(t) - worst(t)} \tag{3}$$

where $best(t)$ and $worst(t)$ correspond to the highest and the lowest fitness value of the set of agents at time t. A heavier inertial mass $m_i(t)$ means that an agent is a better candidate and therefore requires a larger force to move. Using $m_i(t)$, the mass ratio denoted in $M_i(t)$ is calculated for each agent in Eq. (4).

$$M_i(t) = \frac{m_i(t)}{\sum_{i=1}^{I} m_i(t)} \tag{4}$$

The mass ratio $M_i(t)$ divides the mass of an agent by the total mass of the whole set of agents. This represents the relative mass of a particular agent and is used as the input for the calculation of the gravitational force F in Eq. (5)

$$F_{ij}(t) = G(t) \frac{M_i(t) \times M_j(t)}{R_{ij}(t) + \varepsilon} \left(x_j(t) - x_i(t) \right) \tag{5}$$

where $j \in \{1, \dots, I\}$ and $i \neq j$. $R_{ij}(t)$ is the Euclidian distance between the agents i and j and ε is a small constant that prevents the division of zero in the case that two agents appear in the same position within the search space. $F_{ij}(t)$ is a two dimensional vector representing the x and y component of the force acting on the agent. $G(t)$ is the gravitational constant given by Eq. (6).

$$G(t) = G\left(\frac{1}{t}\right)^{\alpha} \tag{6}$$

where G is a pre-defined constant, t is the current iteration and $\alpha < 1$ is a parameter controlling the convergence speed of GSA. $G(t)$ allows the agents to search more aggressively in the search space at early iterations and facilitates convergence at the end. The forces acting on a particular agent is summed in a stochastic manner and is divided by the agent's mass ratio, giving the acceleration vector of the agent at time t:

$$a_i(t) = \frac{\sum_{j=1, i \neq j}^{I} R_j \cdot F_{ij}(t)}{M_i(t)} \tag{7}$$

$a_i(t)$ is then used in the update the position of the agents using Eqs. (8) and (9).

$$v_i(t + 1) = Rv_i(t) + a_i(t) \tag{8}$$

$$x_i(t + 1) = x_i(t) + v_i(t + 1) \tag{9}$$

where R is a random vector in $[0, 1]$ and $v_i(t)$ is the velocity of the agent at time t. The above-mentioned process is iterated and the best fitting agent is output as the solution for a particular frame at the end of the iteration. For the following frame, agents are reinitialized around the solution of last frame for better tracking results.

3   The Standard PSO Algorithm

Particle Swarm Optimization is a search strategy developed from swarm behaviors such as bird flocking and fish schooling [5]. Similar to the system of agents in GSA, PSO requires the initialization of a swarm of particles and the evaluation of the fitness of each particle, followed by velocity updates according to Eq. (10).

$$v_i(t + 1) = w \cdot v_i(t) + R_1 \cdot C_1 \cdot \left(P_{Best} - x_i(t)\right) + R_2 \cdot C_2 \cdot \left(G_{Best} - x_i(t)\right) \tag{10}$$

where P_{Best} t denotes the location of an individual particle that, among past and present iterations, gives the highest fitness value and G_{Best} denotes the location that, among all particles in past and present iterations, gives the highest fitness value. C_1 and C_2 are constant that by convention is set to 2 whereas w is set to 1.

The first random component updates particles based on the difference between their current location and the P_{Best} while the second random component updates particles based on their current location and the G_{Best}. Intuitively, the first component is the 'cognition' part, reflecting the individual thinking of the particles whereas the second component is the 'social' part, giving the particle's behavior within the whole cohort of particles. These two components allow PSO to perform update on particles using past information and to ensure that the best solution found is not lost during the process of iteration. The new velocities of the particles from Eq. (10) are used to update the location of the particles with an equation identical to Eq. (9). This process is repeated and, at the end of all iterations, gives the G_{Best} as the solution.

4   The Proposed HGSA

The standard GSA described above with Eqs. (1)–(9) can be further simplified and modified to gain potential improvements in tracking performance. Since GSA updates the agents based on all current fitness values, it is less likely than PSO to suffer from pre-mature convergence to local optimum and diversity loss. However, the lack of memory in the standard GSA not only reduces its efficiency in finding the best solution, but also induces the possibility of losing a better solution that was found previously. Therefore, the incorporation of PSO memory and fast convergence property into GSA can draw from the best of both worlds.

The derivation of HGSA starts with Eq. (5) which can be reformulated by factorising the term $G \cdot C(t)$ stated in Eq. (11) to give Eq. (12). $C(t)$ keeps the gravitational force high in early iterations to facilitate exploration of the search space.

$$G(t) = G\left(\frac{1}{t}\right)^{\alpha} = G \cdot C(t) \tag{11}$$

$$F_{ij}(t) = G \cdot C(t) \frac{M_i(t) \cdot M_j(t)}{R_{ij}(t) + \varepsilon} \left(x_j(t) - x_i(t)\right) = G \cdot C(t) \cdot f_{ij}(t) \tag{12}$$

By substituting Eq. (12) into Eq. (7) and perform the same factorization, it gives Eq. (13) for $j \neq i$.

$$a_i(t) = \frac{\sum_{j=1}^{l} R_j \cdot C(t) \cdot f_{ij}(t)}{M_i(t)} = G \cdot C(t) \frac{\sum_{j=1}^{l} R_j \cdot f_{ij}(t)}{M_i(t)} = G \cdot C(t) \cdot \dot{a}_i(t) \tag{13}$$

Pioneer work on hybridising PSO and GSA has been accomplished in research in other fields, for instance, function optimization and economic dispatch problems [8–11]. However, they merely involve the addition of the PSO update components to the GSA with little or no modification and are unfit for application in the context of object tracking. In the proposed approach, the velocity update of the agents is re-designed to synthesize the standard GSA component with two extra components from PSO, represented in Eq. (14).

$$\begin{aligned} v_i(t+1) ={}& R_0 \cdot v_i(t) + C(t) \cdot G \cdot \dot{a}_i(t) + (1 - C(t)) \\ & \cdot \left(R_1 \cdot C_1 \cdot \left(P_{Best} - x_i(t)\right) + R_2 \cdot C_2 \cdot \left(G_{Best} - x_i(t)\right)\right) \end{aligned} \tag{14}$$

Equation (14) is a modification of Eq. (10) using $C(t)$ to incorporate the PSO components. $C(t)$ is a function that starts off at 1 and gradually approaches 0 as iteration increases and therefore is used as a weight to synthesize the two algorithms. At early iterations, larger weight is given to the GSA component to search more thoroughly. As more iterations are completed, the weight is shifted from the GSA component to the cognitive and social components of PSO to allow for more rapid convergence. The parameter α which takes value from 0 to 1 in Eq. (12) determines the dynamic rate of shifting towards PSO. A low value of α indicates that the function $C(t)$ decreases more slowly, resulting in a more gradual shift towards PSO. In addition, the parameter G serves as an adjustable weight that fine-tunes the static balance of the GSA and PSO components. In HGSA, P_{Best} and G_{Best} are documented throughout the whole iteration process, thus giving better results by safeguarding the memory of the historically best solutions.

From Eq. (3), for a given set of agents, the one that has the lowest fitness values always results in a zero mass. It follows that the mass ratio of that particular agent will also be zero, leading to a division by zero in Eq. (7). Moreover, the purpose of Eq. (3) in the original GSA is to normalize the fitness values of the agents to between 0 and 1. Since the output of the Bhattacharyya distance is already normalized, it is therefore desirable to equate the output of Eq. (2) directly to the inertial mass $m_n(t)$. The flowchart in Fig. 1 gives an abstract view of HGSA in object tracking.

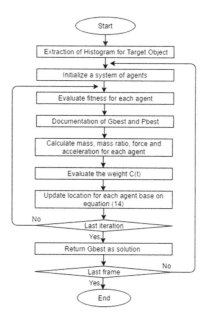

Fig. 1. Flowchart of HGSA

5   Experiment Results

The performance of HGSA was primarily compared with the standard GSA and the standard PSO algorithm. Particle Filter is included as a benchmark for the robustness of our results. The number of agents/particles in HGSA, GSA and PSO are set to 20 and the number of iterations is set to 10, totaling 200 particles per frame. Number of particles in Particle Filter is set to 200. HGSA is found to have the best performance with the parameters $G = 250$, $\alpha = 0.6$, $\varepsilon = 0.1$, $C_1 = 2$ and $C_2 = 2$. Same values of G, α and ε are adopted in GSA and C_1 and C_2 are adopted in PSO. All parameters are chosen at the values that maximizes algorithms' performance from repeated trials.

Many agent/swarm based frameworks impose restriction on particles/agents movement by imposing a sub-search space as a boundary [5, 6]. This sub-search space concentrates particles/agents around the target object and thus increases algorithm efficiency. In testing, the sub-search space is set to 200*200 instead of the standard 30*30 for better visualization of the difference in performance among algorithms.

Test 1 was conducted using the ship video. The target object in this video is a white ship moving linearly in a large frame. Test 2 was conducted using a video of a moving coin. The coin is partially occluded by a human hand and is moved out of and back into frame. The frames of interest of the two tested videos are documented in Tables 1 and 2 while the Euclidean errors of all algorithms for each frame are documented in Figs. 2

and 3. Error during occlusion is set to 0 for better visualization. Link to results: https://www.youtube.com/channel/UCrQAWuVqzbqzvAjkJxo9cZA.

Table 1. Ship video (1920 × 1080)

Frame	Particle Filter	PSO	GSA	HGSA
43				
133				
166				

Table 2. Coin video (720 × 480)

Frame	Particle Filter	PSO	GSA	HGSA
1				
70				
278				

The video in test 1 contains a ship that moves linearly from left to right. This video provides a large search area with many sub-optimums, for instance the cliff and the horizon, which have similar histogram composition to the target ship. It can be observed from Fig. 2 that both HSGA and GSA had very low error and were on track of the target throughout the whole video whereas Particle Filter lost the target object from frame 210 onwards. PSO had a few peaks of errors and lost the target object from frame 144 to frame 230. Since the sub-search space was large relative to the size of the target object, the particles/agents were more likely to miss the target in early iterations. If a solution with similar histogram composition as the target, for instance, the PSO solution in frame 43, 133 and 166 in Table 1 was found instead, particles/agents had a tendency to

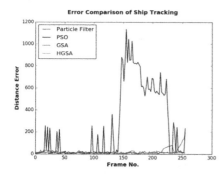

Fig. 2. Error comparison - ship

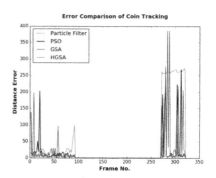

Fig. 3. Error comparison - coin

converge pre-maturely to that local optimum. This explains the poor performance of PSO. On the other hand, HGSA and GSA, with better exploration capability, were able to successfully locate the target object.

In the coin video, a coin was first placed at the center of the frame. As shown in Table 2 frame 70, a finger was positioned on top of the coin to occlude nearly all of the coin's surface. Then the coin was gradually moved towards the edge of the frame. It then disappeared from frame 95 onwards and reappeared in frame 273. Prior to the disappearance of the coin in test 2, HGSA successfully tracked the coin in all frames while the other algorithms had temporary deviations from the target. After the reappearance of the coin, only HGSA was capable in providing satisfactory tracking result. PSO and GSA jumped frequently to local optimums and Particle Filter lost the target permanently as shown in Table 2 frame 278. Testing results showed that HGSA outperforms other algorithms in both accuracy and stability in tracking of occluded and non-linear moving targets.

6   Conclusion

This paper proposes a new Hybrid Gravitational Search Algorithm for object tracking by incorporating PSO into GSA. HGSA retains GSA's capability in fully utilizing all current information while introduces PSO's exploitation of past information and fast convergence property. The choice of the novel weight function ensures the smooth integration of PSO into GSA, allowing HGSA to optimally balance the merits of both algorithms. Experiment results show that, keeping other factors constant, HGSA has a significantly higher accuracy and stability when tracking in a large search space with sub-optimums. An outstanding result is also achieved in occlusion handling and tracking of non-linear moving targets. Future research will aim to implement a weight adjustment function for changing particle inertia in HGSA.

References

1. Li, X., Wang, K., Wang, W., et al.: A multiple object tracking method using Kalman filter. In: 2010 IEEE International Conference on Information and Automation (ICIA), pp. 1862–1866. IEEE (2010)
2. Walia, G.S., Kapoor, R.: Intelligent video target tracking using an evolutionary particle filter based upon improved cuckoo search. Expert Syst. Appl. **41**(14), 6315–6326 (2014)
3. Lee, G., Mallipeddi, R., Jang, G.J., et al.: A genetic algorithm-based moving object detection for real-time traffic surveillance. IEEE Signal Process. Lett. **22**(10), 1619–1622 (2015)
4. Eberhart, R.C., Kennedy, J.: A new optimizer using particle swarm theory. In: Proceedings of the Sixth International Symposium on Micro Machine and Human Science, vol. 1, pp. 39–43 (1995)
5. Sha, F., Bae, C., Liu, G., et al.: A categorized particle swarm optimization for object tracking. In: 2015 IEEE Congress on Evolutionary Computation (CEC), pp. 2737–2744. IEEE (2015)
6. Sha, F., Bae, C., Liu, G., et al.: A probability-dynamic Particle Swarm Optimization for object tracking. In: 2015 International Joint Conference on Neural Networks (IJCNN), pp. 1–7. IEEE (2015)
7. Rashedi, E., Nezamabadi-Pour, H., Saryazdi, S.: GSA: a gravitational search algorithm. Inf. Sci. **179**(13), 2232–2248 (2009)
8. Mirjalili, S., Hashim, S.Z.M.: A new hybrid PSOGSA algorithm for function optimization. In: 2010 International Conference on Computer and Information Application (ICCIA), pp. 374–377. IEEE (2010)
9. Jiang, S., Ji, Z., Shen, Y.: A novel hybrid particle swarm optimization and gravitational search algorithm for solving economic emission load dispatch problems with various practical constraints. Int. J. Electr. Power Energy Syst. **55**, 628–644 (2014)
10. David, R.C., Precup, R.E., Petriu, E.M., et al.: PSO and GSA algorithms for fuzzy controller tuning with reduced process small time constant sensitivity. In: 2012 16th International Conference on System Theory, Control and Computing (ICSTCC), pp. 1–6. IEEE (2012)
11. Gu, B., Pan, F.: Modified gravitational search algorithm with particle memory ability and its application. Int. J. Innovative Comput. Inf. Control **9**(11), 4531–4544 (2013)

Photonic Reservoir Computing Based on Laser Dynamics with External Feedback

Seiji Takeda[1(✉)], Daiju Nakano[1], Toshiyuki Yamane[1], Gouhei Tanaka[2],
Ryosho Nakane[2], Akira Hirose[2], and Shigeru Nakagawa[1]

[1] IBM Research – Tokyo, Kawasaki, Kanagawa 212-0032, Japan
{seijitkd,dnakano,tyamane,snakagw}@jp.ibm.com
[2] Graduate School of Engineering, The University of Tokyo, Tokyo, 113-8656, Japan
gouhei@sat.t.u-tokyo.ac.jp, nakane@cryst.t.u-tokyo.ac.jp,
ahirose@ee.t.u-tokyo.ac.jp

Abstract. Reservoir computing is a novel paradigm of neural network, offering advantages in low learning cost and ease of implementation as hardware. In this paper we propose a concept of reservoir computing consisting of a semiconductor laser subject to external feedback by a mirror, where input signal is supplied as modulation pattern of mirror reflectivity. In that system, non-linear interaction between optical field and electrons are enhanced in complex manner under substantial external feedback, leading to achieve highly nonlinear projection of input electric signal to output optical field intensity. It is exhibited that the system can most efficiently classify waveforms of sequential input data when operating around laser oscillation's effective threshold.

Keywords: Reservoir computing · Recurrent neural network · Sequential data processing · Laser · Silicon photonics · Energy efficiency

1 Introduction

Biological systems possess a wealth of complexity and diversity in its architecture, that sometimes exhibit advanced capability in performing highly complex cognitive tasks. Recent development of artificial intelligence has successfully introduced such biological architectures into their algorithms. One of the most successful examples should be Deep Neural Network (DNN), consisting of multiply connected layers composed of nonlinear elements (neurons), where information propagates unidirectionally from layer to layer. In spite of the fact that DNN profits by its extremely high accuracy in many recognition tasks [1–3], high learning cost is still critical bottle neck, and fundamentally its function is oriented to static recognition. Recurrent Neural Network (RNN), another bio-inspired architecture, consists of connected nonlinear elements but has recurrent loop of information propagation. Owing to its recurrent feedback loop, RNN has internal memory that allows recognition of dynamic (sequential) information. Despite that advantage, however, RNN also suffers from learning cost as same as DNN.

Reservoir Computing (RC) is an emerging computation algorithm, which can overcome drawback of high learning cost that dangles about Neural Network (NN) [4, 5].

© Springer International Publishing AG 2016
A. Hirose et al. (Eds.): ICONIP 2016, Part I, LNCS 9947, pp. 222–230, 2016.
DOI: 10.1007/978-3-319-46687-3_24

RC consists of mainly two parts; RNN with random connections, called reservoir, and linear classifier called output layer. The feature of RC is that randomly given weight values between nodes in reservoir are left untrained, and only weights between reservoir and output layer are updated during training process. That trick allows RC's learning cost to be low compared with other NNs. Provided that the number of reservoir's nodes is large enough, input data is nonlinearly projected into high dimensional space with plenty of random basis, therefore can be linearly classified at output layer, being analogous with kernel method of machine learning. Leveraging its low learning cost and high capability in classification of sequential data, RC has been applied to several classification tasks such as spoken digit recognition [6], hand-written digit recognition [7], phoneme recognition [8], etc.

RC has another great advantage in its ease of implementation as hardware. Due to serial processing and Von-Neumann bottleneck, software implementation of bio-inspired algorithms has been facing critical issue of energy-consumption. While a human brain consumes only 10 W for daily cognitive tasks, today's high-performance computer is anticipated to consume the order of 10 kW in 2020 for the same tasks even on the ideal extrapolation of Moore's Law. In contrast, hardware implementation of bio-inspired algorithms can overcome that issue thanks to its parallel processing and distributed information representation [9]. Owing to small amount of tunable weights, RC is suitable for this purpose. Several hardware implementations of RC have been reported so far by Mackey-Glass electronic circuit [10], soft silicon material [11], connected Semiconductor Optical Amplifier (SOA) [12], laser with time-delayed feedback loop [13], etc. What is remarkable moreover in hardware implementation is that a reservoir is not necessarily a RNN but can be any dynamical physical system, only provided that the system possesses rich nonlinearity. If one can implement reservoir as nonlinear physics with short time constant, processing in reservoir is performed as physical phenomena, therefore high-speed and low-power processing will be achieved.

Among several types of hardware implementations of RC, photonic implementation will be in particular of importance, because photon-electron nonlinear interaction have extremely short time constant around ps to ns, and moreover, it is highly compatible with existing interconnect technology, Silicon Photonics (SiPh) that is highly compatible with matured CMOS process [14]. Multi-integration of electric circuit, SiPh circuit, and photonic RC circuit on the identical silicon chip will tremendously expand the computing capability for cognitive tasks at edge devices, that are indispensable for IoT and cognitive era.

In this paper, we propose a photonic RC consisting of a semiconductor laser with a tunable external mirror, which is suitable for SiPh owing to its simple and compact configuration. The reservoir is tested by a simple task of waveform classification with varying system's parameter, carrier injection rate to the laser and mirror reflectivity. The best operation window as a reservoir is scrutinized.

2 Methods

2.1 Chaotic Dynamics of Laser

Three-variable systems like Lorentz model are well-known to exhibit chaotic behavior under some conditions met. It was 1975 when distinctive analogy between Maxwell-Bloch equations that describe light-matter interaction of laser and Lorenz equations that describe fluid convection was pointed out by Haken. Despite that analogy, in most of lasers including semiconductor lasers, damping of polarization (γ_p) is much faster than other two parameters; population inversion (γ_N) and electric field (κ), then γ_p is adiabatically eliminated, and the lasers operate as stable two-variable systems. Although a semiconductor laser itself oscillates with stable state, external perturbation introduces additional freedom to this two-variable system, and induces chaotic dynamics.

External feedback to a semiconductor laser is one of the most popular methods to bring on chaotic state [15]. Under critical amount of external feedback, intrinsic laser mode with relaxation oscillation frequency f_{RO} and external frequency f_{EC} compete each other and construct complex temporal pattern. The key of RC is to make the system to be highly nonlinear but stay verging on chaotic state, so-called "edge of stability". This delicate state is realized by tuning several laser parameters.

Lang-Kobayashi equation is well-known equations that describe laser system with time-delayed feedback [16, 17]. Electric field $E(t)$ with slowly varying amplitude in a single mode laser and carrier density $N(t)$ are coupled in the form of rate equations,

$$\frac{dE_1(t)}{dt} = \frac{1 + i\alpha}{2}\left[\frac{G_N(N_1(t) - N_0)}{1 + \varepsilon|E_1(t)|^2} - \frac{1}{\tau_p}\right]E_1(t) + \frac{\kappa}{\tau_{in}}E_2(t - \tau)\exp(-i\omega t), \tag{1}$$

$$\frac{dN_1(t)}{dt} = J_1 - \frac{N_1(t)}{\tau_s} - \left[\frac{G_N(N_1(t) - N_0)}{1 + \varepsilon|E_1(t)|^2} - \frac{1}{\tau_p}\right]|E_1(t)|^2, \tag{2}$$

where physical variables J_1, N_1, E_1, E_2, represent carrier injection rate, carrier density, electric field in a laser cavity, electric field reflected from an external mirror respectively. κ represents coupling coefficient between internal mode and external mode, given by $\kappa = (1 - r_1^2)r_2/r_1$, where r_1 and r_2 are laser facet reflectivity and external mirror reflectivity. Other optical parameters are given in Table 1 [16].

We numerically solve those equations to investigate temporal dynamics of the laser-feedback system with varying laser's parameters. And then, we connect a linear classifier to the system and test its capability as a reservoir to nonlinearly transform input signal to allow efficient classification.

Table 1. Optical parameters used in the calculation.

Symbols	Parameters	Values
G_N	Gain coefficient	8.4×10^{-13} m^{3}s^{-1}
α	Linewidth enhancement factor	5.0
N_0	Carrier density at transparency	1.4×10^{24} m^{-3}
N_{th}	Carrier density at threshold	2.02×10^{24} m^{-3}
τ_s	Carrier lifetime	2.04×10^{-9} s
τ_p	Photon lifetime	1.93×10^{-12} s
λ	Optical wavelength	1.54×10^{-6} m
ω	Optical angular frequency	1.23×10^{15} s^{-1}
ε	Gain saturation coefficient	2.5×10^{-23}

2.2   Computational Set-up

We assume the reservoir to consist of a semiconductor laser and an external mirror whose reflectivity is tunable (see Fig. 1). The laser oscillates under the constant carrier injection to the gain medium, being perturbed by external feedback light. Input information to the system is supplied as electric signal pattern of other carrier injection that modulates the mirror reflectivity. The input signal pattern is transported to the laser as modulated optical field pattern, and interacts with intrinsic laser field in the cavity and electrons. Optical output power from the laser is periodically sampled, being regarded as output information from the reservoir. By closely chaotic dynamics in the laser, signal pattern is nonlinearly transformed. Here, each sampling point on temporal axis represents "node" of the reservoir, being connected to output nodes. Sampled optical intensities are weighted-summed and fed into output nodes consisting of nonlinear activation functions such as sigmoid functions. The weights connecting temporal "nodes" and output nodes are updated during training.

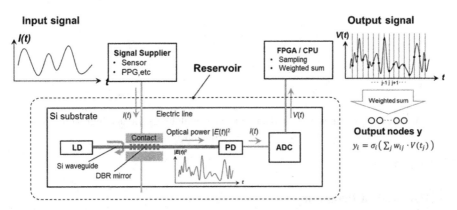

Fig. 1. Conceptual diagram of photonic reservoir. Input signal is supplied to the reservoir as electric signal to modulate DBR mirror reflectivity. Output power from the laser is converted to digital signal, and sampled at FPGA/CPU, then weighted summed.

Considering integration with SiPh, we assume a reservoir consisting of available photonic devices familiar with today's SiPh. For an external mirror, we can leverage Distributed Bragg Reflection (DBR) mirror, which has periodic structure in wavelength scale. Integrating pn junction, electric current is supplied to the DBR to change carrier density inside the structure, causing to modulate the reflectivity. By carefully designing DBR mirror's periodicity to make it's photonic band-edge around wavelength of incident light, the reflectivity can be modulated from 0.0 to 1.0 by small amount of electric current. Optical power is transferred by Si waveguide (SiWG), and detected by Photo Detector (PD), and converted to digital signal by Analogue Digital Converter (ADC), and then its sampling and weighted-summation are performed at CPU/FPGA, that is located outside of the SiPh chip. Compared with looping feedback system, the proposed system has advantage of spatial compactness owing to in-line device configuration free from optical coupler for output monitoring.

To simulate dynamics of the proposed system, we performed calculation by numerically solving Eqs. (1) and (2) by 4th order of Runge-Kutta algorithm.

2.3   Recognition Task

For testing performance of the photonic reservoir, we set a simple task to classify sine/triangular wave forms. Each waveform is supplied to the system independently as a modulation pattern of mirror reflectivity. Weight values of output layer are trained to exhibit 01 and 10 for sine and triangular waveforms respectively. For varying sample's patterns and making the task more complex, frequency ω_n of nth waveform sample is given by $\omega_n = \omega_0 + \alpha_n \omega_0$, where α_n is a random value satisffying $\alpha_n \in [0, A]$. Higher A expands sample's variation and makes the task more difficult. In this paper we set parameters $A = 0.3$ for the first test. The variations of waveforms are shown in Fig. 2.

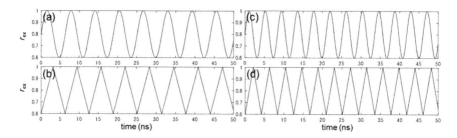

Fig. 2. Waveform variations of input data. Sine wave with lowest (a) and highest (c) frequency and triangular wave with lowest (b) and highest frequency (d).

3   Results

3.1   Dynamics of Reservoir

Fist we investigate fundamental dynamics of the reservoir without injecting input signal patterns, meaning, mirror reflectivity is fixed. We scrutinize temporal dynamics of output optical intensity under different laser parameters; normalized carrier injection rate J/J_0 and reflectivity r_{ex} of the external mirror. It should be noticed that J_0 is laser oscillation threshold without external feedback, but actual threshold slightly decreases under feedback. External cavity length L is fixed to 1 cm.

In order to measure the complexity of laser dynamics, we investigate number of temporal extremal values of output power, N_{ex}, for temporal range from 30 to 50 ns, that is enough after passing transient state required to start laser oscillation. N_{ex} is 2D-plotted as a function of J/J_0 and reflectivity r_{ex} in Fig. 3(a). It is shown that N_{ex} is zero under low carrier injection and mirror reflectivity, meaning the laser oscillates in stable state, but increases with J/J_0 and r_{ex}, and creates variety of temporal patterns. Most representative four patterns, stable oscillation, periodic oscillation, pulse package, and chaos are shown (see Fig. 3(b–e)). With increase of J/J_0 and r_{ex}, those four states appear in seemingly random manner, but statistically pulse package and chaos states occur more frequently. We investigate reservoir's performance as a function of J/J_0 under different r_0, central mirror reflectivity.

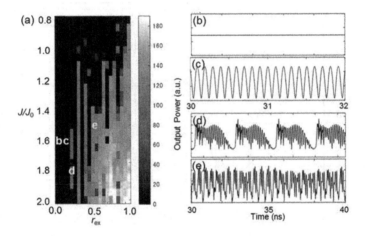

Fig. 3. Number of temporal extremals N_{ex} is color plotted as a function of mirror reflectivity and carrier injection to the laser (a). Typical four temporal dynamics are shown; stable state (b), periodic state (c), pule packages (d), and chaotic state (e), each corresponding to b–e in (a).

3.2 Performance of Classification

To test performance of classification, Eq. (1) and (2) are numerically solved with modulating external mirror reflectivity by $r_{ex} = r_0 + r_{mod}sin(\omega_n t)$ or $r_0 + r_{mod}triangular(\omega_n t)$. We train the system with 400 training samples (200 sine and 200 triangular waveforms), with different J/J_0 ranging from 0.5 to 2.0 under different r_0, central mirror reflectivity. r_{mod} is fixed to 0.2. Output intensity is periodically sampled in 30 ns $< T <$ 50 ns. Sampled 1000 data is fed into output nodes consisting of sigmoid functions and their connection weights are trained. Simple ordinary least square algorithm is used for updating the weights. After training, error rate is measured by 100 test samples independent from training samples.

The error rates as a function of J/J_0 with different r_0 are shown in Fig. 4. Remarkable is that the reservoir exhibits best classification performance when J/J_0 is close to 1 in each r_0. This is the range, according to Fig. 3, where the system transits from periodic to pulse package state by sweeping of r_{ex}. Meanwhile, if the system is fixed around stable state or reaches chaotic state, its classification performance is substantially degraded. That feature will be qualitatively explained by complexity of laser dynamics. Higher carrier injection rate or feedback amount enhance nonlinearity of photon-electron interaction, therefore input signals are transformed in more complex manner, making waveform classification easier because slight but essential differences of waveforms are enhanced. If nonlinearity of the system is too high, inessential trivial differences are also enhanced by "sensitivity to initial condition", and classification becomes even more difficult.

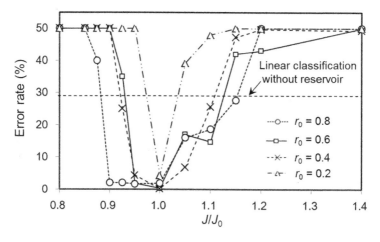

Fig. 4. Error rate of waveform classification task as a function of J/J_0 at different central mirror reflectivity r_0.

Another point to notice is that with increase of mirror reflectivity, "operation window" of reservoir, where error rate is low, broadens and carrier injection rate to achieve that window decreases. From the point of view of low-energy consumption and stability of operation, it will be desired to work with external mirror with high reflectivity.

4   Conclusion

We proposed a photonic RC system consisting of a semiconductor laser with a tunable external mirror, possessing advantage in compactness of integration owing to its in-line device configuration. Complex interaction between internal laser field and reflected laser field realizes highly nonlinear projection of input sequential data to output. Testing by waveform classification task, it is confirmed that the reservoir exhibits best classification performance when laser is tuned to effective oscillation threshold, that is around "edge of stability".

In this paper, we selected J/J_0 and r_0 for reservoir's tuning parameters, but it is also possible to tune laser's parameters more directly, for example Q-factor of laser cavity. It is crucially important as future works in this area to identify the ideal tuning parameters and its operation window as a reservoir, from the viewpoint of practical flexibility after integration.

References

1. Quoc, V.L.: Building high-level features using large scale unsupervised learning. In: 2013 IEEE International Conference on Acoustic, Speech and Signal Processing, pp. 8595–8598, Vancouver, BC (2016)
2. Taigman, Y., Yang, M., Ranzato, M.A., Wolf, L.: Deep face: closing the gap to human-level performance in face verification. In: 2014 IEEE Conference on Computer Vision and Pattern Recognition, pp. 1701–1708 (2014)
3. Wang, W., Arora, R., Livescu, K., Bilmes, J.: On deep multi-view representation learning. In: Proceedings of the 32nd International Conference on Machine Learning, pp. 1083–1092 (2015)
4. Jaeger, H., Haas, H.: Harnessing nonlinearity: predicting chaotic systems and saving energy in wireless communication. Science **304**, 78–80 (2004)
5. Maass, W., Natschläger, T., Markram, H.: Real-time computing without stable states: a new framework for neural computation based on perturbations. Neural Comput. **14**, 2531–2560 (2002)
6. Verstraeten, D., Schrauwen, B., Stroobandt, D.: Isolated word recognition using a liquid state machine. In: Proceedings of the 13th European Symposium on Artificial Neural Networks (ESANN), pp. 435–440 (2005)
7. Jalalvand, A., Wallendael, G.V., Walle R.V.: Real-time reservoir computing network-based systems for detection tasks on visual contents. In: 7th International Conference on Computational Intelligence, Communication Systems and Networks (CICSyN), pp. 146–151 (2015)
8. Triefenbach, F., Jalalvand, A., Schrauwen, B., Martens, J.-P.: Phoneme recognition with large hierarchical reservoirs. Adv. Neural Inf. Process. Syst. **23**, 2307–2315 (2010)
9. Yamane, T., Katayama, Y., Nakane, R., Tanaka, G., Nakano, D.: Wave-based reservoir computing by synchronization of coupled oscillators. In: Arik, S., Huang, T., Lai, W.K., Liu, Q. (eds.) ICONIP 2015. LNCS, vol. 9491, pp. 198–205. Springer, Heidelberg (2015). doi: 10.1007/978-3-319-26555-1_23
10. Appeltant, L., Soriano, M.C., Van der Sande, G., Danchaert, J., Massar, S., Dambre, J., Schrauwen, B., Mirasso, C.R., Fischer, I.: Information processing using a single dynamical node as complex system. Nature Commun. **2**, 468–472 (2011)
11. Nakajima, K., Hauser, H., Li, T., Pfeifers, R.: Information processing via physical soft body. Sci. Rep. **5**, 10487 (2015)
12. Vandoorne, K., Dierckx, W., Schrauwen, B., Verstraeten, D., Baets, R., Bienstman, P., Campenhout, J.V.: Toward optical signal processing using photonic reservoir computing. Opt. Express **16**, 1182–1192 (2008)

13. Brunner, D., Soriano, M.C., Mirasso, C.R., Fischer, I.: Parallel photonic information processing at gigabyte per second data rates using transient state. Nature Commun. **4**, 1364 (2012)
14. Vlasov, Y.: Silicon integrated nanophotonics: From fundamental science to manufacturable technology. http://spie.org/newsroom/pw15_plenary_landing/pw15_plenary_vlasov
15. Sciamanna, M., Shore, K.A.: Physics and applications of laser diode chaos. Nat. Photonics **9**, 151–162 (2015)
16. Kannno, K., Uchida, A.: Complexity analysis in a semiconductor laser with time-delayed optical feedback. Rev. Laser Eng. **39**, 543–549 (2011)
17. Sukow, D.W., Gauhier, D.J.: Entraining power-dropout events in an external-cavity semiconductor laser using weak modulation of the injection current. IEEE J. Quantum Electron. **36**, 175–183 (2000)

FPGA Implementation of Autoencoders Having Shared Synapse Architecture

Akihiro Suzuki$^{(\boxtimes)}$, Takashi Morie, and Hakaru Tamukoh

Graduate School of Life Science and Systems Engineering,
Kyushu Institute of Technology,
2-4 Hibikino, Wakamatsu-word, Kitakyushu-City 808-0196, Japan
suzuki-aKihiro@edu.brain.kyutech.ac.jp

Abstract. Deep neural networks (DNNs) are a state-of-the-art processing model in the field of machine learning. Implementation of DNNs into embedded systems is required to realize artificial intelligence on robots and automobiles. Embedded systems demand great processing speed and low power consumption, and DNNs require considerable processing resources. A field-programmable gate array (FPGA) is one of the most suitable devices for embedded systems because of their low power consumption, high speed processing, and reconfigurability. Autoencoders (AEs) are key parts of DNNs and comprise an input, a hidden, and an output layer. In this paper, we propose a novel hardware implementation of AEs having shared synapse architecture. In the proposed architecture, the value of each weight is shared in two interlayers between input-hidden layer and hidden-output layer. This architecture saves the limited resources of an FPGA, allowing a reduction of the synapse modules by half. Experimental results show that the proposed design can reconstruct input data and be stacked. Compared with the related works, the proposed design is register transfer level description, synthesizable, and estimated to decrease total processing time.

Keywords: Autoencoders · Deep learning · Digital hardware · FPGA · Shared synapse architecture

1 Introduction

In recent years, deep neural networks (DNNs) have been actively studied because they exhibit high performance in the field of machine learning [8]. DNNs consist of stacked neural networks; consequently, they require a huge amount of processing. DNNs are mainly developed via a software implementation with a hardware, especially graphic processing units (GPU). There are several frameworks and libraries for DNNs [1–3]. In contrast, there are few reports of implementations with hardware such as field programmable gate arrays (FPGA) and specialized chips [6,11].

One of the applications of DNNs is embedded systems. Artificial intelligence is expected to apply on latest products such as service robots and automatic

© Springer International Publishing AG 2016
A. Hirose et al. (Eds.): ICONIP 2016, Part I, LNCS 9947, pp. 231–239, 2016.
DOI: 10.1007/978-3-319-46687-3_25

motorists. To realize them, embedded systems should incorporate DNNs. Embedded systems require real-time processing, low power consumption, and flexibility to their purpose. Since hardware can run faster than software, hardware implementation is better for them. Particularly in hardware, FPGA are low power consumption and what reconfigure the circuit to realize a function as user want. Therefore, it is best to implement DNNs with FPGA.

Autoencoders (AEs) are related to DNNs [4]. They are composed of three layers, and multilayered version of them are employed as part of DNNs [5]. To date, a few relevant reports about hardware implements of AEs have been published. One such report discussed the simulation of learning Kyoto pictures with the behavior model sparse AE; however, the model was too large to realize on an actual circuit by logic synthesis [9]. A second report disclosed the implementation of stacked AEs [5] by high-level synthesis (HLS) with Open Computing Language. According to this report, the AEs were implemented onto FPGA, Altera Stratix V GS D5 [10]. Although Stratix V has abundant resources and exhibits a high performance, a circuit by HLS is far from an optimized circuit by register transfer level (RTL) and the performance would be less than that of GPU and mobile GPU.

In this paper, AEs are implemented with a novel hardware architecture, shared synapse architecture. To save the FPGA resources, the proposed architecture halves the number of synapse modules by using the AEs' feature allowing the weights of the hidden layer and output layer to be shared. The proposed design is based on the RTL design to optimize the circuit, and it is intended to be stacked and construct DNNs. Experimental results show that a digital circuit with the proposed design works as an AE, and the circuit can easily change the number of AEs' units and can be stacked. Depending on the design, proposed AE shows better performance than related works at the point of processing speed estimation.

2   Autoencoders

DNNs comprise pre-training and fine-tuning phases. AEs are stacked and form a pre-training phase of DNNs. AEs comprise three layers: an input layer, a hidden layer, and an output layer. Each layer has a unit that is connected via weights to all other layer's units, as shown in Fig. 1. AEs reconstruct the input via two operations, an encode and a decode; therefore, the output and input layers have the same number of units.

The learning algorithm of AEs is as follows. In AEs, vector y of hidden units is encoded by vector x of input units, as shown in Eq. 1, and vector z of output units is decoded by vector y, as shown in Eq. 2, where σ is the sigmoidal function, vector b is the bias of hidden units, b' is the bias of output units, and W is a weight between each layer and is $N \times M$ matrix. Each parameter is updated by Eqs. 3, 4, and 5 with η as learning rate.

$$y_m = \sigma(\sum_{n=0}^{N} W_{mn} x_n + b) \quad m = 0,1 \ldots, \ M-1, M. \tag{1}$$

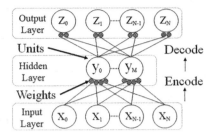

Fig. 1. Autoencoders

$$z_n = \sigma(\sum_{m=0}^{M} W_{nm}^{T} y_m + b') \quad n = 0, 1 \ \ldots, \ N-1, N. \tag{2}$$

$$\Delta W = \eta \left(\left(W (x - z) * y * (1 - y) \cdot x^T \right) + \left(((x - z) y^T)^T \right) \right) \tag{3}$$

$$\Delta b = \eta \left(W (x - z) * y * (1 - y) \right) \tag{4}$$

$$\Delta b' = \eta (x - z) \tag{5}$$

A processing flow of the learning phases of the proposed design AE is as follows.

1. Get the hidden units.
   Calculate Eq. 1. The hidden units are mapped by the input units.
2. Get the output units.
   Calculate Eq. 2. The output units are reconstructed by the hidden units.
3. Update the parameters.
   Reconstruction error is measured by the cross error function. To reduce the error, the update value of each parameter is determined by Eqs. (3, 4, and 5). Each update value is added to the original parameters, W, b, and b'.
4. Repeat operations 1, 2, and 3 until the error is small enough.

3 FPGA Implementation of Autoencoders

3.1 Shared Synapse Architecture

Assuming all processes of evaluating the outputs of AEs are divided into two modules—a neuron module and synapse module—the entire circuit is described in Fig. 2(a) when the input layer has four units and the hidden layer has two units.

We propose a shared synapse architecture where the synapse modules are reduced by half, as shown in Fig. 2(b), because the value of each weight could is shared by two interlayers [7] between the input layer-hidden layer and the hidden layer-output layer.

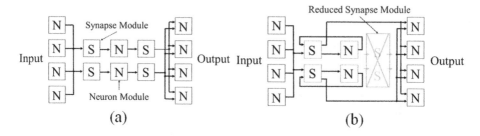

Fig. 2. Architecture of AEs. (a) Flow of evaluating outputs of AEs. (b) Shared synapse architecture

3.2   Digital Circuit of Shared Synapse Architecture

In the proposed circuit, there are three main modules that further contain some submodules. The entire circuit is shown in Fig. 3, and the details of the circuits that include the submodules are shown in Figs. 4 and 5.

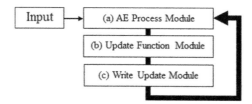

Fig. 3. Entire circuit

All modules have a common interface and, consequently, are easily combined with each other. In addition, the parameters that control the modules' size and function can be externally controlled.

The roles of each circuit is described in the list below.

(a) AE Process Module: The reconstruction from the input to output
   (a-1) Synapse Module: The multiplication and memory for the weights
   (a-2) Neuron Module: The sequence sum and sigmoidal function
(b) Update Function Module: The calculation for the update value of each parameter
   (b-1) b' Update Module: For the update value of b'
   (b-2) b Update Module: For the update value of b
   (b-3) W Update Module: For the update value of W
(c) Write Update Module: The update values are added to each parameter.

4   Experimental Results

To evaluate the proposed architecture, the digital circuits with the proposed architecture were written by Verilog Hardware Description Language and evaluated by logic simulation in Veritak Verilog HDL simulator.

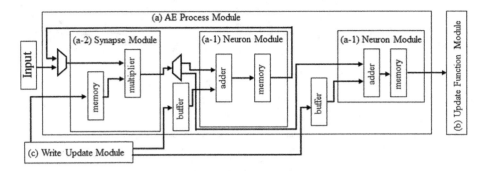

Fig. 4. Detail of AE circuit

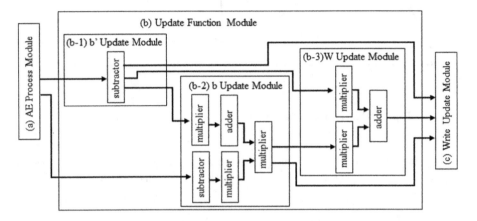

Fig. 5. Detail of update function circuit

4.1 Performance Validation as AEs

To validate the proposed design circuit as AEs, the simulation determined whether the output of the circuit could reconstruct the input. Each parameter was given 0 as an initial value, and the learning rate η was 0.0078125, which was represented by shifting seven bits to the right side.

The input data was sixteen kinds of a set of four binary data, e.g., (0,0,0,0), (0,0,0,1), (1,1,1,1), etc. The output data must be given as fractional number and considered in the following manner below for representing the output data as a reconstruction of the input data.

$$output = \begin{cases} 1 & (z_n \geqq 0.5) \\ 0 & (z_n < 0.5) \end{cases} \quad n = 0,1\ldots,\ N-1,N. \tag{6}$$

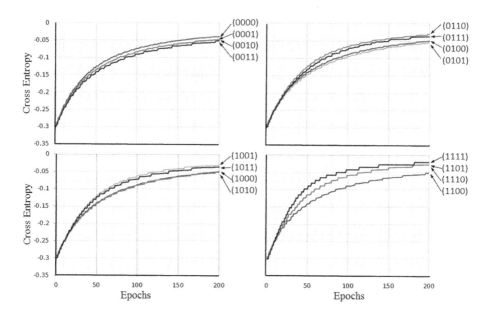

Fig. 6. Observation of learning process by cross entropy errors

The cross entropy error C was calculated by Eq. 7, which we employed as an error function and was calculated every epoch in the learning process.

$$C = \frac{1}{4} \sum_{n=0}^{3} (x \log z + (1 - x) \log (1 - z)) \quad \text{n} = 0,1,2,3. \tag{7}$$

The results of the evaluation of sixteen learning processes are shown in Fig. 6. From the results, it is obvious that the more the epoch increases, the more each unit approximates the target value until, finally, each output may be regarded as a reconstruction of its own input.

We set the FPGA, Xilinx Virtex-6 xc6vlx240t as a target device and synthesized it by Xilinx ISE 14.7. The slice utilization and maximum frequency for the entire circuit and main three parts are shown in Table 1.

4.2   Construction of Stacked AEs

In stacked AE, there is a second AE instead of the hidden layer of the first AE, as shown in Fig. 7. To prove the possibility to change the number of units in each layer, the second AE was constructed with the same parts as the first AE constructed in the previous sections. To shift the first AE to the second AE with the same parts, each part had to be slightly rewritten. The result of logic synthesis is shown in Table 2.

Since the hidden layer of the first AE could be the second AE, the input layer of the second AE was reduced by sharing it with the hidden layer of the

Table 1. Results of logic synthesis of shared synapse AE and its parts

	Registers	LUTs	Freq. (MHz)
Entire circuit	**6407 (2.13 %)**	**5835 (3.87 %)**	**310.366**
AE process module	**2855 (0.95 %)**	**2484 (1.65 %)**	**370.233**
Synapse module	325 (0.11 %)	239 (0.16 %)	453.926
Neuron module	144 (0.05 %)	159 (0.11 %)	395.101
Update function module	**5844 (1.94 %)**	**5490 (3.64 %)**	**388.999**
b' update module	594 (0.20 %)	454 (0.30 %)	419.639
b update module	1141 (0.38 %)	1187 (0.79 %)	402.414
W update module	397 (0.13 %)	410 (0.27 %)	419.639
Write update module	**518 (0.17 %)**	**7 (0.004 %)**	**651.042**

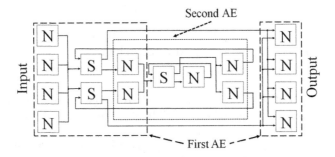

Fig. 7. Stacked AE with shared synapse architecture

first AE. Hence, the total number of stacked AE process module's registers and
LUTs was less than the sum of AE process module's and the second AE's by
sharing the neuron module and any other sub modules.

4.3   Comparison with Other Reports

A comparison of the proposed implementation with others on FPGA is shown
in Table 3.

Table 2. Results of logic synthesis of stacked AE process module and second AE
process module

	Registers	LUTs	Freq. (MHz)
AE process module	2855 (0.95 %)	2484 (1.65 %)	370.233
Stacked AE process module	3133 (1.04 %)	2975 (1.97 %)	370.233
Second AE process module	815 (0.27 %)	819 (0.54 %)	391.083

Table 3. Comparison of the proposed implementation with other implementations

	Algorithm	Method	Logic synthesis
Proposed	AE (Stacked AE)	RTL	Possible
Jin et al. [9]	Sparse AE	Behavior	Impossible
Joao et al. [10]	Stacked AE	HLS	Possible

Table 4. Modules that need an extra clock

Module name	Extra
Synapse	2766
Neuron	5837
b update	5106
Other sub modules	7830

According to Table 3, the proposed design circuit was compared with Ref. [9] in terms of the implementation on FPGA and, with Ref. [10], in terms of processing speed.

Since the design of Ref. [9] needs huge FPGA resources such as memory and multiplier, realizing it in the FPGA is difficult. The synthesis report was not shown in the report.

In HLS, the code of a behavioral model is analyzed to be scheduled to create RTL hardware description language. Generally, there is a difference as a wasteful part between a circuit written by HLS and a circuit optimally written like a proposed architecture. The processing speed of the proposed design was compared with Ref. [10]; since the RTL designed circuit has no wasteful part, it was expected to operate faster than the HLS-designed circuit that has a wasteful part. To expand the network size of the proposed design to the network size of the Ref. [10], the proposed AE needs 21617 clocks, which includes the extra 21539 clocks shown in Table 4. Considering the extra clocks, the total processing time of the proposed design is 2.90[s] while the Ref. [10] design needs 16.87[s] for doing the same thing.

5   Conclusion

In this paper, we proposed a novel hardware shared synapse architecture for AEs. This proposed architecture can save the limited resources of FPGA since the synapse modules are halved because the weight is the same for both the hidden and the output layer. The proposed design was validated and could work as an AE that can reconstruct a set of four binary data. Furthermore, the circuit can diminish the size of network and the different size circuits can be combined with each other to construct stacked AEs.

Future studies will use the proposed design circuit to construct huge scale AEs and stacked AEs. Furthermore, the other neural networks related to DNNs will be implemented with FPGA and we plan to develop a system where any digital circuit of neural networks is dealt with as an object.

Acknowledgement. This research was supported by JSPS KAKENHI Grant Number 26330279 and 15H01706.

References

1. Caffe. http://caffe.berkeleyvision.org/
2. Chainer. http://chainer.org/index.html
3. Theano. http://deeplearning.net/software/theano/
4. Bengio, Y.: Learning deep architectures for AI. Found. Trends Mach. Learn. **2**(1), 1–127 (2009)
5. Bengio, Y., Lamblin, P., Popovici, D., Larochelle, H.: Greedy layer-wise training of deep networks. In: Bernhard, S., John, P., Thomas, H. (eds.) Advances in Neural Information Processing Systems 19, pp. 153–160. MIT Press, Cambridge (2007)
6. Chen, Y., Luo, T., Liu, S., Zhang, S., He, L., Wang, J., Li, L., Chen, T., Xu, Z., Sun, N.: Dadiannao: a machine-learning supercomputer. In: Proceedings of the 47th Annual IEEE/ACM International Symposium on Microarchitecture, pp. 609–622 (2014)
7. Droniou, A., Sigaud, O.: Gated autoencoders with tied input weights. In: International Conference on Machine Learning (2013)
8. Hinton, G.E., Osindero, S., Teh, Y.W.: A fast learning algorithm for deep belief nets. Neural Comput. **18**(7), 1527–1554 (2006)
9. Jin, Y., Kim, D.: Unsupervised feature learning by pre-route simulation of auto-encoder behavior model. Int. J. Comput. Electr. Autom. Control Inf. Eng. **8**(5), 668–672 (2014)
10. Joao, M., Joao, A., Gabriel, F., Luis, A.A.: Stacked autoencoders using low-power accelerated architectures for object recognition in autonomous system. Neural Process. Lett. **43**, 1–14 (2015)
11. Park, S., Bong, K., Shin, D., Lee, J., Choi, S., Yoo, H.J.: A 1.93 TOPS/W scalable deep learning/inference processor with tetra-parallel MIMD architecture for big-data applications. In: IEEE International Solid-State Circuits Conference (ISSCC), pp. 80–82 (2015)

Time-Domain Weighted-Sum Calculation for Ultimately Low Power VLSI Neural Networks

Quan Wang, Hakaru Tamukoh, and Takashi Morie$^{(\boxtimes)}$

Graduate School of Life Science and Systems Engineering,
Kyushu Institute of Technology,
2-4, Hibikino, Wakamatsu-ku, Kitakyushu 808-0196, Japan
morie@brain.kyutech.ac.jp

Abstract. Time-domain weighted-sum operation based on a spiking neuron model is discussed and evaluated from a VLSI implementation point of view. This calculation model is useful for extremely low-power operation because transition states in resistance and capacitance (RC) circuits can be used. Weighted summation is achieved with energy dissipation on the order of 1 fJ using the current CMOS VLSI technology if 1 GΩ order resistance can be used, where the number of inputs can be more than a hundred. This amount of energy is several orders of magnitude lower than that in conventional digital processors. In this paper, we show the software simulation results that verify the proposed calculation method for a 500-input neuron in a three-layer perceptron for digit character recognition.

Keywords: Time-domain computing · Weighted sum · Spike-based computing · Deep neural networks

1 Introduction

In artificial neural networks, weighted summation is an essential and heavy calculation task. Usually, such arithmetic is digitally performed by very-large-scale integration (VLSI) circuits in current computers, but if analog operation of complementary metal-oxide-semiconductor (CMOS) VLSI circuits is used for the task, extremely low-power consumption operation can be achieved. Although the calculation precision is limited because of the non-idealities of analog operation such as noise and device mismatches, neural network circuits and their operation can be designed to be robust to such non-idealities [4].

Time-domain computation based on spiking neuron models was proposed as a realistic mathematical model for biological neurons [1,2]. In contrast to conventional weighted-sum operations using analog voltages or currents, time-domain computation is more suitable for lower-power consumption operation in CMOS VLSI implementation of artificial neural networks.

© Springer International Publishing AG 2016
A. Hirose et al. (Eds.): ICONIP 2016, Part I, LNCS 9947, pp. 240–247, 2016.
DOI: 10.1007/978-3-319-46687-3_26

We have already proposed a device and circuit that performs time-domain weighted-sum calculation [5]. The circuit used consists of plural input resistors and a capacitor, which can lead to extremely low-power operation. However, we have to consider how weighted-sum operations that include both positive and negative weight values are performed.

In this paper, we discuss such a case and show simulation results for a 500-input neuron in a three-layer perceptron for digit character recognition.

2   Time-Domain Weighted-Sum Calculation

An integrate-and-fire-type (IF) neuron model is shown in Fig. 1(a). In this model, the neuron receives spike pulses via synapses from other neurons. The spike pulse represents an analog value by the input timing, and its pulse width and amplitude give no effects on the following processing. A spike generates a temporal voltage change as a response called a post-synaptic potential (PSP), and the internal potential of neuron n, $V_n(t)$, is equal to the spatiotemporal summation of all PSPs. When $V_n(t)$ reaches the firing threshold θ, the neuron outputs a spike, and then $V_n(t)$ settles back to the resting state.

Based on the model proposed in [1], a simplified weighted-sum operation model using IF neurons has been proposed [5]. Only one spike is assumed to be fed from each neuron during the arbitrarily predefined time span T_{in}. Furthermore, the time course of a PSP generated at the input spike timing t_i is assumed to be linear with slope k_i, as shown in Fig. 1(b).

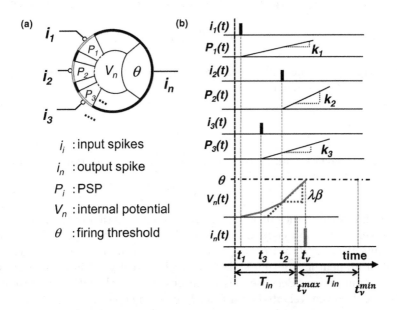

Fig. 1. IF neuron model for weighted-sum operation: (a) schematic of the model and (b) time-domain weighted-sum operation using rise timing of PSPs.

A required weighted-sum operation is that normalized variable x_i ($0 \le x_i \le 1$) is multiplied by a predefined weight coefficient a_i and the multiplication results are summed regarding i ($i = 1, 2, \cdots, N$), where N is the number of inputs. This weighted-sum operation can be performed using the rising slopes of PSPs in the above-mentioned IF neuron model. Input spike timing t_i is determined based on x_i using the relationship $t_i = T_{in}(1-x_i)$. Coefficients a_i are transformed into the PSPs' slopes k_i; $k_i = \lambda a_i$, where λ is a positive transformation constant. If the neuron's firing time is defined as t_ν, the following equation is obtained:

$$\sum_{i=1}^{N} k_i(t_\nu - t_i) = \theta. \tag{1}$$

If we define the following parameter:

$$\beta = \sum_{i=1}^{N} a_i, \tag{2}$$

we obtain

$$\sum_{i=1}^{N} a_i \cdot x_i = \frac{\theta/\lambda + \beta(T_{in} - t_\nu)}{T_{in}}. \tag{3}$$

In this approach, the normalization of sum of a_i is not necessary unlike the previous work [1,2,5].

Here, we assume that all weights of a_i have the same sign. When all inputs are minimum ($\forall i \; x_i = 0$), the left side of Eq. (3) is zero, and therefore the output timing t_ν is given by

$$t_\nu^{min} = \frac{\theta}{\lambda\beta} + T_{in}. \tag{4}$$

On the other hand, when all inputs are maximum ($\forall i \; x_i = 1$), the left side of Eq. (3) is β, and t_ν is given by

$$t_\nu^{max} = \frac{\theta}{\lambda\beta}. \tag{5}$$

The time span when t_ν can be observed is $[t_\nu^{max}, t_\nu^{min}]$, and its interval is

$$T_\nu \equiv t_\nu^{min} - t_\nu^{max} = T_{in}. \tag{6}$$

Thus, the time span of the output spikes are the same as that of the input spikes, T_{in}. However, since the weighted-sum result is given by the expression including the multiplication of β and t_ν, as shown in Eq. (3), and since β are generally different for all neurons, output spikes cannot be directly fed into the next stage circuit as input spikes. Because β is given before the calculation operation, we have to calculate the weighted-sum from t_ν at each stage.

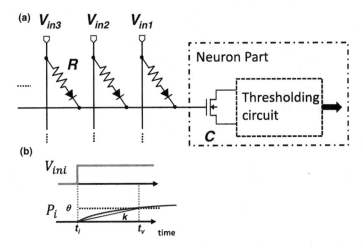

Fig. 2. RC circuit for performing time-domain weighted-sum calculation: (a) circuit diagram in which resistors are connected with the gate of an FET for weighted-sum operation, where each resistor should have rectification function to prevent an inverse current, and (b) step voltage input and its approximate linear response generated by an RC circuit.

3   RC Circuits That Perform Time-Domain Weighted-Sum Calculation

The above-mentioned calculation model can approximately be implemented using a circuit that consists of a field effect transistor (FET) connected to an array of resistors, as shown in Fig. 2(a). The approximate linear slope k is generated by capacitance C of the FET gate and resistance R with step voltage input V_{in}, as shown in Fig. 2(b), where we use step voltages instead of spike pulses as inputs.

In such time-domain calculations, time constant τ for processing should be fairly long, such as $\tau = 1\ \mu$s, to guarantee a sufficient calculation resolution. If C is assumed to be on the order of $1 \sim 10$ fF, which is a typical order for the gate capacitance of a nanoscale MOSFET and parasitic capacitance, R should be around 1 GΩ to obtain $\tau \sim 1\ \mu$s.

Under the above assumption, the energy consumed for a weighted-sum operation is $E_{ws} \sim CV^2 \sim 1 \sim 10$ fJ, where V is assumed to be 1 V. Furthermore, if N resistors are connected with capacitor C for N-parallel multiplication operations, the energy required for one multiplication operation is $E_m = E_{ws}/N \sim 10 \sim 100$ aJ for $N = 100$, which will be more than several orders of magnitude lower than the current highest efficiency digital processors [3]. How many inputs can be effectively implemented in this method from the power consumption point of view depends on how large the parasitic capacitance for each input is.

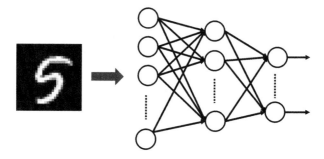

Fig. 3. Network of a perceptron, which has 784 neurons in the input layer, 500 neurons in the hidden layer, and 10 neurons in the output layer, for classifying MNIST digit characters.

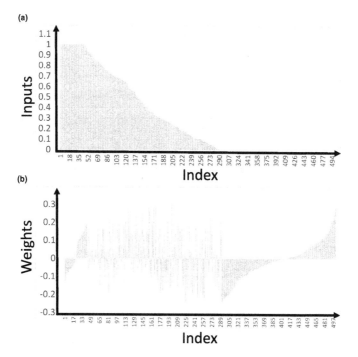

Fig. 4. Inputs and weights of a neuron in the output layer: (a) distribution of 501 inputs, and (b) that of 501 weights corresponding to the 501 inputs.

4   Simulation Results of Time-Domain Weighted-Sum Calculation

To perform a weighted-sum operation with both positive and negative weights, we divide a weighted-sum operation into two operations, in which one is for all the positive weights and the other is for all the negative ones. Then the time-domain weighted-sum calculation in an RC circuit is performed in each of them. According to Eq. (3), the results of the two operations are obtained as spike timing, and the two results are summed as the final result of the original weighted summation. We performed software simulation to verify the proposed method.

In this simulation, we used a multi-layer perceptron (MLP) with a single hidden layer shown in Fig. 3, which is a feed-forward neural network to classify the MNIST digit character set. The size of each digit character pattern is 28×28 pixels. The MLP has 784 input neurons, 500 neurons in the hidden layer, and ten output neurons, so that there exist a large number of multiplications and summations for each neuron in the hidden and output layers to perform the weighted-sum calculation. First, we trained the MLP, and then, we applied the time-domain weighted-sum method described in Sect. 2 to the trained MLP to test the performance of MNIST digit character recognition. Figure 4 showed the

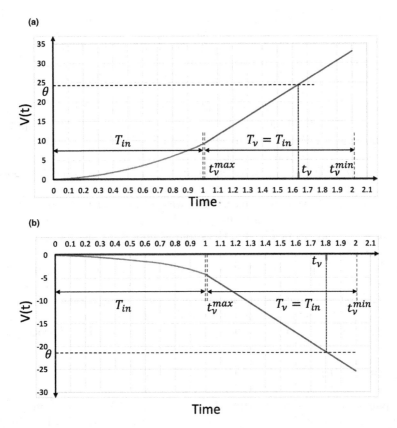

Fig. 5. Simulation results for the proposed approach: (a) PSP of positively weighted-sum operation with 249 inputs in which $T_{in} = 1, \lambda = 1, \beta = 24.01, \theta = 24.25$, and thus $t_\nu = 1.635$, and the weighted-sum result is 8.99, (b) PSP of negatively weighted-sum operation with 252 inputs in which $T_{in} = 1, \lambda = 1, \beta = -21.19, \theta = -21.4$, and thus $t_\nu = 1.81$, and the weighted-sum result is -4.272.

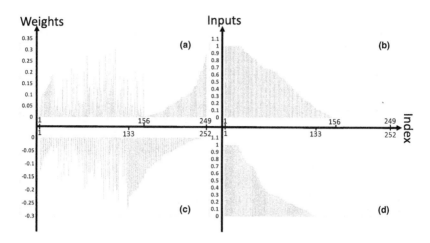

Fig. 6. Distributions of weights a_i and inputs x_i involved in the time-domain weighted-sum calculation: (a) distribution of 249 positive weights, (b) distribution of 249 inputs corresponding to the 249 positive weights, (c) distribution of 252 negative weights, and (d) distribution of 252 inputs corresponding to the 252 negative weights.

distributions of the inputs and the corresponding weights of a neuron, which has 500 inputs from the hidden neurons and one bias input, in the output layer during testing. Note that the well-trained weights consist of both positive and negative values, as shown in Fig. 4(b).

Figure 5 shows the simulation results, in which the neuron has 249 positive weights and 252 negative weights, and a total of 501 weights and inputs. The correct value of the weighted summation was 4.718, and the sum of the two results is equal to that, as shown in Fig. 5. Figure 6 shows the distributions of weights a_i and inputs x_i involved in the simulation of the time-domain weighted-sum calculation.

Considering the energy consumption for this weighted-sum calculation, we have to use two capacitors to complete one precise weighted-sum operation. This means that we will consume only twice as much energy as that discussed in Sect. 2. Even so, the energy consumption will still be more than several orders of magnitude lower than that of the current highest efficiency digital processors [3].

5   Conclusions

In this paper we addressed the time-domain weighted-sum calculation based on a spiking neuron model. To perform weighted-sum operation with both positive and negative weights, we proposed an approach where such an operation is divided into two parts with two RC circuits in which one performs positive weighted-sums and the other performs negative weighted-sums.

We performed simulation and verified that the result of the weighted-sum can be calculated correctly. It can be concluded that the proposed approach is promising to achieve the extremely low-energy consumption for a vastly large number of weighted-sum calculations.

Acknowledgments. This work was supported by JSPS KAKENHI Grant Nos. 22240022 and 15H01706. Part of the work was carried out under the Collaborative Research Project of the Institute of Fluid Science, Tohoku University.

References

1. Maass, W.: Fast sigmoidal networks via spiking neurons. Neural Comput. **9**, 279–304 (1997)
2. Maass, W.: Computing with spiking neurons. In: Maass, W., Bishop, C.M. (eds.) Pulsed Neural Networks, pp. 55–85. MIT Press (1999)
3. Merolla, P.A., Arthur, J.V., Alvarez-Icaza, R., Cassidy, A.S., Sawada, J., Akopyan, F., Jackson, B.L., Imam, N., Guo, C., Nakamura, Y., Brezzo, B., Vo, I., Esser, S.K., Appuswamy, R., Taba, B., Amir, A., Flickner, M.D., Risk, W.P., Manohar, R., Modha, D.S.: A million spiking-neuron integrated circuit with a scalable communication network and interface. Science **345**(6197), 668–673 (2014)
4. Morie, T., Amemiya, Y.: An all-analog expandable neural network LSI with on-chip backpropagation learning. IEEE J. Solid-State Circuits **29**(9), 1086–1093 (1994)
5. Tohara, T., Liang, H., Tanaka, H., Igarashi, M., Samukawa, S., Endo, K., Takahashi, Y., Morie, T.: Silicon nanodisk array with a fin field-effect transistor for time-domain weighted sum calculation toward massively parallel spiking neural networks. Appl. Phys. Express **9**, 034201-1-4 (2016)

A CMOS Unit Circuit Using Subthreshold Operation of MOSFETs for Chaotic Boltzmann Machines

Masatoshi Yamaguchi[1], Takashi Kato[1], Quan Wang[1], Hideyuki Suzuki[2],
Hakaru Tamukoh[1], and Takashi Morie[1(✉)]

[1] Graduate School of Life Science and Systems Engineering,
Kyushu Institute of Technology,
2-4, Hibikino, Wakamatsu-ku, Kitakyushu 808-0196, Japan
morie@brain.kyutech.ac.jp
[2] Graduate School of Information Science and Technology, Osaka University,
1-5, Yamada-oka, Suita, Osaka 565-0871, Japan

Abstract. Boltzmann machines are a useful model for deep neural networks in artificial intelligence, but in their software or hardware implementation, they require random number generation for stochastic operation, which consumes considerable computational resources and power. Chaotic Boltzmann machines (CBMs) have been proposed as a model using chaotic dynamics instead of stochastic operation. They require no random number generation, and are suitable for analog VLSI implementation. In this paper, we describe software simulation results for CBM operation, and propose a CMOS circuit of CBMs using the subthreshold operation of MOSFETs.

Keywords: VLSI implementation · Chaotic Boltzmann machine · Subthreshold operation · MOSFET

1 Introduction

Boltzmann machines (BMs) are considered a useful neural network model not only for deep neural networks with deep learning but also for solving optimization problems [1,2,8,10,11]. Different trials for hardware implementation of BMs have been reported [3,6,7,9,12,15]

The BM is a network of symmetrically connected neuron-like units that make stochastic decisions about whether to be on or off, and therefore we call the original BMs stochastic BMs (SBMs). However, the computation cost for SBMs is very high, and it therefore seems difficult to apply SBMs to real-world problems. To reduce their computation cost, chaotic Boltzmann machines (CBMs) have been proposed [14]. CBMs use chaotic dynamics instead of stochastic operation.

From the hardware implementation point of view, SBMs require random number generators to emulate stochastic operation [4,5]. In contrast, CBMs

© Springer International Publishing AG 2016
A. Hirose et al. (Eds.): ICONIP 2016, Part I, LNCS 9947, pp. 248–255, 2016.
DOI: 10.1007/978-3-319-46687-3_27

require analog dynamical operation, which means that CBMs are suitable for implementation in analog VLSI circuits.

In this paper, we describe software simulation results for the operation of CBMs, and propose a CMOS circuit of CBMs using the subthreshold operation of MOSFETs.

2 Chaotic Boltzmann Machine Model

We first describe the operation of a conventional SBM. Let us define $S_i \in \{0, 1\}$ as the output of unit i in an SBM consisting of N units. Unit i turns on with a probability given by

$$P[S_i = 1] = \frac{1}{1 + \exp(-z_i/T)}, \tag{1}$$

where T denotes the temperature of the system, and z_i represents its total input calculated by

$$z_i = \sum_{j=1}^{N} w_{ij} S_j + \theta_i, \tag{2}$$

where w_{ij} is the symmetric weight of the connection between units i and j, and θ_i is the bias applied to unit i. If the units are updated sequentially, the network of units will eventually reach its equilibrium state. Hardware implementation of SBMs requires random number generation for stochastic operation, which consumes considerable computational resources and power.

In contrast, a CBM is a deterministic system represented by the dynamics of nonlinear oscillators based on a pseudo-billiard model, as shown in Fig. 1. CBMs can emulate BMs without using any random numbers. In CBMs, a unit is associated with state variable $x \in [0, 1]$, which is called the internal state of the unit. Internal state x_i of unit i evolves according to the following differential equation:

$$\frac{dx_i}{dt} = (1 - 2S_i) \left\{ 1 + \exp \frac{(1 - 2S_i)z_i}{T} \right\}, \tag{3}$$

which is determined so that the speed $|dx_i/dt|$ is inversely proportional to the probability $P[S_i]$ given by Eq. (1) (see [13,14] for more details). The states of the units are updated by the deterministic rule in which they change when and only when their internal states reach 0 or 1, as shown in Fig. 1(a). The change speed of a unit's internal state changes when the other units' states change, as shown in Fig. 1(b). In this way, the units interact with each other. In general, the state of BMs is represented with an energy function defined by

$$E = -\sum_{i<j} w_{ij} S_i S_j - \theta_i S_i. \tag{4}$$

The units in CBMs can operate in parallel, and no random number generation is required. Therefore, it is highly anticipated that CBMs can be efficiently implemented as parallel and distributed hardware systems.

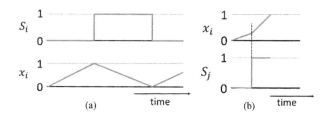

Fig. 1. Dynamics of CBMs: (a) single unit dynamics and (b) slope change in x_i when S_j of another unit changes.

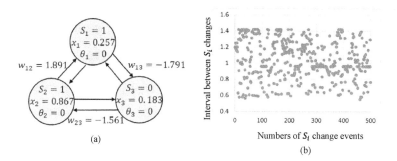

Fig. 2. Software simulation results of a CBM that consists of three units: (a) initial condition and (b) intervals between S_i changes in a unit at $T = 1$.

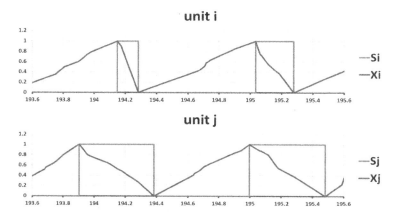

Fig. 3. Software simulation results of the dynamics of units i and j in a CBM composed of ten units.

Figure 2 shows the software simulation results of a CBM that consists of three units. The interval between change events of S_i is plotted as a function of the step of change events of S_i in the time sequence of a unit. The distribution of the interval seems chaotic, although a detailed analysis is needed. The unit dynamics shown in Fig. 3 are consistent with those shown in Fig. 1. According to

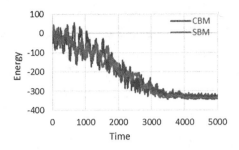

Fig. 4. Software simulation results of changes in energy of chaotic and stochastic Boltzmann machines by lowering temperature T from the same initial condition.

the dynamics shown in Fig. 1, internal state x_i oscillates between 0 and 1 continuously, which is clearly shown in the simulation results in Fig. 3. In addition, in the simulation, temperature T is lowered after each state update of an arbitrary unit in both SBMs and CBMs from the same initial temperature. Figure 4 shows that with the temperature change, the energy of the two models has nearly the same time course.

3   Design of a CMOS Unit Circuit for Chaotic Boltzmann Machines

The proposed unit circuit consists of a switched variable current source (SVCS) circuit, a switched exponential current source (SECS) circuit, and a voltage pulse converter (VPC) circuit, as shown in Fig. 5.

3.1   Operation of SVCS Circuit

The SVCS circuit converts synaptic current $I_{zi} \ (= \sum_{j=1}^{N} w_{ij} S_j + \theta_i)$ to $I_{SVCS} (= (1 - 2S_i) I_{zi}/T)$. Current I_{zi} is changed to voltage V_{zi} with a resistor. The SVCS operates as follows;

(1) If the synapse current is not fed into the SVCS: $I_{zi} = 0 \ (V_{zi} = V_{z0})$, the SVCS outputs no current; $I_{SVCS} = 0$.
(2) If a positive synapse current is fed into the SVCS: $I_{zi} > 0 \ (V_{zi} > V_{z0})$, and if unit state $S_i = 1$, the SVCS outputs a negative current; $I_{SVCS} < 0$.
(3) If unit state S_i turns over or the direction of I_{zi} reverses, the direction of I_{SVCS} reverses.
(4) If voltage V_T is changed, the absolute value of I_{SVCS} changes.

3.2   Operation of SECS Circuit

The SECS converts I_{SVCS} to $I_{SECS} \ (= (1 - 2S_i)(1 + \exp(I_{SVCS})))$. MOSFET M_1 operates in a subthreshold region and continues outputting normalized current I_1 that corresponds to "1" in Eq. 3. Current I_{SVCS} is changed to voltage V_{exp} with a resistor. The SECS operates as follows:

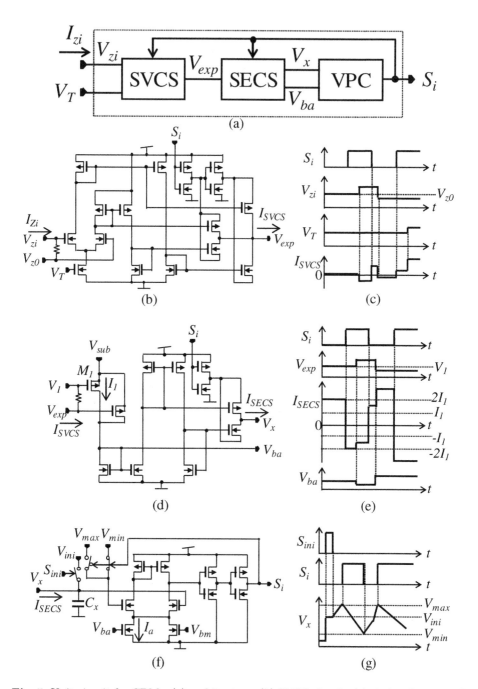

Fig. 5. Unit circuit for CBMs: (a) architecture, (b) SVCS circuit, (c) timing diagram of SVCS, (d) SECS circuit, (e) timing diagram of SECS, (f) VPC circuit, and (g) timing diagram of VPC.

(1) If I_{SVCS} is not fed into SECS: $I_{SVCS} = 0$ ($V_{exp} = V_1$) and $S_i = 0$, SECS outputs $I_{SECS} = 2I_1$.
(2) If unit state S_i turns over, the direction of I_{SECS} reverses.
(3) If V_{exp} increases, the absolute value of I_{SECS} decreases exponentially, and voltage V_{ba} decreases. However, $|I_{SECS}|$ larger than $|I_1|$.
(4) If V_{exp} decreases, $|I_{SECS}|$ increases exponentially, and voltage V_{ba} increases.

3.3   Operation of VPC Circuit

The oscillation of a CBM is achieved by the VPC as follows:

(1) Voltage V_x is initialized at V_{ini} by initialization signal S_{ini} and VPC starts to oscillate.

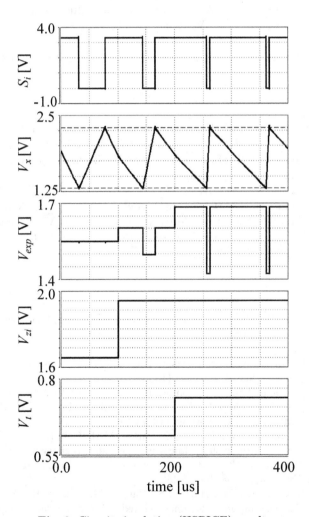

Fig. 6. Circuit simulation (HSPICE) results.

(2) If $S_i = 0$, capacitor C_x is charged by I_{SECS}, and V_x increases.

(3) When V_x exceeds V_{max}, S_i turns over.

(4) If $S_i = 1$, capacitor C_x is discharged by I_{SECS}, and V_x decreases.

(5) When V_x exceeds V_{min}, S_i turns over.

For low power operation, we designed a circuit in which V_{ba} controls a part of the tail current of the differential pair circuit to optimize the power consumption of the circuit depending on the slope of V_x.

3.4 Circuit Simulation Results

We designed a CBM unit circuit using 0.25 μm CMOS technology and performed a circuit simulation (HSPICE) for the operation of this circuit. The simulation results are shown in Fig. 6. We set the following values: $V_{z0} = 1.65$ V, $V_1 = 1.55$ V, $V_{sub} = 2.2$ V, $V_{ini} = 1.9$ V, $V_{max} = 2.3$ V, $V_{min} = 1.3$ V and $V_{bm} = 0.6$ V. We changed V_{zi} from 1.65 V to 1.95 V and V_T from 0.65 V to 0.75 V. These simulation results verified that the designed circuit operated successfully.

4 Conclusion

We verified the operation of chaotic Boltzmann machines (CBMs) by software simulation, designed a CMOS unit circuit for CBMs, and verified its operation. In future research, we will fabricate a CMOS VLSI chip based on the designed results, and evaluate the chip by measuring the circuit.

Acknowledgments. This work was supported by JSPS KAKENHI Grant Nos. 15H01706 and 15K1211. The circuit design was supported by VLSI Design and Education Center(VDEC), the University of Tokyo in collaboration with Cadence Design Systems, Inc., and Synopsys, Inc.

References

1. Aarts, E., Korst, J.: Simulated Annealing and Boltzmann Machines: A Stochastic Approach to Combinatorial Optimization and Neural Computing. Wiley, New York (1989)
2. Ackley, D.H., Hinton, G.E., Sejnowski, T.J.: A learning algorithm for Boltzmann machines. Cognitive Sci. **9**, 147–169 (1985)
3. Alspector, J., Allen, R.B.: A neuromorphic VLSI learning system. In: Losleben, P. (ed.) Advanced Research in VLSI: Proceedings of the 1987 Stanford Conference, pp. 313–349. MIT Press, Cambridge (1987)
4. Alspector, J., Gannet, J.W., Harber, S., Parker, M.B., Chu, R.: Generating mutiple analog noise sources from a single linear feedback shift register with neural network applications. In: IEEE Proceedings of International Symposium on Circuits and Systems (ISCAS), pp. 1058–1061 (1990)
5. Alspector, J., Gannet, J.W., Harber, S., Parker, M.B., Chu, R.: A VLSI-efficient technique for generating multiple uncorrelated noise sources and its application to stochastic neural networks. IEEE Trans. Circ. Syst. **38**, 109–123 (1991)

6. Arima, Y., Murasaki, M., Yamada, T., Maeda, A., Shinohara, H.: A refreshable analog VLSI neural network chip with 400 neurons and 40k synapses. IEEE J. Solid State Circ. **27**, 1854–1861 (1992)
7. Garda, P., Belhaire, E.: An analog chip set for multi-layered synchronous Boltzmann machines. In: International Neural Network Conference, vol. 2, pp. 568–571 (1990)
8. Hinton, G.E.: Deterministic Boltzmann learning performs steepest descent in weight-space. Neural Comput. **1**, 143–150 (1989)
9. Morie, T., Amemiya, Y.: Deterministic Boltzmann machine learning improved for analog LSI implementation. IEICE Trans. Electron. **E76-C**(7), 1167–1173 (1993)
10. Peterson, C., Anderson, J.R.: A mean field theory learning algorithm for neural networks. Complex Syst. **1**, 995–1019 (1987)
11. Salakhutdinov, R., Hinton, G.E.: Deep Boltzmann machines. In: Proceedings of AISTATS, pp. 448–455 (2009)
12. Schneider, C.R., Card, H.C.: Analog CMOS deterministic Boltzmann circuits. IEEE J. Solid State Circ. **28**, 907–914 (1993)
13. Suzuki, H.: Monte carlo simulation of classical spin models with chaotic billiards. Phys. Rev. E **88**, 052144 (2013)
14. Suzuki, H., Imura, J., Horio, Y., Aihara, K.: Chaotic Boltzmann machines. Sci. Rep. **3**, 1610 (2013)
15. Tomberg, J., Raittinen, H., Kaski, K.: VLSI architecture of the Boltzmann machine algorithm. In: International Neural Network Conference, vol. 2, pp. 568–571 (1990)

An Attempt of Speed-up of Neurocommunicator, an EEG-Based Communication Aid

Ryohei P. Hasegawa[✉] and Yoshiko Nakamura

Human Informatics Research Institute, AIST, Umezono 1-1-1 Central 2, Tsukuba,
Ibaraki 305-8568, Japan
r-hasegawa@aist.go.jp

Abstract. We have been developing the "Neurocommunicator", an EEG-based communication aid for people with severe motor disabilities. This system analyzes an event-related potential (ERP) to the sequentially flashed pictograms to indicate a desired message, and predicts the user's choice in the brain. To speed-up of this decoding process, we introduced a special algorithm, the Virtual Decision Function (VDF), which was originally designed to reflect the continuous progress of binary decisions on a single trial basis of neuronal activities in the primate brain. We applied the VDF to the EEG signals, and succeeded in faster decoding of the target.

1 Introduction

Brain-computer/machine interface (BCI/BMI) to provide a direct link between the brain and external devices. There is recent world-wide interest in developing the BMI as advanced assistive technologies, which have the potential to improve the quality of life for individuals with disabilities. Although the recent big movement was motivated by the motor BMI that control prosthetic devices such as a robot arm mainly via invasive multielectrode recording in the brain [13], the cognitive BMI is also expected to support communication for patients with severe motor deficits.

In order to develop the cognitive BMIs, we have been focusing two topics, the decision-making process as well as single trial prediction. It is important to know the decision-making process because it is related to the selection of the internal messages in the brain [4, 6, 9], On the other hand, single-trial prediction of brain activities is also core technique for the real-time control of the external devices [5, 7, 8].

One of the good examples of BMIs is the P300 speller, which is based on the Electroencephalography (EEG) [2, 3]. In the P300 speller system, a user focused on one out of 36 different characters ('target'), which was presented with a 6 by 6 matrix of characters. Six rows and six columns of this matrix were successively and randomly flashed at a specific rate. As one particular row and one particular column contained the target, the row and the column with the target were expected to evoke the stronger P300 responses than those without the target. Therefore, the target was decoded by the combination of the row and the column.

Recently we have extended this technique to develop a cognitive BMI, "Neurocommunicator" (Fig. 1) (reported by a press release at 2010/03/29); Neurocommunicator is

© Springer International Publishing AG 2016
A. Hirose et al. (Eds.): ICONIP 2016, Part I, LNCS 9947, pp. 256–263, 2016.
DOI: 10.1007/978-3-319-46687-3_28

a practical BMI system that interprets the intention in real time based upon EEG data with a small device and enables the users to convey 512 kinds of message (8 * 8 * 8). Although we expect that this system will enables communication by patients with severe motor disabilities, both the accuracy and the speed should be more optimized as the basic specs.

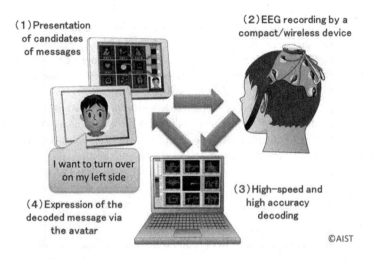

Fig. 1. An EEG-based communication system, Neurocommunicator. This device decodes the user's intentions based upon the brain waves in real time; it can effectively and quickly convey even a long message via his/her CG or Robot avatar.

In this study, we attempted to speed up the decoding of the pictogram that was the "target" in the brain. We have introduced a special algorithm, Virtual Decision Function (VDF). The VDF was originally designed to reflect the continuous progress of binary decisions on a single trial basis of neuronal activities of the superior colliculus in the primate brain. We have applied the VDF to the EEG signals to develop the high-speed version of the Neurocommunicator with the same level of accuracy.

2   Methods

2.1   Behavioral Paradigm

We collected EEG data from 17 normal adult subjects under the protocol approved by the guideline and the committee of our institutes. All subjects were tested in two sessions ('training' and 'test') each. Each session consisted of 8 'games'. In each game, the subject focused attention on one of 8 pictograms ('target') in the matrix, which was prescribed by the investigator. During each game, each pictogram was flashed at 8 Hz, displaying 4 Japanese characters (e.g. "Ko-Re-Ka-Na") with green color [10, 12]. In a block of 8 flashes, all 8 pictograms were selected in a pseudorandom fashion. A block of 8 flashes were consecutively repeated 15 times for 'training' session and 5 times for 'test' session. As completing one block corresponded to 1 s, it took 15 s and 5 s for the training and the test sessions respectively. The subjects had 10 s of resting time between games. The total times were 190 s for the training session and 110 s for the test session. There was 5 min of resting time between sessions.

2.2 Recording

EEG data was obtained by a custom-made recording system, in which a small EEG amplifier was attached on a plastic headgear. The headgear localized the electrode positions around the top of the head; in this study positions of 8 signal electrodes (ID1@FC1, ID2@FC2, ID3@C3, ID4@CZ, ID5@C4, ID6@CP1, ID7@CP2, and ID8@PZ) and one earth (ground) electrode (@CPZ) were selected in the 10 % (10-10) system [1]. A common reference electrode was positioned on a neutral point (earlobe). While conductive gels were used for the 8 signal electrodes and the earth electrode, a disposable electrode with solid gel was used for the reference. Raw EEG data were measured at a sampling late of 256 Hz, bandpass filtered (0.2–30 Hz) and digitized as 16 bit per sample. The digitized data were, in real time, sent to the PC with a wireless transmission method.

2.3 Decoding

In the PC, the original 8 channels of continuous EEG data were downsampled to 21.3 Hz after additional software bandpass filter (1–30 Hz). Then the data were aligned to extract the event-related potential (ERP) associated with the onset of the single flash of each pictogram. As described above, all subjects completed both the training and the test sessions. We performed linear discriminant analysis (LDA) to generate a pattern recognition model after the training session. The optimized LDA model was designated to produce a high score for the target and a low score for the non-target. The test session was conducted to confirm whether the LDA model had enough ability to discriminate the target from the non-targets. The pictogram with the highest total (accumulative) discriminant score was regarded as the target.

In the test session real-time feedback was given to the subject about the prediction of the target on the final (5th) block after each game. After the 8 games of a session, the success rate was computed by dividing the number of the successfully predicted games (0 to 8) by the number of total games (8) in each subject. We mainly focused on the success rate on the final (5th) block of the test session as the index of the accuracy of our system. In order to reveal the progress of the accuracy, we also calculated the success rate not only on all 5 blocks of the test session but also on all 15 blocks of the training session. The success rate of the test session was examined using the single model generated by all data of the training session of each subject. On the other hand, the success rate of the training session was examined by the cross-validation, especially the "leave-one-out" method, in which the prediction of individual games was made using the model generated by the data of remaining games.

In addition to the standard decoding methods above, we have introduced the new algorithm for high-speed decoding, using the virtual decision function (VDF) [7, 8] with some small changes for EEG data such as the use of LDA instead of multiple regression analysis. In order to obtain this function in this study, the accumulative LDA score for each pictogram in each game in the test session was multiplied by the success rate at the corresponding block in the training session (Fig. 2). The post-hoc prediction of the target by the VDF in the test session was made when any of the VDFs reached the threshold, which was set, by the simulation in each subject using the data from the training session so as to keep the same level of the success rate. The advantage of this method is to speed up the decoding by finishing to flash the pictogram when the target was predicted even before the final block.

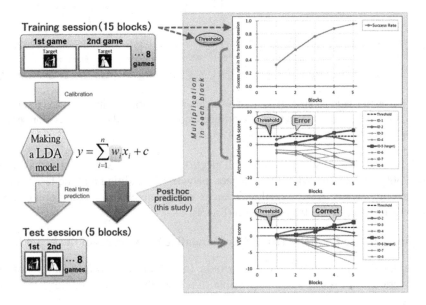

Fig. 2. Schematic drawing of the prediction by the virtual decision function (VDF). The VDF scores (bottom panel) were obtained by multiplication of the success rate (top) and the accumulative LDA scores (middle) in each block. The pictogram with the VDF score that first reached the threshold is regarded as the target (user's choice).

Although this simple idea and the procedure of the calculation could be applied to the test session in real time, we conducted the post hoc analyses with the VDF in order to compere the result by the VDF with the result by the original (currently used) decoding method, in which the pictogram with the highest accumulative LDA score at the final (fifth) block was regarded as the target.

3   Results

3.1   Average EPR Analysis

In this study, it was expected for the flash of the target pictogram to extract the P300, which should be strong enough for the real-time prediction. Therefore, we first made sure whether the P300 was included, at least, in the average ERP. We compared average ERPs between the target and non-target conditions in the training session. The response to the target was typically stronger than that to the non-target, showing a positive peak

Training session   Test session

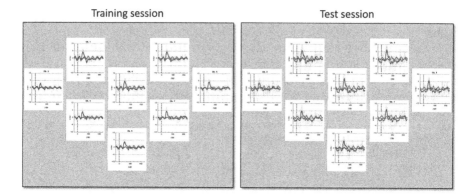

Fig. 3. Event-Related Potentials (ERPs) of a single subject in the training (left) and test (right) sessions. Each panel corresponds to each electrode location. Blue and purple lines represent the response to the target and non-targets respectively.

around 200–400 ms, to some extent, at all electrode positions. Although the waveform of the P300 was similar, within the subject, between the training and the test session, we observed a variety of individual differences among subjects about temporal patterns of waveforms at each electrode location (Fig. 3).

3.2 Single-Game Prediction by the Accumulative LDA Scores of the Fixed Block

We first describe the original decoding method. Instead of the waveform of P300 itself, the converted LDA scores made it possible to predict the target in single games. We

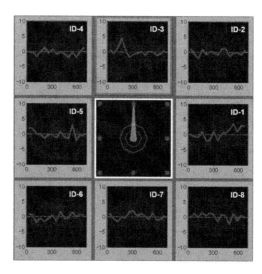

Fig. 4. An example of the decoding in a specific game. The average ERPs of all 8 channels to all 5 flashes for each pictogram (outer panels) and accumulative discrimination scores (central panel) to the pictograms are shown when the pictogram ID-3 was predicted.

compared the accumulative LDA scores among pictograms. In an example of this analysis, the pictogram with the highest LDA score was the correct target (Fig. 4); the target pictogram elicited the P300-like waveform, only showing a positive number. Although the non-target pictograms sometimes showed a positive number too, the correct target was mostly higher than it (if not the prediction was considered to be unsuccessful). The radar chart by LDA scores was useful to visualize how confidently the target was predicted.

It is thought that this method is useful if the appropriate number of the flashes (blocks) to elicit the differential responses between the target and the non-target is consistent. The minimum number to predict the correct target, however, could change game by game.

3.3   Post-hoc Prediction by the VDF

In this study, we attempted to speed up the decoding of the target pictogram using the flexible number of flashing method. For this purpose, we introduced the concept to stop flashing the pictograms depending on the level of confidence of the prediction; we converted the LDA scores to the VDF scores (Fig. 2) as well as determined the level of the threshold to be referred.

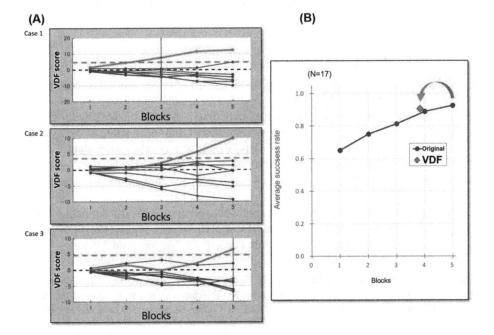

Fig. 5. (**A**) Examples of the prediction by the VDF in specific games. The VDFs for the target (green line) generally reached the threshold (dashed orange line) faster than the non-targets although such a timing was different game by game. (**B**) The reduction of the block made by the VDF prediction. (Color figure online)

We applied this method to 8 games of all 17 subjects. While some predictions were correctly made before the final block (e.g. Fig. 5(A) cases 1–2), others still needed to be waited until the final block (e.g. Fig. 5(A) case 3). While the prediction by the original method resulted in the average success rate of 17 subjects at the fifth block in the test session was 91 %, the VDF made the faster prediction at the 3.84 block, keeping the similar accuracy level (90 %), which corresponded to the 1.16 s reduction (Fig. 5(B)).

4 Discussion

4.1 General

We have been developing our cognitive BMI system, "Neurocommunicator" as a proto-type of a communication aid based upon the EEG. In this study, we administered neural decoding experiments to normal subjects in order to verify the spec of our system as well as our data collection methods, which would be the adjustment processes for individual patients in the future. Our system was originally designed to create a variety of messages by the combinations of pictograms after 3 consecutive predictions of the targets. As a BMI, the neural decoding module was one of the most important aspects in our system. Therefore, we focused on the accuracy of not the 3 consecutive but the single predictions of the target as the indicator of the accuracy. We obtained satisfactory data from the subjects, which encouraged us to start systematic bed-side monitor experiments.

4.2 Speed-up of the Prediction by the VDF

It is useful to obtain the data from the normal subjects before the clinical trials in order to assess our system from the technical aspects independent of specific conditions of individual patients. The success rate of the prediction, especially on the 5th block, was used as the index of the accuracy of our system. In fact, the success rate of our system was about (91 %) in average of 17 subjects, which was far above the chance level (12.5 %). Keeping the similar accuracy level (90 %), the VDF succeeded in speed-up of the decoding from the 5th block to the 3.84 block, which is close to our final goal 95 % at the 3rd block.

We are planning to start systematic monitor experiments for the patients mostly living at home [11]. Toward such a clinical use, these results suggest the VDF might contribute to improve the QOL of the future users of the Neurocommunicator.

Acknowledgments. We thank Drs. Chikashi Fukaya, Tetsuto Minami, Yuki Nakayama and Hiroki Hagiwara for helpful comments, and staff of Neurotechnology research group for technical support. This work was supported by JSPS KAKENHI Grant Number 25293449 and by NEDO Grant Number 15102349-0.

References

1. American Electroencephalographic Society: Guideline thirteen: guidelines for standard electrode position nomenclature. J. Clin. Neurophysiol. **11**, 111–113 (1994). Official Publication of the American Electroencephalographic Society
2. Donchin, E., Spencer, K.M., Wijesinghe, R.: The mental prosthesis: assessing the speed of a P300-based brain-computer interface. IEEE Trans. Rehabil. Eng. **8**, 174–179 (2000). A Publication of the IEEE Engineering in Medicine and Biology Society
3. Farwell, L.A., Donchin, E.: Talking off the top of your head: toward a mental prosthesis utilizing event-related brain potentials. Electroencephalogr. Clin. Neurophysiol. **70**, 510–523 (1988)
4. Hasegawa, R., Sawaguchi, T., Kubota, K.: Monkey prefrontal neuronal activity coding the forthcoming saccade in an oculomotor delayed matching-to-sample task. J. Neurophysiol. **79**, 322–333 (1998)
5. Hasegawa, R.P., Blitz, A.M., Geller, N.L., Goldberg, M.E.: Neurons in monkey prefrontal cortex that track past or predict future performance. Science **290**, 1786–1789 (2000)
6. Hasegawa, R.P., Blitz, A.M., Goldberg, M.E.: Neurons in monkey prefrontal cortex whose activity tracks the progress of a three-step self-ordered task. J. Neurophysiol. **92**, 1524–1535 (2004)
7. Hasegawa, R.P., Hasegawa, Y.T., Segraves, M.A.: Single trial-based prediction of a go/no-go decision in monkey superior colliculus. Neural Netw. **19**, 1223–1232 (2006). The Official Journal of the International Neural Network Society
8. Hasegawa, R.P., Hasegawa, Y.T., Segraves, M.A.: Neural mind reading of multi-dimensional decisions by monkey mid-brain activity. Neural Netw. **22**, 1247–1256 (2009). The Official Journal of the International Neural Network Society
9. Hasegawa, R.P., Matsumoto, M., Mikami, A.: Search target selection in monkey prefrontal cortex. J. Neurophysiol. **84**, 1692–1696 (2000)
10. Minami, T., Inoue, Y., Hasegawa, R.P.: Development of neurocommunicator system: Kansei brain-machine-interface. J. Jpn. Soc. Kansei Eng. **11**, 509–518 (2012)
11. Nakayama, Y., Matsuda, C., Ogura, A., Haraguchi, M., Mochizuki, Y., Nakamura, Y., Hasegawa, R.P.: The usability of an electroencephalography-based communication devices "neuro-communicator" and nursing intervention for severely neurological or neuromuscular disorders. J. Jpn. Intractable Illn. Nurs. Soc. **17**, 187–204 (2013)
12. Takai, H., Minami, T., Hasegawa, R.P.: Efficient methods of presenting visual stimuli for the P300-based cognitive BMI. J. Jpn. Soc. Kansei Eng. **10**, 89–94 (2011)
13. Wessberg, J., Stambaugh, C.R., Kralik, J.D., Beck, P.D., Laubach, M., Chapin, J.K., Kim, J., Biggs, S.J., Srinivasan, M.A., Nicolelis, M.A.: Real-time prediction of hand trajectory by ensembles of cortical neurons in primates. Nature **408**, 361–365 (2000)

Computational Performance of Echo State Networks with Dynamic Synapses

Ryota Mori[1]($\boxtimes$), Gouhei Tanaka[1,2,3], Ryosho Nakane[2], Akira Hirose[2], and Kazuyuki Aihara[1,2,3]

[1] Graduate School of Information Science and Technology,
The University of Tokyo, Tokyo 113-8656, Japan
{mori,gouhei,aihara}@sat.t.u-tokyo.ac.jp
[2] Graduate School of Engineering, The University of Tokyo, Tokyo 113-8656, Japan
nakane@cryst.t.u-tokyo.ac.jp, ahirose@ee.t.u-tokyo.ac.jp
[3] Institue of Industrial Science, The University of Tokyo, Tokyo 153-8505, Japan

Abstract. The echo state network is a framework for temporal data processing, such as recognition, identification, classification and prediction. The echo state network generates spatiotemporal dynamics reflecting the history of an input sequence in the dynamical reservoir and constructs mapping from the input sequence to the output one in the readout. In the conventional dynamical reservoir consisting of sparsely connected neuron units, more neurons are required to create more time delay. In this study, we introduce the dynamic synapses into the dynamical reservoir for controlling the nonlinearity and the time constant. We apply the echo state network with dynamic synapses to several benchmark tasks. The results show that the dynamic synapses are effective for improving the performance in time series prediction tasks.

Keywords: Echo state networks · Reservoir computing · Dynamic synapses · Short-term synaptic plasticity · Time series prediction · Recurrent neural networks

1 Introduction

The echo state network (ESN) is a computational framework for processing time series data, consisting of two parts: the dynamical reservoir (DR) and the readout [1]. The DR is often constructed with sparsely connected recurrent neural networks (RNNs), which play the role to map the input time series into nonlinear spatiotemporal dynamics generated by the DR. The dynamics of the DR is a function of the input history, and therefore, the activation states in the DR contain the information of the input data. The readout is used to make a mapping from the activation states in the DR to the output time series. In the readout, the outputs of the ESN are often created by a linear combination of the activation states of the DR.

The feature of the ESN is that only the connection weights in the readout part are trained. The input connection weights and the internal connection weights

© Springer International Publishing AG 2016
A. Hirose et al. (Eds.): ICONIP 2016, Part I, LNCS 9947, pp. 264–271, 2016.
DOI: 10.1007/978-3-319-46687-3_29

in the DR are all fixed in advance. Therefore, in terms of computation time, the ESN is advantageous compared with the RNNs where all the weights are adjusted [2]. Particularly when the readout is a linear transformation, it is easy to obtain the weights that minimize the difference between the network output sequence and the desired output sequence by any linear regression method.

The ESNs have been successfully applied to a variety of tasks such as time series prediction, system identification, system control, adaptive filtering, noise reduction, function generation, and pattern classification. However, the ESN with the neuron-based reservoir is not good at dealing with slowly changing time series data whose time constant is smaller than that of the neuron unit. A time delay is required to handle the slow dynamics in the sample data. Although a delay line can be realized by connecting neurons in a chain in an unidirectional way, more time delay requires more neurons. Here, to change the nonlinearity and the time constant of the DR in another way, we introduce dynamic synapses into the DR.

Dynamic synapse, also called short-term synaptic plasticity, refers to the synapse in which the efficiency of synaptic transmission changes transiently due to the changes in the calcium concentration and the release of neurotransmitters [3]. The short-term synaptic plasticity persists for only several hundred milliseconds. Mongillo et al. theoretically showed that short-term facilitation in the prefrontal cortex is implicated in working memory [4]. Their simulations indicate that a population activity can be reactivated by weak nonspecific excitatory inputs as long as the synapses remain facilitated. Thus, dynamic synapses can store the history of the past neural activities for about one second as the changes in the synaptic transmission efficiency. Therefore, they can process information in accordance with the past neural activity. Hence dynamic synapses may play an important functional role in time series processing in the brain.

In this study, we propose the ESN with dynamic synapses and investigate its computational performance in time series prediction and memory capacity. We perform numerical experiments on several benchmark tasks to evaluate the effectiveness of the dynamic synapses in information processing. This research is important for understanding the dynamical behavior of the brain as well as realizing bio-inspired / energy-efficient information processing systems.

2   Methods

2.1   Models

(1) **Echo State Network.** In this study, we use the ESN with dynamic synapses. Our ESN consists of K input layer neurons $\mathbf{I}(t) = [I_i(t)]_{1 \leq i \leq K}$, randomly connected N hidden layer neurons $\mathbf{H}(t) = [h_i(t)]_{1 \leq i \leq N}$ (dynamical reservoir), and L output layer neurons $\mathbf{Y}(t) = [y_i(t)]_{1 \leq i \leq L}$. The $N \times K$ input weight matrix $\mathbf{W}^{in} = [w_{i,j}^{in}]$ is created as an uniform random matrix in the range $[-0.5, 0.5]$, and the $N \times N$ internal weight matrix $\mathbf{W} = [w_{ij}]$ is created as an uniform sparse random matrix in the range $[-0.5, 0.5]$. $\mathbf{W}$ is normalized by the spectral radius represented as $\alpha_{sd}(< 1)$. We set the sparseness of the internal

weight matrix $\mathbf{W}$ at 15 %. The reservoir state $\mathbf{H}(t)$ is generated by the following difference equation:

$$\mathbf{H}(t+1) = \mathbf{f}(\mathbf{W}^{in}\mathbf{I}(t+1) + \mathbf{W}\mathbf{H}(t)), \tag{1}$$

where $\mathbf{f}$ denotes the component-wise application of the unit's activation function f. We use the sigmoid function $f(s) = 1/(1 + \exp(-s))$, the hyperbolic tangent $f(s) = \tanh(s)$, and the linear function $f(s) = s$. The output $\mathbf{Y}(t)$ is generated by the following difference equation:

$$\mathbf{Y}(t+1) = \mathbf{f}^{out}(\mathbf{W}^{out}\mathbf{H}(t)), \tag{2}$$

where $\mathbf{W}^{out}$ is an $L \times (K + N + L)$ output weight matrix calculated by using input-output training data pairs [2]. We use the output activation function $f^{out}(s) = \tanh(s)$.

(2) Dynamic Synapse. The short-term plasticity of dynamic synapses is caused by quantitative alteration of the releasable neurotransmitters and the calcium concentration [5,6]. The dynamics of dynamic synapses are described by the following two equations for the variables x_i representing the ratio of the releasable neurotransmitters and u_i representing the calcium concentration of neuron $i(i = 1, ..., N)$:

$$x_i(t+1) = x_i(t) + \frac{1 - x_i(t)}{\tau_D} - x_i(t)u_i(t)h_i(t), \tag{3}$$

$$u_i(t+1) = u_i(t) + \frac{U_{se} - u_i(t)}{\tau_F} + U_{se}(1 - u_i(t))h_i(t), \tag{4}$$

where τ_D and τ_F are time constants for the dynamics of x_i and u_i, respectively. If no action potential comes to the presynaptic terminal, x_i and u_i recover to the steady state level 1 and U_{se}, respectively. Here, the efficiency of synaptic transmission is proportional to $x_j(t)u_j(t)$. Therefore, when we innovate dynamic synapses, the strength of the connection from the jth neuron to the ith neuron is redefined as $D_{ij}(t) = w_{ij}x_j(t)u_j(t)/U_{se}$. This standard dynamic synapse model requires $0 < h_i$. Therefore, this model cannot use tanh units and linear units, and instead we use the sigmoid units in this model.

The standard dynamic synapse model reproduces faithfully the experimental results. However, this model is not necessarily suitable for the neural network. For example, (1) this dynamic synapse model is not superset of the static synapse. (2) Dynamics of synaptic efficacy is not monotonic. (3) Dynamic synapse is complex because it includes two variables. To solve these problem, I devised a new univariate dynamic synapse model as follows:

$$e_i(t+1) = e_i(t) + \frac{1 - e_i(t)}{\tau_i} + a_i(e_i(t) - m_i)(M_i - e_i(t))h_i(t), \tag{5}$$

where $e_i(t)$ is the synaptic efficacy of neuron i, and a_i determines the rate of change of synaptic efficacy, m_i and M_i are the minimum and maximum values of synaptic efficacy, respectively ($0 < m_i < 1$, $1 < M_i$). Now w_{ij} can be described by $D_{ij}(t) = w_{ij}e_j(t)$. In this model, if $a_i > 0$, then the short-term facilitation occurs. If $a_i < 0$, then the short-term depression occurs. If $a_i = 0$, then synapses become static. This model can use tanh units.

2.2   Tasks

(1) Memory Capacity. In order to evaluate the short-term memory capacity of ESN with dynamic synapses, we calculated the Memory Capacity (MC) [2,7] of the network. We consider an ESN with a single input unit $I(t)$ and many output units $\{y_k(t); k = 1, 2, ...\}$. The input $I(t)$ is a random signal generated by sampling from a uniform distribution in the interval $[-0.5, 0.5]$. Training signal $d_k(t) = I(t-k)$ are delayed versions of input signal $I(t)$. Memory Capacity (MC) of an ESN is defined as follows:

$$\text{MC} = \sum_{k=1}^{\infty} \max_{\mathbf{w}^{out}} r^2(I(t - k), y_k(t)), \tag{6}$$

where

$$r^2(I(t - k), y_k(t)) = \frac{cov^2(I(t - k), y_k(t))}{\sigma^2(I(t - k))\sigma^2(y_k(t))} \tag{7}$$

is the determination coefficient (*cov* denotes the covariance and σ^2 denotes the variance).

(2) NARMA 10 Time Series. NARMA (Nonlinear autoregressive moving average) is a generalized version of the autoregressive moving average model, where the regression is nonlinear. NARMA is often used in many studies to evaluate the performance of time series processing of RNNs. The NARMA 10 time series, which includes 10 steps time lag, is generated by the following recurrence relation:

$$y(t) = \alpha y(t - 1) + \beta y(t - 1) \sum_{i=1}^{n} y(t - i) + \gamma I(t - n)I(t - 1) + \delta, \tag{8}$$

where $\alpha = 0.3$, $\beta = 0.05$, $\gamma = 1.5$, $\delta = 0.1$, $n = 10$ [8]. The input $I(t)$ is a signal generated by randomly sampling from a uniform distribution in $[0, 0.5]$. The task is to predict $y(t)$ from $I(t)$.

(3) NARMA 20 Time Series. The NARMA 20 time series includes 20 steps time lag. This is a more difficult task than the NARMA10 because of the longer history. The NARMA 20 has an additional nonlinear transformation by tanh to confine the signal in a finite range. The NARMA 20 is generated by the following recurrence relation:

$$y(t) = \tanh \left(\alpha y(t - 1) + \beta y(t - 1) \sum_{i=1}^{n} y(t - i) + \gamma I(t - n)I(t - 1) + \delta \right), \tag{9}$$

where $n = 20$ and the rest of conditions are the same as NARMA 10.

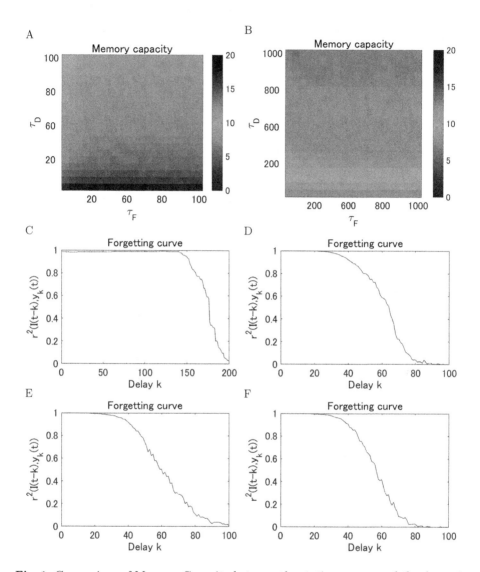

Fig. 1. Comparison of Memory Capacity between the static synapse and the dynamic synapse. We set $N = 400$ and $\alpha_{sd} = 0.99$. We use the standard dynamic synapse model in A–B, and the univariate dynamic synapse model in C–F. **A**, **B**. Dependence of Memory Capacity on the time constant. We ran 20 trials with different initial weights, and we calculate the average value. We use the standard dynamic synapse model **C**. Forgetting curves of linear DR with static synapses. **D**. Forgetting curves of the tanh DR with static synapses. **E**. Forgetting curves of the linear DR with dynamic synapses. **F**. Forgetting curves of the tanh DR with dynamic synapses.

3   Results

First, the memory capacity of the ESN with dynamic synapses was evaluated in comparison with the ESN without static synapses. In this task, we set $\alpha_{sd} = 0.99$, $N = 400$, and $w_{ij}^{in} \in [-0.1, 0.1]$. The activation function of the neuron units in the dynamic reservoir is given by the linear type, the tanh type or the sigmoid type. Figure 1 shows the results of the numerical experiments on the memory capacity task. Figures 1A and B show how the memory capacity depends on the delay parameters τ_D and τ_F in the dynamic synapses when using sigmoid units. These results show that a smaller value of τ_D gives a larger memory capacity and the value of τ_F is not influential on the capacity. Figures 1C–F show the forgetting curves which indicate how much input history can be embedded in the spatiotemporal dynamics of the dynamic reservoir. As the length of the delay k is increased, the determination coefficient tends to decrease. In this task, we use the univariate dynamic synapse model. We set $a_i \in [0, 0.1]$ (random) , $\tau_i \in [1, 20]$ (random), $m_i \in [0, 0.1]$ (random), and $M_i \in [1, 4]$ (random). Figures 1C and D for linear units show that the memory capacity is much decreased by introducing the dynamic synapses. This means that the linearity of the original dynamic reservoir, which is favorable for the transmission of the input data without transformation, is lost by the nonlinearity of the dynamic synapses. In the case of tanh units, the decrease in the memory capacity is relatively small

Table 1. Comparison of the performance between the static synapse and the dynamic synapse. We ran 20 trials with different initial weights, and we calculate the average value and the standard deviation. In these tasks, we use the univariate dynamic synapse model. **A.** Memory Capacity of the ESN. We set $\alpha_{sd} = 0.99$, $N = 400$, $a_i \in [0, 0.1]$ (random), $\tau_i \in [1, 20]$ (random), $m_i \in [0, 0.1]$ (random), and $M_i \in [1, 4]$ (random). **B.** NARMA task. We set $\alpha_{sd} = 0.8$, and use the tanh units. The rest of conditions are the same as those in A.

A

task	activation function	static synapse	dynamic synapse
MC	linear DR	159.3380 ± 8.1009	49.8540 ± 3.0186
	tanh DR	56.1503 ± 2.9915	48.9405 ± 4.7646

B

task	measure		static synapse	dynamic synapse
NARMA10	NRMSE	training	0.13920.0220	0.0768±0.0155
		testing	0.2452±0.0294	0.1541±0.0267
	RMSE	training	0.0204±0.0033	0.01080.0020
		testing	0.0359±0.0049	0.0218±0.0050
NARMA20	NRMSE	training	0.3511±0.0641	0.2997±0.0743
		testing	0.5514±0.0503	0.4966±0.0842
	RMSE	training	0.0148±0.0028	0.0127±0.0033
		testing	0.0232±0.0020	0.0210±0.0040

as shown in Figs. 1E and F. Table 1A summarizes the statistical results for 20 trials in the above four cases. Overall, the dynamic synapses are not effective for improving the memory capacity due to their highly nonlinear property.

Next, NARMA time series prediction performance of the ESN with dynamic synapses was evaluated in comparison with the ESN without static synapses. In this task, we use the univariate dynamic synapse model. We set $\alpha_{sd} = 0.8$, $N = 400$, $w_{ij}^{in} \in [-0.3, 0.3]$, $a_i \in [0, 0.1]$ (random) , $\tau_i \in [1, 20]$ (random), $m_i \in [0, 0.1]$ (random), and $M_i \in [1, 4]$ (random). Table 1B summarizes the statistical results for 20 trials. In the NARMA 10 task, we normalize output to range $[-0.5, 0.5]$. As a result, we found that the dynamic synapse reduce the prediction error about 39.3–47.1 % in NARMA 10 and about 9.5–14.2 % in NARMA 20. Figure 2 shows the time series of training signals (NARMA 10, NARMA 20) and output signals.

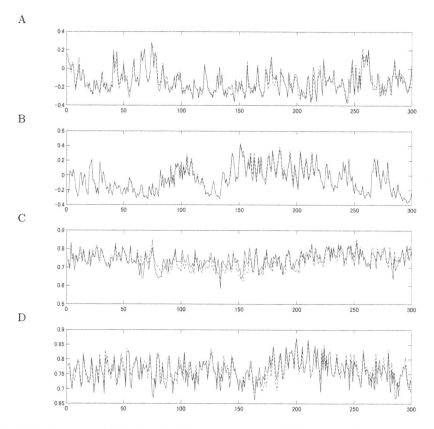

Fig. 2. Training signal (the dotted line) vs Output signal (the solid line). The activation function of the internal units is given by $f = \tanh$. **A.** NARMA10 time series with the static synapses. **B.** NARMA10 time series with the dynamic synapses. **C.** NARMA20 time series with static synapses. **D.** NARMA20 time series with the dynamic synapses

4   Conclusion

We have proposed an echo state network incorporating dynamic synapses which can change the nonlinearity and the time constant of the dynamic reservoir. Numerical experiments were performed to evaluate the effect of dynamic synapses on the computational ability of the echo state network. In the memory capacity task, the dynamic synapses are not effective for improving the performance. This is because the linear dynamics, which is advantageous for the memory capacity task, is broken by the dynamic synapses. In the time series prediction tasks with NARMA 10 and NARMA 20, the dynamic synapses can reduce the prediction error. These tasks require highly nonlinear dynamics, which can be brought about by the dynamic synapses. In this way, the dynamic synapses are suited for relatively difficult tasks that require a dynamical reservoir generating highly nonlinear dynamics. Further numerical experiments for other tasks, however, are necessary to fully reveal the effect of the dynamic synapses on computational ability of the echo state network. In particular, the prediction of time series with slow dynamics is an interesting task to understand how dynamic synapses control the time constant.

Acknowledgments. This work was partially supported by JSPS KAKENHI Grant Number 16K00326 (GT), 26280093 (KA).

References

1. Jaeger, H.: The "echo state" approach to analysing and training recurrent neural networks. Technical Report 148, GMD - German National Research Institute for Computer Science (2001)
2. Jaeger, H.: Tutorial on training recurrent neural networks, covering BPPT, RTRL, EKF and the "echo state network" approach. GMD-Forschungszentrum Informationstechnik (2002)
3. Markram, H., Tsodyks, M.: Redistribution of synaptic efficacy between neocortical pyramidal neurons. Nature **382**, 807–810 (1996)
4. Mongillo, G., Barak, O., Tsodyks, M.: Synaptic theory of working memory. Science **319**, 1543 (2008)
5. Tsodyks, M., Markram, H.: The neural code between neocortial pyramidal neurons depends on neurotransmitter release probability. Proc. Natl. Acad. Sci. USA **94**, 719–723 (1997)
6. Tsodyks, M., Markram, H.: Differential signaling via the same axon of neocortical pyramidal neurons. Proc. Natl. Acad. Sci. USA **95**, 5323–5328 (1998)
7. Jaeger, H.: Short term memory in echo state networks. GMD-Report 152, German National Research Institute for Computer Science (2002)
8. Goudarzi, A., Banda, P., Lakin, M.R., Teuscher, C., Stefanovic, D.: A Comparative Study of Reservoir Computing for Tenporal Signal Processing, arXiv:1401.2224v1 [cs.NE] (2014)
9. Schrauwen, B., Verstraeten, D., Van Campenhout, J.: An overview of reservoir computing: theory, applications and implementations. In: Proceedings of the 15th European Symposium on Articial Neural Networks, pp. 471–482 (2007)

Whole Brain Architecture: Toward a Human Like General Purpose Artificial Intelligence

Whole Brain Architecture Approach Is a Feasible Way Toward an Artificial General Intelligence

Hiroshi Yamakawa$^{(\boxtimes)}$, Masahiko Osawa, and Yutaka Matsuo

Dwango Artificial Intelligence Laboratory, DWANGO Co., Ltd., Kyoto, Japan
ailab-info@dwango.co.jp

Abstract. In recent years, a breakthrough has been made in infant level AI due to the acquisition of representation, which was realized by deep learning. By this, the construction of AI that specializes in a specific task that does not require a high-level understanding of language is becoming a possibility. The primary remaining issue for the realization of human-level AI is the realization of general intelligence capable of solving flexible problems by combining highly reusable knowledge. Therefore, this research paper explores the possibility of approaching artificial general intelligence with such abilities based on meso-scopic connectome.

Keywords: Artificial general intelligence · Computational neuroscience · Biologically inspired cognitive architecture

1 Introduction: Deep Learning Realize Infant Level AI

Conventional AI had strength in intelligence tasks in which adults excel, such as planning and logical inference. In contrast, it was weak in nonlinguistic intelligence such as pattern recognition and motion generation, which could be performed by a three-year old child. Because of this disparity, representations for pattern recognition had to be designed by hand in conventional machine learning technology.

However, in 2012, deep learning [1] in visual information processing over-whelmed conventional technology, and was surpassing humans by 2015. The basis of this change is the possible automatic acquisition of representation from data by deep learning technology. For example, in the Google cat that became famous in the earliest state, the representation for identifying typical cats was generated inside by making the deep learning device read a large volume of data. This representation learning technology greatly resolved the frame issue, symbol grounding issue, and intelligence acquiring bottleneck, which were long-standing problems in AI.

By realizing such "child-like AI," AI became capable of human-like image recognition and improved skillful control. Thus, AI research that is aligned with human development became possible. Current AI is widely realized for specialized learning

The Whole Brain Architecture Initiative, a specified non-profit organization.

© Springer International Publishing AG 2016
A. Hirose et al. (Eds.): ICONIP 2016, Part I, LNCS 9947, pp. 275–281, 2016.
DOI: 10.1007/978-3-319-46687-3_30

already, except for high-level understanding of natural language. However, the versatility for solving various problems by combining multiple types of heterogeneous knowledge has not yet been realized, and this artificial general intelligence (AGI) is described in the next chapter. The whole brain architecture (WBA) approach tries to achieve AGI by using the knowledge of neuroscience that had made significant progress in recent years. Non-profit organizations and Youngers group have been established for promoting research on WBA approach.

2 Artificial General Intelligence [2]

The modern practical AI is a specialized AI that behaves intelligently in individual task areas. In contrast, AGI is the AI that acquires intelligence in a wide range of problem areas autonomously and flexibly solves problems. Therefore, it is an AI that can solve problems beyond the assumptions at the time of design.

Consequently, versatility is the biggest issue toward the realization of human-level AI. Therefore, AGI is a critical technology goal. However, like humans, it does not mean that AGI possesses universal intelligence from the beginning.

2.1 Impact of AGI

As the realization of AGI nears, AI can be designed in various problem areas at a low cost, and it surpasses specialized AI. For example, designing intelligence for work sites such as for a convenience store clerk is too cluttered for a human designer.

Moreover, if AGI had sufficient autonomy, it would consider an enormous amount of hypotheses including failure as if it were playing, and it would perform various trials to the outside world. By supporting the problem solving ability for exceptional situations for which it cannot be designed, the knowledge of the outside world that was acquired through play would obtain robustness and resilience as intelligence. For example, this would be a useful technology for service robots performing domestic chores in various environments such as in a household.

The ability to expand knowledge through play is the basis for creativity. This ability contributes to artistic activities, new business planning, and scientific technology development. If such AI can understand the world, it will help innovate medical technologies and solve global problems of humanity.

2.2 Cognitive Architectures

AGI research is mainly based on the technology of cognitive architecture, which is a static design that draws the placement of components that constitute the intelligent agent, so it is not a part of learning technology. Then, it implements the overall real-time behavior from perception to action.

What is especially important for the realization of AGI is providing various functions through a dynamic combination (interaction) of highly reusable components, enabling some kind of action for unexpected or unknown situations, and having versatility as a result.

3   Whole Brain Architecture (WBA) Approach

In conventional cognitive architecture research, design and evaluation were done based on human behavior. However, the approach from cognitive architecture (BICA: Biologically Inspired Cognitive Architectures) [3] that takes hints from living creatures has been reexamined, since the knowledge in neuroscience has increased rapidly in recent years.

3.1   WBA Approach Become Feasible

Based on such a technological background, the WBA approach was proposed as a strong new course for the realization of AGI. The source hypothesis is the following "Whole Brain Architecture Centric Hypothesis":

   The brain combines modules, each of which can be modeled with a machine learning algorithm, to attain its functionalities, so that combining machine learning modules in the way the brain does enables us to construct a generally intelligent machine with human-level or super-human cognitive capabilities.

Using this hypothesis, research and development in a WBA approach is divided into the following two tasks shown in the Fig. 1:

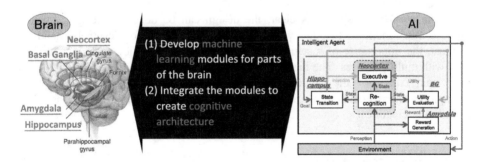

Fig. 1. Whole brain architecture/approach

(1) Developing machine learning modules having the function of brain organs, and
(2) Combining the machine learning modules into a cognitive architecture.

   Until recently, there was a technological difficulty in both tasks, but it has become an increasingly possible approach in recent years because of the following background:

(1) The neocortex has a large role in the versatility of the brain; however, researchers have had success with deep learning studies in recent years, which could be seen as a model for neocortex.

(2) Overall brain connection manner (connectome) can be seen as a model for constructing the cognitive architecture for the flexible combination of machine learning [5]. A connectome includes static nerve connection structure [6] and also dynamic nerve activity connectivity during task execution and resting states.

If the cognitive architecture based on a mesoscopic connectome can be constructed, then this will advance research on the realization of AGI capable of flexible problem solving.

3.2 Why It Is Beneficial to Learn from the Brain

The following reasons outline why it is beneficial to learn from the brain in AGI construction:

- We finally reach the AGI by modeling more detail as needed;
- We can create AI with a high affinity with people;
- We can validate the direction of research;
- We can deal with the difficulty with a general purpose technology design;
- We can glean hints about unsolved issues;
- We can promote collaboration; and
- We can use it as a trunk for aggregating knowledge.

Each of these viewpoints is important, but it is important to highlight the "difficulty of general purpose technology design." There are three methods for designing a system: based on the purpose decomposition, repeating prototypes (or machine learning), and mimicking creatures.

If the purpose of system is clear, the design can be done by breaking up that purpose into functions. In other words, specialized AI is fit for design by hand. In contrast, the importance of machine learning and creature mimicking is increased in AGI with multi purposes. The basic architecture of deep learning has taken hints from the visual cortex of the brain during early development. Therefore, before the asymptotic improvement by engineering, there often exist breakthrough ideas taken from living creatures.

3.3 Research Map of AGI

The organizations working on the realization of AGI can be largely categorized into four quadrants based on two axes: whether they aim to recreate the overall brain function and whether they aim to seek biological plausibility (Fig. 2). Recently, the first quadrant (shown in the upper right in Fig. 2) became crowded with organizations aiming to build an entire brain based on neuroscientific knowledge. Our WBA approach also sits in this position.

Shown in this upper right quadrant, DeepMind made their debut with a research project called the Deep Q Network in which an algorithm that combined deep learning and reinforcement learning is made to learn the play of an arcade game. This technique is applied to Go game and beats world-class player Lee Se-dol March 2016.

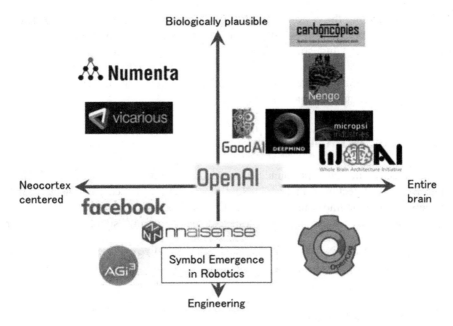

Fig. 2. World developer map of AGIs

These machine-learning technologies have corresponding regions in the brain, and have a good affinity with the WBA approach. Additionally, the CEO of DeepMind, Mr. Hassabis, is interested in the modeling of the brain and is aiming for the realization of AGI. DeepMind is releasing the research results (such as the algorithm) in the form of research paper. GoodAI, a Czech company founded in 2015, aims to realize WBA-like AGI, and they are releasing their tools. The Centre for Theoretical Neuroscience at the University of Waterloo in Canada is developing and releasing the whole brain simulator Nengo[1] [5]. The NPO carboncopies is targeting the reverse engineering of the brain by whole brain emulation. Compared to carboncopies and Nengo, the WBA approach does not necessarily seek strong conformity with biological truth.

In the upper left quadrant of Fig. 2, Numenta and Vicarious are companies trying to create learning machines that learn from the cerebral cortex, but they do not handle the overall brain. However, Numenta has a method of learning from the cerebral cortex, which is a useful reference, and is expected to partially release their algorithm. In the lower right quadrant, OpenCog does not take the approach of referencing the brain, but instead has an open AGI research approach that takes the same stance as WBAI. The NPO OpenAI established December 2015 as research company to build AI beneficial for humanity.

In Japan, new AI projects are being led by the Ministry of Economy, Trade, and Industry; the Ministry of Education, Culture, Sports, Science, and Technology; and the Ministry of Internal Affairs and Communications. Ichisugi, one of the advocates of

---

[1] http://www.nengo.ca/.

WBAI, is conducting the research of BESOM, a machine learning device that learns from the cerebral cortex, at the Artificial Intelligence Research Center of the National Institute of Advanced Industrial Science and Technology. He aims for the gradual, long-term realization of AGI.

As demonstrated by these examples in Japan, alone, research and development that targets AGI has expanded rapidly in the last few years.

4 Promoting WBA Approach

In August 2015, the authors founded the specified nonprofit corporation "WBAI: Whole Brain Architecture Initiative."[2] Now, we believe that a co-creation of 10,000 engineers on brain architecture is a democratic and fastest AGI development scenario, and promote it by indicating direction and by preparing supportive tools for developments.

In addition to this, a youth WBA organization called WBA Future Leaders, which aims to support the future of WBA, was founded a year earlier.

4.1 WBA Future Leaders[3]

WBA Future Leaders was established in August 2014 as an organization that cooperates in WBA study groups. As a community for everyone who is interested in the brain, artificial intelligence, and their effects, two targets are proposed:

(1) Create a place where the technology for modeling the vast knowledge related to WBA can be learned systematically, and
(2) Think about the ideal form of the "coexistence of human and artificial intelligence" that can be widely understood, and heighten the interpretability of artificial intelligence in society.

The main activities performed so far focus on two-hour study groups hosted every 1–2 months with a set lecturer and theme. Lecturers selects themes related to AI and neuroscience regardless of their expertise, and present to the group after a 2–5 month preparation period. The casual talk is hosted every six months, where ideas and individual interests can be offered in a short presentations. A diverse group of people gathers at the exchange meeting focused on the "brain and AI," and they are encouraged to engage in candid discussions regardless of their professional positions. The first WBA Hackathon was co-hosted with the NPO WBAI in September 2015, and another is planned for this year.

The participants include students (from high school, college, and graduate school), researchers (post-doctoral, research lab staff, and faculty members), business people, entrepreneurs, and creators. As of July 2016, there were more than 1,100 participants in the related Facebook group. Through these activities, participants can be matched with

---

[2] http://wba-initiative.org/.

[3] http://wbawakate.jp/.

young researchers, corporations, and other parties interested in contributing to the research and development of the WBA approach.

5 Conclusions

This research paper described the feasibility, advance point, positioning, and driving force of AGI development based on the WBA approach. Today's acceleration of AI technology makes the AGI researches level more realistic soon. So we should promote our development scenario speedily, and we should also consider various contribution and impacts of AGI to humanity from technological view.

The vision of this program is that "Let's build the AGI, that enhance our future happiness, along with a number of engineers and citizens based on understanding of human."

Acknowledgements. Thanks to all members, advisors and supporters of the WBAI and the various members of the WBA Future Leaders.

References

1. Yann, L., Bengio, Y., Hinton, J.: Deep learning. Nature **521**(7553), 436–444 (2015)
2. Goertzel, B.: Artificial general intelligence: concept, state of the art, and future prospects. J. JSAI **29**(3), 228–233 (2014)
3. Goertzel B., et. al.: A world survey of artificial brain projects, Part II: Biologically inspired cognitive architectures. Neurocomputing **74**, 30–49 (2010)
4. Oh, S.W., et al.: A mesoscale connectome of the mouse brain. Nature **508**, 207–214 (2014)
5. Eliasmith, C.: How to Build a Brain: A Neural Architecture for Biological Cognition. Oxford Series on Cognitive Models and Architectures (2013)
6. Petersen, S.E., Sporns, O.: Brain networks and cognitive architectures. Neuron **88**(1), 207–219 (2015)

Learning Visually Guided Risk-Aware Reaching on a Robot Controlled by a GPU Spiking Neural Network

Terence D. Sanger[✉]

Department Biomedical Engineering, University of Southern California,
Los Angeles, CA 90089, USA
terry@sangerlab.net

Abstract. Risk-aware control is a new type of robust nonlinear stochastic controller in which state variables are represented by time-varying probability densities and the desired trajectory is replaced by a cost function that specifies both the goals of movement and the potential risks associated with deviations. Efficient implementation is possible using the theory of Stochastic Dynamic Operators (SDO), because for most physical systems the SDO operators are near-diagonal and can thus be implemented using distributed computation. I show such an implementation using 4.3 million spiking neurons simulated in real-time on a GPU. I demonstrate successful control of a commercial desktop robot for a visually-guided reaching task, and I show that the operators can be learned during repetitive practice using a recursive learning rule.

Keywords: Stochastic control · Spiking network · Reaching · Optimal feedback control · Stochastic Dynamic Operators · Risk-aware control

1 Introduction

Robot reaching in a laboratory or manufacturing environment has many known solutions. However, if the robot kinematics or dynamics are unknown, if there are potential random or malicious perturbations to the movement, if sensory data are unreliable, or if there are fixed or moving high-risk or delicate areas that must be avoided, then current methods are rarely effective. Such problems are effortlessly solved by humans, and this leads us to ask whether simple approximately-correct algorithms exist that can be implemented efficiently on neural-like hardware.

Risk-aware control theory describes how the dynamics of human movement change when the risk of movement changes [4]. In particular, humans will modify their trajectory, response to perturbation, baseline stiffness, and muscle co-contraction in response to perceived changes in risk, where risk is defined as the product of probability of error and cost of error [1,2]. The response is primarily involuntary, although it can be modified by training. Imagine for example, the difference in the kinematics of walking while on a narrow path on a flat road compared to walking on a narrow and high bridge without a handrail.

© Springer International Publishing AG 2016
A. Hirose et al. (Eds.): ICONIP 2016, Part I, LNCS 9947, pp. 282–289, 2016.
DOI: 10.1007/978-3-319-46687-3_31

The theory of Stochastic Dynamic Operators (SDOs) provides a mechanism by which risk-aware control can be implemented. SDOs describe the behavior of stochastic nonlinear optimizing systems when state variables are replaced by probability densities and desired trajectories are replaced by cost functions. SDOs have the particularly useful property that the effect of multiple control inputs is a linear combination of the individual SDOs, and thus flexible control of complicated partially unknown approximate systems in uncertain environments can be implemented using linear operations [3]. SDOs for physical systems are always near-diagonal, which means that if the probability densities for state variables are represented by coarse-coded population behavior, then there will be very little nonlocal interaction. This permits an efficient distributed representation of SDOs for complex systems.

Here I will demonstrate such a system, using it to control visually-guided reaching on a desktop robot. To make the problem more realistic, a variety of visually-identifiable non-target objects will also be visible, including objects that are dangerous or painful to touch, and others that are so delicate that compliance needs to be controlled to avoid damaging contact. I show that a distributed spiking neural network implemented on GPU hardware is able to control the robot in realtime in order to reach toward identified objects while avoiding potentially dangerous obstacles in its path. Furthermore, I show that the network can be learned recursively from multiple samples of reaching behavior.

2   Theory

2.1   Stochastic Dynamic Operators

For a large class of stochastic physical systems, dynamics can be described by the equation:

$$\frac{\partial p(x,t)}{\partial t} = Lp(x,t) \tag{1}$$

where x is the underlying (and only approximately known) state variable, $p(x,t)$ is the time-varying probability density of x, and L is a linear operator. We can write this more simply as $\dot{p} = Lp$. For example, the well-known Fokker-Planck equation describing a drift/diffusion process is given by Eq. 1 where

$$Lp = -a\frac{\partial p}{\partial x} + b^2 \frac{\partial^2 p}{\partial x^2} \tag{2}$$

with drift rate a and diffusion rate b. For physical systems, there is an equivalent Ito stochastic differential equation of the form $dx = f(x,t)dt + g(x,t)dw$, where f and g are often nonlinear.

When $a(x)$ depends on x, the rate of flow or the direction of flow can be different for different values of x. In particular, it is possible to represent stable dynamics with convergence to x_0 if $a(x) > 0$ for $x < x_0$, and $a(x) < 0$ for $x > x_0$. Oscillatory behavior occurs only for second-order systems, which in this formulation would mean that there are coupled dynamics for x and $\dot{x}$ so that $\dot{p}(x, \dot{x}, t) = Lp(x, \dot{x}, t)$.

2.2   Structure of Operators

Since real physical systems have state variables $x(t)$ that are continuous and cannot make instantaneous jumps, $\dot{p}(x, t)$ will be nonzero only near values of x for which $p(x, t)$ is nonzero. Therefore the operators L are "diagonal" in this sense. When x is discretized (the underlying state is discrete or the digital representation is quantized), then L admits a matrix representation and the matrix will have nonzero values only near the diagonal. For example, consider the dynamics described by

$$\dot{p} = -a\frac{\partial p}{\partial x} \tag{3}$$

which describes a deterministic unidirectional movement at rate a. The matrix representation for discrete state x will have $-a$ above the diagonal, 0 on the diagonal, and a below the diagonal for each column. The differential operator will be the limit as the discretization becomes progressively finer, which is the derivative of a delta function centered on the diagonal for each column.

2.3   Control of Dynamics

If we wish to introduce a control vector u, we can do so by parametrizing $L(u)$. Because L is a linear operator even when the underlying stochastic dynamics are nonlinear, it is possible to construct superposition systems using a weighted average of different dynamics:

$$\dot{p} = \left(\sum_i u_i L_k\right) p \tag{4}$$

and this provides a mechanism for control. In particular, the vector $< u_0 \ldots u_N >$ is a control vector that determines the superposition. This is a parametrization $L(u) = \sum u_i L_i$. When $u(t)$ is time-varying, it can be shown that the set of short-time achievable dynamics is given by:

$$p(x, t + \Delta t) = e^{M\Delta t}p(x, t) \tag{5}$$

for short intervals Δt, where M is an element of the Lie algebra generated by the set $\{L_i\}$.

Note that even when each member of the set of operators L_i is unstable, stable dynamics can be achieved by a state-dependent superposition $u(x)$. For example, if L_1 has $a > 0$, L_2 has $a < 0$ and $u(x) =< u_1(x), u_2(x) >$ is such that $u_1(x) = 1$ for $x < x_0$ and 0 otherwise, and $u_2(x) = 1$ for $x > x_0$ and 0 otherwise, then the state-dependent superposition dynamics $L = \sum u_i(x)L_i$ is equivalent to a single operator with stable dynamics around x_0. When $u(x)$ depends on state we refer to this as "feedback control". Note that feedback control significantly expands the set of achievable dynamics, since the linear superposition of operators L_i can be different for each column (each value of x). If $u(x, t)$ is also time-varying, then the resulting Lie algebra of achievable dynamics is much larger than the span of the original set of operators L_i.

2.4   Short-Time Optimal Feedback Control

Suppose we have a value function $v(x)$ which gives the current (or immediate future) value of each state x. Then the expected total value at time t is $Ev = \int v(x)p(x,t)dx$, and the rate of change of expected total value is $\partial Ev/\partial t = \int v(x)\dot{p}(x,t)dt$ which we can write more succinctly as $v^T\dot{p}$ or vLp. This is maximized for superposition dynamics when

$$u_i = v^T L_i p \tag{6}$$

meaning that the weighting for each operator L_i is proportional to the expected increase in value due to that operator. Because $v(x)$ depends on x and $p(x,t)$ depends on t, Eq. 6 describes a time-varying feedback controller and thus makes full use of the Lie algebra of available operators.

2.5   Distributed Representation of Operators

When x takes discrete values $\{x_k\}$, the probability density $p(x,t)$ can be represented by a vector $< p_0(t) \ldots p_M(t) >$, where $p_k(t) = \text{Prob}\{x(t) = x_k\}$. For a tridiagonal matrix operator L with elements $l_{k,m}$, we have

$$\dot{p}_k(t) = l_{k,k-1}p_{k-1}(t) + l_{k,k}p_k(t) + l_{k,k+1}p_{k+1}(t) \tag{7}$$

This is a local representation, since $\dot{p}_k(t)$ depends only on nearby values of $p_k(t)$.

We can perform short-time optimal control in the distributed representation if $v(x) =< v_0 \ldots v_M >$ is also distributed, in which case we obtain the controller

$$u_i(t) = \sum_k u_{ik}(t) = \sum_k v_k \left(l_{k,k-1}^i p_{k-1}(t) + l_{k,k}^i p_k(t) + l_{k,k+1}^i p_{k+1}(t) \right) \tag{8}$$

where $l_{k,m}^i$ is the k,m'th element of matrix L_i. This requires a sum-reduction operation over all the distributed elements, but since only a few local values p_k are nonzero at any time, the sum-reduction itself is local.

2.6   Adjoint Feedback Controller

While it is natural to compute $\dot{p}_i = L_i p$ and then set $u_i = v^T \dot{p}_i$, we could just as easily calculate $\tilde{v}_i = L_i^T v$ and then set $u_i = p^T \tilde{v}_i$, where L_i^T indicates the adjoint operator of L_i, which for real matrices is the transpose. In practice, this is usually more efficient because p tends to change faster than v which means the calculation of $\tilde{v}_i = L_i^T v$ for the adjoint controller can be updated less frequently than the calculation of $\dot{p}_i = L_i p$ for the forward operator.

2.7   Distributed Learning Rule

When the operator L is unknown, observation of $\dot{p}$ and p allows each element to be learned, and because of the locality of the representation elements can be learned using a local learning rule. This type of learning will create an internal representation of observed dynamics. In order to learn a controller, we also need to learn the effect of each control input u_i. Because of linear superposition, we have

$$\dot{p}_k(t) = \sum_i u_i \left(l^i_{k,k-1} p_{k-1}(t) + l^i_{k,k} p_k(t) + l^i_{k,k+1} p_{k+1}(t) \right) \tag{9}$$

which can be learned by approximating the coefficients $l^i_{k,m}$ in the linear network over the second-order polynomial basis $u_i p_m$:

$$\dot{p}_k(t) = \sum_{i,m} l^i_{k,m} (u_i p_m(t)) \tag{10}$$

where m goes from $k-1$ to $k+1$.

For prediction we seek to estimate the forward model $\dot{p} = \mathcal{M}_f(p, u)$ whereas for control we seek to estimate the inverse model $u = \mathcal{M}_i(p, v)$. In other words, estimation of $\dot{p}$ allows the circuit to predict the effect of actions u, whereas for control with a specified cost function v our goal is to calculate u that leads to an increase in the expected value of the state $d/dt(v^T p) = v^T \dot{p}$. This will increase if $\dot{p}(x)$ and $v(x)$ have the same sign for most values of x. In particular, the goal for local learning is to choose u_{ik} such that $\dot{p}_k$ and v_k have the same sign. Therefore we can learn $u = \mathcal{M}_i(p, v)$ by approximating $u \approx \tilde{\mathcal{M}}_i(p, \dot{p})$ where $\dot{p}$ is the change that resulted from exerting command u with state estimate p. Because there may be a delay Δ between the command and the result, we are actually learning: $u(t - \Delta) \approx \tilde{\mathcal{M}}_i(p(t - \Delta), \dot{p}(t))$, which can be done by learning the coefficients a_{imk} for the linear model:

$$u_{ik}(t - \Delta) = \sum_m a_{imk} \dot{p}_k(t) p_m(t - \Delta) \tag{11}$$

where m goes from $k-1$ to $k+1$. Once learned (or during learning), we can control the system by using the coefficients to calculate u from v according to $u(t) \approx \tilde{\mathcal{M}}_i(p(t), v(t))$ using:

$$u_{ik}(t) = \sum_m a_{imk} v_k(t) p_m(t) \tag{12}$$

where $v(t)$ is playing the role of the desired $\dot{p}(t + \Delta)$, and the control output is $u_i = \sum_k u_{ik}$.

2.8   Spike-Based Learning Rule

A common form of rate-coded spike representation is for spikes to occur with Poisson statistics, where the Poisson rate $\lambda(t)$ encodes the underlying variable. In any small time interval Δt the probability of a spike event is $\lambda(t)\Delta t$. To implement Eq. 12, we code $v_k(t)$ and $p_m(t)$ in this way, but allow the coefficient a_{imk} to be a continuous signed variable perhaps represented by the synaptic strength of an excitatory or inhibitory synapse. If spike generators with rates v_k

and p_m are independent, then the probability of the joint event of both firing during interval Δt is the product $v_k p_m \Delta t^2$. So if we set u_k to be proportional to a_{imk} whenever the v_k and p_m spikes both fire, then the expected probability of the output neuron firing will correctly represent Eq. 12 [5]. Note that v_k is allowed to be negative (representing states with negative value, or positive cost). Therefore we use two copies of the implementation of Eq. 12: one for positive values of v_k, and another (with the sign of both v_k and a_{imk} changed) for negative values of v_k.

Equation 11 can be implemented similarly, in this case with $\dot{p}$ being signed so that there are two copies for the two cases with different signs. For recursive learning, only the sign of the change in a matters, so when $a_{imk}\dot{p} > 0$, we increase a_{imk} whenever u_{ik} and p_m both fire, and when $a_{imk}\dot{p} < 0$, we decrease a_{imk} whenever u_{ik} and p_m both fire. It is usually necessary to enforce both positive and negative bounds on a_{imk} which represents realistic limitations on excitatory or inhibitory synaptic efficacy. For each local block k, we need 2 neurons (for positive and negative values) for each of $\dot{p}$, v, and $L^T v$, and 1 neuron for p (which has only positive values). Therefore we need 7 neurons per block.

For each local computation k, $m \in (k - 1 \ldots k + 1)$, and $i \in (1 \ldots N)$, so we need $3N$ different coefficients a_{imk}. For example, if u has two outputs which control leftward or rightward force pulses, then there are 6 coefficients for each local block. If we have additional outputs for horizontal or vertical force pulses, then we need to learn a total of 12 coefficients. Since all the coefficients are learned independently of other blocks, the learning is quite rapid so long as experience brings the robot into the region that activates p_k sufficiently often (persistent excitation).

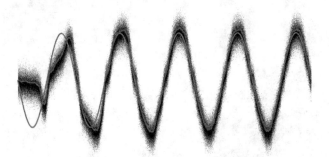

Fig. 1. One-dimensional simulation for 5000 time points t (horizontal axis). Vertical axis is the state index x_k. Black dots show firing of the neurons representing $p(x,t)$, red line is the desired position $x^*(t) = \max_x v(x,t)$, and blue line is the actual position $\hat{x}(t) = \max_x p(x,t)$. (Color figure online)

3   Simulation Results

Figure 1 shows results of simulation of a one-dimensional simulated controller. At each time-point, the position $x(t) \in [0, 1]$ (sensory information) is encoded in $p(t)$ as a Gaussian with standard deviation (SD) of 0.03, the cost function $v(t) \in [-1, 1]$ is encoded as a Gaussian with SD = 0.06, and $\dot{p}(t)$ is encoded from the first difference approximating $\partial p / \partial t$. The cost function is centered on a sinusoidal desired trajectory. All representations use Poisson-coded binary spikes with a population size of 1000 blocks. Two output controls u_{LEFT} and u_{RIGHT} provide forces in each direction, and the sum of u_i over all cell clusters provides the total output force. The true dynamics are thus described by $L_{LEFT} = \partial p / \partial x$ and $L_{RIGHT} = -\partial p / \partial x$ and so it is helpful to compute the spatial derivative $\partial p / \partial x$ as well as the derivative for adjoint control $\partial v / \partial x$. All data elements for each local computation are binary (spikes either fired or didn't), and the learning trains weights using Eq. 11 and at each step exerts control using Eq. 12.

Initial weights are set to random values centered on zero, and Fig. 1 shows poor tracking of the target at the start of learning. Coefficients a are very rapidly learned so that after 1000 time points and 1.5 cycles of the sinusoidal desired trajectory, tracking is reliable. Note that learning and movement are occurring simultaneously; there are no separate learning and performance phases.

Fig. 2. Robot with visually-controlled two-dimensional tracking. The cost function is positive only for green objects seen by the camera, and the robot tracks the objects in realtime. (Color figure online)

Figure 2 is a single frame from a video demonstrating two-dimensional tracking. In this case, there are separate controllers for horizontal and vertical movement, requiring in 14 neurons per block, $640 \times 480 = 307,200$ blocks (one for each

VGA pixel) for a total of 4.3 million neurons. Realtime control is implemented on a GPU running on a laptop computer (NVidia GTX970) programmed using the CUDA environment under Python (Anaconda2, Python 2.7). The desktop robot has a 1 kHz update loop (Geomagic Touch, using Ghost SDK software) and the cost function v is provided from the USB camera at 30 Hz (openCV).

4 Conclusion

The results demonstrate successful adaptive feedback control using an array of Poisson spiking neurons. Because of the probability representation of state, this structure yields risk-aware control, with rapid corrections for perturbations and ongoing adjustment of stiffness and safety margins. This is the first full implementation of Stochastic Dynamic Operators using a population of simulated spiking neurons, and it demonstrates the speed of convergence of learning as well as the flexibility of programming. It is our hope that in the future this will provide a new type of flexible and scalable compliant adaptive controller for use in human-computer interactions, surgery, exoskeletons, or wherever rapid and appropriate response to a compliant yet dangerous environment with unpredictable perturbations is needed.

References

1. Bertucco, M., Bhanpuri, N.H., Sanger, T.D.: Perceived cost and intrinsic motor variability modulate the speed-accuracy trade-off. PloS one **10**(10), e0139988 (2015)
2. Dunning, A., Ghoreyshi, A., Bertucco, M., Sanger, T.D.: The tuning of human motor response to risk in a dynamic environment task. PloS one **10**(4), e0125461 (2015)
3. Sanger, T.D.: Distributed control of uncertain systems using superpositions of linear operators. Neural Comput. **23**(8), 1911–1934 (2011)
4. Sanger, T.D.: Risk-aware control. Neural Comput. (2014)
5. Sanger, T.D.: Probability density methods for smooth function approximation and learning in populations of tuned spiking neurons. Neural Comput. **10**(6), 1567–1586 (1998)

Regularization Methods for the Restricted Bayesian Network BESOM

Yuuji Ichisugi$^{(\boxtimes)}$ and Takashi Sano

Artificial Intelligence Research Center (AIRC),
National Institute of Advanced Industrial Science and Technology (AIST),
Tsukuba Central 1, Tsukuba 305-8560, Japan
y-ichisugi@aist.go.jp

Abstract. We describe a method of regularization for the restricted Bayesian network BESOM, which possesses a network structure similar to that of Deep Learning. Two types of penalties are introduced to avoid overfitting and local minimum problems. The win-rate penalty ensures that each value in the nodes is used evenly; the lateral-inhibition penalty ensures that the nodes in the same layer are independent. Bayesian networks with these prior distributions can be converted into equivalent Bayesian networks without prior distributions, then the EM algorithm becomes easy to be executed.

1 Introduction

One of the remarkable hypotheses in the latest neuroscience is "the cerebral cortex is a kind of Bayesian network [4]." The cerebral cortex plays an important role in human intelligence. The cerebral cortex has many similarities to Bayesian networks [2], from the functional and structural point; this is suggested by a number of neuroscientific phenomena, well-simulated by the models involving Bayesian networks (For example [4–6,8–10]).

Deep Learning, which stems from the Neocognitron [1], is garnering attention for its high recognitive performance. Neocognitron was designed to have the functionality of the visual cortex through the imitation of the hierarchical structure of ventral pathway of the cortex.

Combining the latest neuroscientific insights and the Deep Learning technology will lead to the better performing machine learning technology, which has the more human-like ability. With this goal in mind, we are developing a machine-learning algorithm called BESOM (BidirEctional Self-Organizing Map), a Bayesian network with a layer structure and each node has restricted CPT (Conditional Probability Table) model [6]. Though our BESOM algorithm is under development and lacks accuracy, it already has the ability to show the potential applications in engineering and as a possible computational model of the cerebral cortex [6,8,11].

Deep Learning using a Bayesian network is thought to be promising not only because of its similarity to the human brain but also from a technical viewpoint, particularly with respect to the following points:

© Springer International Publishing AG 2016
A. Hirose et al. (Eds.): ICONIP 2016, Part I, LNCS 9947, pp. 290–299, 2016.
DOI: 10.1007/978-3-319-46687-3_32

- Inference in Bayesian networks can sometime be executed with low computational complexity.
- Because Bayesian networks have top-down information flow in addition to bottom-up, they may be more powerful than feed-forward neural networks.
- It is easy to build in prior knowledge about learning targets.

Despite these advantages, large-scale Bayesian networks like BESOM are not widely used, probably because of their large computational complexity, overfitting and local minima problems.

For the issue of computational complexity, efforts are being addressed by the use of restricted CPTs [3,11].

The problems of overfitting and local minima are thought to arise from the high expressiveness of large-scale Bayesian networks. Assigning an adequate prior distribution to the parameters, this high expressiveness would be reasonably lowered to solve these problems.

In this study, we describe two types of prior distribution: the win-rate penalty and the lateral-inhibition penalty. We also introduce an approximate learning rule for use with these penalties. The two mechanisms can be applied simultaneously. They add biases to the recognition results: the win-rate penalty ensures that each value in the nodes is used evenly; the lateral-inhibition penalty ensures that the nodes in the same layer are independent.

2   The Architecture of BESOM

BESOM is a Bayesian network having a deep hierarchical structure similar to Deep Learning (Fig. 1). Like many Deep Learning architectures, it has connections between layers forming local receptive fields, while there are no connections in the same layer.

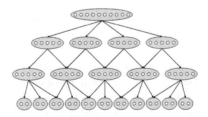

Fig. 1. An example of a BESOM network. Ovals are nodes (random variables) and the white circles inside are units (possible values for the random variables).

In BESOM, variables are called *nodes* and possible values for the variables are called *units*. In general, nodes are multinomial variables. If a black and white image is to be learned, the input pixel values are given as the observed binary variables in the bottom layer.

BESOM can be used in both unsupervised and supervised learning. When used for unsupervised learning, all variables in BESOM are hidden, except those in the bottom layer. In this case, acquired features are expressed in the upper layers. For supervised learning by BESOM, there are ways to assign the supervisory signal. One of the way, for example, is to assign the supervisory signal to a single node in the uppermost layer. In the test phase, the uppermost node becomes a hidden variable, whose inference value (the maximum posterior probability) is taken to be the recognition result.

Another significant feature of BESOM is the limitation placed on CPT, which will be explained in Sect. 4.

3   The Objective Function for Learning

Let $P(\mathbf{h}, \mathbf{i}|\theta)$ be a joint probability model with the a set of hidden variables $\mathbf{h}$ and the set of input variables $\mathbf{i}$, with given parameter θ. By $\mathbf{i}(t)$, we give the set of values of input variables at the time t. Under the assumption that the input data sequence forms i.i.d. (independent and identical distributions) for fixed parameter θ, the probability for the input data sequence $\mathbf{i}(1), \mathbf{i}(2), \cdots, \mathbf{i}(t)$ occurring under the parameter θ, which is likelihood of θ, is calculated like this:

$$P(\mathbf{i}(1), ..., \mathbf{i}(t) \mid \theta) = \prod_{i=1}^{t} P(\mathbf{i}(i) \mid \theta) = \prod_{i=1}^{t} \sum_{\mathbf{h}} P(\mathbf{h}, \mathbf{i}(i) \mid \theta). \tag{1}$$

The objective of learning is to obtain MAP (maximum a posteriori) estimate of the parameter. In other words, the objective is to find maximizing parameter of θ, say θ^*:

$$\theta^* = \underset{\theta}{\mathrm{argmax}} \left[\prod_{i=1}^{t} \sum_{\mathbf{h}} P(\mathbf{h}, \mathbf{i}(i) \mid \theta) \right] P(\theta). \tag{2}$$

To estimate parameter θ, the online EM (Expectation-Maximization) algorithm or its approximation is used. One method of approximation is given as follows.

The approximation algorithm here is combination of two steps, one for recognition and the other for learning. First, in the recognition step, based on current parameter $\theta(t)$ and given the input values $\mathbf{i}(t)$, the maximum posterior probability estimation values of the hidden variables $\hat{\mathbf{h}}(t)$ (i.e., MPE, Most Probable Explanation) are obtained as follows:

$$\hat{\mathbf{h}}(t) = \underset{\mathbf{h}}{\mathrm{argmax}}\, P(\mathbf{h}|\mathbf{i}(t), \theta(t)) = \underset{\mathbf{h}}{\mathrm{argmax}}\, \frac{P(\mathbf{h}, \mathbf{i}(t)|\theta(t))}{P(\mathbf{i}(t))}$$
$$= \underset{\mathbf{h}}{\mathrm{argmax}}\, P(\mathbf{h}, \mathbf{i}(t)|\theta(t)). \tag{3}$$

Approximate calculation of this formula can be efficiently executed by, for example, loopy belief revision algorithm [8].

Next, in the learning step, the marginalization of the hidden variables in Eq. (2) is approximated using the estimated value $\hat{\mathbf{h}}(i)$ and the result is taken to be $\theta(t+1)$.

$$\theta(t+1) = \operatorname*{argmax}_{\theta}\left[\prod_{i=1}^{t} P(\hat{\mathbf{h}}(i), \mathbf{i}(i)|\theta)\right] P(\theta). \tag{4}$$

The prior distribution for parameter θ is defined as the product of two factors, as follows:

$$P(\theta) = P^{\text{WinRate}}(\theta) \; P^{\text{Lateral}}(\theta). \tag{5}$$

The detailed explanation of the win-rate penalty $P^{\text{WinRate}}(\theta)$ and lateral-inhibition penalty $P^{\text{Lateral}}(\theta)$, are given in Sects. 5 and 6, respectively.

4   The Conditional Probability Table Model

One important characteristic of BESOM is in its CPT model. (Note that the win-rate penalty and lateral-inhibition penalty mechanisms, which are the main subject of this paper, are thought to work with the other types of CPT models.)

In a Bayesian network an $O(2^m)$ number of parameters is generally needed with respect to m, the number of parent nodes, to express the CPT for each node. This causes an explosive increase in computational complexity and memory requirements as well as introducing the problem of overfitting and local minima.

To allow CPTs to be expressed with fewer parameters, we limited them in the following manner:

$$P(x|u_1, \cdots, u_m) = \frac{1}{m}\sum_{k=1}^{m} w(x, u_k). \tag{6}$$

As the simplest form of $w(x, u_k)$, we currently use the following:

$$w(x, u_k) = P(x|u_k). \tag{7}$$

In this case, the conditional probability $P(x|u_k)$ is expressed by a single parameter w_{xu_k}.

When this restrictions are introduced, the belief propagation algorithm can be optimized and computational complexity is dramatically reduced [11]. It has also been shown that the information flow between the nodes closely matches the characteristic anatomical structure of the cerebral cortex [6,8].

5   Win-Rate Penalty

5.1   Purpose

If the BESOM network parameters are learned in a naive way, learning progresses for only a small set of units and the other units tend to stay at their initial values. In this case, learning is thought to fall into a local minimum.

An effective way to address this problem is to set the prior distribution for the parameter appropriately, and assign a bias so that the units in each node are used evenly.

Our approach uses the Kullback-Leibler (KL) divergence between the win-rate distribution that is targeted by each unit and the actual distribution. Penalties are imposed when the divergence is large. Here, the win-rate refers to the frequency with which units become the estimated values for the node.

In BESOM, the units are values of random variables, and the unit corresponding to the estimated value in one node (a random variable) is called the *winner unit*.

Using this penalties, it is expected that as learning progresses, the win-rate of each unit will approach the target value. The mechanism is called the *win-rate penalty* because units with larger win-rates are penalized.

5.2   The Problem of Complex Prior Distributions and Its Solution

The maximum likelihood estimate for the parameters of a Bayesian network with hidden variables can be estimated using an EM algorithm which can be executed efficiently using the result of inference [7]. However, when the parameter has a complex prior distribution, it is not obvious to perform the EM algorithm efficiently.

If a Bayesian network with a prior distribution for its parameter can be converted into an equivalent Bayesian network without a prior distribution, then the EM algorithm will become easy to be executed.

Fortunately, a Bayesian network with a prior distribution describe below can be converted into an approximately equivalent Bayesian network with no prior distribution. In the converted Bayesian network, *restriction nodes* are added to give bias to the recognition result.

5.3   Defining a Prior Distribution, and Deriving an Equivalent Bayesian Network

The win-rate penalty $P^{\text{WinRate}}(\theta)$ is defined as follows:

$$P^{\text{WinRate}}(\theta) = \prod_{X \in \mathbf{X}} \exp(-C^{\text{WinRate}} D_{KL}(Q(X)\|P(X;\theta))). \qquad (8)$$

where $\mathbf{X}$ is the set of all nodes and C^{WinRate} is a constant that determines the strength of the win-rate penalty.

$Q(X)$ is the distribution set as the target for the win-rate of node X and the network architect decides the shape of this distribution. For example, if the goal is to make the win-rate of the node units uniform, $Q(X)$ is defined for all units x_i ($i = 1, 2, \cdots, s$) as

$$Q(X = x_i) = 1/s \qquad (9)$$

where s represents the number of units in node X.

The Kullback-Leibler divergence between distributions $Q(X)$ and $P(X; \theta)$ is defined by the following equation:

$$D_{KL}(Q(X)\|P(X;\theta)) = \sum_x Q(x) \log \frac{Q(x)}{P(x;\theta)}. \tag{10}$$

We define a function $R(x; \theta)$ as follows:

$$R(x;\theta) = \frac{Q(x)}{P(x;\theta)} \log \frac{Q(x)}{P(x;\theta)}. \tag{11}$$

We can expect the following approximation holds because $x \sim P(x;\theta)$:

$$\sum_x f(x;\theta) \approx \sum_{i=1}^t \frac{1}{t} \frac{1}{P(x(i);\theta)} f(x(i);\theta). \tag{12}$$

Therefore,

$$D_{KL}(Q(X)\|P(X;\theta)) = \sum_x Q(x) \log \frac{Q(x)}{P(x;\theta)}$$

$$\approx \sum_{i=1}^t \frac{1}{t} \frac{1}{P(x(i);\theta)} Q(x(i)) \log \frac{Q(x(i))}{P(x(i);\theta)} = \sum_{i=1}^t \frac{1}{t} R(x(i);\theta) \tag{13}$$

holds. Then, $P^{\text{WinRate}}(\theta)$ can then be rewritten as follows:

$$P^{\text{WinRate}}(\theta) = \prod_{X \in \mathbf{X}} \exp(-C^{\text{WinRate}} D_{KL}(Q(X)\|P(X;\theta)))$$

$$\approx \prod_{X \in \mathbf{X}} \exp(-C^{\text{WinRate}} \sum_{i=1}^t \frac{1}{t} R(x(i);\theta))$$

$$= \prod_{X \in \mathbf{X}} \prod_{i=1}^t \exp(-\frac{1}{t} C^{\text{WinRate}} R(x(i);\theta))$$

$$= \prod_{i=1}^t \prod_{X \in \mathbf{X}} \exp(-\frac{1}{t} C^{\text{WinRate}} R(x(i);\theta)). \tag{14}$$

Equation (4), which MAP estimates parameter θ in the learning step, can be rewritten as follows (for simplicity, $P(\theta) = P^{\text{WinRate}}(\theta)$ is assumed here):

$$\theta(t+1) = \underset{\theta}{\arg\max} \left[\prod_{i=1}^t P(\hat{\mathbf{h}}(i), \mathbf{i}(i)|\theta) \right] P^{\text{WinRate}}(\theta)$$

$$= \underset{\theta}{\arg\max} \left[\prod_{i=1}^t \prod_{X \in \mathbf{X}} P(x(i)|\text{pa}(x(i));\theta) \right] \left[\prod_{i=1}^t \prod_{X \in \mathbf{X}} \exp(-\frac{1}{t} C^{\text{WinRate}} R(x(i);\theta)) \right]$$

$$= \underset{\theta}{\arg\max} \prod_{i=1}^t \prod_{X \in \mathbf{X}} \left[P(x(i)|\text{pa}(x(i));\theta) \exp(-\frac{1}{t} C^{\text{WinRate}} R(x(i);\theta)) \right] \tag{15}$$

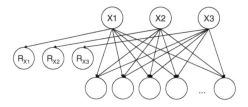

Fig. 2. Restriction nodes representing the win-rate penalty.

where $\text{pa}(x(i))$ represents the values of the parent nodes of node X at time i.

The above equation can be interpreted as that each node X has a corresponding restriction node R_X (Fig. 2) whose conditional probability is defined as follows:

$$P(R_X = 1 | X = x; \theta) = \exp(-\frac{1}{t} C^{\text{WinRate}} R(x; \theta)). \tag{16}$$

(Node R_X always has a observed value 1.) The Bayesian network with restriction nodes no longer has prior, therefore, it is possible to conduct parameter estimation using an EM algorithm.

Because C^{WinRate} is multiplied with the regularization parameter $1/t$, the penalty's influence decreases as time progresses when online learning.

The value $P(x; \theta)$ needed to calculate $R(x; \theta)$ can be simply estimated using statistics of values.

6   Lateral-Inhibition Penalty

6.1   Purpose

When the BESOM network parameters are learned in a naive way, nodes in the same layer that receive inputs from the same child nodes tend to represent similar feature. This phenomenon is also thought to relate to local minima or overfitting problems. Each of the hidden layers in BESOM, similar to those in Deep Learning, is expected to work as a feature extractor. When the same feature are redundantly expressed in many nodes, it is not preferable for recognition in the upper layers.

As in the previous section, this problem is addressed by defining a penalty as a prior distribution, and then, an equivalent Bayesian network without a prior is derived. In the prior distribution, a bias is applied in such a way that by assigning penalties to cases in which two nodes express similar values. Because this mechanism is thought to have a similar role to the lateral inhibition mechanism in the cerebral cortex, we name it the *lateral-inhibition penalty*.

6.2 Defining a Prior Distribution, and Deriving an Equivalent Bayesian Network

The prior $P^{\text{Lateral}}(\theta)$ corresponding to lateral inhibition is defined as

$$P^{\text{Lateral}}(\theta) = \prod_{(U,V) \in L} \exp(-C^{\text{Lateral}} I(U, V; \theta)) \tag{17}$$

where C^{Lateral} is the constant which determines the strength of the lateral-inhibition penalty and L are the set of pairs of nodes conducting lateral inhibition. Usually, each pair of nodes in the same layer that share child nodes should laterally inhibit each other.

$I(U, V; \theta)$ is the mutual information between nodes U and V and is defined as follows:

$$I(U, V; \theta) = \sum_u \sum_v P(u, v; \theta) \log \frac{P(u, v; \theta)}{P(u; \theta) P(v; \theta)}. \tag{18}$$

Here, we define a function $R(u, v; \theta)$ as follows (θ has been omitted):

$$R(u, v) = \frac{P(u, v)}{P(u)P(v)} \log \frac{P(u, v)}{P(u)P(v)} = (P(u|v)/P(u)) \log P(u|v)/P(u). \tag{19}$$

Given these definitions and the approximate Eq. (12), the following holds:

$$I(U, V; \theta) \approx \sum_{i=1}^{t} \frac{1}{t} R(u(i), v(i); \theta). \tag{20}$$

Following the same logic as in Sect. 5, we can derive an equivalent Bayesian network without the prior. In the network, for each node pair $(U, V) \in L$ that displays lateral inhibition, there is a shared binary-valued child node R_{UV} (Fig. 3) whose conditional probability is defined as follows:

$$P(R_{UV} = 1|u, v; \theta) = \exp(-\frac{1}{t} C^{\text{Lateral}} R(u, v; \theta)). \tag{21}$$

Thus, the maximum likelihood value of a parameter can be easily estimated using an EM algorithm.

The values $P(u|v; \theta)$ and $P(u; \theta)$ required to calculate the value of $R(u, v; \theta)$ can be simply estimated using statistics of values.

7 Evaluation

We evaluated the effectiveness of the proposed method using recognition rates of an MNIST handwritten digit database[1].

---

[1] http://yann.lecun.com/exdb/mnist/.

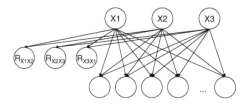

Fig. 3. Restriction nodes representing the lateral-inhibition penalty.

Table 1. Accuracy of recognition results of MNIST hand-written digits. (WR: Win-Rate penaltiy, LI: Lateral-Inhibition penaltiy)

	With WR	Without WR
With LI	80.6 %	81.8 %
Without LI	82.2 %	63.6 %

We used a four-layer BESOM network for this experiment. For the bottom layer, we created a 28×28 layout of two-unit nodes to take binary pixel values from a 28×28 input images. For the uppermost layer, we used a single 10-unit node to provide the supervisory signal. There were two hidden layers: for the layer immediately above the input layer we created a 5×5 array of 20-unit nodes and for the layer above that, a 3×3 layout of 30-unit nodes.

We evaluated the recognition rate by having the network first randomly learn 10,000 pieces of training data from a possible 60,000 pieces and then randomly recognize 1,000 pieces from 10,000 pieces of test data.

For learning, a very rough approximation of an online EM algorithm was used. First, an optimized loopy belief propagation algorithm [11] was applied. This was used to calculate the marginal posterior probabilities for each node. For each node, the value with maximum posterior was taken to be its estimated value, and the parameter was updated using the value.

Table 1 summarizes the results; each value is the average of 10 experiments. For both penalties, the recognition rate was higher than when no penalties were applied. This result also shows that two prior distribution can be applied simultaneously; however, it does not show the best accuracy in this case.

8   Conclusion and Future Work

Two regularization methods for parameter learning of layered Bayesian networks are proposed and an experiment shows that they are promising. We believe they alleviate both overfitting and local minima problems; however, more detailed evaluation and analysis may still be required.

BESOM is beginning to be used as a machine-learning algorithm; however, sufficient recognition precision has not been attained to enable its use in practical applications. The main reason for this is thought to be that the restrictions of

CPTs described in Sect. 4 are too strong. To address this problem, it is necessary to develop a new CPT model and a suitable approximate belief propagation algorithm. This is what we are currently working on.

Acknowledgements. This paper is based on results obtained from a project commissioned by the New Energy and Industrial Technology Development Organization (NEDO).

References

1. Fukushima, K.: Neocognitron: a self-organizing neural network model for a mechanism of pattern recognition unaffected by shift in position. Biol. Cybern. **36**(4), 93–202 (1980)
2. Pearl, J.: Probabilistic Reasoning in Intelligent Systems: Networks of Plausible Inference. Morgan Kaufmann, San Mateo (1988)
3. Heckerman, D.: Causal independence for knowledge acquisition and inference. In: Proceedings of UAI 1993, pp. 122–127 (1993)
4. Lee, T.S., Mumford, D.: Hierarchical Bayesian inference in the visual cortex. J. Opt. Soc. Am. A **20**(7), 1434–1448 (2003)
5. George, D. Hawkins, J., A hierarchical Bayesian model of invariant pattern recognition in the visual cortex. In: Proceedings of IJCNN 2005, vol. 3, pp. 1812–1817 (2005)
6. Ichisugi, Y.: The cerebral cortex model that self-organizes conditional probability tables and executes belief propagation. In: Proceedings of IJCNN 2007, pp. 1065–1070 (2007)
7. Koller, D., Friedman, N.: Probabilistic Graphical Models: Principles and Techniques - Adaptive Computation. The MIT Press, Cambridge (2009)
8. Ichisugi, Y.: Recognition model of cerebral cortex based on approximate belief revision algorithm. In: Proceedings of IJCNN 2011, pp. 386–391 (2011)
9. Hosoya, H.: Multinomial Bayesian learning for modeling classical and nonclassical receptive field properties. Neural Comput. **24**(8), 2119–2150 (2012)
10. Dura-Bernal, S., Wennekers, T., Denham, S.L.: Top-down feedback in an HMAX-Like cortical model of object perception based on hierarchical Bayesian networks and belief propagation. PLoS ONE vol. 7(11) (2012)
11. Ichisugi, Y., Takahashi, N.: An efficient recognition algorithm for restricted Bayesian networks. In: Proceedings of IJCNN 2015, pp. 1–6 (2015)

Representation of Relations by Planes
in Neural Network Language Model

Takuma Ebisu[1,2(✉)] and Ryutaro Ichise[1,2,3]

[1] SOKENDAI (The Graduate University for Advanced Studies), Tokyo, Japan
{takuma,ichise}@nii.ac.jp
[2] National Institute of Informatics, Tokyo, Japan
[3] National Institute of Advanced Industrial Science and Technology,
Tokyo, Japan

Abstract. Whole brain architecture (WBA) which uses neural networks to imitate a human brain is attracting increased attention as a promising way to achieve artificial general intelligence, and distributed vector representations of words is becoming recognized as the best way to connect neural networks and knowledge. Distributed representations of words have played a wide range of roles in natural language processing, and they have become increasingly important because of their ability to capture a large amount of syntactic and lexical meanings or relationships. Relation vectors are used to represent relations between words, but this approach has some problems; some relations cannot be easily defined, for example, sibling relations, parent-child relations, and many-to-one relations. To deal with these problems, we have created a novel way of representing relations: we represent relations by planes instead of by vectors, and this increases by more than 10 % the accuracy of predicting the relation.

1 Introduction

As an approach to achieving artificial general intelligence, brainlike calculation methods are gathering attention, and one such method is known as whole brain architecture (WBA) [1]. An important aspect of this is determining how knowledge is represented in a computer, and one approach is to use distributed representations of words in a neural network. *Distributed representations* use dense vectors in a relatively low-dimensional vector space to represent the targets. A convenient characteristic of this is that the vectors that represent relations between words are obtained by using distributed representations of words. For example, suppose that a vector r represents the relation between the present and past tense of verbs; then, the distributed representation of "ran" is located nearest to the vector, that is, the distributed representation of "run" plus r. However, there are various types of relations, and some cannot be easily defined in this way.

In this paper, we propose a novel approach to this problem: instead of using vectors to represent relations, we propose the use of planes, that is, sets of vectors. The use of planes increases the number of degrees of freedom in the representation of relations.

© Springer International Publishing AG 2016
A. Hirose et al. (Eds.): ICONIP 2016, Part I, LNCS 9947, pp. 300–307, 2016.
DOI: 10.1007/978-3-319-46687-3_33

We evaluated the accuracy of predictions, and we found that the proposed method is more successful than the previous method.

In Sect. 2, we will review past research on distributed representations, and in Sect. 3, we will discuss the problems of representing relations using a distributed representation. Our proposed method will be presented in Sect. 4, and we will present the results of our evaluation in Sect. 5. Our conclusions are presented in Sect. 6.

2 Neural Network Language Model

A neural network language model uses a neural network to learn distributed representations of words from a large, unlabeled corpus. The first neural network language model was introduced by Bengio et al. [2], and Mikolov et al. [3] reduced the computational complexity; the model of Mikolov et al. is shown in Fig. 1. These models use nearby words to predict a given word in a sentence. In particular, in the model of Mikolov et al. distributed representations of words are learned in to order to increase the dot product of words that co-occur. As a result, words that are lexically or grammatically close are mapped to close points in representation space. However, this approach has some problems.

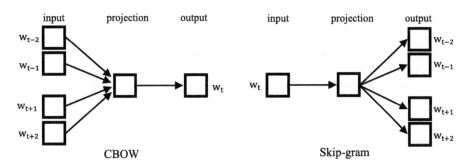

Fig. 1. Two models for distributed representations of words, as proposed by Mikolov et al. [3]. The continuous bag-of-words (CBOW) model predicts the current word from the surrounding words, and the Skip-gram model predicts the surrounding words from the current word.

The first problem is that the existing methods are unable to deal with words that have multiple meanings, so an active area of research is to develop methods that can do this. For example, Jauhar et al. [4] uses ontology with corpus as the input data. Specifically, WordNet, a lexical ontology, is used to separate multiple meanings, and distributed representations of words are learned so that related words in WordNet are at nearby points in representation space. Neelakantan et al. [5] proposed a model in which the average of the distributed representation of co-occurring words is defined as the context, and meanings are then distinguished for a given context.

The second problem is that the previous methods were unable to deal with ambiguity, although there is an existing model that can do so. In this approach, each word is represented by a Gaussian function; for ambiguous words, the variance is high, and thus they are widely distributed in representation space. A word with an unambiguous meaning is thus represented by a low-variance Gaussian function and will be locally distributed.

The last problem is concerned with how relations are represented, and the earlier methods cannot represent certain relations. This will be discussed in the following section.

3   Problems of Current Representation of Relations of Words

It was shown by Mikolov et al. [3] that relationships between words can be represented by vectors. For example, (Tokyo, Japan), (Paris, France), and (Beijing, China) are pairs of words that each have the same relation, a capital and its country. If we let v_w be a distributed representation of word w, then

$$v_{Japan} - v_{Tokyo} \fallingdotseq v_{France} - v_{Paris}$$
$$\fallingdotseq v_{China} - v_{Beijing}$$

holds. Therefore, if (Paris, x) has the same relation as (Tokyo, Japan), we can predict that x represents "France" by searching for the word whose distributed representation is the nearest to $v_{Japan} - v_{Tokyo} + v_{Paris}$

However, if we assume an ideal situation, that is, where equality holds, then we cannot explain relations that follow a reflexive law, such as a sibling relation, those that follow a transitive law, such as a is-a relation or a many-to-one relation. For example, if we let (A, B) be a reflexive law, then (B, A) has the same relation. Hence,

$$v_B - v_A = v_A - v_B$$
$$\therefore v_A = v_B$$

Note that A and B are not the same words, but they have the same distributed representation and the relation vector is thus 0; these are problematic. To deal with this problem, we propose a novel way to represent relations.

4   Representations of Relations Using a Plane

4.1   Summary of Proposed Method

We propose a method for representing relations as planes (sets of vectors), which can then be used to predict the subject, given an object, the distributed representation of the words, and a training set consisting of subject-object pairs of words in the same relation. In other words, a relation is represented by the sum of a certain vector and a certain subspace of representation space V. Let R be the set that represents the relation, then

$$R = r + U$$
$$\overset{\text{def}}{=} \{r + u | u \in U\}$$

where r is a vector and U is a subspace. We now discuss how to determine R. Let $S_{training} = \{(s_i, o_i) | i = 1, \ldots, N$ be a set of pairs of subject s_i and object o_i in the same relation; then, we find the R that minimizes the cost function f, which is defined as the sum of squares of the Euclid distance between $s_i + R$ and o_i.

$$f = \sum_{i=1}^{N} d(v_{s_i} + R, v_{o_i})^2$$

$$= \sum_{i=1}^{N} \min_{x \in v_{s_i} + R} |x - v_{o_i}|^2 \tag{1}$$

Let (s, x) be a pair in the same relation as in $S_{training}$; then, x is predicted by calculating:

$$\underset{w}{\text{argmin}} \, d(v_s + R, v_w)$$

4.2 Calculation Technique

We now explain a technique for implementing the proposed method. Eq. 1 is transformed as follows:

$$f = \sum_{i=1}^{N} d(R, v_{o_i} - v_{s_i})^2$$

If we let $x_i = v_{o_i} - v_{s_i}$, then we can show that $r = \bar{x}$ (*average of x_i*) holds.

Let D be the dimension of V, let E be the dimension of U, and let $(b_1, \ldots, b_D)$ be an orthonormal basis of V, where $b_1, b_2, \ldots, b_E$ spans U; then,

$$d(R, x_i)^2 = d(U, x_i - r)^2$$

$$= |x_i - r|^2 - \sum_{j=1}^{E} \left(b_j^\top (x_i - r)\right)^2$$

$$= \sum_{j=1}^{D} \left(b_j^\top (x_i - r)\right)^2 - \sum_{j=1}^{E} \left(b_j^\top (x_i - r)\right)^2$$

$$= \sum_{j=E+1}^{D} \left(b_j^\top (x_i - r)\right)^2$$

$$\therefore f = \sum_{i=1}^{N} d(R, v_{o_i} - v_{s_i})^2$$

$$= \sum_{j=E+1}^{D} \sum_{i}^{N} \left(b_j^\top (x_i - r) \right)^2$$

Solving this, we find that f is minimized when $r = \bar{x}$. Therefore, by setting $\hat{x}_i = x_i - \bar{x}$, we can determine the value of U that minimizes the following equation:

$$f = \sum_{i=1}^{N} d(U, \hat{x}_i)^2$$

$$= \sum_{j=E+1}^{D} \sum_{i}^{N} \left(b_j^\top \hat{x}_i \right)^2$$

We note that f can be transformed as follows:

$$f = \sum_{j=E+1}^{D} \sum_{i}^{N} \left(b_j^\top \hat{x}_i \right)^2$$

$$= \sum_{j=E+1}^{D} \sum_{i}^{N} b_j^\top \hat{x}_i \hat{x}_i^\top b_j$$

$$= \sum_{j=E+1}^{D} b_j^\top \sum_{i}^{N} \hat{x}_i \hat{x}_i^\top b_j$$

We note that $\sum_{i}^{N} \hat{x}_i \hat{x}_i^\top$ is a symmetric matrix, and hence there exists an orthonormal basis of eigenvectors that can be used to transform it to a diagonal matrix. Therefore, we can choose E eigenvectors in descending order of eigenvalues and use them to define U, which represents the subspace spanned by them.

Let B be a matrix whose columns are normalized eigenvectors that are orthogonal to U. Then, the Euclid distance function d can be simplified as follows:

$$d(v_s + R, v_w) = \left| B^\top (v_w - v_s) \right|$$

4.3   Another Interpretation of the Proposed Method

Our proposed method can be interpreted in another way. Here, $\sum_{i}^{N} \hat{x}_i \hat{x}_i^\top$ is the variance-covariance matrix of $\{x_i\}$ multiplied by a constant value. Hence, the method can be interpreted as changing the basis so that the covariance of $\{x_i\}$ becomes 0, reduce the elements of distributed representations in descending order of variances of $\{x_i\}$, and taking the vector that is the average of $\{x_i\}$ to represent the relations, as in the previous method. This leads to choosing the correct elements to explain the relation.

5 Experiments

5.1 Experimental Settings

We used word2vec[1] to obtain the distributed representations of the words. The dimensionality of the representation space was set to 300, and we used both the continuous bag-of-words (CBOW) model and the Skip-gram model. The training corpus was the English language Wikipedia site[2], which contains a total of 1.8 billion words, 1.7 million of unique words occur more than five times; we obtained a distributed representation for each word in this subset. As in Mikolov [3], we used a normalized distribution.

Pairs of city and country names were used to evaluate how well the proposed method could learn the relation. The data were obtained from GeoNames[3]; 10 pairs were obtained by country in descending order of the population of the cities. We only used pairs in which the city and the country names consist of a single word each, because our method is currently unable to process phrases that consist of two or more words. We obtained 599 pairs of city and country names.

We performed two experiments on this data. In Experiment 1, we evaluated the accuracy rate of predicting the country, given the city; we performed a ten-fold cross-validation for each E, with the dimension of U ranging from 1 to 160. The experiment was performed under two conditions: in the first case, a response was considered to be correct when the country name was the word nearest to $v_{subject} + R$; in the second, a response was considered to be correct when the country name was within the 100 words nearest to $v_{subject} + R$. For comparison, we also performed an experiment in which the relation vector was simply the following average [7]:

$$R = \bar{x}$$
$$= \frac{\sum_{i=1}^{N} (v_{o_i} - v_{s_i})}{N}$$

In the proposed method, it is necessary to determine E in advance, which is why we performed Experiment 2. For each step of the ten-fold cross-validation, we repeated the ten-fold cross-validation for the training set, and we determined E as the number of dimensions that would result in the best prediction of the country; we then evaluated the prediction accuracy using this value for E. To prevent a sudden change in the accuracy rate, the accuracy of a given dimension was defined as the average of the accuracy rates in the range of that dimension plus and minus two. For example, the accuracy for ten dimensions was the average for $D = 8, 9, 10, 11, 12$.

---

[1] https://code.google.com/archive/p/word2vec/.
[2] https://en.wikipedia.org/wiki/Main_Page.
[3] http://www.geonames.org/.

5.2 Experimental Results

The results of Experiment 1 are shown in Table 1. The results of the comparative experiment were equivalent to these results when the dimensionality of U was 0. For CBOW 1, the accuracy rate was improved by about 11 %, compared to the comparative results, and this occurred when the dimensionality of U was 50. For CBOW 100, the accuracy rate was improved by about 5 %. For Skip-gram 1, the accuracy rate was improved by about 12 % when the dimensionality of U was 120, and for CBOW 100, the accuracy rate was improved by more than 6 % when the dimensionality of U was either 70 or 120. Hence, we demonstrated that our proposed method of using a plane to represent a relation is more effective than the previous method. Overall, the skip-gram model yielded better scores than did the CBOW model, but our method improved the accuracy of both. The accuracy rates improved and stabilized as the dimensionality of U increased, and this corresponds to reducing the dimensionality of the distributed representation. Therefore, it is conceivable that some elements causes the representation of the relation to deteriorate, and thus the accuracy rate can be improved by discarding redundant elements.

Table 1. Results of Experiment 1. The values are the obtained accuracy rates, and E is the dimensionality of U. The number n after CBOW and Skip-gram means that the prediction was considered correct when the correct words fell within the n nearest words to the predictive vector. For example, in CBOW 1, the prediction was only considered correct when the correct word was the word nearest to the predictive vector. In each column, the best score is shown in bold. Due to space limitations, we show the results for U in increments of ten.

E	CBOW 1	CBOW 100	Skip-gram 1	Skip-gram 100
0	0.600653	0.905951	0.546843	0.882874
10	0.639042	0.942344	0.606313	0.940493
20	0.662119	0.946226	0.594920	0.934761
30	0.677540	0.948149	0.616038	0.940530
40	0.693033	0.953919	0.621843	0.940530
50	**0.710305**	0.953882	0.616001	0.942453
60	0.698803	0.948149	0.641001	0.946299
70	0.696807	0.953919	0.623730	**0.948186**
80	0.704463	0.953882	0.639151	0.944340
90	0.687192	**0.959652**	0.648766	0.946263
100	0.681459	0.955806	0.648766	0.944376
110	0.693033	0.955806	0.658382	0.946263
120	0.681495	0.946190	**0.662192**	**0.948186**
130	0.687228	0.942380	0.654499	0.946263
140	0.691074	0.940457	0.656495	0.942417
150	0.691074	0.936575	0.648839	0.945417
160	0.666074	0.928919	0.635377	0.940493

The results of Experiment 2 are shown in Table 2. Here, the predictions of both CBOW 1 and Skip-gram 1 were improved by about 10 % compared to the comparative results, and CBOW 100 and Skip-gram 100 were improved by about 5 %. Here, the accuracy was less than the best accuracy rate obtained in Experiment 1, but the proposed method still showed a consistent improvement in the accuracy. Hence, it was demonstrated that our proposed method can be used as a predictor when the dimensionality of U is determined in advance from a training set.

Table 2. Results of Experiment 2.

CBOW 1	CBOW 100	Skip-gram 1	Skip-gram 100
0.704415	0.953935	0.639155	0.944338

6   Conclusions

We investigated the use of planes to represents relations. We found that the use of a plane instead of a vector greatly improved the accuracy of predicting the object in a many-to-one relation, such as cities and their country. There are many other relations that fit this pattern, so a future task will be to evaluate how well the proposed method can represent various other relations. We also intend to perform experiments using other distributed representations with different settings, different models, and using a different corpus.

Acknowledgements. This paper is based on results obtained from a project commissioned by the New Energy and Industrial Technology Development Organization (NEDO).

References

1. Yamakawa, H.: The Whole Brain Architecture Initiative, http://wba-initiative.org/en/
2. Bengio, Y., Ducharme, R., Vincent, P., Jauvin, C.: A neural probabilistic language model. J. Mach. Learn. Res. **3**, 1137–1155 (2003)
3. Mikolov, T., Chen, K., Corrado, G., Dean, J.: Efficient estimation of word representations in vector space. In: Proceedings of Workshop at International Conference on Learning Representations (2013)
4. Jauhar, S.K., Dyer, C., Hovy, E.: Ontologically Grounded multi-sense representation learning for semantic vector space models. In: Proceedings of the 2015 Conference of the North American Chapter of the Association for Computational Linguistics, pp. 683–693, Association for Computational Linguistics (2015)
5. Neelakantan, A., Shankar, J., Passos, A., and McCallum, A.: Efficient Non-Parametric Estimation of Multiple Embeddings per Word in Vector Space, arXiv preprint arXiv:1504. 06662 (2015)
6. Vilnis, A., and McCallum, A.: Word Representations via Gaussian Embedding, arXiv preprint arXiv:1412.6623 (2014)
7. Ichise, R., Arakawa, N.: Relationships Between Distributed Representation and Ontology. In: The 29th Annual Conference of the Japanese Society for Artificial Intelligence, 2I4-OS-17a-5 (2015)

Modeling of Emotion as a Value Calculation System

Takashi Omori[1(✉)] and Masahiro Miyata[2]

[1] College of Engineering, Tamagawa University, Machida, Japan
omori@lab.tamagawa.ac.jp
[2] Graduate School of Engineering, Tamagawa University, Machida, Japan
mytma4re@engs.tamagawa.ac.jp

Abstract. Emotion is a very popular but not well-known phenomenon of animals. Human emotion/feeling is more complex including the emotion features and the intelligent features. Though there are many researches on emotion/feeling, its computational role on self-maintenance is not known well. But it must be important because most of animals look to have similar emotion and there must be a reason for its similarity. Therefore, in this paper, we discuss on a possible component of emotion system, compare their computational model, and propose a possible hypothesis that the emotion is a system of value calculation for a decision making. For a discussion, we show a possible computational model of feeling system in brain.

Keywords: Emotion · Decision making · Value system · Computational model

1 Introduction

Near future, products using AI technology will become popular, and will be requested to have more human like nature. A typical case is a human-AI interaction that is an important application field of AI. The human interaction includes a communication between human and AI in wider sense, and its implementation requires AI a set of ability like an intention estimation and a needs understanding of human and also an ability of expressing intention of self.

However, methods of communication by human is diverse and complex. They include a clear and explicit commands like a language or a sign, an implicit and ambiguous information like a gesture or an action, and their combination on going in parallel. The human interaction is a task in which these various information are read and interpreted in parallel and acted to change other's recognition, thinking and action.

As the task is easy and unconscious for human, we often think it is easy. But it is not easy when we think its internal process and its implementation. The purpose of this paper is a discussion for its understanding and a computational modeling.

In this paper, we focus on an emotion/feeling as a key factor for the understanding of human interaction. The emotion is a phenomenon that contain important function for animals especially in a communication. But most of emotion studies are focusing and analyzing on phenomenological side and few are discussing on its cognitive process, and more, its computational role in a brain information processing. So, in this study, we

© Springer International Publishing AG 2016
A. Hirose et al. (Eds.): ICONIP 2016, Part I, LNCS 9947, pp. 308–315, 2016.
DOI: 10.1007/978-3-319-46687-3_34

discuss on a possibility of its computational modeling based on an idea that the emotion in a wider sense, which we call feeling, is a value calculation system for an action decision.

2 Conventional Models of Emotion/Feeling

2.1 Importance of Mental State Estimation in Communication

There are plenty of phenomenological studies on emotion in physiology and psychology fields [1–3], and on the role of emotion in communication [4, 5]. In human interaction scene, Abe [6] studied an interaction strategy of nursery nurses in a robot-child interaction scene, and suggested an importance of the emotion guiding for a success of communication.

In conventional communication study by language or gesture, the emotion of other person is not considered so well in spite of its importance. The emotion of other is an unobservable internal variable and we encounter many difficulties for its estimation. However, it is obvious that the human communication largely changes based on an emotional state of other. We can't avoid modeling of the emotion when we consider human interaction through a personified agent.

2.2 Conventional Models of Emotion

There are many models for the emotion description [1, 2]. However, these are model of phenomenon description and don't approach to the brain and/or cognitive mechanism, and its computational meaning. For a role of emotion or more complicated human feeling, Toda proposed a qualitative theory for explaining a human complex feeling as a process of value assigning through an inference toward an action decision [7].

Recently, Koelsch proposed a quartet theory of human emotion that divide factors of emotion into four part, self-maintenance, safety, attachment and economic value [8]. In this theory, an economic value is included as a part of the emotion and orbitofrontal area of brain is supposed for its due. This theory is unique as it includes the economic value into the emotion. It is true that our emotion is strongly related to the economic value. For example, we will be pleased when we earn much money. This theory may give us a new concept of the feeling.

3 Hypothesis: Emotion as a Value Calculation System

3.1 Emotion Common in Animals

Some of animals, at least from reptile and later in evolutional course, have brain stem and diencephalon in common, and these part of brain are called as a reptile brain. With the reptile brain, animal can have the emotions, like fear and anger, to protect self. LeDoux studied on fear, and uncovered physiological system of the fear response including a fear condition recognition and its learning [3].

Not limited to the fear, most of basic emotions are common with human and animals suggesting a small change of the emotion system along evolution. It means there should be a reason for the emotions be kept in the evolution process. To discuss on its reason, we have to understand the role of emotions on current animals in more abstracted level before discussing on its evolution.

3.2 Action Decision and Value Calculation

We are making various decisions in our daily life. It is considered that we consciously/ unconsciously calculate value of possible choices for the decision [9]. With this calculation, human sometimes makes useless looking decision when he or she thinks the decision will contribute a future merit. Then, what is the merit?

The Quartet Theory of Emotions (QTE) proposed by Koelsch [8] includes Brainstem, Diencephalon, Hippocampus system and Orbitofrontal to cover the wide range of human emotions. A specific feature of the model is its wide range of its subparts, like a memory in the hippocampus and an inference in the orbitofrontal, to be included in the feeling system. But when we look at its internal process, it looks to be similar to a value calculation and a value representation for the action decision.

3.3 Hypothetical Value Calculation System

In our feeling model in this paper, we think the four brain parts that Koelsch supposed is not enough. We assume the human feeling system includes Amygdala, Basal Ganglia, Medial Prefrontal Cortex and Nucleus Accumbens at least in addition to the areas supposed by Koelsch. Though his paper comments on those areas, a positioning of those area is not clear.

We assume functions of these areas as follows: a maintenance of body by Brainstem, a state value learning by Basal Ganglia, an association between episode and value by Hippocampus, an emotive situation evaluation by Diencephalon, a state value inferenced in Orbitofrontal and a value information integration by Nucleus Accumbens. As a whole, stimulus from an outer world is perceptually processed in Cortex, converted to inherent

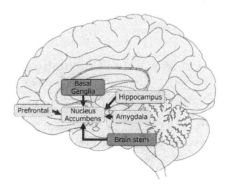

Fig. 1. Overview of Feeling System Model

value information at each brain areas and summed up and distributed at Nucleus Accumbens (Fig. 1).

4   Computational Modeling of the Proposed System

In this model of feeling, we assume an operating cycle of Prefrontal Cortex to implement probabilistic search with a priori knowledge on world events, and to decide a value of the searched results in real-time.

When we assume the feeling system as a value calculation system, the understanding of its computational mechanism is important because it relates the meaning of the computation. Table 1 shows a possible computational theory for each brain area. About the self-maintenance and safety, their process is physically embedded somewhere of the body and is genetically fixed. Though some parts, like Amygdala, shows learning capability, it is rather simple compared to Cortex system and realizable using conventional pattern recognition method.

Table 1. Possible computaional models of value system in brain

Value class	Possible computational algorithm & theory	Brain area
Self-maintenance	Direct detection by biological body state sensors and a pattern recognition of their combined signal.	Brainstem
Safety	Pattern recognition of fixed environment situations with fixed value and learned situations with experienced value.	Diencephalon, Amygdala
Attachment	Episodic association between a scene and value, and their generalization with Reinforcement Learning.	Hippocampus, Basal Ganglia
Economical Value	Probabilistic inference or tree search from current situation, and a value mapping from the basic emotion system.	Orbitofrontal, Medial Prefrontal

In our model, the attachment is assumed as a habitual recollection of past experienced value as a result of accumulated episodes related values. For its realization, we suppose two stages of processing, a recognition of value related object/event that appear in episodes, and a habit formation process as a generalization of those episodes. We can explain and implement the former as a pattern recognition process in Cortex and the latter as RL. Then, we consider the attachment can be approached by a conventional machine learning framework.

The value calculation process with inference could be modeled as below.

(1)   When an agent encounter new situation and can't assign value immediately by past experience.
(2)   The agent iterate predict resulting states of possible actions using a priori knowledge on its current environment, in parallel with a probabilistic way.
(3)   Then, a part of the predicted states reach a state that a value can be assigned.

(4) The agent gives a discounted value by uncertainty to the initial current state and the actions to reach the valued state. This is the inference of value for the action at the initial state.

(5) Assign the values to all the predictable states, and decide an action that maximize the value.

In our model, we use a probabilistic search by Monte Carlo method that can realize tree search in parallel and is possible in brain neural circuit. The probabilistic neural excitation and its associative propagation in Prefrontal Cortex is representing a probabilistically constrained tree search. The predicted neural states are immediately evaluated its value by the emotion system, and affect the next prediction.

5 Simulation of Frontal Value Calculation System

5.1 Navigation Task as a Toy-Model of Value Search in Human Brain

When we decide action, we often can choose optimal one unconsciously. In opposite case that when we have to decide action at a place without any experience, we search our memory for a similar experience and choose a best looking one. In the process, we use a prediction based inference for searching the experience.

As a very simple toy-model of such an experience based and an inference based action decision task, in this paper, we adopted a navigation problem in 2D grid world where an agent thinks to look for a proper action not by the try and error in RL but by a tree search like memory search without an actual action.

In many of grid world action decision and/or optimal route search research, RL is often used and have demonstrated good performances. But RL requires many times of try and error, and is largely different from an intuitive and thinking like action decision that human or animals take.

So, in this model, in place of the try-and-error real action in RL, and in place of a symbolic tree search that is often used in conventional AI algorithm, we adopt multistep Monte Carlo association of neural excitation from a current state representation to a next time step representation implementing the parallel and probabilistic search, and seek for a highest valued action sequence on line.

5.2 Three Layered Model of Distributed Inference

The model has three layered structure, Map layer, Place-Value Association (PVA) layer and Probabilistic Parallel Search (PPS) layer (Fig. 2). Map layer is a representation of spatial map and corresponds to Hippocampus and Prefrontal Cortex in brain. In this study, we assume a priori knowledge of world map. PVA layer has a function to give experience based value to the locations in the world map. Basal Ganglia is supposed to have this function of allocating an expected reward to an experienced situation by RL. In our model, we assume Q-learning for the function using the Eq. (1).

$$Q(s_{t+1}, a_{t+1}) \leftarrow Q(s_t, a_t) + \alpha \left(r + \gamma \max_a Q(s_{t+1}, a) - Q(s_t, a_t) \right) \qquad (1)$$

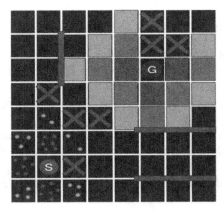

Fig. 2. Navigation task in a toy model simulation

In our simulation, we assumed a part of the world locations have assigned values in advance. If the agent seeks for next action in the locations, it can immediately know what to act by RL mechanism. But if current location of the agent is not assigned, the agent must think what to act next using the iterative inference.

PPS layer represent the neural excitation by the multistep Monte Carlo association in brain. Given a stimulus from an outer world, the agent recognize the current position of self and start searching for a goal position where the value becomes maximum. The iterated activity propagation with a priori knowledge create a wide spread of knowledge constrained neural activity, and enables a reaching of the neural excitation to a value assigned state. We used SoftMax Eq. (2) for the neural excitation propagation direction decision in Monte Carlo method and the value based action choice. We can modulate sensitivity of excitation propagation direction over Q-value by changing a temperature parameter T in Eq. (2). Though we fixed T value, we will be able to control the width of neural excitation by changing the T value in more complex [16].

$$P(i) = \frac{exp\left(\dfrac{Q(s, a_i)}{T}\right)}{\sum_j^n exp\left(\dfrac{Q(s, a_j)}{T}\right)} \qquad (2)$$

5.3 Cost Definition in Our Simulation

In the simulation, an agent is placed at random initial place and started action decision in two ways. One is a conventional RL and another is our method with probabilistic parallel search. As for a cost of search, we supposed cost of actual one step movement of the agent is ten times larger than the neural one step associative propagation in brain.

5.4   Simulation Result

Figure 3 shows a result of the simulation. A vertical line shows the cost till the goal position from a starting position for every trial. The horizontal line indicates the number of each trial. So, the graph shows change of cost for an agent to reach a goal along trial. RL method required large cost for initial part of the learning because many times of physical movement were necessary till Q-values were assigned to each of the location in the map. But with our method, the agent can find without actual moving and the cost became low in most of the map area. It also suggests that the initial location of the agent doesn't affect the searching cost suggesting that a change of goal in a known map, meaning a new task in the same world, doesn't affect searching cost. This is an ability of knowledge reuse.

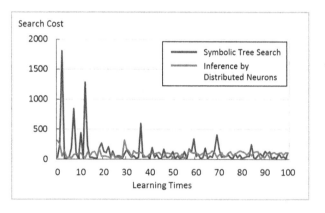

Fig. 3.   Comparison of search cost between RL and the proposed method.

6   Conclusion

In this paper, starting from the discussion on emotion/feeling system in brain, we proposed the hypothesis of an emotion in wide sense is compatible with the value calculation system for action decision, and discussed on the possible neural mechanism of inference for the value estimation. Though the simulation is too simple and the model is rough, whole image of the basic emotion/feeling system has shown and the specific feature of human intuitive inference has reproduced.

For the next study, we need to evaluate the inference model in more realistic situation and implement the whole system in a virtual environment. This study was supported by MEXT KAKENHI Grant Number 15H01622.

References

1. Ekman, P., et al.: What the Face Reveals: Basic and Applied Studies of Spontaneous Expression using the Facial Action Coding System (FACS), Series in Affective Science (1997)
2. Russell, J.A.: A circumplex model of affect. J. Pers. Soc. Psychol. **39**, 1161–1178 (1980)
3. LeDoux, J.: The Emotional Brain: The Mysterious Underpinnings of Emotional Life. Simon & Schuster, New York (1998)
4. Derksa, D., et al.: The role of emotion in computer mediated communication: a review. Comput. Hum. Behav. **24**(3), 766–785 (2008)
5. Ilyasova, A.: Emotional competencies: connecting to the emotive side of engineering and communication. In: IPCC, pp. 1–5 (2015)
6. Abe, K., Iwasaki, A., et al.: Robots that play with a child: application of an action decision model based on mental state estimation. J. Jpn. Robot. Soc. **31**(3), 263–274 (2013)
7. Toda, M.: Emotion. Tokyo University Press, Tokyo (1992)
8. Koelsch, S., et al.: The quartet theory of human emotions: an integrative and neuro-functional model. Phys. Life Rev. **13**, 1–27 (2015)
9. Otake, F., et al.: Economics in Brain, Discover Twenty One (2012). (in Japanese)
10. Gerken, G.M.: Central tinnitus and lateral inhibition: an auditory brainstem model. Hear. Res. **9**(1–2), 75–83 (1996)
11. Kawamura, M.: Organization and function of amygdala. Jpn. J. Clin. Psychiatry **36**(7), 817–828 (2007)
12. Sato, N.: From hippocampus to neocortex: computational models of the hippocampal memory. In: The 29th Annual Conference of the Japanese Society for Artificial Intelligence, 3F4-OS-19b-5 (2015). (in Japanese)
13. Schultz, W., et al.: Reward processing in primate orbitofrontal cortex and basal ganglia. Cereb. Cortex **10**(3), 272–283 (2000)
14. Sutton, R.S., et al.: Reinforcement learning. Cereb. Cortex **10**, 272–283 (2000)
15. Ichisugi, Y.: https://staff.aist.go.jp/y-ichisugi/besom/20060824.pdf (2006)
16. Suzuki, H.: Diversity and evaluation in creative problem solving – finding from insight problem-solving. Trans. Jpn. Soc. Artif. Intell. **19**, 145–153 (2004). (in Japanese)

The Whole Brain Architecture Initiative

Naoya Arakawa[✉] and Hiroshi Yamakawa

The Whole Brain Architecture Initiative, Kabukiza Tower 14F, 4-12-15 Ginza,
Chuo-ku, Tokyo 104-0061, Japan
{naoya.arakawa,ymkw}@wba-initiative.org

Abstract. The Whole Brain Architecture Initiative is a non-profit organization
(NPO) founded in Japan in August 2015, whose purpose is to support research
activities aiming for realizing artificial intelligence with human-like cognitive
capabilities by studying the entire architecture of the brain. It performs educational activities such as holding seminars and hackathons and compiling educational materials, as well as R&D activities such as developing software platforms
to support research in artificial intelligence and facilitating communication among
research communities.

Keywords: Biologically inspired cognitive architecture · Artificial general
intelligence · Open software · Non-profit organization

1 Introduction

The Whole Brain Architecture Initiative (WBAI) is a non-profit organization (NPO)
whose purpose is to support research activities aiming to realize artificial intelligence
with human-like cognitive capabilities by studying the entire architecture of the brain.

1.1 Whole Brain Architecture Seminars

In the summer of 2013, Hiroshi Yamakawa (the current chairperson of WBAI), Yutaka
Matsuo (vice-chairperson), and Yuuji Ichisugi (The National Institute of Advanced
Industrial Science and Technology of Japan (AIST)) met and agreed that it would be
possible to create advanced artificial intelligence having cognitive abilities on par with
human beings by referring to the information processing architecture of the entire brain,
with the increase of computational resources and progress in machine learning, as well
as the rapid accumulation of findings in neuroscience in recent years. They also agreed
that, in order to realize such artificial intelligence, it would be necessary to gather and
foster human resources in various disciplines such as neuroscience, cognitive science,
and machine learning as well as artificial intelligence. They soon decided on using the
term "whole brain architecture (WBA)" to designate the approach to create artificial
general intelligence by referring to the architecture of the entire brain, and held their
first seminar in Tokyo on December 19th in 2013. Since then, whole brain architecture
seminars have been held a few times a year to invite outstanding researchers to speak
about related subjects.

© Springer International Publishing AG 2016
A. Hirose et al. (Eds.): ICONIP 2016, Part I, LNCS 9947, pp. 316–323, 2016.
DOI: 10.1007/978-3-319-46687-3_35

1.2   Up to the Foundation

In 2014, researchers interested in the whole brain architecture approach held meetings and discussed methodology to substantiate the approach. By the end of the year, they agreed that they would create an organization called the Whole Brain Architecture Initiative, to give their programs a concrete form as a research community. At the beginning of 2015, they decided to make the organization non-profit, and after necessary procedures, the NPO was founded on August 21st.

1.3   Financial Support

The operations of WBAI have been financially supported by sponsors including private companies. As of July 2016, it has seventeen sponsors consisting of enterprises and individuals.

2   The Whole Brain Architecture Approach

The WBA approach is an engineering approach that aims to create artificial general intelligence (AGI) by learning from the architecture of the entire brain.

2.1   Artificial General Intelligence

AGI is artificial intelligence that can learn to perform tasks including those not foreseen at the time of its conception, unlike 'narrow' artificial intelligence designed to perform specific tasks. Such general intelligence would be necessary because, for instance, artificial intelligence, sometimes embodied in robots, will be required to cope with unexpected situations when collaborating with human beings in the real world.

2.2   Whole Brain Architecture

Human beings possess general intelligence in the sense that they can learn to perform previously unforeseen tasks. If the human brain instantiates general intelligence, it would be reasonable to endeavor to realize AGI by taking inspiration from the human brain. As intelligence is not a function of part of the brain but of the entire brain, it is also reasonable to seek inspiration from the architecture of the entire brain.

2.3   The Central WBA Hypothesis

Observing the development in machine learning technologies in recent years, WBAI further adopts the following hypothesis:

"The brain combines modules, each of which can be modeled with a machine learning algorithm, to attain its functionalities, so that combining machine learning modules in the way the brain does enables us to construct a generally intelligent machine with human-level or super-human cognitive capabilities".

This is a working hypothesis that constrains the scope of research so that we can concentrate resources.

3   Our Policies

The mission of WBAI is 'to create (engineer) a human-like AGI by learning from the architecture of the entire brain'. WBA deploys educational and R&D businesses to instantiate the mission.

The goal of our educational business is to help people conducting research on the WBA approach on a long-term basis. WBAI conducts educational activities in related areas such as artificial intelligence, neuroscience, cognitive science, and machine learning. In particular, WBAI holds seminars and hackathons; participates and collaborates in academic events; and also collaborates and communicates with related academic societies.

The goal of our R&D business is to support research using the WBA approach. While research projects in general may or may not last for a few years, we are committed to supporting research infrastructures not only for particular projects but also for terms longer than project lifespans. Such infrastructure includes software for supporting research and neuroinformatic databases. However, WBAI itself does not conduct research on the WBA approach and does not compete with researchers in the area.

3.1   WBAI and Open Development of AGI

As an NPO, WBAI aims to make related technical information available to the public to be used in better ways. To this end, WBAI not only publishes the products of its activities, but is also determined to facilitate open research and development. For example, it collaborates and has discussions with other open AGI projects such as OpenCog[1]. It also tries to facilitate research by publishing the previously mentioned research infrastructure so that more people can try out or apply published technologies with ease.

4   Activities in the First Year

Activities in the first year (from September 2015 to July 2016) include educational and R&D endeavors following the previously mentioned policy.

4.1   Educational Business

As noted previously, the goal of the educational business is to help people who can conduct research on the WBA approach on a long-term basis. In the first year, WBAI held the first hackathons and WBA seminars; participated in BICA 2015 in Lyon; and started creating learning material on the Web (in Japanese).

---

[1] http://opencog.org.

WBA Seminars and Symposium. As previously noted, WBA seminars have been held since before the foundation of WBAI. In the first year, WBAI held three seminars and a symposium with the following themes and speakers (mainly in Japanese):

- 11th Seminar: August 26, 2015, *Inside Deep Learning*
  Masayuki Ohzeki (Assistant Professor, Kyoto University), Yoichi Mototake (Univ. of Tokyo), Adam Gibson (CTO, Skymind)
- 12th Seminar: January 14, 2016, *Learning Architecture of the Brain*
  Kenji Doya (Professor, Okinawa Institute of Science and Technology)
- 13th Seminar: March 15, 2016, *Connectome and AI*
  Haruo Mizutani (Harvard University), Hiroki Kurashige (Univ. of Tokyo)
- First WBAI Symposium: May 18, 2016, *Accelerating AI, Accelerating World*
- 14th Seminar: May 18, 2016, *Neocortical Computational Models beyond Deep Learning*
  Takuya Matsuda (NPO Einstein), Manabu Tanifuji (Riken BSI)
- 15th Seminar: June 14, 2016, *Evolution, Development, and Learning in Intelligence*
  Nobuyuki Kawai (Nagoya Univ.), Hiroyuki Okada (Tamagawa Univ.)

The First Hackathon. Together with the Whole Brain Architecture Future Leaders, WBAI held its first hackathon at the Yokohama campus of Keio University for five days from September 19. Seven teams consisting of mainly undergraduate and graduate students participated in this event. Each team set their own task to meet the theme "development of a combined learner" and worked on it while staying in lodging facilities on campus. As one of WBAI's educational activities, this event aimed to improve the knowledge and skill of the participants and provide an opportunity for social networking among students and researchers interested in areas such as neuroscience and machine learning. The event received support from AIST, the University of Electro-Communications, and AlpacaDB, Inc., as well as additional backing from the Dwango AI Lab (Dwango Corporation). The products of the hackathon have been published on GitHub in English.[2]

Participating in BICA 2015. The BICA Society is an academic community on biologically inspired cognitive architectures (BICA) that holds international conferences annually. As WBA is apparently BICA, their theme accords well with ours. The BICA 2015 conference was held in Lyon, France, where five regular members and a WBA Future Leader participated as authors of submitted papers [1, 2]. There was also a WBA special session in the conference.[3] Moreover, WBAI invited three students, as the winners of the first hackathon, to present their work in the WBA session and write a report on the conference.

Compiling Learning Material on the Web. To disseminate basic knowledge for WBA in areas such as artificial intelligence, neuroscience, cognitive science, and machine learning, WBAI has been compiling glossaries (in Japanese) on its Web site.

---

[2] https://github.com/wbap/Hackathon2015.
[3] https://liris.cnrs.fr/bica2015/wiki/doku.php/wba.

Furthermore, WBAI is collaborating with university educators in each area. For the time being, a glossary in machine learning is getting ready.

4.2 R&D Business

As noted in Sect. 3, the goal of the R&D business is to support research activities on the WBA approach.

R&D at WBAI. WBAI is actively working to develop research infrastructure such as software and databases to be used in research and make it public. In particular, WBAI is working on a generic software platform for constructing WBA, evaluation methods of AGI, learning environments for WBA, and infrastructure for neuroinformatics.

- Generic Software Platform [1]
  The generic software platform for constructing WBA supports a mechanism that performs cognitive functions while machine learning modules corresponding to brain parts communicate each other, according to the core WBA hypothesis described in Sect. 2.3. In particular, platform modules communicate with numeric vector values corresponding to signals transmitted in axons. In collaboration with Riken and Keio University, the effort to create this platform began in 2014. It was named BriCA (Brain-inspired Computing Architecture) and implemented in Java (Version 0), then Python (Version 1), and currently in C++ (Version 2).
  Together with the BriCA platform, a language that describes its architecture has also been designed and implemented. This BriCA Language describes modules and connections among them for WBA platforms. Such an architecture description language facilitates the understanding and module modification of architecture implementation, and would work well with open, collaborative development. While the current implementation of BriCA Language uses the BriCA platform, it can also use other platforms such as ROS, Brain Simulator TM, or Nengo, in which modules pass messages to each other. Dwango AI Lab has offered support also for the design and implementation of this BriCA Language.
- Learning Environments
  Human-like AGI is expected to learn and acquire skills in a world similar to that in which human beings live. Though such a learning agent could be implemented as a physical robot, such a design may also require electro-mecharfical engineering besides artificial intelligence. Therefore, in cooperation with the Dwango AI Lab, WBAI is working on robot simulators in the virtual world as learning environment for AGI. Some results from the first quarter of 2016 have been published on the Web. The first result is the creation of environments with the Gazebo robot simulator combined with BriCA, Nengo, or Brain Simulator TM (Fig. 1), and the second shows the Unity game engine combined with machine learning modules from Chainer APIs (Fig. 2). The latter is the first of the software series called *Life in Silico (LIS)*.[4]

---

[4] https://github.com/wbap/lis.

- Infrastructure for Neuroinformatics

Knowledge of the architecture of the entire brain is required to create artificial general intelligence on the WBA approach. In particular, the knowledge (information) on the parts of the human/mammalian brain, microstructure of the parts, and interconnection between parts (connectome) are all required. Much of this knowledge already exists but is scattered across countless academic papers. To use it efficiently, this information should be integrated in one place. WBAI is working to centralize this information for improved knowledge integration.

Fig. 1. Gazebo simulator + BriCA   **Fig. 2.** LIS: unity game engine + Chainer

WBAI is also developing a software prototype of a WBA viewer called BICAmon (Brain-Inspired Cognitive Architecture monitor), which shows the activities of parts of cognitive architecture as if they are parts of a brain.[5] It interactively displays virtual brain parts and connections on a Web browser, while active parts in the corresponding cognitive architecture are highlighted. The viewer has been developed together with the Dwango AI Lab.

R&D on WBA. As noted in Sect. 3, WBAI does not conduct research on whole brain architecture itself, but supports others doing research using the WBA approach. One form of support is sharing discussion among researchers in areas related to WBA; WBAI hold discussion meetings inviting interested researchers and students. Areas of interest include affects and the hippocampus.

4.3 Forming an Open Community for AI Development

In June 2016, WBAI created a community of engineers for open AI development on *Slack*. This occurred after mini-hackathons held in the same month, where participants succeeded in implementing Deep Predictive Coding Networks (Deep PredNet)[6], proposed by William Lotter et al. in May as a promising model of the neocortex [3].

---

[5] https://github.com/kiyomaro927/bicamon.
[6] e.g., https://github.com/quadjr/PredNet.

The community integrates engineering activities around WBAI including those described in Sect. 4.2.

WBAI intends to augment this community in quality and quantity with the support of researchers in related areas to work on issues such as:

- Development of learning environment
- Development and evaluation of machine learning algorithms
- Experiments with AI agents that learns behaviors in interesting environments
- Implementation of new cognitive functions for AI agents
- Data analysis and tool development in neuroscience for the WBA approach

4.4   WBAI Activities and Volunteering

WBAI activities like WBA seminars and hackathons have been conducted with non-paid volunteers, except for two paid part-time workers at the secretariat and honorarium paid for hackathon tutors. PR activities focused on constructing and maintaining the Web site have been performed on a voluntary basis. WBAI has been collaborating with another volunteer organization called WBA Future Leaders for activities including WBA seminars and hackathons.

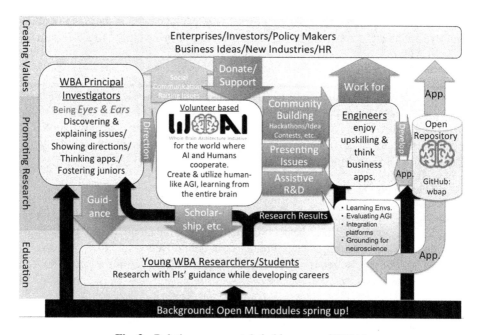

Fig. 3.   Relations among stakeholders around WBAI

5   Future Direction

The current development in AI and machine learning is quite rapid, and the advent of certain AGI is becoming more plausible every day. Thus, WBAI has set a goal to ensure that AGI is beneficial to all of humanity, with the realization of AGI in the near future in mind. To achieve this goal, WBAI is promoting and popularizing the use and development of technology inspired by WBA by holding events such as hackathons and providing tools for related technologies so that AGI development will be democratized. Figure 3 summarizes the relationships among stakeholders around WBAI in this direction.

Acknowledgments. The activities of WBAI have been made possible with support from sponsors and volunteers. We appreciate their support and invite more people to join this promising endeavor.

References

1. Takahashi, K., Itaya, K., Nakamura, M., Koizumi, M., Arakawa, N., Masaru Tomita, M., Yamakawa, H.: A generic software platform for brain-inspired cognitive computing. In: BICA 2015 Proceedings (2015)
2. Omori, T., Kasumi Abe, K., Nagai, T.: Modeling of stress/interest state controlling in robot-child play situation. In: BICA 2015 Proceedings (2015)
3. Lotter, W., Kreiman, G., Cox, D.: Deep Predictive Coding Networks for Video Prediction and Unsupervised Learning. arXiv:1605.08104v2[cs.LG] (2016)

Neural Network for Quantum Brain Dynamics: 4D CP^1+U(1) Gauge Theory on Lattice and Its Phase Structure

Shinya Sakane$^{(\boxtimes)}$, Takashi Hiramatsu, and Tetsuo Matsui

Department of Physics, Kindai University, Higashi-Osaka 577-8502, Japan
shinx0860@gmail.com, matsui@phys.kindai.ac.jp

Abstract. We consider a system of two-level quantum quasi-spins and gauge bosons put on a 3+1D lattice. As a model of neural network of the brain functions, these spins describe neurons quantum-mechanically, and the gauge bosons describes weights of synaptic connections. It is a generalization of the Hopfield model to a quantum network with dynamical synaptic weights. At the microscopic level, this system becomes a model of quantum brain dynamics proposed by Umezawa et al., where spins and gauge field describe water molecules and photons, respectively. We calculate the phase diagram of this system under quantum and thermal fluctuations, and find that there are three phases; confinement, Coulomb, and Higgs phases. Each phase is classified according to the ability to learn patterns and recall them. By comparing the phase diagram with that of classical networks, we discuss the effect of quantum fluctuations and thermal fluctuations (noises in signal propagations) on the brain functions.

Keywords: Hopfield model · Gauge neural network · Quantum brain dynamics

1 Introduction

Various functions of the human brain such as awareness, learning, and recalling patterns have been subjects of intense studies in wide area of science including neuroscience, medical science, psychology. A widely adopted approach in these studies is to model the brain by a neural network (network of neurons) and simulate its static and dynamical properties. A well known example of such network is the Hopfield model [1], which offers us an interesting mechanism of associative memory (recalling memorized patterns of neurons).

In the Hopfield model, each neuron may have two states (fired or not) and the state of the i-th neuron is described by the Ising (Z(2)) variable $S_i(= \pm 1)$ ($i = 1, \cdots, N$). S_i represents the scaled membrane potential as $S_i = 1$ (fired) and $S_i = -1$ (not fired). The information of memorized patterns of S_i is stored here in the parameters J_{ij}, which are called synaptic weights, through the Hebb's rule [2]. The time development of $S_i(t)$ ($t = 0, 1, 2, \cdots$ is a discrete time) is intrinsically

© Springer International Publishing AG 2016
A. Hirose et al. (Eds.): ICONIP 2016, Part I, LNCS 9947, pp. 324–333, 2016.
DOI: 10.1007/978-3-319-46687-3_36

deterministic, but, due to noises in signal propagation, it becomes random. This situation is modeled by introducing the energy $E(S_i(t), J_{ij})$ and considering statistical mechanics with Boltzmann distribution $P(S_i) \propto \exp[-\beta E(S_i, J_{ij})]$ where the effective "temperature" $T \equiv 1/\beta$ starts from zero (no noise) and rises as noise increases. Then the system is regarded as an Ising spin system with random (lomg-range) interactions J_{ij}. The phase diagram is calculable and consists of three phases; spin-ordered phase, spin-disordered phase, spin-glass phase according to the values of J_{ij} and β. The spin-ordered (ferromagnetic) phase corresponds to the state of successful recalling of learned patterns of S_i, while the spin-disorders (paramagnetic) phase to failed recalling, and the spin-glass phase to failed learning due to more patterns than the capacity.

As the next step, by regarding synaptic weights connecting neurons as plastic dynamical variables, various models of learning patterns have been proposed [3]. In Refs. [4,5] a set of new networks for learning have been proposed by promoting synaptic-weight parameters J_{ij} appeared in the Hopfield model to a dynamical gauge field $J_{ij}(t)$ (t is the time). The energy of these gauge neural networks respects gauge symmetry. Introduction of gauge theory as a model of brain functions is motivated from the function of synaptic weight itself. Let us consider the electric signal which starts from the neuron j and arrives at the neuron i. The electric potential transported by this signal is modulated from the initial value S_j at j to $J_{ij}S_j$ at j through the synapse. The synaptic weight J_{ij} is just the conversion factor of propagating potential. That is, J_{ij} expresses relative difference of two frames of potential at j and i. Any quantity having this nature, i.e., a measure of relative orientations of local frames, is to be called a gauge field. The gauge symmetry just implies that observable quantities such as energy should be independent of change of local frames as it should be. By treating these gauge models as models in statistical mechanics, we calculated their phase diagrams. Generally, they consist of three phases; confinement phase, Coulomb phase and Higgs phase. Each phase is characterized by the ability of learning patterns and recalling them (See Table 1).

Our common sense tells us that the brain functions have nothing to do with quantum theory (or quantum effect is negligibly small). However, as long as our brain is made of atoms and molecules at the microscopic level, the microscopic model of the brain itself should be described in terms of these atoms and molecules. If we are involved in the enterprise of describing and understanding the brain functions by a framework of physics, our task should be relating such microscopic quantum model to widely studied neural networks at the macroscopic level and calculating the quantum effect upon them quantitatively. This paper concerns these two points.

In Sect. 2 we briefly explain quantum field theory proposed by Umezawa et al. [6] as a model of brain dynamics at the microscopic level. It consists of two-level quasi-spin variables describing dielectric dipoles of water molecules and bosons describing photons inside the brain which mediate the electromagnetic (EM) forces between dipoles. We respect the U(1) gauge invariance of EM interaction and introduce the $CP^1+U(1)$ lattice gauge theory put on a 4D lattice

Table 1. Three phases of gauge neural network and abilities of learning and recalling patterns of S_i [4,5]. $\langle O \rangle$ is the Boltzmann average of O. $\langle J_{ij} \rangle \neq 0$ implies that J_{ij} has small fluctuations around the average (given by Hebb's law [2]), and the enough information of memorized patterns are stored in J_{ij}, while $\langle J_{ij} \rangle = 0$ implies that strong fluctuations wash out such information. Similarly $\langle S_i \rangle \neq 0$ implies that S_i sustains an almost definite pattern, while $\langle S_i \rangle = 0$ implies S_i is totally unfocused.

Phase	$\langle J_{ij} \rangle$	$\langle S_i \rangle$	Ability of learning	Ability of recalling
Higgs	$\neq 0$	$\neq 0$	Yes	Yes
Coulomb	$\neq 0$	0	Yes	No
Confinement	0	0	No	No

(3 spatial directions and 1 imaginary-time direction for path-integral quantization) as its lattice version. Introduction of a lattice model is to discuss an effective model at semi-macroscopic scales through renormalization. We then discuss that this lattice gauge theory itself may be regarded also as an effective GNN after parameters of the model are renormalized through coarse graining.

In Sect. 3 we calculate the phase diagram of this 4D $CP^1 + U(1)$ lattice gauge theory for general parameters and characterize each phase of Table 1 by measuring electric field, magnetic field, and magnetic monopole density. By considering this model as a GNN, we discuss the ability of learning and recalling patterns in each phase, and the quantum and thermal(noise) effects upon that ability by referring to the results of classical GNN's.

2 Quantum Brain Dynamics and the 4D $CP^1 + U(1)$ Lattice Gauge Theory

Umezawa et al. [6] proposed a quantum spin-boson model that may describe the brain at the microscopic level, and argued that memories may be stored in the ordered ground state and low-energy excitations. They considered a system of N atoms which interact through exchanging bosons. The m-th ago $(m = 1, \cdots, N)$ is described by $s = 1/2$ SU(2) pseudo-spin operators $\boldsymbol{S}_m = (S_{m1}, S_{m2}, S_{m3})$, and a boson having a 3D momentum k and energy $E(k)$ is described by canonical annihilation operator C_k and creation operator $C_k^\dagger$. Its Hamiltonian H is given by

$$H = \sum_k E_k C_k^\dagger C_k + \epsilon \sum_m S_{m3} - f \sum_m (C_m S_{m+} + \text{H.c.}), \qquad (1)$$

where $S_{m+} = S_{m1} + iS_{m2}$ is the spin rising operator and C_m is the Fourier transform of C_k. Each term expresses energy of bosons, level splitting of spins by external field, and emission and absorption of bosons and associated spin flips. Jibu and Yasue [7] argued that the quasi-spins and bosons in Eq. (1) have explicit counterparts in the human brain; each quasi-spin $\boldsymbol{S}_m$ describes an electric dipole moment of each molecule of bound water (water molecules stand almost still) and the bosons C_k describe evanescent photons mediating short-range interaction among dipoles.

To pursue this interpretation further and improve a couple of points of the model (1), we introduce a model with the following properties; (i) manifest U(1) local gauge invariance of EM interaction; (ii) self-consistently determined photon energy $E(k)$ (massive or massless); (iii) a lattice model with a cut-off scale to make renormalization-group transformation straightforward. It is a $CP^1+U(1)$ lattice gauge theory defined on the 4D hyper-cubic lattice, a variation of Wilson's lattice gauge theory [8] by replacing fermonic quark variables by the CP^1 spin variables. We shall work in the path-integral representation of the partition function, $Z = \mathrm{Tr}\exp(-\beta H)$. The imaginary time $\tau(\in [0,\beta])$ is also discretized with the lattice spacing a_0. We use $x = (x_0, x_1, x_2, x_3)$ as the site index of the 4D hypercubic lattice, and $x_1, x_2, x_3 = 0, 1, \cdots, N-1$ and $x_0 = 0, 1, \cdots, N_0 - 1$ and $\beta = N_0 \times a_0$. We use $\mu = 0, 1, 2, 3$ as the direction index and also as the unit vector in the μ-th direction. The lattice spacing $a_\mu = (a_0, a, a, a)$ is regarded as a parameter to set the scale of the model in the sense of renormalization group. The $s = 1/2$ spins are described by the so-called CP^1 (complex projective) variables $z_{x\sigma}(\sigma = 1, 2)$ on each site x, a two-component complex variables satisfying $|z_{x1}|^2 + |z_{x2}|^2 = 1$. On each link $(x, x+\mu)$ (straight path between two nearest-neighbor (NN) sites), we have a U(1) gauge variable, $U_{x\mu} = \exp(i\theta_{x\mu})$ $[\theta_{x\mu} \in (-\pi, +\pi)]$. In the naive continuum limit $(a_\mu \to 0)$, it is expressed as $U_{x\mu} = \exp(igaA_\mu(x))$ where $A_\mu(x)$ is the vector potential and g is the gauge coupling constant [8]. $U_{x\mu}$ measures the relative orientation of the two internal coordinates which measure the wave function of charged particles at x and $x+\mu$ [8]. Then Z is written as

$$Z = \int [dU][dz] \exp(A[U, z]),$$

$$[dU] \equiv \prod_{x,\mu} dU_{x\mu} = \prod_{x,\mu} \frac{d\theta_{x\mu}}{2\pi}, \quad [dz] \equiv \prod_x dz_{x1} dz_{x2} \delta(|z_{x1}|^2 + |z_{x2}|^2 - 1). \quad (2)$$

$A[U, z]$ is the action of the model given by

$$A = \frac{c_1}{2} \sum_{x,\mu,\sigma} \left(\bar{z}_{x+\mu,\sigma} U_{x\mu} z_{x\sigma} + \text{c.c.} \right) + \frac{c_2}{2} \sum_{x,\mu<\nu} \left(\bar{U}_{x\nu} \bar{U}_{x+\nu,\mu} U_{x+\mu,\nu} U_{x\mu} + \text{c.c.} \right), \quad (3)$$

where c_1 and c_2 are real parameters of the model. These parameters are regarded to characterize each brain, i.e., each person has his(her) own values of c_1 and c_2 (and the other parameters for (irrelevant) interactions not included here). The action A is invariant under the following U(1) local (x-dependent) gauge transformation;

$$z_{xa} \to z'_{xa} = e^{i\Lambda_x} z_{xa}, \quad U_{x\mu} \to U'_{x\mu} = e^{i\Lambda_{x+\mu}} U_{x\mu} e^{-i\Lambda_x}. \quad (4)$$

Here we note that the partition function Z of (2) is a function of βc_1 and βc_2. Below we set $\beta = 1$ in the most of expressions for simplicity. The β-dependence is easily recovered by replacing $c_{1(2)} \to \beta c_{1(2)}$. In the continuum limit $a, a_0 \to 0$, the c_1-term of (3) becomes the kinetic term of z_x, while the c_2-term becomes the electomagnetic action $\propto \boldsymbol{EE} + \boldsymbol{BB}$.

By applying the renormalization-group transformation to the model (2), one may obtained an effective theory at the lattice spacings $a'_\mu = \lambda a_\mu$. The analysis made for the related models of lattice gauge theory [9] shows that the relevant interactions at larger distances are the c_1 and c_2 terms and next-NN terms such as $\bar{z}UUUz$, $\bar{z}UUUUz$, and no qualitatively different terms emerge. Thus we think that the model (2) may be worth to study as an approximation of the effective model of neural network for the brain. In this viewpoint, the meaning of variables are as follows; (i) the CP^1 variable $z_{x\sigma}$ is the probability amplitude of quantum neuron state $|S_x\rangle = z_{x1}|1\rangle_x + z_{x2}|2\rangle_x$ where $|1\rangle$ and $|2\rangle$ are two independent states, such as fired or unfired, and (ii) the $U(1)$ variable $U_{x\mu} = \exp(i\theta_{x\mu})$ is the phase part of wave function of the synaptic connection weight between NN pair $(x, x + \mu)$. Therefore, by replacing $z_{x\sigma}$ and $U_{x\mu}$ by the neuron variable S_x and the synaptic weight variable $J_{x\mu}$ respectively, the action A of Eq. (3) is viewed as the action of GNN at macroscopic level;

$$A = \frac{c_1}{2} \sum \bar{S}_{x+\mu} J_{x+\mu,x} S_x + \frac{c_2}{2} \sum J_{x,x+\nu} J_{x+\nu,x+\mu+\nu} J_{x+\mu+\nu,x+\mu} J_{x+\mu,x} + \text{c.c.} \quad (5)$$

We note that its first term $c_1 SJS$ corresponds to the Hopfield energy [1] and the second term $c_2 JJJJ$ describes the reverberating current of signals explained in Ref. [2], which runs along a closed loop $(x \to x + \mu \to x + \mu + \nu \to x + \nu \to x)$.

Of course we recognize that the brain itself is far more complicated than this effective model; e.g., the network is multilayer with column structure and the synaptic connections are long-range and asymmetric (J_{ij} and J_{ji} are independent) with various strengths ($J_{ij} \in \mathbf{R}$). However, these points can be incorporated systematically into the present model (2) in the framework of quantum gauge theory as inputs in the stage of model building, and we leave them as future problems.

3   Phase Structure of the 4D CP^1 + U(1) Gauge Theory

In this section we study the phase structure of the model (2) by Monte Carlo simulation (MCS) and mean field theory (MFT) and discuss the effect of quantum and thermal fluctuations upon the ability of learning and recalling patterns.

3.1   Phase Diagram

In our MCS, we consider a hypercubic lattice of size L^4 with periodic boundary condition. This implies the corresponding "temperature" T tends to zero $T \to 0$ as the thermodynamic limit $L \to \infty$ is taken [9]. We use standard Metropolis algorithm to generate Markov process and present the results of $L = 16$ with typical sweep number for single run as $50000 + 10 \times 5000$. Errors are estimated as standard deviation of 10 samples taken in the last half of each run. To locate the phase transition point, we calculate the internal energy U and the specific heat C defined as the thermodynamic averages as

$$\langle O \rangle \equiv \frac{1}{Z} \int [dU][dz] O[U,z] e^{A[U,z]}, \quad U = \frac{1}{L^4}\langle -A \rangle, \quad C = \frac{1}{L^4}\langle (A - \langle A \rangle)^2 \rangle, \quad (6)$$

where Z and A are given in Eqs. (2) and (3). We measure U and C as a function of c_1 for a fixed value of c_2 (and vice versa). Location of phase transition point is determined from their behavior as follows;

(i) If $U(c_1)$ shows hysteresis while c_1 makes a round trip, it exhibits a first-order transition. Such hysteresis effect should diminish as MC runs more sweeps and leaves a gap $\Delta U(c_1)$ at the transition point $c_1 = c_{1c}$ $(\Delta U(c_1) \equiv \lim_{\epsilon \to 0_+}[U(c_1 + \epsilon) - U(c_1 - \epsilon)])$.
(ii) If $U(c_1)$ shows no hysteresis, but $C(c_1)$ has a peak developing as L increases and/or a gap of $\Delta C(c_1)$ appears at $c_1 = c_{1c}$, a second-order transition takes place there.

As typical examples of these transitions, we show the following three figures; In Fig. 1 we show U and C for $c_2 = 0.9$. U exhibits a hysteresis curve around $c_1 \sim 0.9$ and a first-order transition takes place. In Fig. 2 we show U and C for $c_2 = 0.4$. C exhibits a peak around $c_1 \simeq 1.63$ at which a second-order transition takes place. In Fig. 3 we show U and C for $c_2 = 2.0$. C exhibits a small jump which we take as a sign of a gap ΔC implying a second-order transition.

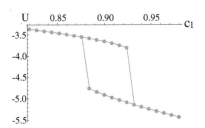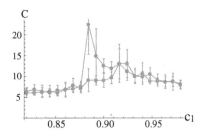

Fig. 1. $U(c_1)$ (left) and $C(c_1)$ (right) for $c_2 = 0.9$. U shows a hysteresis between $c_1 \simeq 0.88 \sim 0.93$. C shows double peaks near the edges of hysteresis.

In Fig. 4 we show the phase diagram in the c_2-c_1 plane. There are three phases as indicated. To identify each phase as shown there we measured squared electric field W_E, squared magnetic field W_B, and the magnetic monopole density Q [10] defined as follows;

$$W_E \equiv \frac{1}{3L^4} \sum_{x,i} \langle (E_{x,i} - \langle E_{x,i} \rangle)^2 \rangle = \frac{1}{3} \sum_{x,i} \left[c_2 \langle \cos \theta_{x,0i} \rangle - c_2^2 \langle \sin^2 \theta_{x,0i} \rangle \right],$$

$$W_B \equiv \frac{1}{3} \sum_{x,i<j} \langle \sin^2 \theta_{x,ij} \rangle,$$

$$Q \equiv -\frac{1}{2} \sum_{i,j,k} \epsilon_{ijk} \langle n_{x+i,jk} - n_{x,jk} \rangle = \frac{1}{4\pi} \sum_{i,j,k} \epsilon_{ijk} \langle \tilde{\theta}_{x,jk} - \bar{\theta}_{x,jk} \rangle, \quad (7)$$

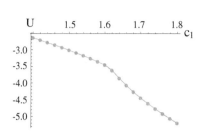 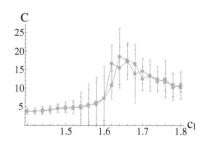

Fig. 2. $U(c_1)$ (left) and $C(c_1)$ (right) for $c_2 = 0.4$. $C(c_1)$ shows a peak at $c_1 \simeq 1.64$, at which a second-order transition takes place.

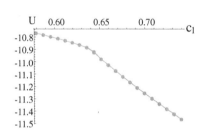 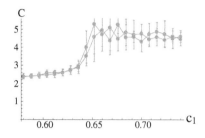

Fig. 3. $U(c_1)$ (left) and $C(c_1)$ (right) for $c_2 = 2.0$. $C(c_1)$ has a jump at $c_1 \simeq 0.65$ which we judge as a gap $\Delta C \neq 0$, implying a second-order transition.

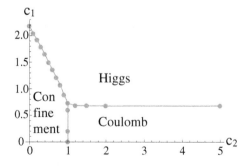

Fig. 4. Phase diagram of the 4D $CP^1 + U(1)$ model (2) in the c_2-c_1 plane determined by the MCS for the lattice $L = 16$. The transition between Coulomb and Higgs phases is of second-order. The confinement-Coulomb transition is of weak first order (almost second order), The confinement-Higgs transition is (i) first-order near the triple point, i.e., for $0.6 \lesssim c_2 \lesssim 1.0$, and (ii) second-order for $c_2 \lesssim 0.6$.

where i, j, k takes $1, 2, 3$ and we decompose $\theta_{x,ij} = \nabla_i \theta_{xj} - \nabla_j \theta_{xi}$ as $\theta_{x,ij} = 2\pi n_{x,ij} + \tilde{\theta}_{x,ij}$, $(-\pi < \tilde{\theta}_{r,ij} < \pi)$. $n_{x,ij} \in \mathbf{Z}$ describes nothing but the Dirac string (quantized magnetic flux). In short, W_E measures the magnitude of fluctuations of electric field $\boldsymbol{E}$, and W_B and Q measure fluctuations of magnetic field $\boldsymbol{B} = \mathrm{rot}\boldsymbol{A}$. Because vector potential $\boldsymbol{A}$ and $\boldsymbol{E}$ are canonically conjugate each other, uncertainty principle $\Delta A \Delta E \sim \Delta B \Delta E \gtrsim \mathrm{const.}$ holds. In confinement phase, $\Delta E \simeq 0$ and ΔB is large. In the deconfinement phase such as Coulomb and Higgs phases, ΔE is large and ΔB is small. ΔB is smaller in the Higgs phase than the Coulomb phase. We show these quantities for three values of c_2 shown in Figs. 1, 2 and 3; $c_2 = 0.9$ in Fig. 5, $c_2 = 0.4$ in Fig. 6, $c_2 = 2.0$ in Fig. 7. In general, in the small-c_1 phase, W_B is large and W_E small, and in the large-c_1 phase, other way around. From these properties, it is straightforward to identify three phases as shown in Fig. 4.

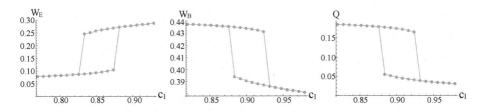

Fig. 5. $W_E(c_1)$ (left), $W_B(c_1)$ (middle), $Q(c_1)$ (right) for $c_2 = 0.9$.

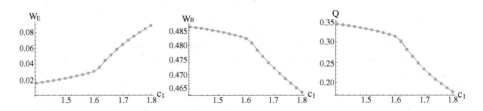

Fig. 6. $W_E(c_1)$ (left), $W_B(c_1)$ (middle), $Q(c_1)$ (right) for $c_2 = 0.4$.

3.2 Effect of Quantum and Thermal Fluctuations

To discuss the effect of quantum fluctuations, we introduce a classical model corresponding to the present quantum model (2). It is the 4D Z(2) gauge theory defined by the action of Eq. (5) with the choice $S_x = \pm 1$ and $J_{x,x+\mu} = J_{x+\mu,x} = \pm 1$. These Z(2) variables are discrete and express thermal fluctuations but no quantum fluctuations. In Fig. 8 we show the phase diagrams of these two models obtained by MCS. It shows that the region of Higgs phase is smaller in the

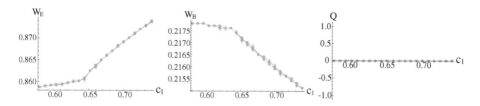

Fig. 7. $W_E(c_1)$ (left), $W_B(c_1)$ (middle), $Q(c_1)$ (right) for $c_2 = 2.0$. Q almost vanishes here due to strong suppression of monopoles due to large c_2, while fluctuations in zero-monopole sector generate small but finite W_B.

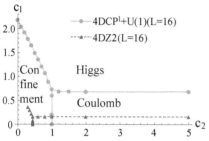

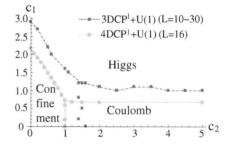

Fig. 8. Phase diagrams by MCS for 4D Z(2) model and 4D CP^1+U(1) model. Higgs region is smaller in the CP^1+U(1) model. The transition line of the Z(2) model terminates at $c_2 \simeq 0.28$.

Fig. 9. Phase diagrams by MCS for 3D and 4D CP^1+U(1) lattice gauge models. Higgs region is smaller in the 3D model. The 3D model has no Coulomb phase and the marks at $c_2 \simeq 1.4 \sim 1.6$ show the crossover.

CP^1+U(1) model than in the Z(2) model. Therefore we conclude that the quantum fluctuations in the present model generally reduce both abilities of learning patterns and recalling them (see Table 1).

So far we considered the case of no noises ($T = 0$). In contrast with $T = 0$, the high-temperature limit $T \to \infty$ implies $N_0 \to 1$ in $\beta = N_0 a_0$; i.e., the CP^1+U(1) model put on the 3D cubic lattice. Therefore, the effect of noises in signal propagations is estimated by comparing the results of the 4D model and 3D model with the same set of variables and action. In Fig. 9 we show the phase diagrams of the 3D model obtained by MCS [11] together with that of the 4D model in Fig. 4. In the 3D model, the confinement-Coulomb transition becomes a crossover and Coulomb phase disappears. Furthermore, the region of Higgs phase is smaller than that of the 4D model. Therefore we conclude that the thermal fluctuations in the present model generally reduce both abilities of learning patterns and recalling them.

4   Conclusion

We introduced the $CP^1+U(1)$ gauge theory on a 4D lattice as a microscopic model of quantum brain dynamics. It describes a system of molecules of bound water and photons in the brain and respects $U(1)$ gauge symmetry of the electro-magnetism. This model may be regarded also as a neural network of the brain after coarse graining. We calculated its phase diagram and compared it with related models. We found that both quantum fluctuations and thermal fluctuations by noise reduce the ability of learning and recalling patterns. We plan to confirm this point by an explicit simulation of learning processes.

Finally, we comment on the network structure of the $CP^1+U(1)$ model. To be realistic, the human brain has complicated network structures, such as left and right hemispheres, multilayer-structure, column-structure, small-world network, etc. Because the way to coase-grain the microscopic model is not unique by itself, additional argument is required to explain the realistic brain structure. On this point, it is interesting to define the coase-grained $CP^1+U(1)$ models on these networks and study their phase diagrams. Although we expect the basic three phases appeared in Table 1, the details should be structure-dependent and shed some light on the study of brain architecture.

Acknowledgment. The authors thank Prof. K. Sakakibara and Dr. Y. Nakano for discussion. This work was supported by JSPS KAKENHI Grant No. 26400412.

References

1. Hopfield, J.J.: Proc. Nat. Acd. Sci. USA **79**, 2554 (1982)
2. Hebb, D.O.: The Organization of Behavior: A Neuropsychological Theory. Wiley, New York (1949)
3. Haykin, S.: Neural Networks; A Comprehensive Foundation. Macmillan Pub. Co., New York (1994)
4. Matsui, T.: Fluctuating Paths and Fields, Janke, W., et al. (eds.). World Scientific (2001) 271 (arXiv:cond-mat/0112463); Kemuriyama, M., Matsui, T., Sakakibara, K.: Physica A **356**, 525 (2005) (arXiv:cond-mat/0203136); Takafuji, Y., Nakano, Y., Matsui, T.: Physica A **391**, 5258 (2012)
5. Fujita, Y., Hiramatsu, T., Matsui, T.: Proceedings of International Joint Conference on Neural Networks (Montreal, Canada, 2005) 1108. There is a similar model describing each neuron state by $U(1)$ variables instead of CP1 variable; Fujita, Y., Matsui, T.: Proceedings of 9th International Conference on Neural Information Processing, Wang, L., et al. (eds.) (2002) 1360 (arXiv:cond-mat/0207023)
6. Stuart, C., Takahashi, Y., Umezawa, H.: J. Theor. Biol. **71**, 605 (1978); Found. Phys. **9**, 301 (1979); Ricciardi, L.M., Umezawa, H.: Kybernetik **4**, 44 (1967)
7. Jibu, M., Yasue, K.: Informatica **21**, 471 (1997); Quantum Brain Dynamics. An Introduction. John Benjamins, Amsterdam (1995); See also Vitiello, G.: Int. J. Mod. Phys. B **9**, 973 (1995)
8. Wilson, K.: Phys. Rev. D **10**, 2445 (1974)
9. Rothe, H.J.: Lattice Gauge Theories: An Introduction. World Scientific, Singapore (2005)
10. DeGrand, T.A., Toussaint, D.: Phys. Rev. D **22**, 2478 (1980)
11. Takashima, S., Ichinose, I., Matsui, T.: Phys. Rev. B **72**, 075112 (2005)

BriCA: A Modular Software Platform for Whole Brain Architecture

Kotone Itaya[1,2], Koichi Takahashi[1,2,3](✉), Masayoshi Nakamura[3],
Moriyoshi Koizumi[4], Naoya Arakawa[3], Masaru Tomita[1],
and Hiroshi Yamakawa[3]

[1] Keio University Graduate School of Media and Governance, Minato, Japan
[2] RIKEN QBiC, Suita, Japan
ktakahashi@riken.jp
[3] Dwango AI Laboratory, Minato, Japan
[4] Open Collector Inc., Tokyo, Japan

Abstract. Brain-inspired Computing Architecture (BriCA) is a generic software platform for modular composition of machine learning algorithms. It can combine and schedule an arbitrary number of machine learning components in a brain-inspired fashion to construct higher level structures such as cognitive architectures. We would like to report and discuss the core concepts of BriCA version 1 and prospects toward future development.

Keywords: Software platform · Cognitive architecture · Machine learning · Modularity · The Whole Brain Architecture

1 Introduction

Recent advancements in computational neuroscience has driven the development of machine learning algorithms based on the neurological characteristics of the brain: one of the most prominent being deep learning [8]. This has motivated research of implementing complex machine learning systems by combining machine learning algorithms of different paradigms to achieve performance and functions which were unaccomplished with conventional machine learning systems [7,10,15]. The Whole Brain Architecture (WBA) project has set up a central hypothesis which claims that the brain attains its functionality by combining modules which can be modeled as machine learning algorithms, thus combining machine learning modules according to the brain will result in at least human-level artificial intelligence. The goal of the WBA project is to constructively test this hypothesis by developing machine learning modules which represent specific brain components and combining those modules to build cognitive architectures. The WBA hypothesis contains at least three major points of discussion; if the brain is modular, if brain components are representable as machine learning algorithms, and if machine learning algorithms are non-additive.

© Springer International Publishing AG 2016
A. Hirose et al. (Eds.): ICONIP 2016, Part I, LNCS 9947, pp. 334–341, 2016.
DOI: 10.1007/978-3-319-46687-3_37

In order to test the WBA hypothesis it is required to collect neurological knowledge of the brain, develop novel machine learning algorithms, design cognitive architectures, and implement software to execute agents. As research in a wide variety of domains is a necessity, it is essential to construct an environment to encourage an open community driven cooperative development. To gather as many collaborators as possible, it would be helpful to create a reference architecture for WBA and software for editing, sharing, and executing such architecture. Brain-inspired Computing Architecture (BriCA) is an integrated software platform for implementing, hierarchically connecting, and executing multiple machine learning algorithms, which provides a domain specific language (DSL) for editing architectures as well as learning curricula construction.

Implementation of WBA requires BriCA to be able to execute machine learning systems of distinct paradigms while providing interfaces for efficient scheduling, synchronization, and communication. Some existing software which may be applicable for such use case include robot middleware, data analysis platforms, and simulation software. Robot OS (ROS) [11] and Middleware for Robotic Applications (MIRA) [2] are robot middleware capable of executing heterogeneous software on distributed platforms, however they allow very high communication latency and do not offer a DSL for architecture definition. Data analysis platforms (Weka [4], Garuda [3], Jubatus, TensorFlow [1]) and simulation software (Simulink, LabVIEW, E-Cell [13] are not designed for execution in real time and have limited functionalities. From our requirements analysis of BriCA V0 [14], the minimal specifications for BriCA are currently as follows:

1. Provide a module library of novel and existing machine learning algorithms.
2. Support hierarchical connection of machine learning modules to compose cognitive architectures.
3. Implement a unified messaging protocol to connect machine learning modules.
4. Provide a unified sensor/actuator interface to interact with an environment.
5. Provide real time scheduling for asynchronous calling and controlling of machine learning modules.
6. Be scalable in terms of software and performance for N modules.
7. Support learning curricula creation and execution to deal with the combined learning system problem.
8. Support a community based distributed development.

In BriCA, algorithms are chosen and used at the level of brain components instead of that of neurons, which will most likely result in modules having different loads during execution according to the chosen algorithm. As modules with heavy loads will potentially postpone or block the execution of other modules from the given interval in synchronous execution, modules must be executed in a concurrent manner. Message passing is our current model for communication for two reasons; it is scalable in terms of implementation and it is more likely to be the communication model between brain components than the shared memory model. However, as architectures including the blackboard model may be required in future development, a module dedicated for the purpose may be implemented.

2   Implementation

The main functionalities of BriCA V1 core are provided by five classes, namely *Unit, Component, Module, Agent,* and *Scheduler.* The *Unit* is a class which is meant to represent a unit of functionality and is inherited to implement the *Component* and *Module* classes. A *Unit* has a set of input and output *Port*s which holds some value in its buffer which may be interchanged with another *Port* connected to itself. *Port*s can be aliased to other *Port*s, in which case the internal buffers and its connections are shared and acts as the exact same *Port.* *Component*s represent a unit of implementation, and has three extra buffers called inputs, states, and outputs. Its design is based on BriCA V0 *Module*s [14], with an exception of allowing any type of data to be passed through the *Port*s. Like the V0 *Module*, a *Component* also additionally owns a method named *fire()*, which performs the following operation.

```
outputs, states <- fire(inputs, states)
```

The behavior of a *Component* is specified by implementing an algorithm within the *fire()* function which must be overridden by inheriting from the *Component* base class. In practice, because these *Component*s must be executed concurrently, the *fire()* can only access the buffers from the input and output *Port*s indirectly. Therefore the following three step execution takes place as from V0 *Module*s [14].

1. `inputs <- In Ports`
2. `outputs, states <- fire(inputs, states)`
3. `Out Ports <- outputs`

BriCA V0, only with the *Module* class, had no way of creating nested *Modules* which for the development of WBA is a shortfall. To support hierarchical structuring of cognitive architectures developed with BriCA, the *Module* class implements an interface to contain an arbitrary number of *Modules* and *Components* within itself. Unlike *Components*, *Modules* have no methods for defining an implementation which makes structuring its sole purpose. The *Agent* class is a subclass of the *Module* and serves as the top level *Module* of a cognitive architecture. A *Scheduler* will execute the *Components* within a given *Agent* starting at the offset time set for each *Component* and then periodically given the interval specific to the *Component*. There are three categories of *Schedulers* planned for implementation; virtual time, real time, and external time. Virtual time schedulers will execute as if the given time has passed, which may be faster or slower than the actual time. Real time and external time schedulers execute synchronously with a given time system, where in real time *Schedulers* the reference will be the actual time and in external time *Schedulers* it will be another software which may or may not be in sync with the actual time. There are two types of *Schedulers* for each category, either calling all of the *fire()* methods from every *Component* synchronously, or asynchronously.

3   Results

BriCA V1 is implemented in Python language, available freely under the terms of the Apache License. The source code is currently hosted online as a GitHub repository (https://github.com/wbap/V1) and its documentation is available at a GitHub pages site for the WBAP/V1 repository (http://wbap.github.io/V1/). The NumPy Python library is required as a dependency when installing BriCA V1.

To demonstrate the application of the BriCA architecture, a stacked autoencoder consisting of three individual autoencoders and a single layer perceptron has been implemented to perform a categorization task of hand-written digits. An autoencoder is a multi-layered artificial neural network which is optimized to reconstruct the input from the hidden representation [6] which can result in reduction of dimensionality, extracting important features from the input data.

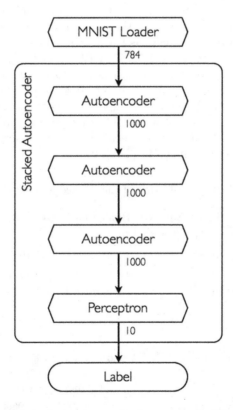

Fig. 1. The BriCA based stacked autoencoder is composed of three autoencoder *Components* and a simple perceptron *Component*. Arrows represent the connection between *Components* and the labeled numbers are the dimensionality of vectors being communicated.

Autoencoders can be "stacked" to create stacked autoencoders. The hidden layers of the preceding autoencoder is fed to the input of the following autoencoder, allowing the second autoencoder to extract features from the features extracted by the first autoencoder. By training each autoencoder step by step it is possible to construct a very deep neural network. For experimental purposes we have encapsulated an autoencoder into a BriCA V1 *Component* and connected the autoencoder *Components* to create a stacked autoencoder (Fig. 1). The autoencoder *Components* take the input vector and transforms the values to a hidden representation which is exposed to an Out *Port*. At the same time, a single iteration of unsupervised learning is performed to train on the fed data. A single layer perceptron that also exposes its transformed data to an Out *Port* and performs a single iteration of supervised learning has been attached to the end of the stacked autoencoder so the model can be used to perform classification tasks. The stacked autoencoder, along with the perceptron has been wrapped inside a *Module* to create a single classifier *Module*.

The MNIST dataset [9] has been used to train a stacked autoencoder with a composition identical to that from the stacked autoencoder section of Deep Learning Tutorials (Table 1) which has been trained for 20 epochs on a machine with 12 cores Intel(R) Core(TM) i7-5930K CPU @ 3.50 GHz and GeForce GTX TITAN X. The source code for the stacked autoencoder and its training is posted as a GitHub gist (https://gist.github.com/ktnyt/455694506ee6595c92e4). A stacked autoencoder with the same composition written only with the Chainer framework has also been implemented and trained for comparsion. The classification accuracy for the test dataset and execution time of the script in microseconds for 100 iterations have been listed in (Table 2).

Table 1. Architecture of the stacked autoencoder

Network	Input layer size	Hidden layer size	Output layer size
Autoencoder 1	768 (28 * 28)	1000	768
Autoencoder 2	1000	1000	1000
Autoencoder 3	1000	1000	1000
Perceptron	1000	No Hidden Layer	10

Table 2. Comparison of BriCA and chainer stacked autoencoders

Criteria	Chainer	BriCA V1
Accuracy (Test data)	0.965987	0.966094
Execution time (microseconds)	194,354,796	201,845,639

For both cases, the prediction accuracy for the test dataset reached 96.6 %. Results from F-test on the samples have revealed that both results have equal variance (P $\gg$ 0.05) and Student T-test showed no significant change in its mean (P $\gg$ 0.05). Execution time showed a significant difference in its variance (P = 0.004593) and Welch T-test indicated a significantly longer execution time for the BriCA implementation (P = 2.2e–16). The overhead for communication between components compared to the Chainer implementation was 3.85 %.

4   Discussion

In BriCA V1 there are two types of pluggable *Units*: *Components* and *Modules*, while there was only one in V0 with characteristics of both classes. The base class was split into two classes in order to loosen the restriction of the ports to allow the communication of more flexible structures which could not be encoded as distributed representations (e.g. trees and stacks). However, this partition raises a question as a component is a unit of machine learning algorithm and not necessarily a unit of brain function abstraction. Therefore scheduling of components may not be the equivalent to the scheduling of brain regions depending on the architecture. Although the current design of the library makes BriCA V1 simple in terms of software design it may not be intuitively suited for the use as a WBA platform. This aspect, though, may be wrapped away from users by implementing a domain specific markup language.

One aspect to take into account for future development is the support of training curricula to solve the combined learning problem. There are a series of technical issues in machine learning systems which were not present in classical computing, as pointed out by a paper published by a team at Google [12]. The problem that most attracts the attention of the BriCA development group is boundary erosion. In machine learning systems it is difficult to draw strict abstraction boundaries between two machine learning algorithms. For example, in the case of the three layer SDA implementation from this paper, the second autoencoder depends on the output of the first autoencoder, whose internal representations of the raw signals may change with the introduction of a new training datum. Such nature makes it difficult to add changes to a single component as Changing Anything Changes Everything (CACE) [12]. This issue is controversial, as (quoted in [12]) "the desired behavior cannot be effectively implemented in software logic without dependency on external data". However, as a running WBA agent will have a static architecture these considerations must be made during the design of the cognitive architecture. Ordering and scheduling of the training curriculum must be taken into account during execution. An implementation of a trainer class for supervising the training curricula for an architecture is being planned.

As the application of BriCA is assumed in a wide variety of environments including robotics, gaming, and data analysis, there is a need for a sensor/actuator interface capable of connecting to platforms widely used in these areas. In the case of robotics, the widely used ROS and MIRA platforms provide interfaces to add modules and have an interface for Gazebo, a robot simulation software. Integration into popular game engines including Unity, Irrlicht

Engine, and cocos2D will enable working with games. Some data analysis platforms include Jubatus and Garuda. The current BriCA implementation supports connection with ROS Indigo through the RosPy interface.

It is also important to implement popular algorithms from open source machine learning libraries as BriCA V1 *Components* for use in machine learning system development. Some examples of general libraries include Chainer, Scikit-learn, PyBrain, Theano, Pylearn2, Apache Spark, Weka, MALLET, Dlib-ML, shogun, and Stuttgart Neural Network Simulator (SNNS). Specialized libraries include a number of implementations for simultaneous localization and mapping (SLAM) (RatSLAM, RT-SLAM, LSD-SLAM), a model of the cerebral neocortex designated BESOM, and a cerebrum feedback model MOSAIC [5].

Finally, the performance is a key issue in BriCA V1 as the library is implemented in Python to serve as a proof of concept for the architecture. The current communication overhead is large at 10 mis compared to the BriCA V0 Java counterpart which is at 100 ns. The stacked autoencoder benchmarks showed a significant increase in execution time when implemented in BriCA V1, which suggests that the current implementation is not suited as the final product for WBA implementation. As the most frequent neuron activation is at about 1 kHz, assuming that 1 % of the execution time may be accepted as communication overhead, it must be at maximum 10 ms, which cannot be overcome by Python. We are currently developing V2 in C++ which will support reference passing in an attempt to reducing the overhead. Another problem of Python is the global interpreter lock (GIL) which restricts the number of threads executing under a single Python interpreter instance to only one. BriCA V2 also seeks to provide concurrency by implementing the *Scheduler* in C++ and allow asynchronous scheduling of *Components*.

Acknowledgments. We would like to thank Yuji Ichisugi, Makoto Taiji, Shinji Nishimoto, Hidemoto Nakada, and the members of the Whole Brain Architecture Initiative, especially Ryutaro Ichise, Takashi Omori, Hideki Kashioka, Satoshi Kurihara, Takeshi Sakurada, Takeshi Sato, and Yutaka Matsuo, along with the Whole Brain Architecture Future Leaders for their support, comments, and discussion. This research was supported in part by funds from Yamagata Prefectural Government and Tsuruoka City.

References

1. Abadi, M., Agarwal, A., Barham, P., Brevdo, E., Chen, Z., Citro, C., Corrado, G.S., Davis, A., Dean, J., Devin, M., Ghemawat, S., Goodfellow, I., Harp, A., Irving, G., Isard, M., Jia, Y., Jozefowicz, R., Kaiser, L., Kudlur, M., Levenberg, J., Mané, D., Monga, R., Moore, S., Murray, D., Olah, C., Schuster, M., Shlens, J., Steiner, B., Sutskever, I., Talwar, K., Tucker, P., Vanhoucke, V., Vasudevan, V., Viégas, F., Vinyals, O., Warden, P., Wattenberg, M., Wicke, M., Yu, Y., Zheng, X.: TensorFlow: Large-scale machine learning on heterogeneous systems (2015), http://tensorflow.org/, software available from tensorflow.org
2. Einhorn, E., Langner, T., Stricker, R., Martin, C., Gross, H.M.: Mira - middleware for robotic applications. In: 2012 IEEE/RSJ International Conference on Intelligent Robots and Systems, pp. 2591–2598, October 2012

3. Ghosh, S., Matsuoka, Y., Asai, Y., Hsin, K.Y., Kitano, H.: Software for systems biology: from tools to integrated platforms. Nat. Rev. Genet. **12**(12), 821–832 (2011). http://www.ncbi.nlm.nih.gov/pubmed/22048662
4. Hall, M., Frank, E., Holmes, G., Pfahringer, B., Reutemann, P., Witten, I.H.: The weka data mining software: an update. SIGKDD Explor. Newsl. **11**(1), 10–18 (2009). http://doi.acm.org/10.1145/1656274.1656278
5. Haruno, M., Wolpert, D.M., Kawato, M.: Mosaic model for sensorimotor learning and control. Neural Comput. **13**(10), 2201–2220 (2001). http://www.ncbi.nlm.nih.gov/pubmed/11570996
6. Hinton, G.E., Salakhutdinov, R.R.: Reducing the dimensionality of data with neural networks. Science **313**(5786), 504–507 (2006). http://science.sciencemag.org/content/313/5786/504
7. Karpathy, A., Li, F.: Deep visual-semantic alignments for generating image descriptions. CoRR abs/1412.2306 (2014). http://arxiv.org/abs/1412.2306
8. LeCun, Y., Bengio, Y., Hinton, G.: Deep learning. Nature **521**(7553), 436–444 (2015). http://dx.doi.org/10.1038/nature14539
9. Lecun, Y., Cortes, C.: The MNIST database of handwritten digits. http://yann.lecun.com/exdb/mnist/
10. Mnih, V., Kavukcuoglu, K., Silver, D., Rusu, A.A., Veness, J., Bellemare, M.G., Graves, A., Riedmiller, M., Fidjeland, A.K., Ostrovski, G., Petersen, S., Beattie, C., Sadik, A., Antonoglou, I., King, H., Kumaran, D., Wierstra, D., Legg, S., Hassabis, D.: Human-level control through deep reinforcement learning. Nature **518**(7540), 529–533 (2015). http://dx.doi.org/10.1038/nature14236
11. Quigley, M., Conley, K., Gerkey, B., Faust, J., Foote, T., Leibs, J., Wheeler, R., Ng, A.Y.: ROS: an open-source robot operating system. In: ICRA Workshop on Open Source Software, vol. 3, p. 5
12. Sculley, D., Holt, G., Golovin, D., Davydov, E., Phillips, T., Enber, D., Chaudhary, V., Young, M.: Machine learning: the high interest credit card of technical debt. In: SE4ML: Software Engineering for Machine Learning (NIPS Workshop) (2014)
13. Takahashi, K., Kaizu, K., Hu, B., Tomita, M.: A multi-algorithm, multi-timescale method for cell simulation. Bioinformatics **20**(4), 538–546 (2004). http://www.ncbi.nlm.nih.gov/pubmed/14990450
14. Takahashi, K., Itaya, K., Nakamura, M., Koizumi, M., Arakawa, N., Tomita, M., Yamakawa, H.: A generic software platform for brain-inspired cognitive computing. Procedia Comput. Sci. **71**, 31–37 (2015). http://www.sciencedirect.com/science/article/pii/S1877050915036467, 6th Annual International Conference on Biologically Inspired Cognitive Architectures, BICA 2015, 6–8 November Lyon, France
15. Vinyals, O., Toshev, A., Bengio, S., Erhan, D.: Show and tell: a neural image caption generator. CoRR abs/1411.4555 (2014). http://arxiv.org/abs/1411.4555

An Implementation of Working Memory Using Stacked Half Restricted Boltzmann Machine

Toward to Restricted Boltzmann Machine-Based Cognitive Architecture

Masahiko Osawa[1,2(✉)], Hiroshi Yamakawa[2], and Michita Imai[1]

[1] Department of Information and Computer Science, Keio University,
3-14-1 Hiyoshi, Kohoku, Yokohama 223-8522, Japan
{mosawa,michita}@ailab.ics.keio.ac.jp
[2] Dwango AI Laboratory, Tokyo, Japan
hiroshi_yamakawa@dwango.co.jp

Abstract. Cognition, judgment, action, and expression acquisition have been widely treated in studies on recently developed deep learning. However, although each study has been specialised for specific tasks and goals, cognitive architecture that integrates many different functions remains necessary for the realisation of artificial general intelligence. To that end, a cognitive architecture fully described with restricted Boltzmann machines (RBMs) in a unified way are promising, and we have begun to implement various cognitive functions with an RBM base. In this paper, we propose new stacked half RBMs (SHRBMs) made from layered half RBMs (HRBMs) that handle working memory. We show that an ability to solve maze problems that requires working memory improves drastically when SHRBMs in the agent's judgment area are used instead of HRBMs or other RBM-based models.

Keywords: Restricted Boltzmann machine · Cognitive architecture

1 Introduction

Cognitive architectures are blueprints for artificial general intelligence that models the behavior of humans and other organisms. Although many cognitive architectures have been proposed, to the best of our knowledge, none describe the entire architecture with a unified computational theory. However, architecture described by a single computational theory would likely have a superior architectural view, extendibility, and module connectivity.

If we adopt the core techniques of deep learning as a model for a unified description of the entire cognitive architecture, then the results of recent research can be easily incorporated. Furthermore, because many modules are complexly connected in computing cognitive architecture, architecture based on unsupervised learning named pretraining is more appropriate than that with a technique based on backpropagation or end-to-end learning, as in [1].

© Springer International Publishing AG 2016
A. Hirose et al. (Eds.): ICONIP 2016, Part I, LNCS 9947, pp. 342–350, 2016.
DOI: 10.1007/978-3-319-46687-3_38

The primary techniques used in pretraining are autoencoder- and restricted Boltzmann machine (RBM)-type techniques. Among the several differences between the types, of particular importance is that RBMs are stochastic models, which in cognitive architecture look promising for stabilizing the whole system [2].

From that perspective, RBM-based cognitive architectures are promising as standard computational models. In response, this paper proposes stacked half RBMs (SHRBMs) that carry out working memory functions as a step toward realizing RBM-based cognitive architecture. Of all action test battery maze problems, we model and use the eight-arm radial maze problem for our evaluative experiments. Such problems rank among those often used in real life to investigate cognitive functions concerning the working memory of mice.

In Sect. 2 of the paper, we summarize RBMs and the basis of half RBMs (HRBMs). In Sect. 3, we explain the details of HRBMs and, in Sect. 4, the evaluative experiments. We close the paper with a summary in Sect. 5.

1.1 Related Works

Since Soar [3] and ACT-R [4] use symbolic information expressions, they cannot perform stochastic behaviors. At the same time, although OpenCogPrime [5, 6] partially incorporates stochastic functions, because the entire architecture is constructed using diverse modules, it experiences major problems with connections among modules. Moreover, though Nengo [7] has a highly united cognitive architecture, it does not include learning functions.

RBM-based cognitive architecture, if actualized, might be able to realize stochastic behaviors and learning functions, given the advantage of simple connections among modules, many of which are constructed from RBMs and their extension modules.

At the same time, several proposed extended RBMs can handle time series data. Temporal RBMs (TRBMs) [8] use both hidden and visible layers from several steps earlier, and as some authors have shown, another method can perform unsupervised learning of time series data using RBMs [9]. Though similar to that of TRBMs, the method differs insofar as uses only hidden layer expressions from one moment earlier. However, learning becomes difficult when the number of visible units is smaller than the number of hidden units. Plus, as literature on TRBMs has shown, learning also becomes difficult when it is multilayered.

On the contrary, HRBMs and SHRBMs can mitigate the tendency for learning to become difficult when recursive rates are introduced, even if there is a large difference in the number of visible and hidden units. Furthermore, when the primary focus is working memory, performances improve with multilayering.

Although recurrent TRBMs (RTRBMs) [10] and RNN-RBMs [11] have been proposed as improved models, RTRBMs are mitigated in terms of their stochastic activity, and RNN-RBMs, which involve techniques combining RBMs and recurrent neural networks (RNNs), constitute undesirable optimizations for use as base cognitive architecture.

2 Restricted Boltzmann Machines and Extended Models

2.1 Restricted Boltzmann Machines

The probability that a unit's value is 1 is described in the following equations:

$$p\left(h_j = 1|v\right) = \sigma\left(c_j + \sum_i^V v_i w_{ij}\right) \tag{1}$$

$$p(v_i = 1|h) = \sigma\left(b_i + \sum_j^H h_j w_{ji}\right) \tag{2}$$

in which $\sigma(x)$ is a sigmoid function, V and H are the numbers of visible and hidden units, respectively, v_i and h_j are the visible and hidden units, b and c are biases of the visible and hidden units, and w_{ij} is the weight between units v_i and h_j.

RBMs update their weights in order to minimise the following energy function against the training data:

$$E(v, h) = -\sum_i^V b_i v_i - \sum_j^H c_j h_j - \sum_i^V \sum_j^H v_i w_{ij} h_j \tag{3}$$

The probability of all possible pairs of visible layers and hidden layer can be described as follows, in which Z is a normalisation coefficient:

$$p(v, h) = \frac{1}{Z} e^{-E(v,h)} \tag{4}$$

The gradient of the weight is expressed by the following equation, in which the parameter lr is the learning rate:

$$\Delta w_{ij} = lr\left(\langle v_i h_j \rangle_{data} - \langle v_i h_j \rangle_{model}\right) \tag{5}$$

2.2 Weight Determination Method for Echo State Networks Using RBMs [9]

Some authors have proposed a weight determination method for echo state networks (ESNs) using RBMs with hidden layer vectors from the previous moment connected to visible layer vectors. Accompanying that, (1) and (2) are modified as (6) through (8), energy function is expressed in (9), and the updated rules are (10):

$$p\left(h_j^t = 1|v^t, h^{t-1}\right) = \sigma\left(\sum_i^V v_i^t w_{ij} + \sum_k^H h_k^{t-1} w_{kj}\right) \tag{6}$$

$$p(v_i^t = 1|h) = \sigma\left(c_i + H \sum_j h_j^t w_{ji}\right) \tag{7}$$

$$p\left(h_k^{t-1} = 1 \mid \boldsymbol{h}^t\right) = \sigma\left(\sum_j^H h_j^t w_{jk}\right) \tag{8}$$

$$E\left(\boldsymbol{v}^t, \boldsymbol{h}^t, \boldsymbol{h}^{t-1}\right) = \sum_i^{V+H} \sum_j^H comb\left(\boldsymbol{v}^t + \boldsymbol{h}^{t-1}\right)_i w_{ij} h_j^t \tag{9}$$

$$\Delta w_{ij} = \alpha\left(\left\langle comb\left(\boldsymbol{v}^t, \boldsymbol{h}^{t-1}\right)_i, h_j\right\rangle_{data} - \left\langle comb\left(\boldsymbol{v}^t, \boldsymbol{h}^{t-1}\right)_i, h_j\right\rangle_{model}\right) \tag{10}$$

Where, $\boldsymbol{h}^t$ represents the expression of hidden layers during time t, and $comb(\boldsymbol{x}, \boldsymbol{y})$ represents the connection of vector $\boldsymbol{x}$ and $\boldsymbol{y}$.

Trained RBM is converted into an ESN. Weights between $\boldsymbol{v}^t$ and $\boldsymbol{h}^t$ and between $\boldsymbol{h}^{t-1}$ and $\boldsymbol{h}^t$ in the RBM becomes the weights between the input layer and the reservoir and among the reservoir in the ESN, respectively.

However, in that method, when the number of hidden units is larger than the number of visible units, the learning result of the hidden layers' expression constantly becomes the same expression. That result is arguably due to the fact that when the number of units of the hidden layers is larger than that of the visible layers, $W_{\boldsymbol{v}^t \boldsymbol{h}^t}$ learning is neglected, because the priority of learning with $W_{\boldsymbol{h}^{t-1} \boldsymbol{h}^t}$ and $W_{\boldsymbol{v}^t \boldsymbol{h}^t}$ depends on the number of units of both the visible and hidden layers.

3 Stacked Half Restricted Boltzmann Machines

In this section, we explain the model that [9] has extended even further. Because that RBM is thought to be one with the hidden layer restriction ultimately removed, it is called an HRBM. When that HRBM is further multilayered, it is called a stacked HRBM.

Figure 1 shows the schematic diagram of the HRBM, with two points of improvement from the method in [9].

The first improvement is that a recursive rate r_{rec} has been introduced and a restriction on recursive connections established. In Fig. 1, the number of units that connect to a visible unit is limited to two by introducing the recursive rate $r_{rec} = 1/3$

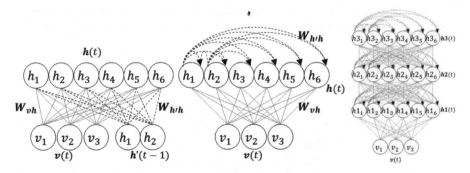

Fig. 1. Structure of half restricted Boltzmann machines, with learning time construction (left), deduction time (centre), and a stacked half restricted Boltzmann machine (right)

for six hidden units. Thanks to this improvement, the network structure can be flexibly adjusted, even when the number of units differs greatly.

The other improvement comes with changing the energy function. The HRBM divides the energy function into two parts, and the energy is minimised accordingly:

$$E\left(v^t, h^t\right) = -\sum_i^V v_i^t w_{ij} h_j^t \qquad (11)$$

$$E\left(h^{t-1}, h^t\right) = -\sum_i^{H*r_{rec}} h_i^{t-1} w_{ij} h_j^t \qquad (12)$$

In the same way, the weights' volume of change is expressed by the following:

$$\Delta w_{ij} = \begin{cases} \alpha\left(\left\langle v_i^t h_j^t \right\rangle_{data} - \left\langle v_i^t h_j^t \right\rangle_{model}\right) & \text{if } \Delta w_{ij} \in W_{vh}. \\ \alpha\left(\left\langle h_i^{t-1} h_j^t \right\rangle_{data} - \left\langle h_i^{t-1} h_j^t \right\rangle_{model}\right) & \text{if } \Delta w_{ij} \in W_{h'h}. \end{cases} \qquad (13)$$

Through not renewing W_{vh} and $W_{h'h}$ at the same time, the learning of visible unit groups with different properties can be adequately controlled.

A SHRBM is shown in Fig. 1 (right). When normal RBMs are multilayered, it is an unsupervised learning model with greedy layer-wise training from layers close to the input.

4   Evaluative Experiments

In our evaluative experiments, we modelled and used the eight-arm radial maze problem familiar in a behavioural test battery learning and memory experiments.

4.1   Eight-Arm Radial Maze Problem Outline and Modelling

Figure 2 (left) depicts an image of an eight-arm radial maze. In an eight-arm radial maze problem, tasks are learned in an environment in which food, so to speak, lies at the tips of mazes that fan out radially, with a mouse, so to speak, placed in the centre of the maze. The mouse is required to obtain as much of the food as possible. Because the mouse needs to remember where it has already visited, it tests working memory.

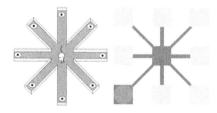

Fig. 2. Eight-arm radial maze problem (left) and its model (right)

To explain the modelling of the eight-arm radial maze, Fig. 2 (right) presents a picture of the modelled maze. The mouse, or agent, operates in the environment according to the following steps:

1. Set the initial value to the middle of the maze;
2. Receive the current status;
3. Select the destination;
4. Receive a reward; and
5. Repeat Steps 2–4 15 times.

Here, the selected destination and current status are expressed respectively by a nine-dimension symbolic vector. The agent receives a point when it acquires food, and 0 points in all other cases. After the above steps are followed, the test score is (Number of foods acquired by the agent) - (Number of illegal state transitions attempted).

In this case, the ideal behaviour without sequential training is: (A) if it is outside the centre of the maze, then move to the centre, and (B) if its own status is in the middle of the maze, then move in the maze at random. It is particularly important that the place selected in (B) is unbiased. When we assume that ideal status and request an expectation with a computer simulation, the result is 5.25 points. We call that value the *chance level*.

4.2   Agent Outline

Figure 3 is a schematic diagram of the agent.

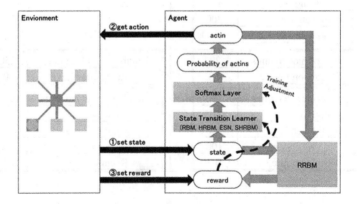

Fig. 3. Agent block diagram

Reward RBMs (RRBMs) apply unlearned data detection methods [12]. It is possible for RRBMs to judge situations in which receiving rewards are easy. When the energy for the current status value is lower than the past average energy level, it gives an anticipatory reward.

The parameters of the softmax layer and RRBM common to every state transition learner (STL) are shown in Table 1.

Table 1. Common parameters of the softmax layer and reward restricted Boltzmann machine

	Softmax layer	Reward RBM
Number of input units	20	18
Number of output units	9	20
Learning rate	0.01	0.01
Mini batch size	100	1
Training epochs	1,000	1

The STL and softmax layer learn only the rewarding behaviour. The RBM, ESN, HRBM, and their multilayered model are used as a STL and compared. However, the SRBM is omitted because its improved accuracy via multilayering is undesirable. Table 2 shows each parameter of the STL. For parameters shown here, the calculations of the best results for each model within the range of the preliminary examination results are used.

Table 2. Parameters for the state transition learners

				Single-layer models		Multilayer models	
RBM	Number of hidden units	20		Number of hidden units	20		
	Learning rate	0.05				-	
	Mini batch size	200					
	Training epochs	1,000					
ESN	Number of hidden units	20		Number of hidden units	20,50,100		
	Recursion rate	0.3		Recursion rate	0.3		
HRBM	Number of hidden units	100		Number of hidden units	20,50,100		
	Learning rate	0.05		Learning rate	0.05		
	Mini batch size	200		Mini batch size	200		
	Training epochs	100		Training epochs	100		
	Recursion rate	0.3		Recursion rate	0.3		

4.3   Results

Figure 4 shows the results of each STL as the moving average of each 100 trials. The smaller box illustrates the intervals of trial scores less than 0.

A comparison of the single-layer RBM, ESN, and HRBM show that all three have scores of roughly five. However, only the HRBM has intervals that exceed the chance level. The stacked ESN has multiple layers of reservoirs with random connection weights, and it has a high possibility of irregular expression with a complicated highest

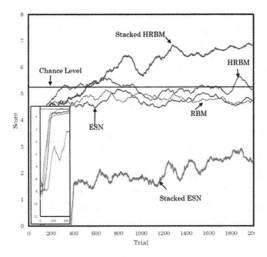

Fig. 4. Trial scores of each state transition learner (moving average of each 100 trials)

layer. Conversely, the SHRBM performs learning that enables it to ultimately acquire the food seven times.

5 Conclusion

In this paper, we have proposed SHRBMs formed from multiple layers of HRBMs that can learn time series and HRBMs, in order to aid the realisation of a RBM-based cognitive architecture.

In our evaluative experiment, we modelled and used the eight-arm radial maze problem used in real life to investigate the cognitive functions of agents. The results suggested that HRBMs can display stronger working memory functions by being stacked in multiple layers.

References

1. Mnih, V., et al.: Human-level control through deep reinforcement learning. Nature **518** (7540), 529–533 (2015)
2. Miri, H.: CernoCAMAL: a probabilistic computational cognitive architecture. Ph.D. thesis, University of Hull (2012)
3. Laird, J.E.: The Soar Cognitive Architecture. MIT Press, Cambridge (2012)
4. Anderson, J.R., Lebiere, C.: The newell test for a theory of cognition. Behav. Brain Sci. **26** (05), 587–601 (2003)
5. Goertzel, B.: The Hidden Pattern. Brown Walker, Boca Raton (2006)
6. Goertzel, B.: Opencog prime: a cognitive synergy based architecture for embodied artificial general intelligence. In: 8th IEEE International Conference on Cognitive Informatics, pp. 60–68 (2009)

7. Eliasmith, C., Stewart, T.C., Choo, X., Bekolay, T., DeWolf, T., Tang, Y., Rasmussen, D.: A large-scale model of the functioning brain. Science **338**(6111), 1202–1205 (2012)
8. Sutskever, I., Hinton, G.E.: Learning multilevel distributed representations for high-dimensional sequences. In: International Conference on Artificial Intelligence and Statistics (2007)
9. Yu, Y., Masahiko, O., Masafumi, H.: A learning method for echo state networks using RBM. In: International Symposium on Advanced Intelligent Systems (2015)
10. Sutskever, I., Hinton, G.E., Taylor, G.W.: The recurrent temporal restricted Boltzmann machine. In: Advances in Neural Information Processing Systems (2009)
11. Boulanger-Lewandowski, N., Bengio, Y., Vincent, P.: Modeling temporal dependencies in high-dimensional sequences: application to polyphonic music generation and transcription. In: Proceedings of the 29th International Conference on Machine Learning (2012)
12. Osawa, M., Hagiwara, M.: A proposal of novel data detection method and its application to incremental learning for RBMs. IEICE Technical report, ME and Bio Cybernetics, vol. 114, no. 259, pp 283–288 (2015). (In Japanese)

A Game-Engine-Based Learning Environment Framework for Artificial General Intelligence

Toward Democratic AGI

Masayoshi Nakamura[1,2(✉)] and Hiroshi Yamakawa[1,2]

[1] Dwango Artificial Intelligence Laboratory, DWANGO Co., Ltd.,
Kabukiza Tower 14F, 4-12-15 Ginza, Chuo-Ku, Tokyo 104-0061, Japan
{masayoshi_nakamura,hiroshi_yamakawa}@dwango.co.jp
[2] The Whole Brain Architecture Initiative, Kabukiza Tower 14F, 4-12-15 Ginza,
Chuo-Ku, Tokyo 104-0061, Japan

Abstract. Artificial General Intelligence (AGI) refers to machine intelligence that can effectively conduct variety of human tasks. Therefore AGI research requires multivariate and realistic learning environments. In recent years, game engines capable of constructing highly realistic 3D virtual worlds have also become available at low cost. In accordance with these changes, we developed the "Life in Silico" (LIS) framework, which provides virtual agents with learning algorithms and their learning environments with game engine. This should in turn allow for easier and more flexible AGI research. Furthermore, non-experts will be able to play with the framework, which would enable them to research as their hobby. If AGI research becomes popular in this manner, we may see a sudden acceleration towards the "Democratization of AGI".

Keywords: Artificial general intelligence · Simulation-Based learning environment · Machine learning

1   Introduction

As the ability of AI draws closer to that of humans, the prospective large-scale effects on society and technological innovation are stimulating efforts to advance Artificial General Intelligence (AGI) via open source, joint development ventures.

One example of such projects is OpenCog, which started in 2008 with "Building better minds together…" as its motto[1]. OpenCog's AGI architecture is dubbed CogPrime and consists of an assembly of heterogeneous modules. For its external environment, CogPrime makes use of Minecraft, the game which provides an open 3D environment[2]. OpenCog has also developed virtual pets that use imitation learning and provides a learning environment using the game engine Unity[3].

---

[1] http://opencog.org/.
[2] https://github.com/OC2MC/opencog-to-minecraft.
[3] http://wiki.opencog.org/w/Setting_up_the_Unity3D_world.

© Springer International Publishing AG 2016
A. Hirose et al. (Eds.): ICONIP 2016, Part I, LNCS 9947, pp. 351–356, 2016.
DOI: 10.1007/978-3-319-46687-3_39

As another example, OpenAI has also released OpenAI Gym[4], a toolkit for developing and comparing reinforcement learning algorithms.

These examples show that, in recent years, learning from environments has been occupying an important space in the development of AGI. Two fundamental parts are needed here - AGI agents and the external environments for their learning. In this particular example, CogPrime is the AGI agent and Minecraft is the external learning environment.

Because of the pace at which technological development is rushing ahead, each of the main components is being replaced rapidly. For example, for AGI agents, as the interest in components that combine deep learning and reinforcement learning has grown rapidly, many standalone techniques are now being open sourced. For external environments, a powerful 3D simulator game engine has been freely available since 2015. This change has made it possible for anyone to freely create virtual environments to reproduce a variety of existing environments. In February and March of 2016, the US companies DeepMind, Facebook, and Microsoft proposed to use a 3D simulation-based learning environment [1–3].

Such game engine innovation brings massive momentum to AGI development. However, from the perspective of developers attempting to start new AGI research, there is a risk of overcommitting to specific components. This is a barrier researchers and developers hoping to enter to the field have to overcome.

We addressed this problem by developing the LIS Framework, which provides a way to approach AGI learning in a flexible and easy-to-use manner by combining multiple, interchangeable components. The framework allows AGI workers including beginners to combine pre-installed LIS Framework components and begin AGI

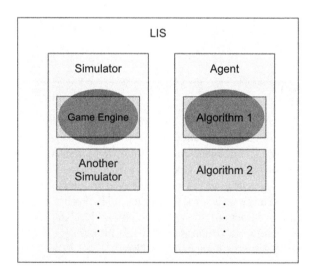

Fig. 1. LIS Framework Conceptual diagram. Users can select one or more components from each of the columns (marked with red circles) to create original modules. (Color figure online)

---

[4] https://gym.openai.com/.

development with ease. Furthermore, non-experts can set up learning environments and develop AI to solve problems of their own choosing. If the basis for AGI research continues to expand with this framework helped by the present deep learning boom, we can look forward to realizing that everyone can create their AGI. We call it "Democratization of AGI" (Fig. 1).

2   LIS Framework

The LIS Framework[5] is a simulation-based learning environment framework equipped with state-of-the-art machine-learning methods and learning environment simulators. The framework is designed to allow for off-the-shelf AGI research and development.

As shown in Fig. 2, the components in the same column will be switchable in the future to allow a quick response to new technology, such that users can select one or more components from each of the columns and combine them in a functional manner.

- Environment: The agent's learning environment. Users select a preset learning environment or use the simulator to create a custom environment.
- Simulator: The simulator simulates the learning environment. In the future, users will be able to choose simulators based on their needs. The simulator currently runs on the Unity game engine.
- Agent: Agent varieties. In the future, users will be able to choose an agent based on their needs.

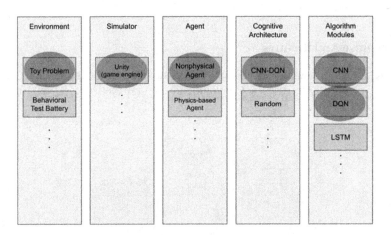

Fig. 2. Detailed AGI learning LIS framework organizational chart A system built with a premise of interchangeable components. Presently implemented components are circled in red. In the cognitive architecture column, "CNN-DQN" consists of AlexNet [4] and Deep Q-Network [5]. (Color figure online)

---

[5] https://github.com/wbap/lis.

- Cognitive Architecture: Cognitive architecture provides the agent with cognitive logic. Users can switch between architectures based on their needs and construct their own architecture.
- Algorithm Modules: Algorithmic components for the cognitive architecture. Users can select modules based on their needs.

3   Implementations

The currently implemented components on the LIS Framework are explained below (i.e., those circled in red in Fig. 2).

3.1   3D Simulation-Based Learning Environment

AGI may require interaction with the environment via sensors and actuators. However, when trying to implement it with real physical robots, spatial, and temporal limitations exist. In a virtual space, many of these limitations are removed. Recently, 3D graphics for games have advanced to the point where their graphics can be mistaken for photographs. Moreover, user bases exist at a scale of several million. The systems are usable with strong community supports.

For these reasons, we decided to adopt Unity Technology's "Unity" game engine, which is equipped with 3D graphics and a simple physics simulator. Agents inside the Unity engine are equipped with socket transmission functionality and are controlled directly via an external program. Unity sends camera images, depth images, and reward information to the agent's external program. The agent receives that information and sends action information back to Unity.

3.2   Cognitive Architectures

The current agent's cognitive architecture consists of cutting-edge machine learning methods. Specifically, it uses a Convolutional Neural Network (CNN) for image

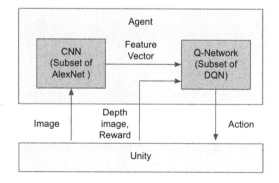

Fig. 3. System configuration

processing and a Deep Q-Network (DQN) for reinforcement learning. The system configuration is shown in Fig. 3.

3.3 Performance

We conducted a benchmark evaluation of this current implementation of the LIS Framework. As this functionality will change as development continues, the newest benchmarks will be listed on the open source page. A standard, commercially available desktop PC (Intel Core i7-4790 K, 32 MB memory, GeForce GTX TITAN X, Ubuntu 14.04) was used for testing.

In the learning environment, a large apple was placed on top of a path set above a river. The agent was "rewarded" with a fail when the apple fell into the water and a success upon taking the apple. The agent was allowed to go forward, jump, and turn right or left. Within this environment, the agent conducted trial-and-error learning to be able to obtain the apple without falling into the river. The agent's 154587-dimension ($227 \times 227 \times 3$) RBG image vision was exchanged by a 9216-dimension ($6*6*256$) vector, which is AlexNet's poop5 convolutional layer output. Afterwards, it was changed to a 10240-dimension image by combining with depth imaging 1024-dimension (32×32) vector. This 10240-dimension was input into a Q-Network that depicts the policy with a neural net. For learning, the Q-Learning reinforcement-learning framework was used.

Our test allowed us to confirm that learning was qualitatively possible within our framework. The decisions occurred every 0.15 s of Unity's time. This allows approximately 20 frames to be sent every second. Simulation can be accelerated by 2 to 3 times compared to the normal speed.

Useable sensors are an RGB camera, depth camera, microphone (sound speed settings are infinite for game engines), collision sensor, ray sensor, GPS sensor, IMU sensor, and sonar sensor. A torque sensor can be used if Unity is connected with a

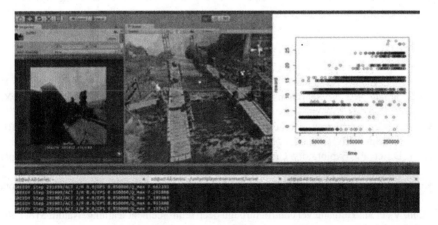

Fig. 4. Screenshot of the LIS interface

database such as the Open Dynamics Engine (ODE) Library. This will, of course, incur implementation costs (Fig. 4).

4   Results of AGI Research Foundational Expansion

We attempted to determine whether LIS could have a real user base by holding a Hackathon in Tokyo, Japan on April 9, 2016 in which 151 attendees participated. LIS was received favorably by the attendees as a form of "play using AI." Partly because of the synergistic effects from the deep learning boom of recent years, we were able to confirm a high level of appeal to a general audience. From this, we conclude that this framework can also be expected to be effective in spreading AGI research and development to the general public.

Additionally, one of the aims of the LIS Framework is to make it possible for non-experts to set up a learning environment and freely design AI. In approximately 8 h, beginners of machine learning and Unity were able to build games such as "3D Pong" and "an agent that runs away from an approaching wall as in a 2D Super Mario game," and each person was able to learn the skills needed to actualize such tasks. Although these were toy problems, they show the practicality of realizing the "democratization of AI."

5   Conclusion

As we made the LIS Framework as open-source, the threshold for AGI research and development within a 3D environment was lowered greatly. Additionally, we confirmed that non-experts in AGI were able to set up an environment and implement intellectual functions. There were also people who felt AGI was a form of play - this has potential for expanding the AGI research base. We intend expanding the framework and grow the community around it. It will bring AGI to everyone as a form of play. Therefore we will be able to see "Democratization of AGI".

Acknowledgements. Thanks to all members of the WBAI and the members of the WBA Future Leaders.

References

1. Mnih, V., Badia, A.P., Mirza, M., Graves, A., Lillicrap, T.P., Harley, T., Silver, D., Kavukcuoglu, K.: Asynchronous methods for deep reinforcement learning (2016). arXiv preprint arXiv:1602.01783
2. Lerer, A., Gross, S., Fergus, R.: Learning physical intuition of block towers by example (2016). arXiv preprint arXiv:1603.01312
3. Abel, D., et al.: Exploratory Gradient Boosting for Reinforcement Learning in Complex Domains (2016). arXiv preprint arXiv:1603.01312
4. Krizhevsky, A., Sutskever, I., Hinton, G.: ImageNet classification with deep convolutional neural networks. In: NIPS (2012)
5. Mnih, V., et al.: Humanlevel control through deep reinforcement learning. Nature **518**(7540), 529–533 (2015)

Neurodynamics

Modeling Attention-Induced Reduction of Spike Synchrony in the Visual Cortex

Nobuhiko Wagatsuma[1]([✉]), Rüdiger von der Heydt[2],
and Ernst Niebur[2]

[1] School of Science and Engineering, Tokyo Denki University, Saitama, Japan
nwagatsuma@rd.dendai.ac.jp
[2] Krieger Mind/Brain Institute, Johns Hopkins University, Baltimore, USA
{von.der.heydt,niebur}@jhu.edu

Abstract. The mean firing rate of a border-ownership selective (BOS) neuron encodes where a foreground figure relative to its classical receptive field. Physiological experiments have demonstrated that top-down attention increases firing rates and decreases spike synchrony between them. To elucidate mechanisms of attentional modulation on rates and synchrony of BOS neurons, we developed a spiking neuron network model: BOS neurons receive synaptic input which reflects visual input. The synaptic input strength is modulated multiplicatively by the activity of Grouping neurons whose activity represents the object's location and mediates top-down attentional projection to BOS neurons. Model simulations agree with experimental findings, showing that attention to an object increases the firing rates of BOS neurons representing it while decreasing spike synchrony between pairs of such neurons. Our results suggest that top-down attention multiplicatively emphasizes synaptic current due to bottom-up visual inputs.

Keywords: Border-ownership · Selective attention · Spike synchrony · Modulatory input

1 Introduction

Neural mechanisms underlying figure–ground segregation have been studied for decades. Particularly, the most fundamental process to perceive and understand objects and their locations is the determination of the figure direction. Reports of physiological studies have described that most neurons in monkey V2 and V4 have selectivity to border-ownership: The mean firing rate of a border-ownership selective (BOS) neuron changes depending on where a foreground figure is located relative to its classical receptive field [1]. It is particularly interesting that when animals attend to the foreground figure, the firing rates of these neurons are increased [2].

Martin and von der Heydt [3] recorded from pairs of BOS neurons responding either to contours of different objects, or contours of the same object (Fig. 1A). In their experiments, keystone-like stimuli were presented on the classical receptive fields of BOS neurons. They reported that stimulation by the "Bound" condition produced enhanced spike synchrony between the pairs of BOS neurons compared to "Unbound".

© Springer International Publishing AG 2016
A. Hirose et al. (Eds.): ICONIP 2016, Part I, LNCS 9947, pp. 359–366, 2016.
DOI: 10.1007/978-3-319-46687-3_40

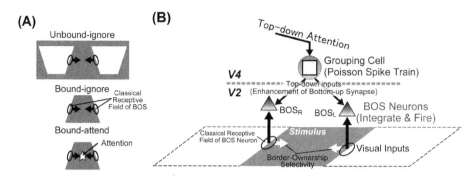

Fig. 1. (**A**) Examples of conditions for visual inputs and attention for physiological experiments [3]. Black ellipses on the borders of the keystone-like stimuli represent classical receptive fields of BOS neurons. In the "Unbound" condition, a pair of BOS neurons responded to contours of different objects. In the "Bound" condition, the two BOS receptive fields lie on the borders of the same object. On "attend" trials, monkeys attended the object, as shown by the star. (**B**) Model architecture. Grouping (G) cells in V4 represent a grouping structure and top-down attention on BOS$_L$ and BOS$_R$ neurons in V2. Details of the effects of feedback signals from G-cells are presented in Fig. 2.

However, while attention to an object increased the firing rates of these neurons, synchrony significantly decreased from the "ignore" to the "attend" condition.

To elucidate the mechanisms for paradoxical attentional modulation between mean rates and synchrony of BOS neurons, we developed a network model of spiking neurons (Fig. 1B). In the model, BOS neurons receive synaptic input from non-BOS feature-selective neurons which reflect visual input. The synaptic input strength is modulated multiplicatively by the activity of Grouping cells (G) which receive their input from BOS neurons and whose activity represents the location of visual objects in the scene [4]. Furthermore, the mean firing rate of a G-cell is increased when attention is directed to an object that is represented by the G-cell. Model simulations agree with experimental findings [3], showing that attention to an object increases the mean firing rates of BOS neurons representing the object while decreasing spike synchrony between pairs of such neurons. Our results therefore suggest that top-down attention multiplicatively emphasizes synaptic currents due to bottom-up visual input. Results further suggest that attention exerts its influence on BOS cells by boosting the firing rates of G-cells, rather than directly influencing the activity of BOS or other feature-selective neurons.

2 Proposed Model

2.1 Network Architecture

Figure 1B displays the network model architecture, which consists of two BOS populations (BOS$_L$ and BOS$_R$) and one Grouping cell population (G). The grouping hypothesis assumed that G-cells integrate the responses of BOS neurons to generate a representation of a fast sketch of the location and rough shapes of objects in the scene. The G-cell function requires very little specificity in their responses; they are fully characterized by the center location and receptive field size [4]. Although BOS neurons are observed in cortical area V2 and neighboring areas V1 and V4, it remains unknown where G-cells reside. The BOS neuron whose receptive field is presented by the left (right) circle has right (left) side-of-figure preference (Fig. 1). It is therefore called as BOS_R (BOS_L). These BOS neurons are driven by two external sources of input corresponding to bottom-up visual inputs and top-down G-cell signals. For simplification, these external inputs, including G-cell, are given as a Poisson spike train. Bottom-up visual inputs are independent processes, whereas top-down G-cell signals are common to BOS_R and BOS_L neurons representing the same object. In our model and simulations, the firing rate of a G-cell v_G represented the conditions of visual stimuli and attention (Fig. 1A). Details of settings for mean rates of these external inputs are presented in Simulation Results. To elucidate the fundamental mechanisms of attentional modulation of BOS neurons, we included in our current model only the minimum number of neurons and synaptic connections. There were solely two types of synaptic connections in this work: feedforward connections representing visual stimuli to BOS and feedback from G to BOS cells. Note that, for simplifying the model network and exploring the mechanism of top-down attentional modulation in BOS neurons, we excluded the feedforward connections from BOS to G-cells (Fig. 1B).

In our model, BOS neurons are described as integrate-and-fire neurons. We used specific values of the parameters [5, 6]. We assumed the firing threshold as $\theta = -50$ mV, and the reset potential as $V_{reset} = -65$ mV. Membrane capacitance C_m was 0.5 nF. The membrane time constant τ_m was 20 ms. In mathematical terms, the dynamics of the sub-threshold membrane potential V of a BOS neuron is given as the following equation.

$$\frac{dV(t)}{dt} = -\frac{V(t)}{\tau_m} + \frac{I_{syn}(t)}{C_m} \tag{1}$$

In our proposed model, the bottom-up excitatory postsynaptic currents from visual stimuli I_{syn} are simply modeled by an instantaneous rise of the synaptic conductance [5, 6].

$$I_{syn}(t) = w_{vis}^{BOS} exp\left(-t/\tau_{syn}\right) \tag{2}$$

In that equation, the postsynaptic current time constant is $\tau_{syn} = 0.5$ ms and w_{vis}^{BOS} represents the excitatory synaptic weight.

For BOS_L and BOS_R neurons, G-cells provide common modulatory inputs [4]. For simplification of the model, the modulatory feedback from G-cells transiently boosts the bottom-up synaptic currents from visual inputs. In our model, when BOS neurons receive signals from G-cells, the bottom-up excitatory synaptic weight w_{vis}^{BOS} is increased to 1.4-fold during 50 ms (Fig. 2). Actually, the G-cell signals themselves do not induce a spike of BOS neurons.

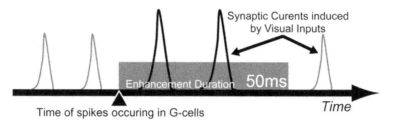

Time of spikes occuring in G-cells

Fig. 2. Feedback projections from G-cells temporally and multiplicatively enhanced the synaptic currents induced by visual inputs. The black triangle shows the time of spikes occurring in G-cells. Strength of the synaptic input is modulated multiplicatively by the signals from G-cells whose activity was increased when attention is directed to an object represented by it.

2.2   Computation of Synchrony Between BOS Neurons

We computed the correlation between BOS neurons to see whether our proposed model exhibits sufficient synchrony. Our methods for the analysis of spike synchrony were based on physiological studies [3, 7, 8]. To analyze the simulation data more accurately, we conducted the simulations of the model using 20 trials having length of 252 s each.

To compute the spike correlation, we divided time into bins of biological width 1 ms, each containing either 0 or 1 spike. The spike train, $S_j^i(n)$, is a binary vector in which each component takes on either a value 0 if no spike is present in the interval $(n, n+1]$ ms or 1 if a spike exists. Here, n stands for the bin index, i represents the trial number, and j denotes a BOS_L or BOS_R neuron. The cross-correlation is defined as

$$CC^i(\tau) = \sum_{\mu = t_0 - 250}^{t_{end} + 250} \left(S_{BOS_L}^i(\mu + \tau) - f_{BOS_L}^i \right) \left(S_{BOS_R}^i(\mu) - f_{BOS_R}^i \right) \qquad (3)$$

$$f_j^i = \frac{1}{t_{end} - t_0} \sum_{n = t_0}^{t_{end}} S_j^i(n), \qquad (4)$$

where τ signifies the time lag between two spike trains (-250 ms $\leq \tau \leq 250$ ms). The interval of the spike train is defined as t_0 and t_{end}. Also, f_j^i represents the mean spike count per bin of spike train $S_j^i(n)$ in the interval t_0 to t_{end}. The magnitude of synchrony between BOS_L and BOS_R neurons, M^i, is defined as the integral of the correlation, Eq. 3, in the range ± 40 ms as the following;

$$M^i = \sum_{\tau = -40}^{40} CC^i(\tau) \cdot binsize \qquad (5)$$

where binsize = 1 ms.

3   Simulation Results

To examine the mechanism of synchrony between BOS neurons, we conducted numerical simulations of our proposed model with the stimulus corresponding to Martin and von der Heydt [3] (Fig. 1(A)). The firing rate of a G-cell v_G in the "Bound" condition is higher than in the "Unbound" condition. In the latter situation (top of Fig. 1(A)), the grouping hypothesis assumes that two objects (gray shapes) are located left and right in the scene and different two G-cells are activated [4]. In addition to this G-cell activity corresponding to the geometry of the scene, we assume that attention to an object increases the firing rate of the corresponding G-cell further, as in the "Bound-attend" condition (bottom of Fig. 1(A)). We used mean rates of G-cells v_G of 2, 15, and 25 Hz to represent the "Unbound-ignore", "Bound-ignore", and "Bound-attend" conditions, respectively. The visual inputs have the same statistics in all three conditions, modeled as a Poisson spike train with mean rates of 675 Hz. With these settings, the mean firing rates of BOS neurons for the "Unbound-ignore" condition are about 10 Hz.

Figure 3(A) presents a summary of the mean firing rates of BOS neurons (v_{BOS}) for three conditions. The firing rates of BOS neurons for the "Bound" condition (gray bar)

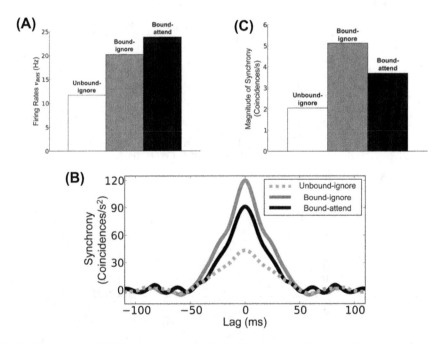

Fig. 3. Responses of BOS neurons to stimuli for the corresponding physiological experiment [3]. **(A)** Firing rates of BOS neurons. White, gray and black bars show "Unbound-ignore", "Bound-ignore" and "Bound-attend" conditions, respectively. **(B)** Simulated synchrony between BOS_L and BOS_R neurons. Synchrony is highest in the "Bound-ignore" condition (gray solid), lowest in the "Unbound-ignore" (gray dashed), and intermediate for "Bound-attend" (black). **(C)** Magnitude of synchrony. Same conventions as those of (A).

were significantly higher than that for the "Unbound" condition (white bar) (*t*-test, $p < 0.01$). Furthermore, a significant difference in v_{BOS} was found between "Bound-ignore" (gray bar) and "Bound-attend" (black bar) conditions (*t*-test, $p < 0.01$). These results show good agreement with physiological results [2, 3].

Important results of [3] were the observations that binding increased the spike synchrony in the absence of attention compared to the unbound case, and that top-down attention to an object decreased synchrony between BOS neurons representing that object. We simulated this experiment by computing the spike train correlations between BOS_L and BOS_R neurons in the "Unbound-ignore", "Bound-ignore" and "Bound-attend" conditions (Fig. 3(B)). In the ignored conditions ("Unbound-ignore" and "Bound-ignore"), the synchrony for the "Bound-ignore" condition (gray solid line) was markedly higher than for the "Unbound-ignore" condition (gray dashed line). In contrast, the "Bound-attend" condition (black line) was much lower than for the "Bound-ignore" condition (gray line). To quantify our simulation results, we computed the magnitude of the synchrony based on Eq. (5) (Fig. 3(C)). A significant difference was found in the magnitude of the synchrony between "Unbound-ignore" (white bar) and "Bound-ignore" (gray bar) conditions (*t*-test, $p < 0.01$). However, we found a

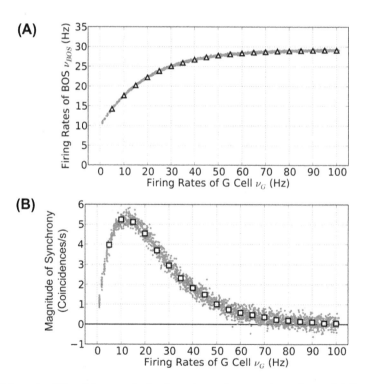

Fig. 4. Firing rates (A) and the magnitude of synchrony (integrated over interval ±40 ms around lag 0) (B) for BOS neurons as function of the mean rates of G-cells ($v_G = 1$–100 Hz). Small gray dots represent the rate and the magnitude of synchrony for each simulation trial. **(A)** Mean rates of BOS neurons are shown as white triangles in steps of 5 Hz. **(B)** White squares represent the mean magnitude of synchrony of 20 simulated trials, in steps of 5 Hz of mean v_G.

significant decrease in the magnitude of synchrony from the "Bound-ignore" to the "Bound-attend" (black bar) condition.

To ascertain the effects of top-down modulatory projections from G-cells on the mean rates of BOS neurons and their synchrony, we parametrically varied the mean rates of G-cells in the range $v_G = 1$–100 Hz. The firing rates of BOS neurons monotonically increased with increasing top-down signals (Fig. 4(A)). In contrast, the magnitude of synchrony between BOS_L and BOS_R neurons indicated non-monotonic modulation patterns, rising to a peak at about 15 Hz and then decreasing (Fig. 4(B)). These simulation results show good agreement with the physiologically observed changes both in firing rates and in synchrony between neural pairs [3].

4 Discussion and Conclusion

We have investigated the neural mechanisms of attentional modulation on the mean firing frequency and spike synchrony of BOS neurons through computational simulations of a network model of spiking neurons. In our proposed model, the strength of the synaptic input is transiently and multiplicatively enhanced by the feedback signals of G-cells whose activity is increased when attention is directed to an object that is represented by the G-cell [4] (Figs. 1 and 2). Simulation results of our model indicate that attention to an object increased the mean firing rates of BOS neurons while decreasing spike synchrony between the pairs of these neurons, which shows agreement with physiological responses [3].

In our proposed model, G-cells provided common inputs for BOS_L and BOS_R neurons, which transiently boosted the bottom-up synaptic currents from visual inputs (Fig. 2). In the absence of common inputs ($v_G = 0$), the synchrony between BOS neurons cannot exceed that of chance. At the other extreme, if the common feedback inputs from G-cells are dense, then BOS neurons constantly receive the feedback signals. As a consequence of constant steady input, the synaptic weight is apparently consistently enhanced throughout simulation, which corresponds to the model consisting of only BOS neurons and bottom-up visual inputs with synaptic weight $1.4w_{vis}^{BOS}$. The marked activation of the G-cells increases the mean firing rates, as shown in Fig. 4(A), but it generates no spike–spike synchrony beyond chance. Therefore, synchrony vanished both for very small and for very high firing rates of G-cells, with a peak location at some intermediate point, as shown in Fig. 4(B).

Our model reproduced the recent physiological report for the behaviors of BOS neurons [3]: selective attention to an object increased the firing rates, while decreased spike-spike synchrony. However, little is known about the function of attention-induced reduction of synchrony for visual processing and perception. These paradoxical attentional modulations between mean rates and spike synchrony will provide important insights for understanding the perceptual mechanism of the camouflaged/occluded object. Perhaps, in order to understand the function of these attentional modulations, it is important to discuss about the noise redundancy and correlation of the BOS mechanism. Further studies are necessary to clarify the attentional mechanism for the modulation in BOS neurons.

We implemented the transient enhancement of the bottom-up synaptic currents as the modulatory feedback from G-cells. Recent physiological work has reported that the cortical feedback projections use slow, modulatory NMDA receptors rather than fast, driving AMPA receptors [9]. Interestingly, a computational model with the feedback projections mediated by synaptic kinetics of NMDA receptor [10] indicated the similar behaviors of this current model.

In this work, we only used the integrate-and-fire neuron model for simulating the BOS neurons. However, various types of neuron model have been proposed such as leaky integrate-and-fire, Hodgkin-Huxley and Izhikevich neuron. Further studies are needed to investigate whether paradoxical attentional modulations between mean rates and spike synchrony depended on the types of neuron model.

Our model predicts that spike–spike synchrony is useful to infer common modulatory input and that attention exerts its influence on BOS neurons by boosting the responses of G-cells, rather than directly influencing the activity of BOS neurons or other feature-selective neurons. It is expected to examine these predictions from physiological perspectives. Our results provide useful and testable predictions for fundamental problems related to figure–ground segregation and object perception.

Acknowledgements. This work was partly supported by KAKENHI (no. 26880019). We thank Japanese Neural Network Society for supporting English proofreading.

References

1. Zhou, H., Friedman, H.S., von der Heydt, R.: Coding of border ownership in monkey visual cortex. J. Neurosci. **20**, 6594–6611 (2000)
2. Qiu, F.T., Sugihara, T., von der Heydt, R.: Figure-ground mechanisms provide structure for selective attention. Nat. Neurosci. **10**, 1492–1499 (2007)
3. Martin, A.B., von der Heydt, R.: Spike synchrony reveals emergence of proto-objects in visual cortex. J. Neurosci. **35**, 6860–6870 (2015)
4. Craft, E., Schütze, H., Niebur, E., von der Heydt, R.: A neural model of figure-ground organization. J. Neurophysiol. **97**, 4310–4326 (2007)
5. Wagatsuma, N., Potjans, T.C., Diesmann, M., Fukai, T.: Layer-dependent attentional processing by top-down signals in a visual cortical microcircuit model. Front. Comp. Neurosci. (2011). doi:10.3389/fncom.2011.00031
6. Wagatsuma, N., Potjans, T.C., Diesmann, M., Sakai, K., Fukai, T.: Spatial and feature-based attention in a layered cortical microcircuit model. PLoS ONE (2013). doi:10.1371/journal.pone.0080788
7. Roelfsema, P.R., Lamme, V.A., Spekreijse, H.: Synchrony and covariation of firing rates in the primary visual cortex during contour grouping. Nat. Neurosci. **7**, 982–991 (2004)
8. Dong, Y., Mihalas, S., Qiu, F., von der Heydt, R., Niebur, E.: Synchrony and the binding problem in macaque visual cortex. J. Vis. (2008). doi:10.1167/8.7.30
9. Self, M.W., Koojimans, R.N., Super, H., Lamme, V.A., Roelfsema, P.R.: Different glutamate receptors convey feedforward and recurrent processing in Macaque V1. Proc. Natl. Acad. Sci. USA **109**, 11031–11036 (2012)
10. Wagatsuma, N., von der Heydt R., Niebur, E.: Spiking synchrony generated by modulatory common input through NMDA-type synapses. J. Neurophysiol. (in press). doi:10.1152/jn.01142.2015

A Robust TOA Source Localization Algorithm Based on LPNN

Hao Wang, Ruibin Feng, and Chi-Sing Leung$^{(\boxtimes)}$

Department of Electronic Engineering, City University of Hong Kong,
Kowloon Tong, Hong Kong
eeleungc@cityu.edu.hk

Abstract. One of the traditional models for finding the location of a mobile source is the time-of-arrival (TOA). It usually assumes that the measurement noise follow a Gaussian distribution. However, in practical, outliers are difficult to be avoided. This paper proposes an l_1-norm based objective function for alleviating the influence of outliers. Afterwards, we utilize the Lagrange programming neural network (LPNN) framework for the position estimation. As the framework requires that its objective function and constraints should be twice differentiable, we introduce an approximation for the l_1-norm term in our LPNN formulation. From the simulation result, our proposed algorithm has very good robustness.

Keywords: Source location · Time-of-arrival · Outliers · LPNN

1 Introduction

Estimating the position of a mobile source is very important in many applications, such as, emergency rescue, intelligent transport and resource management. The time of arrival (TOA) [1] is the most popular measurement model for source localization. In this model, three or more sensors are used to measure the time of signal transmission from the mobile source to the sensors, as shown in Fig. 1(a). Multiplying the measurement time by the signal propagation speed gives the distances between the mobile source and the sensors. Under the noiseless situation, the exact position of the mobile source can be calculated from those distances. In fact, the measurements usually contain noise. When the noise is Gaussian, the maximum likelihood (ML) concept can be applied to estimate the coordinate of the mobile source. However, the ML function is nonlinear. Using some linearization techniques, we can formulate the TOA problem as a least squares (LS) [2,3] problem. Several numerical methods [2,3] were proposed for TOA source localization. In [4], an analog neural network technique was used for solving the TOA ML problem. However, the mentioned methods do not consider outliers. When there are some outliers in the TOA measurements, the estimated position of the mobile source may have a very large error.

This paper proposes an robust algorithm based on the Lagrange programming neural network (LPNN) framework [5]. As the l_1-norm has a much better

© Springer International Publishing AG 2016
A. Hirose et al. (Eds.): ICONIP 2016, Part I, LNCS 9947, pp. 367–375, 2016.
DOI: 10.1007/978-3-319-46687-3_41

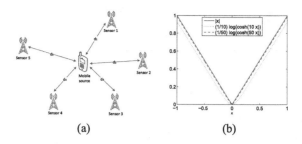

(a)   (b)

Fig. 1. (a) The source localization problem. The distance between the ith sensor and the mobile source is denoted as d_i. (b) The approximation for the l_1-norm.

ability to handle outliers, we replace the l_2-norm in the objective function with the l_1-norm. However, the LPNN requires that its objective function and constraints are twice differentiable. But the l_1-norm term in our objective function is non-differentiable at 0 point. Hence, we introduce an approximation for the l_1-norm term.

The rest of this paper is organized as follows. Section 2 provides an introduction of LPNN and the problem formulation. Section 3 presents our l_1-norm based LPNN positioning approach. Besides, theoretical analysis on the stability of the neural model is also discussed. Simulation results are included in Sect. 4. We then conclude the paper in Sect. 5.

2   Background

TOA-Based Source Localization. Let the unknown position of the mobile source be $\boldsymbol{c} = [c_1, c_2]^{\mathrm{T}}$, and let the coordinates of the sensors be $\{\boldsymbol{u}_1 = [u_{1,1}, u_{1,2}]^{\mathrm{T}}, \cdots, \boldsymbol{u}_m = [u_{m,1}, u_{m,2}]^{\mathrm{T}}\}$. At time τ_o the mobile source emits a signal and the ith ($i = [1, \cdots, m]$) sensor obtains the signal at time τ_i. The travelling time $\tau_i - \tau_o$ can be converted to a distance measurement. Let $d_i = \|\boldsymbol{c} - \boldsymbol{u}_i\|_2$ be the true distance between the ith sensor and the mobile source. The collection of d_i's is denoted as $\boldsymbol{d} = [d_1, \cdots, d_m]^{\mathrm{T}}$. We use $\boldsymbol{r} = [r_1, \cdots, r_m]^{\mathrm{T}}$ to denote the measurement distances, given by

$$r_i = d_i + \psi_i, \text{ for } i = 1, \cdots, m, \tag{1}$$

where ψ_i is the measurement noise.

In the TOA localization problem, given the measurements $\boldsymbol{r}$, the sensor coordinates $\boldsymbol{u}_i$ ($i = [1, \cdots, m]$), and the statistics of the noise $\boldsymbol{\psi}$, our aim is to estimate the source coordinates $\boldsymbol{c}$. Usually, we assume that ψ_i's follow the Gaussian distribution with zero mean, and that ψ's are independent. The objective function for estimating the mobile source position can be formulated as $\mathcal{E} = \frac{1}{2} \sum_{i=1}^{m} (r_i - \|\boldsymbol{c} - \boldsymbol{u}_i\|_2)^2$. The traditional gradient descent like algorithm may not be appropriate for minimizing $\mathcal{E}$ because the gradient vector of $\mathcal{E}$ contains some factors that are in terms of $(1/\|\boldsymbol{c} - \boldsymbol{u}_m\|_2)$'s. When the current estimate of c is close to one of u_m's, we have the ill posed problem. Hence many algorithms [2–4] formulated the problem as a constrained problem, given by

$$\min_{c,d} \frac{1}{2}\|r - d\|_2^2, \tag{2a}$$

$$\text{s.t. } d_i^2 = \|c - u_i\|_2^2, \text{ and } d_i \geq 0, \quad i = 1, \cdots, m. \tag{2b}$$

Lagrange Programming Neural Networks. Applying analog neural networks for optimization has received considerable attention [6–8]. The neural approach allows real-time computation because the neural circuits can be realized by the VLSI or optical technologies. However, many existing models are designed for solving a particular form of the constrained optimization problems. For example, the method in [8] addresses the quadratic programming problem only. While, the LPNN [4,5,9] provides a general approach for solving nonlinear constrained optimization problems. It considers the following optimization problem:

$$\min_{x} f(x), \text{ s.t. } h(x) = 0. \tag{3}$$

where $x = [x_1, \cdots, x_{n_1}]^T$ is the variable vector being optimized, $f : \mathbb{R}^{n_1} \rightarrow \mathbb{R}$ is the objective function, $h : \mathbb{R} \rightarrow \mathbb{R}^{n_2} (n_2 < n_1)$ represents n_2 equality constraints, and f and h should be twice differentiable. In LPNN approach, we first define the Lagrangian function L,

$$L(x, \lambda) = f(x) + \lambda^T h(x) \tag{4}$$

where $\lambda = [\lambda_1, \cdots, \lambda_{n_2}]^T$ is the Lagrangian multiplier vector. The LPNN model has two kinds of neurons: variable neurons and Lagrangian neurons. The n_1 variable neurons are used to hold the decision variable vector x, while the n_2 Lagrangian neurons are used to hold the Lagrangian multiplier vector λ. In the LPNN framework, the dynamics of the neurons are given by

$$\kappa \frac{dx}{dt} = -\frac{\partial L(x, \lambda)}{\partial x} \text{ and } \kappa \frac{d\lambda}{dt} = \frac{\partial L(x, \lambda)}{\partial \lambda}, \tag{5}$$

where the variable κ is called the characteristic time, whose value depends on the resistance and capacitance of the analog circuit. Without loss of generality, $\kappa = 1$ is assigned in this work. The differential equations in (5) are used to govern the state transition of the neurons. After the neurons settle down, the solution is obtained by measuring the neuron outputs at this stable equilibrium point. The dynamics $\frac{dx}{dt}$ are to seek for a state with the minimum objective value, while the dynamics of $\frac{d\lambda}{dt}$ are to restrict the system state within the feasible region. With (5), the network will settle down at a stable state [5,9] if the network satisfies some conditions.

3   Development of Proposed Algorithm

The problem defined in (2) has been studied many years, a number of algorithms [2–4] were developed. However, in (2), we assume that the noise follow a Gaussian distribution. Unfortunately, outliers may exist in the TOA measurements. Hence, to reduce the influence of the outliers, we consider the l_1-norm in the objective function, given by

$$\min_{\boldsymbol{c},\boldsymbol{d}} \frac{1}{2} \|\boldsymbol{r} - \boldsymbol{d}\|_1, \quad \text{s.t. } d_i^2 = \|\boldsymbol{c} - \boldsymbol{u}_i\|^2, \text{ and } d_i \geq 0, \quad i = 1, \cdots, m. \quad (6)$$

Before we apply the LPNN framework to solve the problem stated in (6), we need to resolve two issues. First, there are a number of inequality constraints in (6), but the LPNN framework can handle equality constraints only. Another issue is that the objective function is non-differentiable.

For the first issue, the following theorem shows that the inequality constraints can be removed.

Theorem 1: The optimization problem stated in (6) is equivalence to

$$\min_{\boldsymbol{c},\boldsymbol{d}} \|\boldsymbol{r} - \boldsymbol{d}\|_1, \quad \text{s.t. } d_i^2 = \|\boldsymbol{c} - \boldsymbol{u}_i\|^2, \quad i = 1, \cdots, m. \quad (7)$$

Proof: Suppose $\{\boldsymbol{c}^*, \boldsymbol{d}^*\}$ is the optimal solution of (6). From the properties of absolute value operation, or saying the l_1-norm, we have

$$\|\boldsymbol{r} - \boldsymbol{d}^*\|_1 = \sum_{i=1}^{m} |r_i - d_i^*| \geq \sum_{i=1}^{m} |\text{abs}(r_i) - \text{abs}(d_i^*)| = \sum_{i=1}^{m} |r_i - \text{abs}(d_i^*)| \quad (8)$$

Since r_i's are the distance measurements, we have $r_i \geq 0$. Thus, the last equation in (8) is satisfied. From (8), we have

$$\sum_{i=1}^{m} |r_i - d_i^*| \geq \sum_{i=1}^{m} |r_i - \text{abs}(d_i^*)| \quad (9)$$

Inequality (9) means that the objective function value $\sum_{i=1}^{m} |r_i - d_i^*|$ at the optimal solution $\{\boldsymbol{c}^*, \boldsymbol{d}^*\}$ is greater than or equal to the objective function value at the point $\{\boldsymbol{c}^*, \text{abs}(\boldsymbol{d}^*)\}$. Note that $\{\boldsymbol{c}^*, \text{abs}(\boldsymbol{d}^*)\}$ is also a feasible point. Since $\{\boldsymbol{c}^*, \boldsymbol{d}^*\}$ is the optimal solution, it is not possible to have another feasible point with a smaller objective function value. Hence the equality in (9) is hold. That means, $\boldsymbol{d}^* = \text{abs}(\boldsymbol{d}^*)$. Therefore, the inequalities in (6) can removed. The proof is complete. ∎

To resolve the second issue, we consider the following approximation:

$$|x| \approx \frac{\ln(\cosh(ax))}{a}, \quad (10)$$

where $a > 1$. Figure 1(b) shows the approximation with different a. It can be seen that the approximation is quite accurate when a is large.

With the approximation, the problem (7) can be modified as

$$\min_{\boldsymbol{c},\boldsymbol{d}} \frac{1}{a} \sum_{i=1}^{m} \ln(\cosh(a(r_i - d_i))) \quad \text{s.t. } d_i^2 = \|\boldsymbol{c} - \boldsymbol{u}_i\|^2, \quad i = 1, \cdots, m. \quad (11)$$

The above formulation has several good properties. First, the objective function is twice differentiable, hence we can apply the LPNN framework to solve it. Second, the derivative of $(1/a)\sum_{i=1}^{m}\ln(\cosh(a(r_i - d_i)))$ is equal to $\tanh(a(r_i - g_i))$, namely, the hyperbolic tangent function, which is frequently used as an activation function in artificial neural networks. According to (11), we can construct its Lagrangian function as

$$L_o(\boldsymbol{c}, \boldsymbol{d}, \boldsymbol{\lambda}) = \frac{1}{a}\sum_{i=1}^{m}\ln(\cosh(a(r_i - d_i))) + \sum_{i=1}^{m}\lambda_i(d_i^2 - \|\boldsymbol{c} - \boldsymbol{u}_i\|_2^2). \tag{12}$$

where the vector $\boldsymbol{\lambda} = [\lambda_1, \cdots, \lambda_m]^{\mathrm{T}}$ contains Lagrange multipliers. Afterward, we can use (12) to drive the neural dynamics. However, according to our preliminary simulation results, we find that an equilibrium point may not be stable. To improve the convexity and the stability, we introduce an augmented term $C_0/2\sum_{i=1}^{m}(d_i^2 - \|\boldsymbol{c} - \boldsymbol{u}_i\|_2^2)^2$ into the objective function. The augmented Lagrangian function becomes

$$L(\boldsymbol{c}, \boldsymbol{d}, \boldsymbol{\lambda}) = \frac{1}{a}\sum_{i=1}^{m}\ln(\cosh(a(r_i - d_i))) + \sum_{i=1}^{m}\lambda_i(d_i^2 - \|\boldsymbol{c} - \boldsymbol{u}_i\|_2^2)$$
$$+ \frac{C_0}{2}\sum_{i=1}^{m}(d_i^2 - \|\boldsymbol{c} - \boldsymbol{u}_i\|_2^2)^2. \tag{13}$$

Note that at an equilibrium point the constrains are satisfied, i.e., $d_i^{*2} = \|\boldsymbol{c}^* - \boldsymbol{u}_i\|_2^2$, $i = 1, ..., m$. Thus, the augmented term $C_0/2\sum_{i=1}^{m}(d_i^{*2} - \|\boldsymbol{c}^* - \boldsymbol{u}_i\|_2^2)^2$ is equal to zero. Introducing the augmented term does not influence the objective function value at equilibrium. Moreover, the augmented term is very useful in accelerating the convergence [5].

According to (5) and (13), the dynamics are given by

$$\frac{d\boldsymbol{c}}{dt} = 2\sum_{i=1}^{m}[\lambda_i + C_0(d_i^2 - \|\boldsymbol{c} - \boldsymbol{u}_i\|_2^2)](\boldsymbol{c} - \boldsymbol{u}_i) \tag{14}$$

$$\frac{dd_i}{dt} = \tanh(a(r_i - d_i)) - 2\lambda_i r_i - 2C_0 d_i(d_i^2 - \|\boldsymbol{c} - \boldsymbol{u}_i\|_2^2) \tag{15}$$

$$\frac{d\lambda_i}{dt} = d_i^2 - \|\boldsymbol{c} - \boldsymbol{u}_i\|_2^2. \tag{16}$$

For LPNN, its circuit complexity depends on the complexity to compute the time derivatives. From (14)–(16), the complexity to obtain the time derivatives is $O(m)$ only.

Another issue that we need to investigate is the local stability. Local stability means that a minimum point should be stable. Otherwise, the network never converges to the minimum [5]. Let $\{\boldsymbol{x}^*, \boldsymbol{\lambda}^*\}$ be a minimum point of the problem, where $\boldsymbol{x}^* = \left[\boldsymbol{d}^{*\mathrm{T}} \boldsymbol{c}^{*\mathrm{T}}\right]^{\mathrm{T}}$. There are two sufficient conditions for local stability of LPNN. The first one is convexity, i.e., the Hessian matrix of the

Lagrangian at $\{x^*, \lambda^*\}$ should be positive definite. It is achieved by introducing the augmented term. Note that if C_0 is chosen sufficiently large, then the Hessian matrix $\partial^2 \mathcal{L}(x, \lambda) / \partial x^2$ at the minimum is positive definite under mild conditions [5]. The second one is that at the minimum point, the gradient vectors of the constraints with respect to x should be linearly independent. The gradient vectors of them at the equilibrium point $x^* = \left[d^{*T} c^{*T} \right]^T$ are given by

$$\left\{ \left. \frac{\partial h_1(x)}{\partial x} \right|_{x=x^*}, \cdots, \left. \frac{\partial h_m(x)}{\partial x} \right|_{x=x^*} \right\} = \left\{ \begin{array}{cccc} 2d_1^* & 0 & \cdots & 0 \\ 0 & 2d_2^* & \cdots & 0 \\ \vdots & \vdots & \ddots & \vdots \\ 0 & 0 & \cdots & 2d_m^* \\ k_{1,1}^* & k_{2,1}^* & \cdots & k_{m,1}^* \\ k_{1,2}^* & k_{2,2}^* & \cdots & k_{m,2}^* \end{array} \right\}, \quad (17)$$

where $k_{i,j}^* = -2(c_j^* - u_{i,j})$, $i = 1, \cdots, m$, $j = 1, 2$. We can see that the m gradient vectors are linear independent with each other, if for all i $d_i^* \neq 0$. In other word, as long as the estimated coordinates of the mobile source are not equal to the coordinates of one of the sensors, the m gradient vectors are linear independent.

4 Simulation

4.1 Settings

This section conducts two experiments to test the robustness and efficiency of our proposed algorithm. As a comparison, we also consider the LLS algorithm [2], the TSWLS algorithm [3], and the l_2-norm LPNN. The sensors are uniformly distribution on a circumference of a circle with centre at $(0,0)$ and radius equal to $R = 10$. Figure 2 shows the sensor position for $m = \{4, 5, 6, 7, 8\}$. The TOA measurement errors are Gaussian random variables combining with some outliers. The outliers follow uniform distribution with mean value ranging from 1 to 19, and the width is equal to 2. For instance, if the mean of the outlier is 5, then the outlier can equal to any real value between 4 to 6. Hence we can control the intensity of the outlier by changing the mean of the outlier.

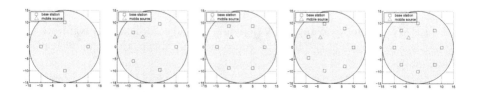

Fig. 2. The configuration of the source and sensors.

4.2   Experiment 1: Fixed Mobile Position

In this experiment, we fix the coordinates of mobile source at point $(-4, 4)$. We vary the outlier level by changing the mean value of outliers. We repeat this experiment 100 times. Figure 3 shows the mean square errors (MSEs) of the estimated mobile source coordinates under different settings.

First of all, we can see that no matter in which setting, the performance of our proposed algorithm is much better than other algorithms.

Second, as shown in Fig. 3(a),(b),(c),(e),(f), if sufficient sensors are used, then the performance of our proposed algorithm is insensitive to the outlier intensity. For example, from Fig. 3(f), when there are two outliers, with eight sensors the MSE of algorithm is lower than -30 dB, regardless of the outlier intensity. While, for the other algorithms, even we increase the number of sensors, their performances are still poor.

Lastly, from Fig. 3(c),(d) our proposed algorithm can work under both low level and high level Gaussian noise environment.

4.3   Experiment 2: Random Mobile Position

This section tests the performance of our proposed algorithm under the situation that the position of the mobile source is randomly chosen at each repeated experiment. The positions of the mobile source are uniformly distributed on a

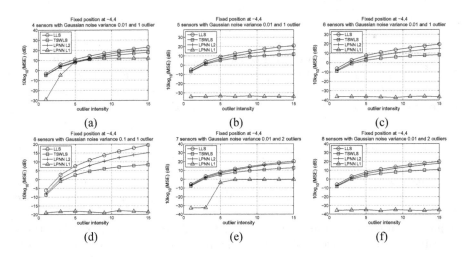

Fig. 3. The performance of various algorithms under different setting. All of them we randomly choose one or two sensors and add outliers into TOA measurements. (a) 4 sensors with Gaussian noise variance 0.01 and 1 outlier; (b) 5 sensors with Gaussian noise variance 0.01 and 1 outlier; (c) 6 sensors with Gaussian noise variance 0.01 and 1 outlier; (d) 6 sensors with Gaussian noise variance 0.01 and 1 outlier; (e) 7 sensors with Gaussian noise variance 0.01 and 2 outliers; (f) 8 sensors with Gaussian noise variance 0.01 and 2 outliers.

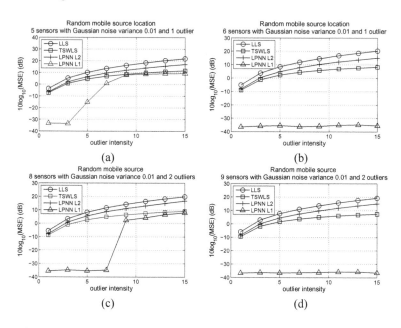

Fig. 4. The performance of various algorithms under different settings when the position of mobile source is randomly chosen. The variance of Gaussian noise level equal to 0.01. (a) 5 sensors with 1 outlier; (b) 6 sensors with 1 outlier; (c) 8 sensor with 2 outliers;(c) 9 sensor with 2 outliers.

circle centered at original point with radius 15. The simulation results are given by Fig. 4. From the figure, we get a conclusion similar to that of Experiment 1. That is, the performance of our proposed algorithm is much better than LLS, TSWLS, l_2-norm LPNN. Besides, if sufficient sensors are used, then the performance of our proposed algorithm is insensitive to the outlier intensity.

5   Conclusion

This paper considers the source localization problem under the situation that outliers exist in the TOA measurements. We propose an robust algorithm, based on LPNNs and an approximation function, to solve this problem. Although the original formulation contain some inequality constraints, we show that those inequality constraints can be removed. To improve convexity of the network, we introduce an augment term into the original objective function. From the simulation results, we can see that our proposed algorithm greatly improves the robustness against outliers. The performance of the proposed approach is much better than the performances of the comparison methods, including the LLS algorithm [2], the TSWLS algorithm [3], and the l_2-norm LPNN.

References

1. So, H.C.: Source localization: algorithms and analysis. In: Zekavat, S.A., Buehrer, R.M. (eds.) Handbook of Position Location: Theory, Practice, and Advances. Wiley, New York (2011)
2. Chen, J.C., Hudson, R.E., Yao, K.: Maximum-likelihood source localization and unknown sensor location estimation for wideband signals in the near field. IEEE Trans. Signal Process. **50**(8), 1843–1854 (2002)
3. Chan, Y.T., Ho, K.C.: A simple and efficient estimator for hyperbolic location. IEEE Trans. Signal Process. **42**(8), 1905–1915 (1994)
4. Leung, C.-S., Sum, J., So, H.C., Constantinides, A.G., Chan, F.K.W.: Lagrange programming neural networks for time-of-arrival-based source localization. Neural Comput. Appl. **24**(1), 109–116 (2014)
5. Zhang, S., Constantinides, A.G.: Lagrange programming neural networks. IEEE Trans. Circ. Syst. II: Analog Digital Signal Process. **39**(7), 441–452 (1992)
6. Tank, D., Hopfield, J.: Simple neural optimization networks: an A/D converter, signal decision circuit, and a linear programming circuit. IEEE Trans. Circ. Syst. **33**(5), 533–541 (1986)
7. Xia, Y.: An extended projection neural network for constrained optimization. Neural Comput. **16**(4), 863–883 (2004)
8. Liu, Q., Wang, J.: A one-layer recurrent neural network with a discontinuous hard-limiting activation function for quadratic programming. IEEE Trans. Neural Netw. **19**(4), 558–570 (2008)
9. Liang, J., Leung, C.-S., So, H.C.: Lagrange programming neural network approach for target localization in distributed MIMO radar. IEEE Trans. Signal Process. **64**(6), 1574–1585 (2016)

Reward-Based Learning of a Memory-Required Task Based on the Internal Dynamics of a Chaotic Neural Network

Toshitaka Matsuki$^{(\boxtimes)}$ and Katsunari Shibata

Oita University, 700 Dannoharu, Oita, Japan
{matsuki,shibata}@oita-u.ac.jp

Abstract. We have expected that dynamic higher functions such as "thinking" emerge through the growth from exploration in the framework of reinforcement learning (RL) using a chaotic Neural Network (NN). In this frame, the chaotic internal dynamics is used for exploration and that eliminates the necessity of giving external exploration noises. A special RL method for this framework has been proposed in which "traces" were introduced. On the other hand, reservoir computing has shown its excellent ability in learning dynamic patterns. Hoerzer et al. showed that the learning can be done by giving rewards and exploration noises instead of explicit teacher signals. In this paper, aiming to introduce the learning ability into our new RL framework, it was shown that the memory-required task in the work of Hoerzer et al. could be learned without giving exploration noises by utilizing the chaotic internal dynamics while the exploration level was adjusted flexibly and autonomously. The task could be learned also using "traces", but still with problems.

Keywords: Chaotic neural network · Reservoir computing · Reward-Modulated Hebbian Learning · Traces · Dynamic higher functions

1 Introduction

In recent years, Deep Learning, in which a large-scale neural network (NN) with many layers learns to process raw sensor signals in parallel, has surpassed existing systems in various fields. That suggests the difficulty in understanding the phenomenal performance of our parallel brain through our sequential consciousness and then developing an appropriate program by hand for such massively parallel processing. For a long time, our group has pointed out this difficulty and has suggested the necessity to develop a system in which the whole process from sensors to motors consists of a NN and necessary functions or useful internal representations emerge through reinforcement learning (RL) with explorations and rewards [1,2]. Recently, a recurrent NN (RNN) has been employed to deal with dynamics, and it was confirmed that the function of "memory" or "prediction" emerges in a simple task [3,4]. However, there seems to be a limitation

© Springer International Publishing AG 2016
A. Hirose et al. (Eds.): ICONIP 2016, Part I, LNCS 9947, pp. 376–383, 2016.
DOI: 10.1007/978-3-319-46687-3_42

for a non-chaotic "silent" RNN to form multi-stage state transitions through learning [5].

Thus, we thought that the complex dynamics is not formed from scratch in a non-chaotic "silent" RNN but is reformed from rich and chaotic internal dynamics in a chaotic NN. The dynamics is used also for exploration in RL, and that eliminates the need for giving exploration noises from outside. It is expected to become purposeful through learning reflecting the causal relations of the world and finally reach dynamic higher functions such as "thinking". In this new RL framework, since exploration components cannot be separated from the outputs, training signals cannot be derived. Therefore, instead of using error back propagation as in the conventional RL, a special learning method, in which "traces" are introduced, was proposed and confirmed to work in an easy task [6,7].

On the other hand, recently, reservoir computing such as Echo State Network [8] and Liquid State Machine [9] has been focused on. In this trends, Sussillo et al. trained reservoir networks by the new learning procedure called FORCE Learning [10]. In this procedure, the outputs are returned to the network along its feedback pathway and only readout weights are modified to match the network outputs with target patterns. Then, the network can learn to generate complex dynamic patterns amazingly easily and rapidly.

Hoerzer et al. showed that a reservoir network can learn through Reward-Modulated Hebbian Learning in which instead of explicit teacher signals, exploration noises and the reward to show the improvement of the performance derived from the error between outputs and targets were given [11]. In this research, it was shown that the network learned various dynamic patterns or a working memory task.

From the above, we think that it is essential to introduce the learning ability of dynamical patterns into our new RL framework to realize dynamic higher functions such as "thinking". In this paper, as the first step of this attempt, we examine whether the working memory task, which a reservoir network learned from reward signals in Hoerzer's work, can be learned without giving external exploration noises by utilizing the internal chaotic dynamics of the network as well as in our new RL framework. Next, we also examine whether the network can be trained with "traces", which are used in our new RL.

2   Method

2.1   Network

In this paper, we use the network as in Fig. 1, which has basically the same structure and connection weights in the previous researches [10,11]. The network is composed of $N = 1000$ neurons, and they are sparsely and recurrently connected (connection probability $p = 0.1$). There are four external inputs each of which is fed to all the neurons. There are two output units called readout units, and are connected by all the network neurons. Each output from the corresponding readout unit is returned to all the network neurons along its feedback pathway.

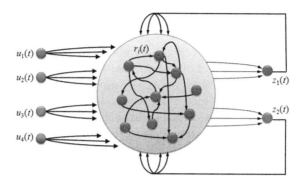

Fig. 1. The network model. It has 4 inputs, u_1 (green), u_2 (orange), u_3 (cyan), u_4 (brown) and 2 outputs, z_1 (red), z_2 (purple). In the network, 1000 neurons (blue) are recurrently connected (connection probability $p = 0.1$). (Color figure online)

The model of each network neuron is a dynamical firing-rate model. The internal activity (membrane potential) of the j-th network neuron at time t is given as

$$x_j(t) = \left(1 - \frac{\Delta t}{\tau}\right)x_j(t - \Delta t) + \frac{\Delta t}{\tau}\left(\lambda \sum_{i=1}^{N} w_{ji}^{rec}r_i(t) + \sum_{i=1}^{I} w_{ji}^{in}u_i(t) + \sum_{i=1}^{O} w_{ji}^{fb}z_i(t)\right),$$
(1)

where the step size $\Delta t = 1$ [ms] and the time constant $\tau = 10$ [ms]. λ is the parameter that gives the scale of recurrent connection weights between the network neurons, whose value is 1.8 or 1.5 (the latter is used in the training with "traces"). Larger λ produces more chaotic activities of the network. w_{ji}^{rec} is the weight of recurrent connection from the i-th neuron to the j-th neuron. These are set to a value generated randomly from a Gaussian distribution with zero mean and variance $1/pN$. I is the number of inputs. w_{ji}^{in} is the weight from the i-th input to the j-th neuron. u_i is the i-th input value. O is the number of readout units. w_{ji}^{fb} is the weight from i-th readout unit to j-th neuron. These are set to a value generated randomly from a uniform distribution between -1 and 1. $z_i(t)$ is the output of the j-th readout unit. The output of network neuron $r_j(t)$ is computed from its internal activity $x_j(t)$ as

$$r_j(t) = tanh\left(x_j(t)\right).$$
(2)

$z_j(t)$ at time t is derived from $r_i(t)$ and the corresponding readout weight w_{ji} as

$$z_j(t) = \sum_{i=1}^{N} w_{ji}r_i(t).$$
(3)

Initially, w_{ji} is set to a value generated randomly from a Gaussian distribution with zero mean and variance $1/N$.

2.2   Learning

In this paper, only readout weights w_{ji} are trained. Excepting that exploration noises are not added, we basically followed the learning procedure by Hoerzer et al. in [11]. The network is trained with reward or penalty which is given dependently on whether the current performance of the network $P(t)$ is improved as compared to its running average $\overline{P}(t)$ with time constant $5\,\mathrm{ms}$. $P(t)$ is defined as

$$P(t) = -\sum_{j=1}^{O}\Big(z_j(t) - f_j(t)\Big)^2, \tag{4}$$

where f_j is the target for the j-th output of the network.

We use two learning methods in this research. First one is Reward-Modulated Hebbian Learning in [11] with a little modification. The modulatory signal $M(t)$ is defined using $P(t)$ and $\overline{P}(t)$ as

$$M(t) = \begin{cases} 1 & P(t) > \overline{P}(t) \\ -1 & P(t) \le \overline{P}(t). \end{cases} \tag{5}$$

In [11], $M(t)$ took the value of 1 or 0, but here 1 or -1 is used. The readout weights are modified with $M(t)$ as

$$\Delta w_{ji} = \eta\Big(z_j(t) - \overline{z}_j(t)\Big)M(t)r_i(t). \tag{6}$$

where η is a learning constant and here $\eta = 0.0005$. $\overline{z}(t)$ is the running average of z with time constant $5\,\mathrm{ms}$.

Second, we use a learning method with traces that are used in our new RL. In this learning, to limit the value range, the output $z(t)$ is derived as

$$z_j(t) = tanh\Big(\sum_{i=1}^{N} w_{ji}r_i(t)\Big). \tag{7}$$

The readout weights are modified with $P(t)$ and $\overline{P}(t)$ as

$$\Delta w_{ji} = \eta\Big(P(t) - \overline{P}(t)\Big)c_{ji}(t), \tag{8}$$

where $c_{ji}(t)$ is the trace which expresses the correlation between the output increase and the i-th input in the j-th readout unit and $\eta = 0.05$ here. $c_{ji}(t)$ is given by

$$\Delta z_j(t) = z_j(t) - z_j(t - \Delta t). \tag{9}$$

$$c_{ji}(t) = \Big(1 - \frac{|\Delta z_j(t)|}{2}\Big)c_{ji}(t - \Delta t) + \frac{\Delta z_j(t)}{2}r_i(t), \tag{10}$$

where 2 is the value range of each readout unit. This equation computes the value similar to the running average of the input, but the time constant is derived from the output change of the unit. Therefore, when the output change is large, the signal $r_i(t)$ is considered to be important and taken into the trace largely. When the output does not change, the past value is kept in the trace.

2.3 Task

The network learns the task that requires working memory [11]. The network has four inputs and two outputs. Input pulses with the average rate of 0.5 Hz are given on each input signal independently. It goes up to 1.0 taking 50 ms and then goes down with time constant 50 ms. Each signal has a different meaning. u_1 and u_2 are respectively ON and OFF signals for the output z_1, and u_3 and u_4 are for the output z_2. An ON or OFF signal makes the corresponding output to be 1.0 or -1.0 respectively with time constant 20 ms, and the value is kept until the opposite signal for the corresponding output comes in.

3 Results

Figure 2 shows the network activity: outputs, inputs and activities of some neurons. Figure 2(a) shows the activities for the first 30 s of learning. At first, the network did not know its desired behavior. However, noise-like fluctuations originated from the chaotic internal dynamics appeared in the output even without external noises. The chaotic internal dynamics performs the role of exploration, and the outputs looks to follow the targets with a lag. However, when learning was stopped at this timing, the output could not follow the target.

Figure 2(b) shows the activities of the network for 30 s of testing after 250 s of learning. The outputs almost match the target with no lag. This result shows that the task can be learned without exploration noises. It is interesting that as the learning progresses with successively given reward and penalty ($M(t)$), the sharp change disappeared gradually. The activity of the network seems to transit from exploration mode to stable mode, and the exploration component from the internal dynamics decreases autonomously.

To observe whether the network can adjust the exploration level autonomously when encountering unknown situation, the rule of learning task was changed suddenly. Figure 2(c) shows the network activities when the ON signal and OFF signal were swapped between u_1 and u_2 and between u_3 and u_4 after 250 s of learning. The chaotic activities appeared again and exploration was resumed even without any direction from outside. The network activities for 300 s of learning after the rule change are shown in Fig. 2(d) in a compressed time scale, and Fig. 2(e) shows 30 s of testing after that. It shows that the network could resume to explore and successfully learn the task even though the environment was changed suddenly in the middle of learning.

To show how the chaotic activities of network neurons change during learning, the output errors and the output change of network neurons were recorded as in Fig. 3. These values are the mean absolute error of the network output and the mean absolute one-step change in each neuron output over all neurons over every 10000 time steps. In Fig. 3, it is seen in the term of 250 s from the start of learning when the network learned under the first rule, the output error decreased and the network activities decreased gradually. As soon as the rule was changed at 250 s, the output error increased and, a little later, the output change in the network neurons also increased. Then the output change was large though it

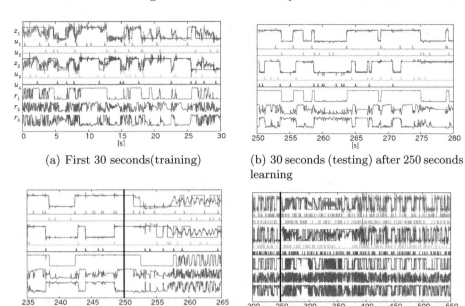

(a) First 30 seconds(training)

(b) 30 seconds (testing) after 250 seconds learning

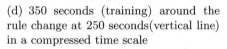

(c) 30 seconds (training) around the rule change at 250 seconds(vertical line)

(d) 350 seconds (training) around the rule change at 250 seconds(vertical line) in a compressed time scale

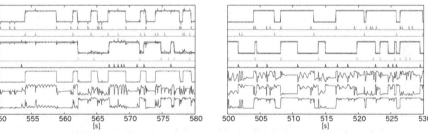

(e) 30 seconds (testing) after 300 seconds from the rule change

(f) Learning with traces: 30 seconds (testing) after 500 seconds of learning

Fig. 2. Network activities. Output z_1, z_2 are in red, purple respectively. The target value is in black. Input u_1, u_2, u_3, u_4 are in green, orange, cyan, brown respectively. The activities of 3 sample neurons from the network are in blue. (Color figure online)

decreased sometimes, before the change decreased again as the error decreased. It shows that the network can make the internal dynamics chaotic autonomously to explore in unknown situation.

The result after learning procedure with traces is shown in Fig. 2(f). In this experiment, we used 0.9 and −0.9 as the maximum and minimum value of targets to prevent reaching a limit of the outputs. Figure 2(f) shows that the reservoir network could be trained with the traces. It seems that the outputs follow the target precisely at a glance, but its values are close to −1.0 or 1.0 that is the

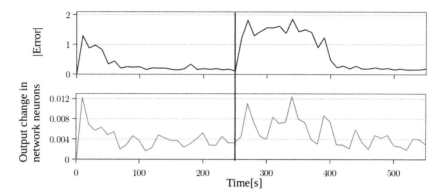

Fig. 3. The mean absolute error (upper) and the mean absolute output change of network neurons (lower) during learning. The vertical line is the timing of rule change.

upper or lower limit of the output. That means that the output values before the transformation by *tanh* are very large due to large readout weights w_{ji}. In addition, the network parameters needed very sensitive adjustment to learn successfully, and occasionally the output deviated largely. In this case, because the outputs stick upper or lower limit, it was difficult to output intermediate values and was impossible to resume to explore when the rule of task was changed during learning. There still remain problems to be solved.

4 Conclusion

In the Reward-Based Learning of Memory Required Task in a reservoir network, it was confirmed that the internal chaotic dynamics can perform the role of exploration on behalf of the exploration noises added from the outside. As the learning progressed, noise-like fluctuations in the outputs originated from the internal dynamics decreased and the network activities autonomously transited from exploration mode to stable mode gradually. It was also shown that when the task setting was changed during learning, the network adaptively resumed exploration and learned appropriately after that. Using the traces, which is used to train a chaotic NN in the newly-proposed novel RL, the same task could be trained as well, but further investigations are necessary. From these results, it is expected that the learning ability of the reservoir computing can be taken into our approach, and that enables the emergence of higher functions as the result of developing internal dynamics through RL.

Acknowledgement. The authors wish to thank Prof. Hiromichi Suetani for introducing FORCE Learning and the work of Hoerzer et al. to us. This work was supported by JSPS KAKENHI Grant Number 15K00360.

References

1. Shibata, K., Okabe, Y.: Reinforcement learning when visual signals are directly given as inputs. In: Proceedings of ICNN 1997, vol. 3, pp. 1716–1720 (1997)
2. Shibata, K.: Emergence of intelligence through reinforcement learning with a neural network. In: Mellouk, A. (ed.) Advances in Reinforcement Learning, pp. 99–120. InTech, Rijeka (2011)
3. Shibata, K., Utsunomiya, H.: Discovery of pattern meaning from delayed rewards by reinforcement learning with a recurrent neural network. In: Proceedings of IJCNN, pp. 1445–1452 (2011)
4. Shibata, K., Goto, K.: Emergence of flexible prediction-based discrete decision making and continuous motion generation through actor-q-learning. In: Proceedings of ICDL-Epirob, ID 15 (2013)
5. Sawatsubashi, Y., et al.: Emergence of discrete and abstract state representation in continuous input task. In: Robot Intelligence Technology and Applications, pp. 13–22 (2012)
6. Shibata, K., Sakashita, Y.: Reinforcement learning with internal-dynamics-based exploration using a chaotic neural network. In: Proceedings of International Joint Conference on Neural Networks (IJCNN) (2015). 2015.7
7. Goto, Y., Shibata, K.: Emergence of higher exploration in reinforcement learning using a chaotic neural network. In: Akira, H., Seiichi, O., Kenji, D., Kazushi, I., Minho, L., Derong, L. (eds.) ICONIP 2016. LNCS, pp. 40–48. Springer, Heidelberg (2016)
8. Jaeger, H.: The "echo state" approach to analysing and training recurrent neural networks. GMD report 148, p. 43 (2001)
9. Maass, W., Natschlger, T., Markram, H.: Real-time computing without stable states: a new framework for neural computation based on perturbations. Neural Comput. 14(11), 2531–2560 (2002)
10. Sussillo, D., Abbott, L.F.: Generating coherent patterns of activity from chaotic neural networks. Neuron 63(4), 544–557 (2009)
11. Hoerzer, G.M., Legenstein, R., Maass, W.: Emergence of complex computational structures from chaotic neural networks through reward-modulated Hebbian learning. Cereb. Cortex 24(3), 677–690 (2014)

Roles of Gap Junctions in Organizing Traveling Waves in a Hippocampal CA3 Network Model

Toshikazu Samura[1,2]([✉]), Yutaka Sakai[2], Hatsuo Hayashi[3], and Takeshi Aihara[2]

[1] Graduate School of Sciences and Technology for Innovation,
Yamaguchi University, 2-16-1 Tokiwadai, Ube, Yamaguchi 755-8611, Japan
samura@yamaguchi-u.ac.jp
[2] Tamagawa University Brain Science Institute,
6-1-1 Tamagawa Gakuen, Machida, Tokyo 194-8610, Japan
{sakai,aihara}@eng.tamagawa.ac.jp
[3] Kyushu Institute of Technology,
1-1 Sensui-cho, Tobata-ku, Kitakyushu, Fukuoka 804-8550, Japan

Abstract. Directional traveling waves are organized in a hippocampal CA3 recurrent network model composed of biophysical pyramidal cells and inhibitory interneurons with gap junctions. The network spontaneously organizes neuronal activities traveling in a particular direction and the organized traveling waves are modified by repetitive local inputs. We found that the distributions of inter-spike intervals (ISIs) of pyramidal cells and interneurons are involved with spontaneous traveling waves that can be modified by local stimulation. Similar ISI distributions emerge in a network that has no gap junctions, but strong mutual connections between pyramidal cells and interneurons. These results suggest that interaction between interneurons through gap junctions contributes to enhancing the inhibition of pyramidal cells for organizing traveling waves.

Keywords: Hippocampus · CA3 · Traveling waves · Gap junctions

1 Introduction

Theta oscillations travel along the longitudinal axis of the hippocampus [1,2]. A possible mechanism of the traveling theta waves is the propagation of neuronal activities through recurrent connections of the hippocampal CA3 [1,2]. It has been demonstrated that radially propagating neuronal activities (non-directional traveling wave) are spontaneously and input-dependently organized in a hippocampal CA3 recurrent network model composed of biophysical pyramidal cells and inhibitory interneurons [8]. On the other hand, we have demonstrated that directionally propagating neuronal activities are input-dependently organized in a simple recurrent network model with anisotropic inhibitory structure, which does not cause propagating neuronal activities spontaneously [3]. A biophysical CA3 recurrent network model with an anisotropic inhibitory structure, however, spontaneously organizes directionally propagating neuronal activities [4].

© Springer International Publishing AG 2016
A. Hirose et al. (Eds.): ICONIP 2016, Part I, LNCS 9947, pp. 384–392, 2016.
DOI: 10.1007/978-3-319-46687-3_43

The inhibitions play an important role in selecting the direction of the propagation. In the hippocampal CA3, an activity of inhibitory interneurons is affected by interactions through gap junctions [7]. It has been demonstrated that gap junctions enhance the propagation of firings in a feedforward network [5]. Here, we investigated the roles of gap junctions in organizing traveling waves in the biophysical CA3 recurrent network model. We show that the biophysical CA3 recurrent network model with gap junctions can not only spontaneously but also input-dependently organize directional propagation of neuronal activities and gap junctions have a role in enhancing inhibition for the organization of traveling waves.

2   Methods

2.1   Hippocampal CA3 Model

We modified CA3 recurrent network model developed by Yoshida and Hayashi [8]. The CA3 network model is composed of $2,304$ pyramidal cells and 288 inhibitory interneurons. Pyramidal cells were placed on 48×48 (vertical $\times$ horizontal) lattice points and inhibitory interneurons were uniformly distributed in the network (Fig. 1). Edge neurons were positioned next to neurons in the opposite side to remove the ununiformity of their connections (i.e. torus structure). Each pyramidal cell (e.g. $\bullet$ in Fig. 1) was connected to 20 pyramidal cells in its surrounding 7×7 region (e.g. solid rectangle in Fig. 1) and to all interneurons in 11×11 region (e.g. dashed rectangle in Fig. 1). On the other hand, each interneuron (e.g. $\bigcirc$ in Fig. 1) was connected to 60 pyramidal cells and 5 interneurons in its surrounding 13×9 region (e.g. gray solid rectangle in Fig. 1). In the region, interneurons have slightly long connections in the downward direction. Indeed, the axon projection of an inteneuron is biased in the hippocampus [6]. Furthermore, each interneuron was mutually connected with up to 5 interneurons in its surrounding 9×9 region (e.g. dashed-dotted rectangle in Fig. 1).

The membrane potential V_i of i-th pyramidal cell was updated as follows:

$$
\begin{aligned}
CdV_i/dt = {} & g_{\mathrm{Na}}m^2h(V_{\mathrm{Na}} - V_i) + g_{\mathrm{Ca}}s^2r(V_{\mathrm{Ca}} - V_i) \\
& + g_{\mathrm{Ca(low)}}s_{\mathrm{low}}^2 r_{\mathrm{low}}(V_{\mathrm{Ca}} - V_i) + g_{\mathrm{K(DR)}}n(V_{\mathrm{K}} - V_i) \\
& + g_{\mathrm{K(A)}}ab(V_{\mathrm{K}} - V_i) + g_{\mathrm{K(AHP)}}q(V_{\mathrm{K}} - V_i) \\
& + g_{\mathrm{K(C)}}c\,min(1, \chi/250)(V_{\mathrm{K}} - V_i) + g_{\mathrm{L}}(V_{\mathrm{L}} - V_i) \\
& + g_{\mathrm{af}}(V_{\mathrm{syn(e)}} - V_i) + I_{\mathrm{syn}},
\end{aligned}
\tag{1}
$$

where g_x and V_x of the 1st–7th term are the conductance and the equilibrium potential for the respective ion channels (x): Na, Ca, Ca(low), K(DR), K(A), K(AHP), K(C). g_{L} and V_{L} are the conductance and the equilibrium potential for leakage, respectively. g_{af} and V_{af} are the conductance and the equilibrium potential for afferent excitatory synapse, respectively. I_{syn} is total synaptic currents from postsynaptic neurons.

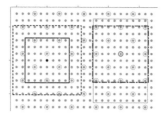

Fig. 1. The position of neurons in the hippocampal CA3 recurrent network (•: pyramidal cell; ◯: interneuron) and connection rage from each type of neuron.

We introduced gap junctions [5] into interneurons. The membrane potential of i-th interneuron was updated as follows:

$$CdV_i/dt = g_{Na}m^3h(V_{Na} - V_i) + g_{K(DR)}n^4(V_K - V_i) + g_L(V_L - V_i) + I_{syn} \quad (2)$$
$$+ \Sigma_j^{N_{gap}} g_{gap}(V_j - V_i),$$

where g_{gap} is the coupling conductance through gap junction between interneurons. N_i^{gap} is the number of the gap connection with other interneurons ($6 \leq N_i^{gap} \leq 8$). The details of the network model are described in ref. [8]; however a part of parameters was changed as follows: $g_{af} = 0.004$, $V_{ip} = -75$, $g_{gap} = 0.0015$. The synaptic conductance between pyramidal cells was updated by asymmetric STDP. We set the synaptic conductance for each synapse as follows: $C_{pp} = 0.0004-0.001, C_{pi} = 0.002, C_{ip} = 0.001, C_{ii} = 0.001$. We set different maximum modification rates for LTP and LTD: $M_{LTP} = 0.05, M_{LTD} = 0.0525$ because stronger modification rate for LTD than that for LTP are required for organizing directional traveling waves stably in the network (data not shown).

2.2 Simulation Conditions

Here, we used network model with gap junctions between interneurons as a control. Additionally, we defined four conditions using a different network model without gap junctions as shown in Table 1. In the first condition, gap junctions are simply removed from interneurons (no-gap). In the three other conditions, C_{pi} (pyramidal neuron → interneuron), C_{ip} (interneuron → pyramidal neuron), or both are strengthened (strPI, strIP, and strPI+strIP). We ran a simulation ten times in each condition. In a simulation trial, we updated the membrane potential of neurons for 250 s. Synaptic inputs and synaptic weights were updated from 1 s and 10 s, respectively. Burst inputs with doublet spikes (inter-spike interval: 10 ms; inter-burst interval: 125 ms) were applied into 3×3 neurons in the center of the network periodically for 100 s from 110 s. We evaluated traveling waves organized in the networks before and after inputs.

2.3 Evaluation of Traveling Waves

Neuronal activities propagate in the vertical direction in the network where interneurons elongate their axons in the vertical direction (e.g. Fig. 2a) [4].

Table 1. Conditions of the network without gap junctions between interneurons

Conditions	no-gap	strPI	strIP	strPI+strIP
C_{pi}	0.002	0.0025	0.002	0.0025
C_{ip}	0.001	0.001	0.0011	0.0011

Therefore, we evaluated synchronized activities of pyramidal cells at location on a vertical axis to confirm whether directional propagation of neuronal activities emerges in the network. We calculated the ratio of co-firing neuron at the same location on the vertical axis for each 20 ms bin (e.g. Fig. 2c). Stripes indicate that traveling waves with horizontal wave fronts propagate in the vertical direction (Fig. 2a, c).

We evaluated the direction, in which a pyramidal cell acquired strong connections. Such a weight direction reflects the direction of neuronal activities passing through each pyramidal cell. The weight direction of the j-th pyramidal cell at time t is calculated as follows:

$$R_j(t) = \frac{1}{N_{\mathrm{trial}} N_{\mathrm{post}}} \sum_k^{N_{\mathrm{trial}}} \sum_l^{N_{\mathrm{post}}} (w_{lj}^k(t) - w_{lj}^k(0)) e^{i\theta_{lj}^k}, \tag{3}$$

where N_{trial} and N_{post} are the number of trials and the number of postsynaptic pyramidal cells, respectively. $w_{lj}^k(t)$ is a synaptic weight to the l-th postsynaptic pyramidal cell from the j-th pyramidal cell at time t in the k-th trial. θ_{lj}^k indicates the angle between the j-th pyramidal cell and the l-th postsynaptic pyramidal cell in the k-th trial. Larger $R_j(t)$ indicates a stronger tendency of traveling waves to pass through the j-th neuron in the evaluated direction across trials (e.g. long arrows in Fig. 2e, f).

3   Results

3.1   Traveling Waves Organized in a Network with Gap Junctions

Directional traveling waves were spontaneously organized in the network model with gap junctions. Figure 2a shows the neuronal activities for 20 ms from 109 s in the 3rd trial. Most pyramidal cells and inhibitory interneurons fired simultaneously around 5 and 30 on the vertical axis. Upward traveling waves with horizontal wave fronts were organized spontaneously in the network in 9 of 10 trials (Fig. 2c). The direction can be confirmed by the weight directions of pyramidal cells (Fig. 2e). On the other hand, traveling waves with horizontal wave fronts were broken at 209 s by repetitive inputs into the center of the network (Fig. 2b, d). The weight directions indicate that neuronal activities around the center of the network rather propagate in the left- and rightward direction (Fig. 2f). In other words, input-dependent traveling waves were newly organized.

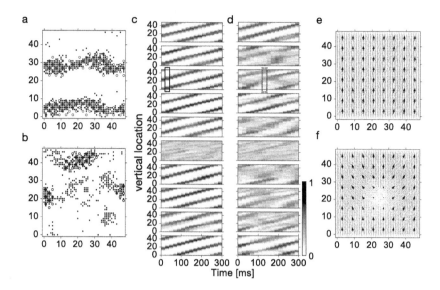

Fig. 2. Neuronal activities and weight directions organized in the hippocampal CA3 recurrent network model with gap junctions before (a, c, e) and after (b, d, f) inputs. (a, b) Neuronal activities for 20 ms at 109 s (a) and 209 s (b) in the 3rd trial. Figures (a) and (b) correspond to the activities in solid and dashed rectangles in (c) and (d), respectively. (c, d) Ratio of co-firing neuron at each location on the vertical axis of the network for 300 ms from 109 s (c) and 209 s (d). Each panel shows the results of each trial. Two panels in the same row show neuronal activities occurring in the same trial. (e, f) Weight directions of pyramidal cells (gray arrows) and the average weight directions within a 6 × 6 region (black arrows) at 110 s (e) and 210 s (f).

3.2   Traveling Waves Organized in a Network Without Gap Junctions

Figure 3 shows neuronal activities of the network without gap junctions before applying inputs. Although traveling waves with horizontal wave fronts were not organized clearly in the network under the no-gap condition (Fig. 3a), the weight directions show that neuronal activities tend to propagate upward at each cell position (Fig. 3e). On the other hand, traveling waves with horizontal wave fronts were frequently organized under the other conditions (Fig. 3b–d), especially in the strPI and strPI+strIP conditions (Fig. 3b, d). Upward weight directions were obviously obtained in the network (Fig. 3f–h). The direction of propagation was consistent across trials, but traveling waves without clear horizontal wave fronts occurred in some trials (Fig. 3b–d). Figure 4 shows neuronal activities of the network without gap junctions after applying inputs. Spontaneous traveling waves were broken in most trials (Fig. 4a–d), and new propagation of neuronal activities was organized by the inputs. Neuronal activities around the center of the network propagated in the left- and rightward direction (Fig. 4e–h).

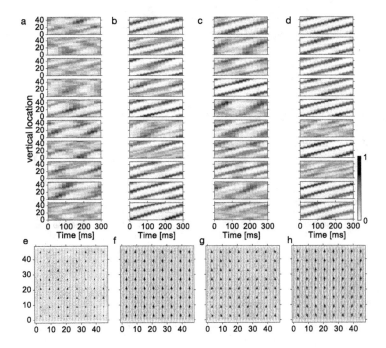

Fig. 3. Spontaneous neuronal activities and weight directions of the hippocampal CA3 recurrent network model without gap junctions. (a–d) Ratio of co-firing neuron at each location on the vertical axis of the network for 300 ms from 109 s under no-gap (a), strPI (b), strIP (c) and strPI+strIP (d) conditions. (e–h) Weight directions of pyramidal cells (gray arrows) and the average weight directions within a 6×6 region (black arrows) at 110 s under each condition.

3.3 Distributions of Inter-Spike Intervals of Pyramidal Cells and Interneurons

Here, we compared the distributions of inter-spike intervals (ISIs) obtained from activities of the network under the five conditions (Fig. 5). In all conditions, the largest and the second largest peaks exist around 20 and 180 ms, respectively. Neuronal activities periodically pass through each cell position. The largest peak around 20 ms corresponds to the intervals between firings within a traveling wave and the second peak around 180 ms corresponds to the intervals between wavefronts of traveling waves.

The peak around 180 ms depends on the conditions (Insets in Fig. 5a–e). Under the control condition (Fig. 5a), the ISI distributions of pyramidal cells and interneurons showed peaks around 180 and 140 ms, respectively. The ISIs of interneurons were shorter than those of pyramidal cells. Under the no-gap condition, the ISI distribution of pyramidal cells increased in the range of 100 to 150 ms, while that of interneurons decreased (Fig. 5b). Under the conditions of strong C_{pi} (strPI and strPI+strIP), the difference between peak timings of the distributions was obvious and the ISI distribution of pyramidal cells in the

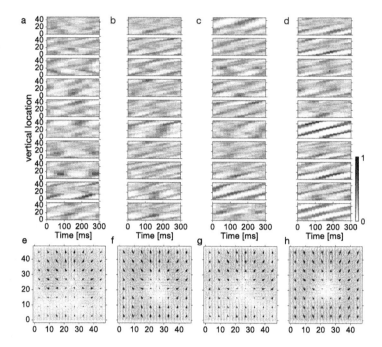

Fig. 4. Neuronal activities and weight directions of the hippocampal CA3 recurrent network model without gap junctions organized by local stimulation. (a–d) Ratio of co-firing neuron at each location on the vertical axis of the network for 300 ms from 209 s under no-gap (a), strPI (b), strIP (c) and strPI+strIP (d) conditions. (e–h) Weight directions of pyramidal cells (gray arrows) and the average weight directions within a 6 × 6 region (black arrows) at 210 s under each condition.

range of 100 to 150 ms decreased compared to no-gap condition (Fig. 5c, e). The ISI distribution of pyramidal cells in the range of 100 to 150 ms also decreased under the strIP condition (Fig. 5d). Consequently, the ISI distribution of pyramidal cells in the rage was prominently reduced under the strPI+strIP condition (Fig. 5e). The similar distribution of ISIs in the control condition can be acquired under the strPI+strIP condition.

Additionally, we evaluated the on-going activity of interneurons for 1 s from 109 s. Interneurons more frequently fire under the conditions in which traveling waves with horizontal wave fronts are organized (control: 18.8 ± 2.58 Hz, strPI: 19.0 ± 2.55 Hz, strPI+strIP: 19.1 ± 2.68 Hz) than under the other conditions (no-gap: 14.7 ± 3.49 Hz, strIP: 14.5 ± 3.20 Hz) (mean $\pm$ SD). Under the control condition, the activities of interneurons were kept at the same level with the activities in the strPI and strPI+strIP conditions.

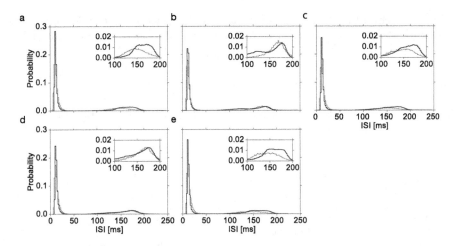

Fig. 5. The distributions of inter-spike intervals of pyramidal cells (solid line) and interneurons (dashed line) obtained from spontaneous neuronal activities at 109 s. For each neuron types, the distributions were obtained from all spike emitted by a neuron in 10 trials under control (a), no-gap (b), strPI (c), strIP (d) and strPI+strIP (e) conditions. Inset of each panel shows the magnification of the peak around 180 ms.

4 Conclusion

We demonstrated that the CA3 recurrent network model with gap junctions organizes traveling waves with horizontal wave fronts not only spontaneously but also input-dependently. Local stimulation produced new traveling waves from the stimulus site in the network causing spontaneous traveling waves. These results suggest that traveling waves propagating rhythmically may encode information applied into the hippocampal CA3 by changing its wave front.

We also found that the network with gap junctions causes specific ISI distributions of pyramidal cells and interneurons in the range of 100 to 200 Hz. Similar ISI distributions were shown in the network where synaptic connections between pyramidal cells and interneurons were strengthened (strPI+strIP condition) even if there were no gap junctions in the network. Since new traveling waves were caused by local stimulation in the network causing spontaneous traveling waves under the strPI+strIP condition, the ISI distributions of pyramidal cells and interneurons would be involved with causing such traveling waves.

These results suggest that the role of gap junctions is the enhancement of inhibition of pyramidal cells for organizing traveling waves as if the synaptic connections from pyramidal cells (interneurons) to interneurons (pyramidal cells) would be enhanced. Under the condition with gap junctions, the connections from inhibitory interneurons to pyramidal cells were weak compared to those under strPI+strIP condition. The high-frequent firing of interneurons is required for causing inhibition at the same level with strPI+strIP condition. However, the on-going activities of interneurons are similar to those under the strPI+strIP condition. The enhancement of inhibitory output to pyramidal cells by gap junctions is not derived just from strong inhibitory connections and high-frequent activities of interuenorns. It is supposed that the interaction between interneurons through gap junctions improves the effect of inhibition by causing inhibition at the proper timing for organizing traveling waves.

Acknowledgement. This study was supported by MEXT -Supported Program for the Strategic Research Foundation at Private Universities, 2009–2013 and JSPS KAK-ENHI Grant Number 15K21193.

References

1. Lubenov, E.V., Siapas, A.G.: Hippocampal theta oscillations are travelling waves. Nature **459**(7246), 534–539 (2009)
2. Patel, J., Fujisawa, S., Berényi, A., Royer, S., Buzsáki, G.: Traveling theta waves along the entire septotemporal axis of the hippocampus. Neuron **75**(3), 410–417 (2012)
3. Samura, T., Hayashi, H.: Directional spike propagation in a recurrent network: dynamical firewall as anisotropic recurrent inhibition. Neural Netw. **33**, 236–246 (2012)
4. Samura, T., Yutaka, S., Hatsuo, H., Takeshi, A.: Localized anisotropic inhibition for self-organized directional traveling waves in the hippocampal CA3. In: The Proceedings of 24th Annual Conference of Japanese Neural Network Society, pp. 80–81 (2014)
5. Shinozaki, T., Naruse, Y., Câteau, H.: Gap junctions facilitate propagation of synchronous firing in the cortical neural population: a numerical simulation study. Neural Netw. **46**, 91–98 (2013)
6. Sik, A., Penttonen, M., Ylinen, A., Buzsáki, G.: Hippocampal CA1 interneurons: an in vivo intracellular labeling study. J. Neurosci. **15**(10), 6651–6665 (1995)
7. Yang, Q., Michelson, H.B.: Gap junctions synchronize the firing of inhibitory interneurons in guinea pig hippocampus. Brain Res. **907**(1–2), 139–143 (2001)
8. Yoshida, M., Hayashi, H.: Regulation of spontaneous rhythmic activity and organization of pacemakers as memory traces by spike-timing-dependent synaptic plasticity in a hippocampal model. Phys. Rev. E **69**(1), 011910:1–011910:15 (2004)

Towards Robustness to Fluctuated Perceptual Patterns by a Deterministic Predictive Coding Model in a Task of Imitative Synchronization with Human Movement Patterns

Ahmadreza Ahmadi and Jun Tani[(✉)]

KAIST, Daejeon, South Korea
ar.ahmadi62@gmail.com, tani1216@gmail.com

Abstract. The current paper presents how performance of a particular deterministic dynamical neural network model in predictive coding scheme differ when it is trained for a set of prototypical movement patterns using their modulated teaching samples from when it is trained using unmodulated teaching samples. Multiple timescale neural network (MTRNN) trained with or without modulated patterns was applied in a simple numerical experiment for a task of imitative synchronization by inferencing the internal states by the error regression, and the results suggest that the scheme of training with modulated patterns can outperform the scheme of training without them. In our second experiment, our network was tested with naturally fluctuated movement patterns in an imitative interaction between a robot and different human subjects, and the results showed that a network trained with fluctuated patterns could achieve generalization in learning, and mutual imitation by synchronization was obtained.

Keywords: Neuro-robotics · Recurrent neural networks · Imitative synchronization · On-line adaptation

1 Introduction

The idea of the predictive coding [1–3] is based on the hypotheses that by accumulated learning of the perceptual experience, our brains become able to predict perceptual outcomes of own intention for acting to the external environment [1, 2]. Some dynamical neural network models such as recurrent neural network with parametric biases (RNNPB) [1] applied the idea of predictive coding. The current intention represented by the PB vectors are modulated by means of prediction error minimization in order to predict perceptual input sequences in RNNBP. Perceptual sequences, for example proprioception and visual input sequences, can be generated by the RNNPB as corresponding to given intention state in terms of the PB vector. Actual motor movement can be generated by feeding predicted proprioception state at each time step into the motor controller as a target posture. Moreover, in RNNPB, an optimal intention can be inferred

---

The original version of this chapter was revised. The name of the first author was mispelled. It should read as "Ahmadreza Ahmadi". An erratum to this chapter can be found at DOI: 10.1007/978-3-319-46687-3_70

© Springer International Publishing AG 2016
A. Hirose et al. (Eds.): ICONIP 2016, Part I, LNCS 9947, pp. 393–402, 2016.
DOI: 10.1007/978-3-319-46687-3_44

for a given perceptual sequence pattern by means of inferencing the internal states by the error regression of the PB vector towards minimizing the prediction error.

This corresponds to recognition of the perceptual sequence pattern. In this manner, the RNNPB mechanized by the predictive coding principle can account for the mirror neuron functions [4] in terms of pairing generation and recognition of movement patterns [1].

One flaw of implementing the idea of the predictive coding in deterministic dynamic models such as RNNPB is their vulnerability to noisy and fluctuated patterns [5]. To solve this problem, Bayesian predictive coding scheme endowed by free-energy principle was introduced [2]. As inspired by [2], so-called the stochastic continuous time RNN (S-CTRNN) in which next step perceptual state can be predicted by its expected mean and variance (inverse precision) instead of its actual value was proposed by Murata and colleagues [5]. However, it has been shown that deterministic RNN models can also learn to extract probabilistic structures latent in observed sequential patterns by embedding them into deterministic chaos or transient chaos [6–8]. This study aims to examines such characteristics of a particular RNN model incorporated with the scheme of predictive coding.

The central ideas to be examined in the current study are as follows. (i) Mean values of periods and amplitudes for a set of prototypical cyclic patterns can be learned from their fluctuated teaching samples as embedded in deterministic dynamic structure of multiple timescale recurrent neural network (MTRNN) by using the initial sensitivity of the internal neural states. The intention for generating the corresponding prototypical patterns as well as variance to account for deviation of each teaching sample are represent by the initial values of the internal neural states. (ii) The network can achieve a sort of generalization in learning accounting for possible fluctuation when mean values of periods and amplitudes for a set of prototypical patterns are learned from such fluctuated teaching samples. (iii) Robustness in recognizing novel fluctuated test patterns by means of inferencing the internal states by the error regression can be obtained when a network achieved generalization in learning.

We evaluated the aforementioned ideas by conducting a simple simulation experiment and an embodied experiment on imitative interaction between a robot and human subjects. The paradigm of synchronized imitation [9] in which the network models trained for a set of prototypical sequential pattern can "recognize" a given target sequence pattern by predicting its next step value as synchronized with the input sequence was employed in both experiments. This recognition by "predictive synchronization" can be performed by inferencing the internal neural states in different timescale levels in the MTRNN model by means of the error regression.

Our first experiment showed that how mean values of periods and amplitudes for two different prototypical cyclic patterns can be learned from multiple cyclic teaching sample patterns which had different amplitudes and periods, obtained from a normal distribution, in each cycle of them. Furthermore, it was shown how better the network model trained with such modulated teaching samples can show synchronized imitation with the test target input than the one trained with unmodulated teaching samples. Moreover, it was revealed that the synchronized imitation using inferencing the internal states by the error regression outperformed the conventional input entrainment scheme where the feeding of cyclic inputs to the network model could entrain the internal neural dynamics for synchronization. In our last experiment, first, an imitation game [9] in which a set of prototypical movement patterns were learned by a humanoid robot was conducted. Then, several subjects were asked to explore those prototypical patterns learned by the robot by means of the imitative synchronization. Our experimental results showed how mutual imitation by synchronization between the robot and the human subjects can be achieved in our proposed scheme in facing with naturally fluctuated movement patterns of the human subjects when the network model in the robot have been trained with modulated teaching samples.

2   MTRNN Model

2.1   Overview

MTRNN uses neurons with different timescales in order to develop the self-organization of a functional hierarchy in which neurons with a small time constant are called fast context (FC) units, and ones that have a bigger time constant are called slow context (SC) units. In our robotic experiments, we also use another type of neurons that own an intermediate time constant called middle context (MC) unit. There are no connections between FC and SC units in this case. Similar to [10], input and output units are only connected to FC neurons although all output units have a time constant of 1 in this paper (there is no recurrency in output units). The current external input states can be received by the input units, and their predicted states are generated by the output units. Neural activities are calculated based on a conventional firing rate model in which each neuron's activity entails average firing rate of other neurons and its own decayed internal value from the previous time step as shown in the following equation.

$$u_{i,t+1} = (1 - \frac{1}{\tau_i})u_{i,t} + \frac{1}{\tau_i}(\sum_j w_{ij}c_{j,t} + \sum_k w_{ik}x_{k,t} + b_i) \qquad (1)$$

where $u_{i,t}$ is the internal state value of ith neuron at time t, w_{ij} is the connectivity weight from ith context neuron to jth context neuron, w_{ik} is the connectivity weight from ith neuron unit to kth input unit, $c_{j,t}$ is the context activation value of jth neuron at time t, $x_{k,t}$ is the external input of kth input unit at time t, b_i is the bias of the ith neuron, and τ_i is the time constant of the ith neuron.

2.2   Generation and Training Method

The equations for the forward dynamics of the MTRNN can be seen in [10, 12]. A conventional back-propagation through time (BPTT) scheme [11] is used to train the networks. Details about the calculation of softmax transformation for the input units and gradients of learnable parameters (weights, biases, and initial states) can be seen in [10, 12]. We do not include these equations here because of lack of space). In all of our experiments, the open-loop generation approach is used in the training mode in which MTRNN receives the current external inputs and generates one or multiple look-ahead prediction steps of the outputs. Open-loop generation can also be called as the entrainment but we often use the first term in this paper. The closed-loop mode is defined as giving the current prediction outputs to the next time inputs. This can also be referred as the mental simulation of actions. A schematic of open-loop and closed-loop generations are shown in [12].

2.3   Training with Fluctuated Teaching Samples

In many real-time neuro-robotics experiments, it is expected to have modulations and fluctuations in signals received by sensors due to plenty of reasons. Consequently, it is crucial that the trained network should be robust enough in controlling the robots successfully against such fluctuated perceptual input patterns. In the training phase of MTRNN, we exploited several modulated teaching samples for each prototypical pattern using the normal distribution in order to provide robustness of the trained network at the moment of on-line recognition of untrained fluctuated patterns. It means that we generated several target sequences in which amplitudes and periods were modulated randomly by means of the normal distribution. Standard deviations were different for each target sequence of a prototypical pattern. It is expected that prototypical patterns with mean values of amplitudes and periods can be learned from their modulated teaching samples as embedded in deterministic MTRNN by using the initial sensitivity of the internal neural states. The intention for generating the corresponding prototypical patterns and variance to account for deviation of each sample are represented by the initial values of the internal neural states. The network can achieve a sort of generalization in learning accounting for possible fluctuations when a set of prototypical patterns are learned from fluctuated teaching samples.

2.4   Inferencing the Internal States by the Error Regression

After a network is trained, it can achieve the imitative synchronization with the test target inputs by inferencing the internal states by using the error signal between the test input and the output. In the testing phase, by passing the output generation errors in a bottom-up manner from the output units to all neural states of MTRNN, the intention can be changed in a top-down manner from the higher neural states to the output units within a temporal window (W) of the immediate past. More in detail, the generation errors from the t-W time step to the current t time step can modulate the internal states at the t-W time step, which result in updating of all context units inside the window, by means of the BPTT scheme while the connectivity weights and biases obtained in the training phase are fixed. This process is called inferencing the internal states by the error regression. Unlike [8] who just updated the SC units, we updated the other contexts units such as MC and FC units also in order to facilitate adaptation to the detail profiles in the target patterns. Inferencing in lower-level internal states by the error regression can facilitate modifications in higher-level internal states for both recognition of actual behavior through the bottom-up pass way and intentional predictions through the top-down pass way. Inferencing the internal states by the error regression was applied on all time steps during the testing mode and in a closed-loop manner. More specifically, there were no external inputs to MTRNN, and inferencing the internal states by the error regression scheme was used in order to achieve the imitative synchronization with test target inputs by minimize the prediction error (closed-loop output patterns compared with target patterns) inside the temporal window. The equations for updating the internal states are similar to updating equations of the initial states in [12]. However, the error regression adaptation rate (α_{ER}) was set to 0.001, and we did not use the momentum term (μ) in this stage. The window length was set to 15 and inferencing the internal states by the error regression was performed for 100 regression steps at each time step within this temporal window in which all former internal states were overwritten from the t-15 time step to the current t time step. By inferencing the internal states inside the temporal window, the prediction of internal states and output values after the window referred as the future plan can be also modified.

3   Numerical Experiment

To examine our proposed method for imitative synchronization with multiple modulated test samples, we conducted one experiment in the simulation in which eleven MTRNN models were trained to generate two-dimensional sinusoidal patterns. The equations to generate sine curves are shown as followings

$$
\begin{cases}
y_1 = A \sin\left(B\dfrac{\pi}{p}x\right) \\[2mm]
y_2 = 1 + A \sin\left(B\dfrac{\pi}{p}x\right) \; if\,(y_1 < 0) \\[2mm]
y_2 = 1 - A \sin\left(B\dfrac{\pi}{p}x\right) \; else
\end{cases}
\tag{2}
$$

$$
\begin{cases}
y_1 = -A \sin\left(B\dfrac{\pi}{p}x\right) \\[2mm]
y_2 = -1 - A \sin\left(B\dfrac{\pi}{p}x\right) if\,(y_1 > 0) \\[2mm]
y_2 = -1 + A \sin\left(B\dfrac{\pi}{p}x\right) \; else
\end{cases}
\tag{3}
$$

where A, B are variables that modulate amplitude and period of the sine curves and p is set to 30.

First, one MTRNN was trained by giving only two prototypical patterns to the input units by setting $A = B = 1$ in both Eqs. (2) and (3). The variable x is an integer which changes from 0 to 300 in order to produce 5 cycles of y_1 and y_2. The MTRNN consisted of 30 fast and 15 slow context neurons with time constants of 2 and 50, respectively. There were 11 softmax units for each real input and output dimensions, which means that MTRNN models had 22 softmax input, and output units in this experiment.

The network was trained to generate 5 look-ahead prediction step (l = 5) of the input sequences in their output units. Then, other ten MTRNN models were trained by the same parameter setting as the first MTRNN model but 8 teaching patterns were given to each network. To generate 4 teaching signals for each prototypical pattern, A and B in both Eqs. (2) and (3) were chosen randomly from a normal distribution with a mean (μ) of 1 and standard deviations (σ) of 0.05, 0.1, 0.15, and 0.2 (σ is 0.05 for the first teaching signal and 0.1 for the second teaching signal, and so on). The outputs (y_1 and y_2) consisted of 5 cycles in all teaching signals, and each cycle of them had different A and B values. For example, the first teaching signal for the first MTRNN was generated by picking 5 pairs of A and B values (each pair for one cycle) randomly from a normal distribution with a mean of 1 and standard deviation of 0.05. For the other MTRNN models, the same procedure was applied but all teaching signals were different because of random selection of A and B values. All eleven networks were trained with the same initial weights and biases for 3000 epochs to generate 5 look-ahead prediction step ($l = 5$) of the input sequences in their output units.

After the training was finished for all networks, the closed-loop generation of all 8 modulated patterns were executed for 50000 time steps by starting from their corresponding initial states of the slow context neurons. In 10 MTRNN models with modulated training patterns, the closed-loop generation outputs were synchronizing with the teaching signals. However, after a few hundred steps, A and B values of all modulated patterns for the both prototypical patterns converged to the same values around the mean ($\mu = 1$), which means that only one stable attractor was generated for each prototypical pattern.

To test our MTRNN models for imitative synchronization, different modulated test patterns were used for both closed-loop generation while inferencing the internal states by the error regression and open-loop generation schemes. The fluctuated test patterns were different in the amplitude, and period from the fluctuated teaching samples by randomly picking different values of A and B in both Eqs. 2 and 3 from a normal distribution with mean of 1 and standard deviation of 0.3. Outputs and the mean square error (MSE) of the first MTRNN model which was trained with unmodulated patterns and one of the tenth MTRNN models which was trained with modulated patterns are depicted in Fig. 1 given the first test target signal (obtained by Eq. 2). By looking at MSEs of two networks, it can be observed that our proposed method significantly outperforms another network. Results of other 9 MTRNN models, open-loop results, and results of fluctuated test patterns obtained by Eq. 3 are not depicted because of lack of space. However, one can compare all results in Table 1 that shows the average mean square error (MSE) of the prediction outputs for all MTRNN models (with or without fluctuated teaching samples) using closed-loop generation while inferencing the internal states by the error regression and open-loop generation. MSE was summed up for all cycles of the prediction outputs except their transient response at the beginning (first 50 time steps) and divided by the time step length to compute the average value. For 10 MTRNN models with modulated patterns, 10 MSE values were obtained and average of them are represented in Table 1 (third column). As can be seen, the MTRNN trained with fluctuated teaching samples showed better performance in the imitative synchronization driven by inferencing the internal states by the error regression.

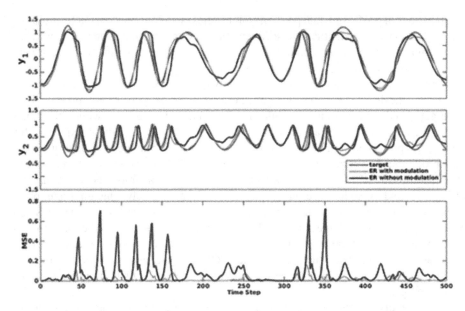

Fig. 1. Outputs and MSE results of two networks (one trained with modulated patterns and one trained with unmodulated patterns) given modulated test pattern with different pairs of A and B obtained from a normal distribution with mean of 1 and standard deviation 0.3.

Table 1. Average MSE of prediction outputs

	MSE			
	ER W/O Mod	ER W/Mod	O-L W/O Mod	O-L W/Mod
Pattern 1	0.0715	**0.0202**	0.2994	0.0456
Pattern 2	0.0467	**0.0097**	0.1695	0.0157

ER W/O Mod: error regression without modulated patterns
ER W/Mod: error regression with modulated patterns
O-L W/O Mod: open-loop generation without modulated patterns
O-L W/Mod: open-loop generation with modulated patterns.

4   Robot Experiments

We designed an imitative interaction game between a robot and human subjects for investigating how well mutual imitation by synchronization can be achieved with naturally modulated patterns of the human subjects using the proposed scheme to control the robot. One experiment was conducted for generation/recognition of multiple prototypical movement patterns. We employed a NAO humanoid robot (developed by Aldebaran Robotics) and a Kinect sensor (developed by Microsoft) for imitative interaction tasks. The Kinect SDK and OpenNI framework were used to track the 3-D (X, Y, Z) coordinates of a human's arm joints. The human-user arms' 3-D positions were mapped to the NAO's arms' 3-D positions with respect to the robot coordinate system. Next, the

3-D positions of NAO were mapped to its joint angles (shoulder roll, and pitch and elbow roll, and yaw) by applying the inverse kinematics. In our experiment, an imitation game was designed for an imitative interaction game in which several human subjects were asked to participate in an exploratory assignment.

First, the training data were collected by only one experimenter who interacted with NAO using the direct method (without any networks). Three cyclic prototypical patterns were generated by using only shoulder roll and pitch of both arms while other joint angles were fixed. To generate the training data, we collected 5 training sequences for each cyclic prototypical pattern. This means that we had 15 movement patterns (5 for each prototypical pattern) that were used to train a MTRNN model, which consisted of 30 FC, 20 MC, 10 SC, 44 softmax input and 44 softmax output units. The time constants of FC, MC, and SC units were set to 5, 25, and 150, respectively. One important point is that there was around 520 ms delay between actual human movement patterns and perceiving them by the Kinect sensor. To overcome this delay, the prediction step was set to 7 because one-time step corresponded to 75 ms in our robot experiments. After the training was completed successfully, the MTRNN model endowed with inferencing the internal states by the error regression was used as the NAO's brain in an imitative interaction game.

10 university students (2 females and 8 males) participated in our experiment in which they interacted with NAO without any prior knowledge about the movement patterns trained. Their first task was to explore all of the prototypical patterns memorized by the robot by achieving imitative synchronization for each pattern. They were given 20 min first for the exploration. All participants were able to interact with NAO successfully. However, two of them could not figure out one of the movement patterns. The second task for the subjects who successfully explored all patterns was to synchronize with the robot. They had to repeat the movement pattern 1 first until they felt that they were synchronized with NAO well, and then they could switch to the movement pattern 2 and do the same and finally, they could switch to the movement pattern 3 and synchronize with the robot.

Figure 2 displays the dynamic for inferencing the internal states by the error regression for one of the participants during the synchronization stage. In Fig. 2, switching of the movement patterns from 1 to 2 (MP1 → MP2) and from 2 to 3 are shown on the top of the sensory prediction panel. The temporal windows of the error regression at a particular moment of switching from movement pattern 2 to 3 are depicted by gray areas in all panels, and the right line of the windows display the current time. The neural internal states cannot be modified before the temporal window but they can be revised within the window to minimize the prediction error by means of BPTT to achieve the imitative synchronization with the human subject's movement patterns. The prediction states after the window can be obtained by continuing the closed loop generation to the end for each time step. As can be seen in the third row, at the transition point of the movement pattern 2 to the movement pattern 3, a large error occurs. As a result, neural internal states in both SC and FC units are revised in order to reduce the error, and the intentional behavior generation is changed.

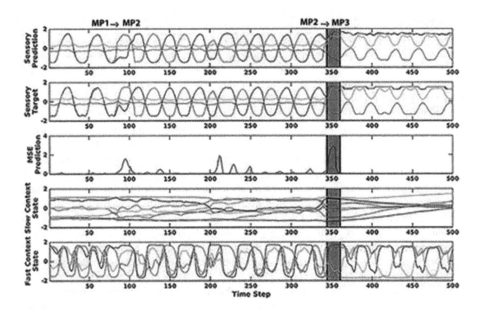

Fig. 2. The dynamic for inferencing the internal states by the error regression for a human subject in the synchronization stage of the imitation game.

5   Conclusion

It was shown in both numerical and robotic experiments that generalization in learning accounting for possible fluctuations was achieved for the deterministic MTRNN by using fluctuated samples for the training. By the aid of this generalization capability, network exhibited robustness in recognizing novel fluctuated test patterns by inferencing the internal states by the error regression. In our first experiment, the network trained with teaching patterns in which the periods and amplitudes obtained from a normal distribution with different standard deviations showed significantly better performance than the one trained with unmodulated patterns. Furthermore, the superior performance of the scheme of inferencing the internal state by error regression compared to conventional entrainment scheme reveled the effectiveness of the on-line modulation of the internal states by the error regression for achieving imitative synchronization.

The second experiment showed that aforementioned advantageous characteristics of the employed model could appear even in quite natural setting of human-robot imitative interaction during spontaneous mutual exploration between two sides. Possible scaling of the proposed scheme such as introducing much more number of more complex human movement patterns with replacing the current Kinect input to pixel level visual inputs should be considered in future study.

Acknowledgement. This work was supported by the ICT R&D program of MSIP/IITP. [2016(R7117-16-0163), Intelligent Processor Architectures and Application Softwares for CNN-RNN].

References

1. Tani, J., Ito, M., Sugita, Y.: Self-organization of distributedly represented multiple behavior schemata in a mirror system: reviews of robot experiments using RNNPB. Neural Netw. **17**(8), 1273–1289 (2004)
2. Friston, K.: The free-energy principle: a unified brain theory? Nat. Rev. Neurosci. **11**(2), 127–138 (2010)
3. Rao, R.P., Ballard, D.H.: Predictive coding in the visual cortex: a functional interpretation of some extra-classical receptive-field effects. Nat. Neurosci. **2**(1), 79–87 (1999)
4. Rizzolatti, G., et al.: Premotor cortex and the recognition of motor actions. Cogn. Brain. Res. **3**(2), 131–141 (1996)
5. Murata, S., et al.: Learning to reproduce fluctuating time series by inferring their time-dependent stochastic properties: application in robot learning via tutoring. IEEE Trans. Auton. Mental Dev. **5**(4), 298–310 (2013)
6. Tani, J., Fukumura, N.: Embedding a grammatical description in deterministic chaos: an experiment in recurrent neural learning. Biol. Cybern. **72**(4), 365–370 (1995)
7. Namikawa, J., Nishimoto, R., Tani, J.: A neurodynamic account of spontaneous behaviour. PLoS Comput. Biol. **7**(10), e1002221 (2011)
8. Murata, S., Yamashita, Y., Arie, H., Ogata, T., Sugano, S., Tani, J.: Learning to perceive the world as probabilistic or deterministic via interaction with others: a neuro-robotics experiment. IEEE Trans. Neural Networks Learn. Syst. **PP**(99), 1–19 (2015)
9. Ito, M., Tani, J.: On-line imitative interaction with a humanoid robot using a dynamic neural network model of a mirror system. Adapt. Behav. **12**(2), 93–115 (2004)
10. Yamashita, Y., Tani, J.: Emergence of functional hierarchy in a multiple timescale neural network model: a humanoid robot experiment. PLoS Comput. Biol. **4**(11), e1000220 (2008)
11. Rumelhart, D.E., Hinton, G.E., Williams, R.J., Learning internal representations by error propagation. DTIC Document (1985)
12. Park, G., Tani, J.: Development of compositional and contextual communicable congruence in robots by using dynamic neural network models. Neural Netw. **72**, 109–122 (2015)

Image Segmentation Using Graph Cuts Based on Maximum-Flow Neural Network

Masatoshi Sato[1]([✉]), Hideharu Toda[2], Hisashi Aomori[3], Tsuyoshi Otake[4],
and Mamoru Tanaka[5]

[1] Faculty of System Design, Tokyo Metropolitan University,
6-6 Asahigaoka, Hino, Tokyo 191-0065, Japan
sato-masatoshi@tmu.ac.jp
[2] Graduate School of Information Cognitive Science,
Chukyo University, 101-2 Yagoto-honmachi, Showa, Nagoya, Aichi 466-8666, Japan
[3] Department of Electrical and Electronic Engineering,
Chukyo University, 101-2 Yagoto-honmachi, Showa, Nagoya, Aichi 466-8666, Japan
[4] Department of Software Science, College of Engineering,
Tamagawa University, 6-1-1, Tamagawa-Gakuen, Machida, Tokyo 194-8610, Japan
[5] Department of Information and Communication Sciences,
Sophia University, 7-1 Kioi-cho, Chiyoda, Tokyo 102-8554, Japan

Abstract. Graph Cuts has became increasingly useful methods for the image segmentation. In Graph Cuts, given images are replaced by grid graphs, and the image segmentation process is performed using the minimum cut (min-cut) algorithm on the graphs. For Graph Cuts, the most typical min-cut algorithm is the B-K algorithm. While the B-K algorithm is very efficient, it is still far from real-time processing. In addition, the B-K algorithm gives only the single min-cut even if the graph has multiple-min-cuts. The conventional Graph Cuts has a possibility that a better minimum cut for an image segmentation is frequently overlooked. Therefore, it is important to apply a more effective min-cut algorithm to Graph Cuts. In this research, we propose a new image segmentation technique using Graph Cuts based on the maximum-flow neural network (MF-NN). The MF-NN is our proposed min-cut algorithm based on a nonlinear resistive circuit analysis. By applying the MF-NN to Graph Cuts instead of the B-K algorithm, image segmentation problems can be solved as the nonlinear resistive circuits analysis. In addition, the MF-NN has an unique feature that multiple-min-cuts can be find easily. That is, it can be expected that our proposed method can obtain more accurate results than the conventional Graph Cuts which generates only one min-cut. When the proposed circuit model is designed with the integrated circuit which can change graph structure and branch conductance, a novel image segmentation technique with real-time processing can be expected.

1 Introduction

In recent years, Graph Cuts is used as an effective technique for the image segmentation [1]. The accuracy of the image segmentation by Graph Cuts is superior

© Springer International Publishing AG 2016
A. Hirose et al. (Eds.): ICONIP 2016, Part I, LNCS 9947, pp. 403–412, 2016.
DOI: 10.1007/978-3-319-46687-3_45

to other several state of the art interactive tools such as Magic Wand, Intelligent Scissors, Level Sets and for matting [2]. In Graph Cuts, given images are replaced by grid graphs with defined edge weights, and the image segmentation is realized by finding a minimum cut (min-cut) of the graph using the min-cut algorithm. Therefore, the most of a performance of Graph Cuts is affected by the min-cut algorithms. For Graph Cuts, the most typical min-cut algorithm is the B-K algorithm proposed by Y. Boykov and V. Kolmogorov [4]. While the B-K algorithm is very efficient, with the execution time of only a few seconds for a typical problem, it is still far from real-time processing. In general, real-time processing is the very important issue in vision applications. In addition, the B-K algorithm gives only the single min-cut even if the graph has multiple min-cuts. Multiple min-cuts show that plural cuts with the same min-cut capacity exist in the same graph. The conventional Graph Cuts has a possibility that a better minimum cut for an image segmentation is frequently overlooked. Therefore, a novel Graph Cuts that corresponds to real-time processing and multiple min-cuts is required.

In this paper, we propose a new image segmentation technique using Graph Cuts based on the maximum-flow neural network (MF-NN). The MF-NN is our proposed min-cut algorithm based on a nonlinear resistive circuit analysis [6]. The MF-NN in which each neuron nonlinearity has a saturation characteristic defined by the piecewise linear function can be realized by using a simple nonlinear resistive circuit, and has the advantage of being suitable for hardware implementation. The hardware implementation of the MF-NN can be associated with "network flow optimization by a resistor circuit" by M. Hasler and J.-E. Nussbaumer [7]. By applying the MF-NN to Graph Cuts instead of the B-K algorithm, image segmentation problems can be solved as the nonlinear resistive circuits analysis. In addition, we previously reported that the MF-NN has an unique feature that multiple min-cuts can be find easily [8]. That is, it can be expected that our proposed method can obtain more accurate results than the conventional Graph Cuts which generates only one min-cut. Moreover, when the proposed circuit model is designed with integrated circuits such as analog type Programmable Logic Device (PLD), Memristor or Phase Change Memory (PCM) which can change graph structure and branch conductance, the novel Graph Cuts based on the MF-NN with real-time processing can be expected.

2   Image Segmentation by Graph Cuts

In this section, we explain about the conventional Graph Cuts for the image segmentation [3]. Let each pixel in an image $\mathcal{P}$ be considered as $p \in \mathcal{P}$, each label on the pixel is set to $L = \{L_p \mid p \in \mathcal{P}\}$. Each L_p belongs to either an object label ("obj") or a background label ("bkg"), and adjacent pixels of p is set to $q \in \mathcal{N}$ where $\mathcal{N}$ is a set of all pairs of neighboring pixels. Then, the soft constraints that we impose on boundary and region properties of L are described by the cost function $E(L)$:

$$E(L) = \lambda \cdot R(L) + B(L), \tag{1}$$

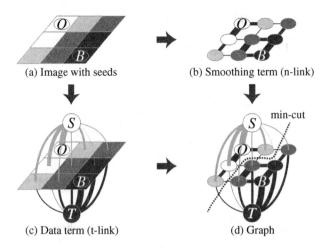

(a) Image with seeds   (b) Smoothing term (n-link)

(c) Data term (t-link)   (d) Graph

Fig. 1. Weighted graph made from input image.

where

$$R(L) = \sum_{p \in \mathcal{P}} R_p(L_p), \qquad (2)$$

$$B(L) = \sum_{(p,q) \in \mathcal{N}} B_{\{p,q\}} \cdot \delta(L_p, L_q), \qquad (3)$$

and

$$\delta(L_p, L_q) = \begin{cases} 1, & if \, L_p \neq L_q, \\ 0, & otherwise. \end{cases} \qquad (4)$$

The coefficient $\lambda \geq 0$ in (1) specifies a relative importance of the region properties term $R(L)$ versus the boundary properties term $B(L)$. The regional term $R(L)$ assumes that the individual penalties for assigning pixel p to "object" and "background", correspondingly $R_p(obj)$ and $R_p(bkg)$, are given. The image segmentation is performed by finding the label L which minimizes the cost function $E(L)$ defined by $R(L)$ and $B(L)$ using Graph Cuts.

In Graph Cuts, the image is replaced by a graph G as shown in Fig. 1. The graph G consists of a source S, a terminal T and nodes corresponding to each pixel of the image. An edge connected to each node is called "n-link", and an edge connected form each node to S and T is called "t-link" as shown in Fig. 1(b) and (c), respectively. Each edge cost of n-link and t-link is set as shown in Table 1.

Table 1. Each edge cost in G.

Edge		Cost	For
n-link	$\{p, q\}$	$B_{\{p,q\}}$	$\{p, q\} \in \mathcal{N}$
t-link	$\{p, S\}$	$\lambda \cdot R_p(bkg)$	$p \in \mathcal{P},\ p \notin \mathcal{O} \cup \mathcal{B}$
		K	$p \in \mathcal{O}$
		0	$p \in \mathcal{B}$
	$\{p, T\}$	$\lambda \cdot R_p(obj)$	$p \in \mathcal{P},\ p \notin \mathcal{O} \cup \mathcal{B}$
		K	$p \in \mathcal{O}$
		0	$p \in \mathcal{B}$

The functions $R_p(obj)$, $R_p(bkg)$, $B_{\{p,q\}}$ and K are defined by

$$B_{\{p,q\}} \propto exp\left(-\frac{(I_p - I_q)^2}{2\sigma^2}\right) \cdot \frac{1}{dist(p, q)}, \tag{5}$$

$$R_p(obj) = -\ln\Pr(I_p|\mathcal{O}), \tag{6}$$

$$R_p(bkg) = -\ln\Pr(I_p|\mathcal{B}), \tag{7}$$

$$K = 1 + \max_{p \in \mathcal{P}} \sum_{q:\{p,q\} \in \mathcal{N}} B_{\{p,q\}}, \tag{8}$$

where $B_{\{p,q\}}$ penalizes a lot for discontinuities between pixels of similar intensities when $|I_p - I_q| < \sigma$. However, if pixels are very different, $|I_p - I_q| > \sigma$, then the penalty is small. $dist(p, q)$ is the Euclidean distance between p and q. $\mathcal{O}$ and $\mathcal{B}$ denote the subsets of pixels marked as "object" and "background" seeds, respectively. The seeds of $\mathcal{O}$ and $\mathcal{B}$ must be specified by user in advance. I_p is a luminosity value of the pixel p. $Pr(I_p|\mathcal{O})$ and $Pr(I_p|\mathcal{B})$ are histograms for $\mathcal{O}$ and $\mathcal{B}$ intensity distributions, and these histograms are used to set the regional penalties $R_p(obj)$ and $R_p(bkg)$ as negative log-likelihoods.

The image segmentation is realized by finding a min-cut of the graph G using the min-cut algorithm as shown in Fig. 1(d). In the image segmentation by Graph Cuts, the most typical min-cut algorithm is the B-K algorithm proposed by Boykov and Kolmogorov [4]. While the B-K algorithm is very efficient, with the execution time of only a few seconds for a typical problem, it is still far from real-time processing. In addition, since the B-K algorithm gives only the single min-cut even if the graph has multiple min-cuts, there is a possibility that the conventional Graph Cuts frequently misses a better min-cut for the image segmentation.

3   Grash Cuts Based on Nonlinear Resistive Circuits

In our previous study, we proposed the Maximum-Flow Neural Network (MF-NN) which is the effective min-cut algorithm based on nonlinear resistive circuits, and we reported that the MF-NN has an unique feature that multiple min-cuts

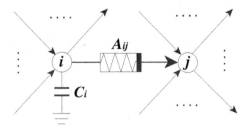

Fig. 2. Association between neuron i and j.

can be find easily [8]. In this research, the graph G as shown in Sect. 2 is replaced by the nonlinear resistive circuit, and image segmentation problems are solved by the nonlinear resistive circuits analysis.

3.1 Maximum-Flow Neural Network

Now, we explain about the MF-NN. Each neuron is connected by the nonlinear resistance which has a saturation property, and the state equation of the MF-NN is described by the simultaneous differential equation on the node voltage. Figure 2 shows the connection between a neuron v_i and a neuron v_j in the nonlinear resistive network. A_{ij} is a nonlinear resistance that exists between v_j and v_i. The MF-NN has the saturation characteristic such that the entire network converges to the equilibrium state if a certain amount of the current in the S-point v_S goes out. The I-V characteristic from v_i to v_j is described by

$$I_{ij} = A_{ij}f(u_i - u_j), \tag{9}$$

where

$$f(x) = \begin{cases} 1, & for \ V_{th} \leqq x, \\ x/V_{th}, & for \ 0 < x < V_{th}, \\ 0, & for \ x \leqq 0. \end{cases} \tag{10}$$

A_{ij} is defined by the branch capacity c_{ij} in given flow network. If a directional branch from v_i to v_j does not exist, A_{ij} is defined by $A_{ij} = 0$. I_{ij} is a current value that flows from v_i to v_j, and I_{ij} is equal to the flow f_{ij} in a flow network. u_i and u_j are node voltages of the neuron v_i and v_j respectively. V_{th} is an important positive threshold which determines the range of the potential difference that shows the state of each branch in the equilibrium state. Now, the state equation of the MF-NN is given by

$$C_i \frac{du_i}{dt} = - \sum_{v_j \in \Gamma(v_i)}^{n} A_{ij}f(u_i - u_j)$$

$$+ \sum_{v_k \in \Gamma^{-1}(v_i)}^{n} A_{ki}f(u_k - u_i), \tag{11}$$

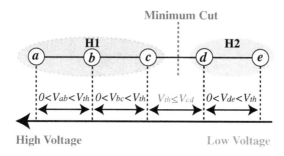

Fig. 3. Minimum cut by MF-NN.

where

$$0 \leqq A_{ij}f(u_i - u_j) \leqq A_{ij} = c_{ij} \tag{12}$$

$$\Gamma(v_i) = v_j \mid (v_i, v_j) \in B(N),$$
$$\Gamma^{-1}(v_i) = v_j \mid (v_j, v_i) \in B(N). \tag{13}$$

C_i is a capacitor that exists between v_i and the ground (0 V). The S-point v_S and the T-point v_T do not have these capacitors. A constant initial voltage u_S is given to v_S, and v_T is grounded ($u_T = 0$ V). When the network goes to the equilibrium state, solutions of the state equation converge as follows,

$$\lim_{t \to \infty} C_i \frac{du_i}{dt} = 0. \tag{14}$$

Then the maximum flow F_{max} which is the sum of currents from v_S is given by

$$F_{max} = \sum_{v_i \in \Gamma(v_S)}^{n} A_{si}f(u_S - u_i). \tag{15}$$

The min-cut which is a cut with the smallest cut capacity can be obtained from node voltages u_i in equilibrium state. The cut is a partition of the vertices of the graph into two disjoint subsets that are joined by at least one edge, and the total of branch capacities contained in the cut is called the cut capacity. In the MF-NN, the min-cut can be defined by a boundary of potential difference which is V_{th} and more in (10) between two disjoint node-subsets. For example, when node voltage is arranged according to high voltage $(v_a, v_b, \cdots, v_e)$ as shown in Fig. 3, the boundary between v_c and v_d which is "$V_{th} \leqq V_{cd}$" becomes the min-cut. When multiple min-cuts exist in the same graph, we can easily obtain all min-cuts by finding all boundary lines with the potential difference over V_{th}. The unique feature of the node information (node voltages u_i) in the MF-NN enables to obtain multiple min-cuts.

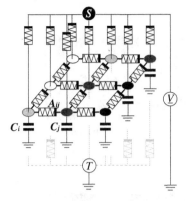

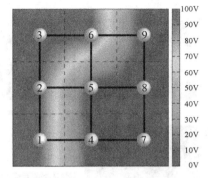

Fig. 4. Nonlinear circuit for grash cuts   **Fig. 5.** Voltage distribution

3.2 Nonlinear Resistive Circuits for Grash Cuts

We apply the MF-NN to Graph Cuts as an alternative to the B-K algorithm, and solve image segmentation problems by the nonlinear resistive circuits analysis. The graph G as shown in Fig. 1(d) is replaced by the nonlinear resistive circuit as shown in Fig. 4. Each A_{ij} is the edge cost defined as shown in Table 1. The threshold of (10) is set to $V_{th} = 1\,\text{V}$. A voltage $u_S = 100\,\text{V}$ is constantly given to S, and T is grounded ($u_T = 0\,\text{V}$). Then, the voltage distribution is generated as shown in Fig. 5. From this circuit analysis, it is observed that the boundary line in which potential difference is over V_{th} corresponds with the min-cut. That is, in the proposed method, image segmentation results can be obtained by finding the boundary line in the voltage distribution.

4 Simulation Result

In this simulation, we uses three color images as shown in Figs. 6(a), 7(a) and 8(a). Each image size was 481×321, 249×196, and 256×256, respectively. The object label ("obj") and the background label ("bkg") were indicated in red line and blue line as shown in Figs. 6(b), 7(b) and 8(b). From nonlinear resistive circuits analysis by our proposed method, each voltage distribution was generated as shown in Figs. 6(c), 7(c) and 8(c). Figures 6(d), 7(d) and 8(d) showed each voltage distribution from a different angle where the horizontal axis is voltage [V]. All min-cuts in each figures were marked with dot-lines. Image segmentation results by the conventional Graph Cuts were obtained as shown in Figs. 6(e), 7(e) and 8(e). The best image segmentation results by our proposed Graph Cuts were obtained as shown in Figs. 6(f), 7(f) and 8(f). Since our method could obtain image segmentation results with the same number of min-cuts which include the result by the conventional Graph Cuts, we selected the best image segmentation result on our judgement in this simulation.

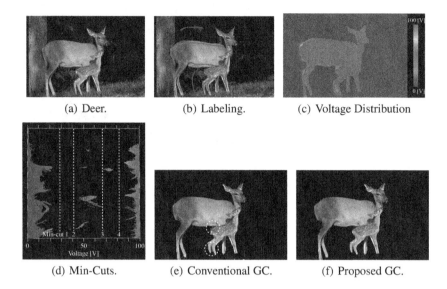

(a) Deer.   (b) Labeling.   (c) Voltage Distribution

(d) Min-Cuts.   (e) Conventional GC.   (f) Proposed GC.

Fig. 6. Deer simulation.

(a) Starfish.   (b) Labeling   (c) Voltage Distribution

(d) Min-Cuts.   (e) Conventional GC.   (f) Proposed GC.

Fig. 7. Starfish simulation.

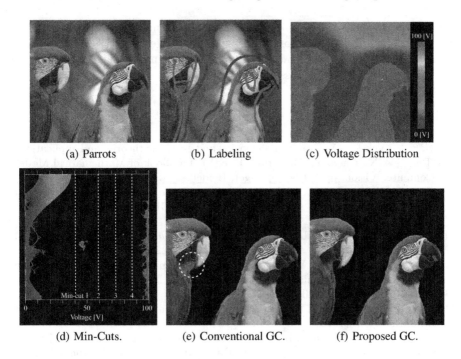

(a) Parrots   (b) Labeling   (c) Voltage Distribution

(d) Min-Cuts.   (e) Conventional GC.   (f) Proposed GC.

Fig. 8. Parrots Simulation

The different between the conventional Graph Cuts and the proposed Graph Cuts were marked with dotted line circles on Figs. 6(e), 7(e) and 8(e). The conventional method could not remove a portion of background area where a segmentation is difficult in particular. On the other hands, our proposed method successfully removed theses areas. The way to select one best min-cut automatically from all min-cuts will be considered in our future work.

5   Conclusion

In this research, we proposed a new image segmentation technique using Graph Cuts based on the maximum-flow neural network (MF-NN). By applying the MF-NN to Graph Cuts, image segmentation problems can be solved as the nonlinear resistive circuits analysis. Moreover, we showed that the proposed method can obtains more accurate results than the conventional Graph Cuts. A higher precision segmentation which has been overlooked in the conventional method became possible by getting various output images from multiple min-cuts. The circuit analysis of the proposed model was performed by computer simulation. Since there is the premise that real-time processing is realized by the hardware implementation, we did not make a comparison of the computation time by computer simulation in this experiment. The hardware implementation of our proposed method is an issue in the future. When the proposed circuit model can be designed in future by the integrated circuit like analog type Programmable Logic Device (PLD), Memristor or Phase Change Memory (PCM) which can change graph structure and branch conductance, the novel image segmentation technique with real-time processing can be expected.

References

1. Boykov, Y., Veksler, O.: Graph cuts in vision and graphics: theories and applications. In: Paragios, N., Chen, Y., Faugeras, O. (eds.) Handbook of Mathematical Models in Computer Vision, pp. 79–96. Springer, Heidelberg (2005)
2. Rother, C., Kolmogorv, V., Blake, A.: GrabCut: interactive foreground extraction using iterated graph cuts. ACM Trans. Graph. SIGGRAPH 2004 **23**(3), 309–314 (2004)
3. Boykov, Y., Veksler, O., Zabih, R.: Fast approximate energy minimization via graph cuts. IEEE Trans. Pattern Anal. Mach. Intell. **23**, 1222–1239 (2001)
4. Boykov, Y., Kolmogorov, V.: An experimental comparison of min-cut/max-flow algorithms for energy minimization in vision. IEEE Trans. Pattern Anal. Mach. Intell. **26**(9), 1124–1137 (2004)
5. Goldberg, A.V., Tarjan, R.E.: A new approach to the maximum-flow problem. J. ACM **35**, 921–940 (1988)
6. Sato, M., Aomori, H., Tanaka, M.: Maximum-flow neural network: a novel neural network for the maximum flow problem. IEICE Trans. Fundam. Electron. Commun. Comput. Sci. **E92–A**(4), 945–951 (2009)
7. Hasler, M., Nussbaumer, J.E.: Network flow optimization by a resistor circuit. In: ECCTD 1989, pp. 644–646, Brighton, U.K., September 1989
8. Sato, M., Aomori, H., Tanaka, M.: Node voltages in nonlinear resistive circuits enable new approach to the minimum cut problem. In: 2014 IEEE International Symposium on Circuits and Systems (ISCAS2014), pp. 2788–2791, Melbourne, Australia, June 2014

Joint Routing and Bitrate Adjustment for DASH Video via Neuro-Dynamic Programming in SDN

Kunjie Zhu[✉], Junchao Jiang, Bowen Yang, Weizhe Cai, and Jian Yang

University of Science and Technology of China, Hefei, China
zkj4873@mail.ustc.edu.cn

Abstract. This paper considers the joint routing and bitrate adjustment optimization for DASH (Dynamic Adaptive Streaming over HTTP) video service using neuro-dynamic programming (NDP) in software-defined networking (SDN). We design an open optimization architecture based on OpenFlow based SDN. Following this architecture, we formulate the joint routing and bitrate adjustment problem as a Markov Decision Process (MDP) for maximizing the average reward. In order to solve the curses of dimensionality, we employ neuro-dynamic programming method to conceive an online learning framework and develop a NDP based joint routing and bitrate adjustment algorithm for DASH video service. At last, an emulation platform based on POX and Mininet is constructed to verify the performance of the proposed algorithm. The experimental results indicate our algorithm has more excellent performance compared with OSPF based algorithm.

Keywords: DASH · Neuro-dynamic programming · Bitrate adjustment · Routing algorithm · Software-defined networking

1 Introduction

With the significant advances of the video compression technology, video services are widely provided via Internet. Dynamic Adaptive Streaming over HTTP (DASH) is an adaptive bitrate streaming technique that enables high quality streaming of media content to be delivered from conventional HTTP web servers. DASH works by breaking the content into a sequence of small HTTP-based file segments. Each segment contains a short part of the content and the content is compressed with different bit rates. Users can autonomously select video segments based on current network conditions. However, with the expansion of the scale of DASH video service, it is unpractical to handle the rapidly increasing bandwidth requirements in the conventional network by increasing physical bandwidth. Software-defined networking (SDN) [1] has been fast emerging as a promising network technology. The architecture of OpenFlow-based [2] SDN provides the programmability of multiple network layers to improve the utility of network resources and the network agility. Against this network, we conceive a joint routing and bitrate adjustment algorithm for DASH video service.

© Springer International Publishing AG 2016
A. Hirose et al. (Eds.): ICONIP 2016, Part I, LNCS 9947, pp. 413–420, 2016.
DOI: 10.1007/978-3-319-46687-3_46

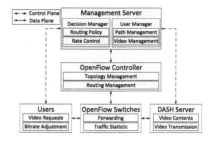

Fig. 1. SDN experimental environment framework

Fig. 2. Architecture of DASH video transmission service

There are some related researches on routing and bitrate adaptation algorithm. [3,4] propose routing algorithms respectively from two aspects of user perception and power allocation in wireless networks. An online routing algorithm using neuro-dynamic programming (NDP) method is proposed in [5]. In [6], a DASH video routing algorithm in SDN is presented based on segment flows. And a dynamic adaptive bitrate algorithm is proposed in [7,8] according to playback buffer state of DASH client. However, there are very few researches discussing a centralized management algorithm for both DASH video routing and bitrate adjustment.

In this paper, we formulate the problem of routing and bitrate adjustment for DASH video transmission in SDN as a Markov decision process [12]. In consideration of the large scale in our network, it is unpractical to solve our problem by traversing the state space. NDP [9,13,14] is an approximate dynamic programming methodology which could provide an approximate value function for MDP solution. Hence, we employ NDP method to solve the curses of dimensionality of our problem. The main contributions of this paper are enumerated as follows. Firstly, an open architecture for joint routing and bitrate adjustment optimization is constructed based on OpenFlow. Based on the observable network state from the network view provided by OpenFlow controller, we formulate the routing and bitrate adjustment into Markov decision problem (MDP) of maximizing the average reward. To solve the curses of dimensionality of the problem, we employ NDP method to construct an online learning framework. At last, in order to verify the performance of the proposed algorithm, we develop an emulation platform and the experimental results show that the proposed strategy significantly improves performance compared with a commonly used OSPF based algorithm.

2   System Model

The OpenFlow based network for the DASH video service is shown in Fig. 1. The network consists of OpenFlow switches as a software-defined substrate layer. A global OpenFlow controller controls all the switches as a control layer, while the management and decision is in the charge of the management server. In addition, a DASH video server is deployed to provide DASH video contents.

We conceive an open architecture for joint routing and bitrate adjustment optimization as depicted in Fig. 2. It is composed of Management Server, Open-Flow Controller, OpenFlow Switches, DASH Server and Users. The basic serving process of the system is described as below. Users request video transmissions from the DASH Server. The User Manager in the Management Server authorize the access to the video service if there are available paths for the transmission. Then the request is further forwarded to the Decision Manager in the Management Server to select an optimal path for the transmission. The Decision Manager will notify the OpenFlow Controller to map the routing decision into forwarding rules and flush these rules into the flow tables in the OpenFlow switches. When there is a bitrate adjustment request from the access user, the Decision Manager in the Management Server will agree the bitrate adjustment request if there is an increase in the average reward. Otherwise, the bitrate adjustment request will be rejected.

3   Dynamic Programming Formulation

The problem of routing and bitrate adjustment for DASH video transmission could be formulated as a Markov decision process. We consider an OpenFlow network consisting of a set of node $\mathcal{N} = \{1, \cdots, N\}$ and a set of unidirectional links $\mathcal{L} = \{1, \cdots, L\}$, where each link l has a total capacity of $B(l)$ units of bandwidth. Let us define $b_t(l)$ to denote the traffic of the link l at the time instant t, which satisfies the capacity constraints, i.e., $b_t(l) \leq B(l)$. Then, the state $\mathbf{s}_t$ of the network at time t consists of a list of the link traffics $b_t(l)$, i.e., $\mathbf{s}_t = [b_t(1), b_t(2), \cdots, b_t(L)]^T$. The set of all possible states is referred to as the state space S. Let $\mathcal{M} = \{1, \cdots, M\}$ be the set of the video identifiers, where each video m has n layers with different bitrates and bandwidth requirements. Let $c(m, n)$ be the immediate reward obtained whenever the request of the layer n of the video m is accepted.

When a DASH video transmission request arrives or a bitrate adjustment request arrives or a transmission is completed, we say that an event $e \in \mathcal{E}$ occurs. When e is a departure event, there is no action to taken. If a DASH video transmission request of the layer n of the video m arrives, the action space $A(\mathbf{s}, e)$ is defined as the set of the rejection action and the possible routes. The rejection action takes place when the bandwidth requirement does not meet the capacity constraints. When a bitrate adjustment request of the access user arrives, the path for the transmission will not be changed, the only action is accepting the bitrate adjustment request or rejecting it. The instantaneous reward after taking the action a is defined as $f(\mathbf{s}, e, a)$. If e represents the video transmission request of the layer n of the video m and a is an action to admit the video m along a route, we have $f(\mathbf{s}, e, a) = c(m, n)$, or e represents the bitrate adjustment request of changing the layer of the video m to n', then $f(\mathbf{s}, e, a) = c(m, n') - c(m, n)$. Otherwise, $f(\mathbf{s}, e, a) = 0$.

We define a policy μ to map the state space and the event space to the action space, *i.e.*, $\mu(\mathbf{s}, e) \in A(\mathbf{s}, e), \forall \mathbf{s} \in S, e \in \mathcal{E}$. Then the average reward associated with the policy μ could be defined as

$$v(\mu) = \lim_{N \to \infty} \frac{1}{t_N} \sum_{k=0}^{N-1} f(\mathbf{s}_{t_k}, e_k, a_{t_k}), \tag{1}$$

where t_k is the time of the kth event e_k occurring, and $a_{t_k} = \mu(\mathbf{s}_{t_k}, e_k)$ is the corresponding action. Then, the target of the problem is maximizing $v(\mu)$. This could be solved as a dynamic programming problem [11]. Let $J^*(\mathbf{s})$ be the optimal (maximal) expected long-term reward (cost-to-go function) starting from state $\mathbf{s}$. Then Bellman's equation takes the following form

$$v^* E_\tau \{\tau | \mathbf{s}\} + J^*(\mathbf{s}) = E_e\{ \max_{a \in A(\mathbf{s}, e)} [f(\mathbf{s}, e, a) + J^*(\mathbf{s}')]\}, \mathbf{s} \in S. \tag{2}$$

Here, v^* denotes the optimal average reward and τ is the time from current state $\mathbf{s}$ to the next state $\mathbf{s}'$. $E_\tau\{\tau | \mathbf{s}\}$ stands for the expectation of τ under the current state $\mathbf{s}$. On the right side of the equation, $E_e\{\cdot\}$ is the expectation reward associated with event e. According to Bellman's equation, the optimal policy could be given by

$$\mu^*(\mathbf{s}, e) = \arg \max_{a \in A(\mathbf{s}, e)} [f(\mathbf{s}, e, a) + J^*(\mathbf{s}')]. \tag{3}$$

When a bitrate adjustment request of changing the layer of the video m to n' arrives, the corresponding reward by the layer changing can be defined as

$$\Delta Q = c(m, n') - c(m, n) + J^*(\mathbf{s}') - J^*(\mathbf{s}). \tag{4}$$

The bitrate adjustment request will be accepted if the bandwidth requirement meets the capacity constraints and $\Delta Q > 0$.

It is easy to solve the problem and make the optimal action while we know the long-term reward $J^*(\mathbf{s})$ of each state. However, the state in our system is defined as the traffic of all links. Then the cardinality of state space is very large. It is too hard to compute the long-term reward with traversing the state space. Hence, we employ the method of NDP, which could provide an approximate reward function to obtain an approximate optimal policy.

4   Neuro-Dynamic Programming Solution

4.1   Approximation Architecture

Neuro-dynamic programming is a simulation based approximate dynamic programming methodology. The central idea of NDP is an approximate architecture shown in Fig. 3, which is easy to calculate to replace the long-term reward function. In order to select an approximation architecture for $\tilde{J}(\mathbf{s}', \alpha)$ to replace $J^*(\mathbf{s}')$ in (3), we need to define a feature vector $\mathbf{r}(\mathbf{s}')$ [10], which is understood

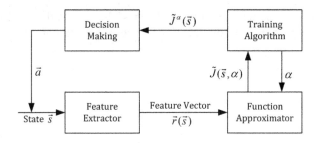

Fig. 3. Approximating architecture of NDP system.

as feature extractor in Fig. 3. The feature vector is extremely important to indicate the basic characteristic of the current state. And it is usually handcraft based on insight or experience. In our system, the traffic of links is used as the feature which shows the information of the current state and the future capacity of the links. We consider a linear architecture, and let $\tilde{J}(\mathbf{s}', \alpha) = \alpha^T \mathbf{r}(\mathbf{s}')$, where the superscript T stands for transpose. The dimension of the parameter vector α is equal to the dimension of the feature vector. Then the approximation architecture could be the form of

$$\tilde{J}(\mathbf{s}', \alpha) = \sum_{l \in \mathcal{L}} [\alpha(l) + \alpha(l, b)b(l) + \alpha(l, b, b)b(l)^2]. \tag{5}$$

The dimension of parameter α is equal to $3L$, where L is the number of links in our system. Then the optimal policy is changed to the form

$$\mu^\alpha(\mathbf{s}, e) = \arg \max_{a \in A(\mathbf{s}, e)} [f(\mathbf{s}, e, a) + \tilde{J}(\mathbf{s}', \alpha)]. \tag{6}$$

And the reward by the layer changing can be given by

$$\Delta Q^\alpha = c(m, n') - c(m, n) + \tilde{J}(\mathbf{s}', \alpha) - \tilde{J}(\mathbf{s}, \alpha). \tag{7}$$

4.2 Iteration Method

In this section, we employ the simplest version of Temporal-Difference (TD) learning algorithm, which is called the TD(0) algorithm, to train the parameter α. The central idea of TD(0) is to improve the approximation of the optimal value function as more state transitions are observed. The iteration rules are given as follow:

$$\alpha_k = \alpha_{k-1} + \gamma_k d_k \nabla_\alpha \tilde{J}(\mathbf{s}_{t_{k-1}}, \alpha_{k-1}), \tag{8}$$

where d_k is the temporal difference, which is defined by

$$d_k = f(\mathbf{s}_{k-1}, e, a_{t_{k-1}}) + \tilde{J}(\mathbf{s}_{t_k}, \alpha_{k-1}) - \tilde{J}(\mathbf{s}_{t_{k-1}}, \alpha_{k-1}) - \tilde{v}_{k-1} \Delta t_k. \tag{9}$$

Then the optimal average reward changes with iteration as follow:

$$\tilde{v}_k = \tilde{v}_{k-1} + \eta_k (f(\mathbf{s}_{k-1}, e, a_{t_{k-1}}) - \tilde{v}_{k-1} \Delta t_k). \tag{10}$$

Here, $\Delta t_k = t_k - t_{k-1}$, which is due to $E_\tau\{\tau|\mathbf{s}\}$ in (2). And γ and η are small step size parameters, which should satisfy the following conditions generally: $\sum_{k=0}^{\infty} \gamma_k = \infty, sum_{k=0}^{\infty} \gamma_k^2 < \infty, \sum_{k=0}^{\infty} \eta_k = \infty, \sum_{k=0}^{\infty} \eta_k^2 < \infty.$

4.3   Decision Process

After the steps presented above. The specific steps of the decision process are shown as follow:

Step (1) With an initial state $\mathbf{s}_0$ and a parameter vector α_0. Establish the approximation architecture $\tilde{J}(\mathbf{s}_0, \alpha)$.

Step (2) When a new video transmission request arrives, make approximate optimal action $\mu^\alpha(\mathbf{s}, e)$ by (6). When a btirate adjustment request arrives, agree the bitrate adjustment request if the bandwidth requirement meets the capacity constraints and $\Delta Q^\alpha > 0$

Step (3) Update the parameter vector α under previous action by (8). The approximate optimal average reward υ is obtained by (10).

Step (4) Return to step (2) if a new video transmission request or a bitrate adjustment request arrives. Otherwise, update the state if a transmission is completed.

5   Experimental Results

In this section, we developed an emulation platform by integrating POX and Mininet, where POX is a real python-version OpenFlow controller, and Mininet is a network emulation orchestration system. For the performance comparison, we also implemented OSPF routing strategy. OSPF is a widely applied routing algorithm which preferentially selects the shortest path by using Dijkstra algorithm. In this benchmark algorithm, the DASH video transmission request and the bitrate adjustment request are accepted if the available bandwidth could accommodate the video stream.

As shown in Fig. 4, our experiment system includes 7 OpenFlow switches and 9 links. The bandwidth of each link is set to 100 Mbps. The DASH server is connected to switch 2 and users are attached to the switch 1 and switch 3 which are regarded as the access points. Each access point has 100 users to randomly request the DASH video content. Three kinds of video are encoded to different bitrate and the corresponding information of the videos is shown in Table 1. In the experiment, each DASH client randomly requests a video transmission with the lowest bitrate, and will select a bitrate version for the next segment based on the average bitrate, which is calculated by gathering statistics of the total bytes and time of the transmission.

We compare our algorithm with OSPF algorithm under different arrival intervals of users. The average reward of two algorithms is shown in Fig. 5. In most of the time, OSPF algorithm causes a more loss of average reward, because OSPF algorithm cannot provide reasonable path management and bitrate management

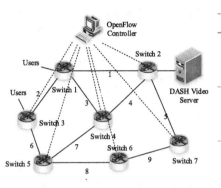

Fig. 4. Topology of the experiment.

Table 1. DASH videos.

Video	Bitrate	Bandwidth	Reward
Video 1[a]	2000	3.91	10
	5000	5.68	12
	8000	8.49	14
Video 2[b]	2000	3.57	11
	5000	7.63	14
	8000	9.55	17
Video 3[c]	2000	4.22	10
	5000	8.19	15
	8000	9.7	17

[a]Big Buck Bunny
[b]Elephants Dream
[c]Red Bull Playstreets

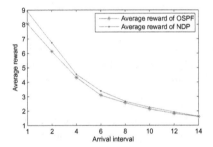

Fig. 5. Average reward of different arrival interval of DASH client

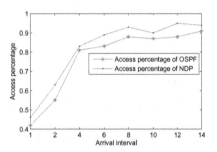

Fig. 6. Access rate of different arrival interval of DASH client

for DASH users. By comparison, our algorithm based on NDP can provide reasonable path management for DASH video transmission according to global network state. Meanwhile, our algorithm can also provide the bitrate adjustment management and reject the bitrate adjustment request when the bitrate adjustment cannot increase the average reward. The percentage of the access users in Fig. 6 further confirms the decision by our algorithm. Therefore, the results of the experiment indicate the joint routing and bitrate adjustment algorithm for DASH video transmission proposed by this paper can effectively increase the average reward and realize the efficient utilization of network bandwidth resources.

6   Conclusion

In this paper, we considered the joint routing and bitrate adjustment optimization for DASH video service in SDN. We conceived an open architecture based on OpenFlow. Based on the observed network state provided by OpenFlow controller, we formulated the joint routing and bitrate adjustment problem into Markov decision problem of maximizing the overall average reward. The neuro-dynamic programming solution is employed to handle the curse of dimensionality in the context of MDP. In order to verify the performance of the proposed algorithm, we developed an emulation platform by integrating POX and Mininet. The experimental results show that the proposed method has higher achievable performance compared with the OSPF based algorithm.

References

1. Sezer, S., Scott-Hayward, S., Chouhan, P.K., Fraser, B., Lake, D., Finnegan, J., Viljoen, N., Miller, M., Rao, N.: Are we ready for sdn? Implementation challenges for software-defined networks. IEEE Commun. Mag. **51**(7), 36–43 (2013)
2. Mckeown, N., Anderson, T., Balakrishnan, H., et al.: OpenFlow: enabling innovation in campus networks. ACM Sigcomm Comput. Commun. Rev. **38**(2), 69–74 (2008)
3. Wu, D., Ci, S., Wang, H., Katsaggelos, A.K.: Application-centric routing for video streaming over multihop wireless networks. IEEE Trans. Circ. Syst. Video Technol. **20**(12), 1721–1734 (2010)
4. Neely, M.J., Modiano, E., Rohrs, C.E.: Dynamic power allocation and routing for time-varying wireless networks. IEEE J. Sel. Areas Commun. **23**(1), 89–103 (2005)
5. Mellouk, A., Hoceini, S., Amirat, Y.: Adaptive quality of service-based routing approaches: development of neuro-dynamic state-dependent reinforcement learning algorithms. Int. J. Commun. Syst. **20**(10), 1113–1130 (2007)
6. Cetinkaya, C., Karayer, E., Sayit, M., Hellge, C.: SDN for segment based flow routing of DASH. In: IEEE Fourth International Conference on Consumer Electronics Berlin (ICCE-Berlin), pp. 74–77 (2014)
7. Liu, C., Bouazizi, I. Gabbouj, M.: Rate adaptation for adaptive HTTP streaming. In: Proceedings of the Second Annual ACM Conference on Multimedia Systems, pp. 169–174 (2011)
8. Tian, G., Liu, Y.: Towards agile and smooth video adaptation in dynamic HTTP streaming. In: Proceedings of the 8th International Conference on Emerging Networking Experiments and Technologies, pp. 109–120 (2012)
9. Bertsekas, D.P., Ioffe, S.: Temporal differences-based policy iteration and applications in neuro-dynamic programming. Lab. for Info. and Decision Systems Report LIDS-P-2349, MIT, Cambridge, MA (1996)
10. Huang, D., Chen, W., Mehta, P., Meyn, S., Surana, A.: Feature selection for neuro-dynamic programming. In: Lewis, F. (ed.) Reinforcement Learning and Approximate Dynamic Programming for Feedback Control, pp. 535–559. Wiley, New York (2011)
11. Bertsekas, D.P.: Dynamic Programming and Optimal Control, vol. 1, no. 2. Athena Scientific, Belmont (1995)
12. Puterman, M.L.: Markov Decision Processes: Discrete Stochastic Dynamic Programming. Wiley-Intersicence, New York (2005)
13. Bertsekas, D.P., Tsitsiklis, J.N.: Neuro-Dynamic Programming (Optimization and Neural Computation Series, 3). Athena Sci. **7**, 15–23 (1996)
14. Marbach, P., Mihatsch, O., Tsitsiklis, J.N.: Call admission control and routing in integrated services networks using neuro-dynamic programming. IEEE J. Sel. Areas Commun. **18**(2), 197–208 (2000)

Stability of Periodic Orbits in Dynamic Binary Neural Networks with Ternary Connection

Kazuma Makita, Ryuji Sato, and Toshimichi Saito$^{(\boxtimes)}$

Hosei University, Koganei, Tokyo 184-8584, Japan
tsaito@hosei.ac.jp

Abstract. This paper studies dynamic binary neural networks that can generate various periodic orbits. The networks is characterized by signum activation function and ternary connection parameters. In order to analyze the dynamics, we present two simple feature quantities that characterize plentifulness of transient phenomena and superstability of the periodic orbits. Calculating the feature quantities for a class of networks, we investigate transient and superstability of the periodic orbits.

Keywords: Dynamic binary neural networks · Periodic orbits · Stability

1 Introduction

Applying a delayed feedback to the binary feed forward network [1,2], the dynamic binary neural network (DBNN) is constructed [3–5]. The DBNN is characterized by signum activation functions, ternary connection parameters, and integer threshold parameters. Depending on parameters and initial conditions, the DBNN can generate various binary periodic orbits in the steady state. The DBNN is based on the logical/sequential circuits [1] and the dynamics can be integrated into the digital return map (Dmap) defined on a set of points [3]. The Dmaps are regarded as a digital version of analog return maps, important study objects in nonlinear dynamics [6]. The Dmaps are related deeply to the cellular automata that can generate various spatiotemporal patterns [7]. Engineering applications of the cellular automata and DBNN include information compressors [8], image processors [9], and control of switching circuits [3,10]. Analysis of the DBNN is important not only as fundamental nonlinear problems but also for engineering applications. However, systematic analysis is not easy because the dynamics is complex even in a simple class of DBNNs.

In order to analyze the dynamics, this paper presents two simple feature quantities. The first quantity characterizes plentifulness of transient phenomena. The second quantity characterize superstability such that all the initial values fall instantaneously into a periodic orbit. Using the feature quantities, we construct a feature plane that is useful in visualization and classification of the stability.

This work is supported in part by JSPS KAKENHI#15K00350.

A. Hirose et al. (Eds.): ICONIP 2016, Part I, LNCS 9947, pp. 421–429, 2016.
DOI: 10.1007/978-3-319-46687-3_47

We then analyze a simple class of DBNNs: 6-dimensional DBNNs having several periodic orbits. Calculating the feature quantities, it is shown that the simple DBNNs can generate a variety of periodic/transient phenomena and cutting connection can reinforce stability of the periodic orbits.

2   Dynamic Binary Neural Networks

Let us begin with introducing the dynamic binary neural network (DBNN) and digital return map (Dmap) presented in [3–5]. As shown in Fig. 1, The DBNN is constructed by applying a delayed feedback to a feed-forward network with the signum activation function. The dynamics is described by

$$x_i^{t+1} = \text{sgn}\left(\sum_{j=1}^{N} w_{ij} x_j^t - T_i\right) \tag{1}$$

$$\text{sgn}(x) = \begin{cases} +1 \text{ for } x \geq 0 \\ -1 \text{ for } x < 0 \end{cases} \quad i = 1 \sim N$$

ab. $x^{t+1} = F_1(x^t)$, $x^t \equiv (x_1^t, \cdots, x_N^t) \in B^N$

where x^t is a binary state vector at discrete time t and $x_i^t \in \{-1, +1\} \equiv B$ is the i-th element. The dimension N is a finite positive integer. The connection parameters are ternary $w_{ij} \in \{-1, 0, +1\}$ and the threshold parameters are integers $T_i \in \{0, \pm1, \pm2, \cdots\}$. Let $w = (w_{ij})$ denote the connection matrix. The "0" elements in w can be effective to suppress spurious memories as discussed in [5]. When an initial state vector x^1 is given as an input, the DBNN outputs x^2, the x^2 is fed back as the next input. Repeating in this manner, the DBNN generates a binary sequence. Since the number of the N-dimensional binary vector is finite, the DBNN must exhibits a binary periodic orbit (BPO) in the steady state.

In order to analyze the DBNN, we introduce the Dmap. Since the set of all the binary vectors B^N is equivalent to a set of 2^N points L_{2^N}, the dynamics of the DBNN can be integrated into the Dmap

$$x^{t+1} = F_D(x^t), \ x^t \in \{C_1, \cdots, C_{2^N}\} \equiv L_{2^N}.$$

The 2^N points C_1 to C_{2^N} are expressed by decimal value of the binary code. x^t denotes a variable of Dmap corresponding to x^t of DBNN. Figure 1 illustrates the DBNN and its Dmap for $N = 3$ where $L_{2^3} = \{C_1, \cdots, C_8\}$; $C_1 = 0 \equiv (-1, -1, -1)$, $\cdots$, $C_8 = 7 \equiv (+1, +1, +1)$; It should be noted that, since the number of points in L_{2^N} is 2^N, direct memory of all the inputs/outputs becomes hard/impossible as N increases. However, for relatively small value of N, the DBNN can exhibit a variety of periodic/transient phenomena some of which are applicable to engineering systems. For examples, the cases $N = 6$ and $N = 9$ have been applied to control signal of switching power converters [3–5].

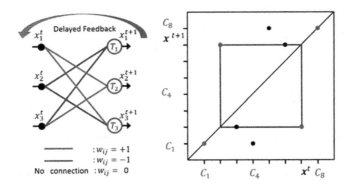

Fig. 1. DBNN and Dmap. $w_{ij} = 1$: red branch. $w_{ij} = -1$: blue branch. $w_{ij} = 0$: no connection. Red points: PEPs. Blue orbit: a PEO with period 2. (Color figure online)

3   Feature Quantities

Here we present two feature quantities and a feature plane in order to analyze various BPOs of the DBNN. As preparations, we give basic definitions.

Definition 1. A point $x_p \in L_{2^N}$ is said to be a periodic point (PEP) with period p if $F_D^p(x_p) = x_p$ and $F_D(x_p)$ to $F_D^p(x_p)$ are all different where F_D^p is the p-fold composition of F_D. A periodic point with period 1 is referred to as a fixed point. A sequence of the PEPs, $\{F_D(x_p), \cdots, F_D^p(x_p)\}$, is said to be a periodic orbit (PEO). Note that the steady state must be a PEO (corresponding to a BPO of the DBNN) because the domain L_{2^N} consists of a finite number of the points. Note also that at least one PEP must exist in the Dmap. Figure 1 shows a Dmap and a PEO with period 2.

Definition 2. A point $x_e \in L_{2^N}$ is said to be an eventually periodic point (EPP) with step q if x_e is not a PEP but falls into some PEP x_p after q iterations: $F^q(x_e) = x_p$. The EPPs are basic to consider stability of a PEO. An EPP with step 1 is said to be a direct eventually periodic point (DEPP). A PEO (PEP) is said to be *superstable* if all the points are DEPPs (except for the PEP).

Here we present two feature quantities. The first quantity is the rate of EPPs that can characterize plentifulness of transient phenomena:

$$\alpha = \frac{\text{The number of EPPs}}{2^N}, \quad \frac{1}{2^N} \leq \alpha \leq \frac{N-1}{2^N}. \tag{2}$$

The second quantity is the rate of DEPPs that can characterize similarity to superstability:

$$\beta = \frac{\text{The number of DEPPs}}{2^N}, \quad \frac{1}{2^N} \leq \beta \leq \alpha. \tag{3}$$

As β increases, a PEO approaches to be superstable. A PEO with k is superstable if $\beta = \alpha = (1-k)/2^N$.

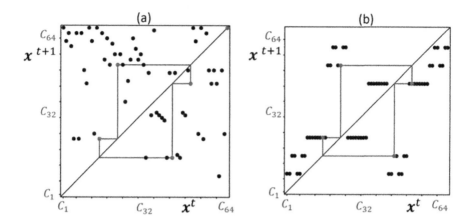

Fig. 2. Dmap examples. (a) w_{ij} in Table 1 and $T_i = 0$. One PEO with period 4 and one fixed point. $\alpha = 59/64$, $\beta = 17/64$. (b) w_{ij} in Table 2 and $T_i = 0$. One PEO period 4. $\alpha = 60/64$, $\beta = 44/64$

Table 1. Connection matrix example 1

j	1	2	3	4	5	6
w_{1j}	-1	$+1$	0	-1	0	$+1$
w_{2j}	-1	0	-1	$+1$	$+1$	0
w_{3j}	$+1$	-1	0	-1	0	$+1$
w_{4j}	$+1$	-1	0	$+1$	0	-1
w_{5j}	$+1$	0	$+1$	-1	-1	0
w_{6j}	-1	$+1$	0	$+1$	0	-1

Table 2. Connection matrix example 2

j	1	2	3	4	5	6
w_{1j}	$+1$	$+1$	-1	0	0	0
w_{2j}	-1	$+1$	$+1$	0	0	0
w_{3j}	$+1$	0	-1	0	$+1$	0
w_{4j}	-1	-1	$+1$	0	0	0
w_{5j}	$+1$	-1	-1	0	0	0
w_{6j}	-1	0	-1	0	-1	0

Figure 2(a) shows the first example of Dmap from the DBNN with connection parameters in Table 1. The connection matrix includes two "0" elements in each row where the "0" element means cutting connection. The Dmap has one PEO with period 4 and one fixed point hence $\alpha = 59/64$. The number of DEPPs is not so large and $\beta = 17/64$. Figure 2(a) shows the second example of Dmap from the DBNN with connection parameters in Table 2. The connection matrix includes three "0" elements in each row. The Dmap has one PEO with period 4 and $\alpha = 60/64$. The number of DEPPs is larger than the first example and $\beta = 44/64$. These results suggest that the PEO approaches to be superstable as the number of "0" elements (cutting connection) increases appropriately.

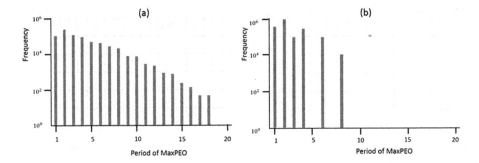

Fig. 3. The histogram of MaxPEOs. (a) For 729000 Dmaps in Case 1. (b) For 1728000 Dmaps in Case 2.

Table 3. RPEO6 for control signal of a dc/ac inverter.

z^1	$(+1, -1, -1, -1, +1, +1)$
z^2	$(+1, +1, -1, -1, -1, +1)$
z^3	$(+1, +1, +1, -1, -1, -1)$
z^4	$(-1, +1, +1, +1, -1, -1)$
z^5	$(-1, -1, +1, +1, +1, -1)$
z^6	$(-1, -1, -1, +1, +1, +1)$
$z^7 = z^1$	$(+1, -1, -1, -1, +1, +1)$

4   Numerical Experiments

Since the DBNN has many parameters and can generate a variety of PEOs and EPPs, general analysis of the dynamics is extremely hard. In this paper, we analyze a simple class of DBNNs. For simplicity, we fix the dimension of the binary vectors and the threshold parameters: $N = 6, T_i = 0$. We then consider the two cases of connection parameters.

Case 1: Each row of the connection matrix consists of two "+1", two "−1", and two "0" elements as shown in Table 1. The first 3 rows are inverse pattern of the second 3 rows: $w_{ij} = -w_{(i+3)j}$ for $i = 1 \sim 3$. The number of connection matrices (the number of the objective DBNNs) is

$$(_6C_2 \times_4 C_2)^3 = 90^3 = 729,000.$$

Case 2: Each row of the connection matrix includes three 0 elements as shown in Table 2. The others are two "+1" and one "−1" elements, or one "+1" and two "−1" elements. The first 3 rows are inverse pattern of the second 3 rows: $w_{ij} = -w_{(i+3)j}$ for $i = 1 \sim 3$. The number of connection matrices (the number of the objective DBNNs) is

$$(_6C_3 \times_3 C_2)^3 = 120^3 = 1,728,000.$$

Table 4. Connection matrix example 3

j	1	2	3	4	5	6
w_{1j}	+1	−1	−1	0	0	+1
w_{2j}	+1	0	0	−1	−1	+1
w_{3j}	+1	+1	0	0	−1	−1
w_{4j}	−1	+1	+1	0	0	−1
w_{5j}	−1	0	0	+1	+1	−1
w_{6j}	−1	−1	0	0	+1	+1

Table 5. Connection matrix example 4

j	1	2	3	4	5	6
w_{1j}	0	+1	−1	0	−1	+1
w_{2j}	0	−1	−1	−1	0	+1
w_{3j}	−1	+1	0	+1	0	−1
w_{4j}	0	−1	+1	0	+1	−1
w_{5j}	0	+1	+1	+1	0	−1
w_{6j}	+1	−1	0	−1	0	+1

Table 6. Connection matrix example 5

j	1	2	3	4	5	6
w_{1j}	+1	0	−1	0	0	+1
w_{2j}	+1	0	0	0	−1	+1
w_{3j}	0	+1	+1	0	−1	0
w_{4j}	−1	0	+1	0	0	−1
w_{5j}	−1	0	0	0	+1	−1
w_{6j}	0	−1	−1	0	+1	0

Table 7. Connection matrix example 6

j	1	2	3	4	5	6
w_{1j}	+1	−1	−1	0	0	0
w_{2j}	+1	+1	−1	0	0	0
w_{3j}	0	+1	+1	0	−1	0
w_{4j}	−1	+1	+1	0	0	0
w_{5j}	−1	−1	+1	0	0	0
w_{6j}	0	−1	−1	0	+1	0

Since it is hard to consider all the BPOs of each Dmap, we consider one PEO with the longest period for each Dmap and refer the representative PEO as MaxPEO. We construct a histogram of the MaxPEOs in Case 1 and Case 2. as shown in Fig. 3. In Case 1, the DBNN can generates a variety of MaxPEOs. In Case 2, kinds of MaxPEOs are reduced, however, stability of the MaxPEOs seems to be reinforced as suggested in examples in Fig. 2.

Using the feature plane, we consider stability of one typical PEO: a PEO with period 6 as shown in Table 3. This PEO corresponds to a control signal of a dc-ac inverter that is one of the most important switching power converters [5]. We refer to this PEO as RPEO6 hereafter. If the stability of the RPEO6 is reinforced, robustness can be reinforced in operation of the inverter.

Figure 4 shows two examples of Dmaps in Case 1 that can generate the RPEO6. The Dmaps have several PEOs except for the RPEO6 and stability is not so strong. In Case 1, 42,248 Dmaps can generate the RPEO6 and Fig. 6 shows their feature quantities on the feature plane. We can see that the two feature quantities distribute in relatively wider region in the feature plane.

Figure 5 shows two typical examples of Dmaps in Case 2 that can generate the RPEO6. Except for the RPEO6, they have no PEO and stability of RPEO6 is stronger than the Dmaps in Case 1. Especially, RPEO6 in Fig. 5(a) is superstable. In Case 2, 99,312 Dmaps can generate the RPEO6 and Fig. 6 shows their feature

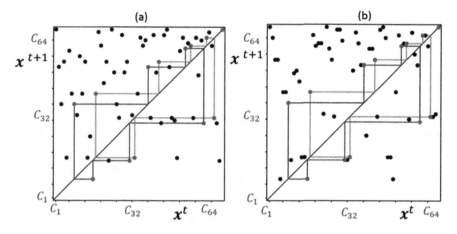

Fig. 4. Dmap examples in Case 1. Blue orbit denotes RPEO6 (a PEO with period 6). (a) w_{ij} in Table 4. $\alpha = 51/64$, $\beta = 27/64$ (b) w_{ij} in Table 5. $\alpha = 51/64$, $\beta = 31/64$ (Color figure online)

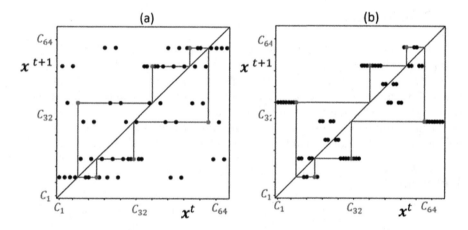

Fig. 5. Dmap examples in Case 2. Blue orbit denotes RPEO6. (a) w_{ij} in Table 6. $\alpha = 58/64$, $\beta = 58/64$ (b) w_{ij} in Table 7. $\alpha = 58/64$, $\beta = 50/64$ (Color figure online)

quantities on the feature plane. We can see that the plots concentrate onto the line $\alpha = (64 - 6)/64$ and the stability is strong. Each connection matrix in Case 2 includes six "0" elements more than Case 1 and the stability is much stronger than the Case 1. We have confirmed that a number of DBNNs can have superstable RPEO6 in Case 2 but no DBNN has superstable RPSO6 in Case 1.

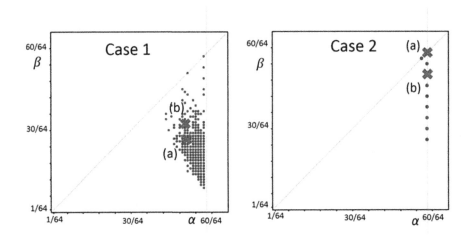

Fig. 6. Feature quantity plane. Case 1 for 42,248 RPEO6. Case 2 for 99,312 RPEO6.

5 Conclusions

In order to analyze stability of the PEO, two feature quantities are presented in this paper. The first and second quantities can characterize plentifulness of transient phenomena and superstability, respectively. Calculating the feature quantities of a simple class of DBNNs, stability of PEOs is investigated. Especially, superstable RPEO6s are confirmed in Case 2.

Future problems include more detailed stability analysis of various PEOs and engineering applications including control of various switching circuits.

References

1. Gray, D.L., Michel, A.N.: A training algorithm for binary feed forward neural networks. IEEE Trans. Neural Netw. **3**(2), 176–194 (1992)
2. Chen, F., Chen, G., He, Q., He, G., Xu, X.: Universal perceptron and DNA-like learning algorithm for binary neural networks: Non-LSBF implementation. IEEE Trans. Neural Netw. **20**(8), 1293–1301 (2009)
3. Kouzuki, R., Saito, T.: Learning of simple dynamic binary neural networks. IEICE Trans. Fundam. **E96–A**(8), 1775–1782 (2013)
4. Moriyasu, J., Saito, T.: A cascade system of dynamic binary neural networks and learning of periodic orbit. IEICE Trans. Inf./Syst. **E98–D**(9), 1622–1629 (2015)
5. Sato, R., Saito, T.: Simple feature quantities for learning of dynamic binary neural networks. In: Arik, S., Huang, T., Lai, W.K., Liu, Q. (eds.) ICONIP 2015. LNCS, vol. 9489, pp. 226–233. Springer, Heidelberg (2015). doi:10.1007/978-3-319-26532-2_25
6. Ott, E.: Chaos in Dynamical Systems. Cambridge University Press, Cambridge (1993)
7. Chua, L.O.: A Nonlinear Dynamics Perspective of Wolfram's New Kind of Science, vol. I and II. World Scientific, Singapore (2005)

8. Wada, W., Kuroiwa, J., Nara, S.: Completely reproducible description of digital sound data with cellular automata. Phys. Lett. A **306**, 110–115 (2002)
9. Rosin, P.L.: Training cellular automata for image processing. IEEE Trans. Image Process. **15**(7), 2076–2087 (2006)
10. Rodriguez, J., Rivera, M., Kolar, J.W.: A review of control and modulation methods for matrix converters. IEEE Trans. Ind. Electron. **59**(1), 58–770 (2012)

Evaluation of Chaotic Resonance by Lyapunov Exponent in Attractor-Merging Type Systems

Sou Nobukawa[1(✉)], Haruhiko Nishimura[2], and Teruya Yamanishi[1]

[1] Department of Management Information Science, Fukui University of Technology,
3–6–1 Gakuen, Fukui, Fukui 910–8505, Japan
nobukawa@fukui-ut.ac.jp
[2] Graduate School of Applied Informatics, University of Hyogo,
7–1–28 Chuo-ku, Kobe, Hyogo 650–8588, Japan

Abstract. Fluctuating activities in the deterministic chaos cause a phenomenon that is similar to stochastic resonance (SR) whereby the presence of noise helps a non-linear system to amplify a weak (under-barrier) signal. In this phenomenon, called chaotic resonance (CR), the system responds to the weak input signal by the effect of intrinsic chaotic activities under the condition where no additive noise exists. Recently, we have revealed that the signal response of the CR in the spiking neuron model has an unimodal maximum with respect to the degree of stability for chaotic orbits quantified by maximum Lyapunov exponent. In response to this situation, in this study, focusing on CR in the systems with chaos-chaos intermittency, we examine the signal response in a cubic map and a chaotic neural network embedded two symmetric patterns by cross correlation and Lyapunov exponent (or maximum Lyapunov exponent). As the results, it is confirmed that the efficiency of the signal response has a peak at the appropriate instability of chaotic orbit in both systems. That is, the instability of chaotic orbits in CR can play a role the noise strength of SR in not only spiking neural systems but also the systems with chaos-chaos intermittency.

Keywords: Cubic map · Chaotic neural network · Chaos · Chaos-chaos intermittency · Lyapunov exponent · Chaotic resonance

1 Introduction

Fluctuating activities in the deterministic chaos cause a phenomenon that is similar to stochastic resonance (SR) whereby the presence of noise helps a non-linear system to amplify a weak (under-barrier) signal [3, 7–11]. In this phenomenon, called chaotic resonance (CR), the system responds to the weak input signal by the effect of intrinsic chaotic activities under the condition where no additive noise exists.

At first, CR has been investigated in one-dimensional cubic map and Chua's circuit [4–6,12,21]. Furthermore, the study of CR has been proceeding in the neural systems [13,14,16,19,20,22]. Recently, we have revealed that the signal

© Springer International Publishing AG 2016
A. Hirose et al. (Eds.): ICONIP 2016, Part I, LNCS 9947, pp. 430–437, 2016.
DOI: 10.1007/978-3-319-46687-3_48

response of CR in a spiking neural system has an unimodal maximum with respect to the degree of stability for chaotic orbits quantified by maximum Lyapunov exponent [15,17]. That is, the appropriate chaotic behavior leads to the generation of spikes (exceeding the threshold) not at specific times, but at varying scatter times for each trial as input signals. This frequency distribution of these spike timings against the input signal becomes congruent with the shape of the input signal. Thus, it can be interpreted that the instability of the chaotic orbit in CR plays a role of the noise strength in SR.

On the other hand, in the systems having a chaotic attractor with chaos-chaos intermittency as typified by one-dimensional cubic map and Chua's circuit, CR is produced by synchronization of chaos-chaos switching to the weak input signal [2]. However, at this stage, the relationship between the signal response and the instability of the orbit has not been revealed yet in this type of CR. Thus, in this paper, we focus on an one-dimensional cubic map and a chaotic neural network with two stored patterns and evaluate their signal response by Lyapunov exponents.

2   Cubic Map

In this study, to evaluate the signal response in CR by chaos-chaos intermittency, we utilize a cubic map driven by a periodic signal $S(t) = A \sin \pi \Omega t$ $(t = 1, 2, \cdots)$ as follows:

$$x(t + 1) = (ax(t) - x^3(t)) \exp(-x^2(t)/b) + S(t). \tag{1}$$

Here, the exponential term is inserted to prevent $x(t)$ from diverging. In our simulation, we set the parameter b to 10.

At first, we investigate the structure of attractor in cubic map under signal-free $(A = 0)$ condition. To evaluate chaos and the instability of the orbit in the system, we use the Lyapunov exponent [18]:

$$\lambda = \frac{1}{\tau M} \sum_{k=1}^{M} \ln(\frac{d^k(t_l = \tau)}{d^k(t_l = 0)}). \tag{2}$$

Here, $d^k(t_l = 0) = d_0$ $(k = 1, 2, \cdots, M)$ are M perturbed initial conditions to $x(t)$ applied at $t = t_0 + (k - 1)\tau$. Their time evolutions for $t_l \in [0 : \tau]$ are $d^k(t_l = \tau) = (x(t) - x'(t))|_{t=t_0+k\tau}$. $x'(t)$ indicates an orbit applied perturbation. Figure 1 shows dependence of bifurcation diagram for $x(t)$ ((a)) and λ ((b)) on parameter a. In $a < a_{cr} = 2.839 \cdots$, this system has two symmetric chaotic attractors $(\lambda > 0)$ divided by $x(t) = 0$. The trajectory is trapped inside one attractor depending on the initial condition for $x(0)$. Note that this bifurcation diagram is constructed by positive and negative initial conditions of $x(0)$. Then, both attractors merge in $a > a_{cr}$, and $x(t)$ moves between these attractors, intermittently.

(a)

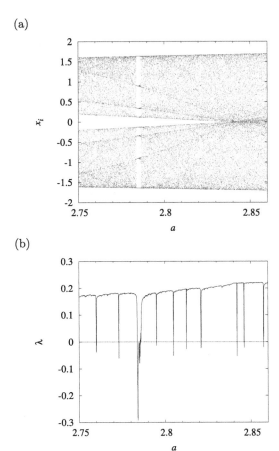

(b)

Fig. 1. Dependence of system behavior on parameter a in cubic map. (a) Bifurcation diagram for $x(t)$. (b) Lyapunov exponent λ. ($b = 10, A = 0$)

Next, we investigate the response of $x(t)$ against the weak signal ($A = 0.005, \Omega = 0.005$). To quantify this signal response, we used the mutual correlation $C(\tau)$ between the time series of $x(t)$ and the signal $S(t)$ is given by

$$C(\tau) = \frac{C_{sx}(\tau)}{\sqrt{C_{ss}C_{xx}}}, \tag{3}$$

$$C_{sx}(\tau) = < (S(t+\tau) - < S(t) >)(x(t) - < x(t) >) >, \tag{4}$$

$$C_{ss} = < (S(t) - < S(t) >)^2 >, \tag{5}$$

$$C_{xx} = < (x(t) - < x(t) >)^2 > . \tag{6}$$

For the time delay factor τ, we check $\max_\tau C(\tau)$ (i.e., the largest $C(\tau)$ between $0 \leq \tau \leq T_0 = 1/\Omega$). Figure 2 shows the scatter plot of λ and $\max_\tau C(\tau)$.

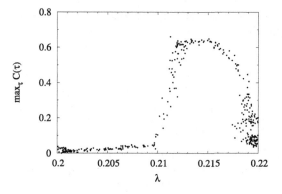

Fig. 2. Scatter plot of λ and $\max_\tau C(\tau)$ in cubic map. ($b = 10, A = 0.005, \Omega = 0.005$)

In $0.2 \leq \lambda \lesssim 0.21$, the signal response efficiency is low ($\max_\tau C(\tau) \approx 0.02$), but increases accompanied with λ, and $\max_\tau C(\tau)$ has the peak at $\lambda \approx 0.215$. However, further enhancing the instability of the chaotic orbit ($\lambda \lesssim 0.215$), $\max_\tau C(\tau)$ decreases.

3   Chaotic Neural Network

In the chaotic neural network [1] consisting of N neurons, the output of neuron i ($i = 1, 2, \cdots, N$) driven by external signal $S_i(t)$ is described by

$$x_i(t + 1) = f(y_i(t) + S_i(t)), \tag{7}$$

where f and y_i indicate the output function defined by $f(x) = \tanh \frac{x}{2\varepsilon}$ with the steepness parameter ε and the internal potential, respectively. The state of the network is exhibited by $\{x_i\}$. The internal potential is given by

$$y_i(t) = \eta_i(t) + \zeta_i(t), \tag{8}$$

$$\eta_i(t) = k_f \eta_i(t - 1) + \sum_{j=1}^{N} w_{ij} x_j(t - 1), \tag{9}$$

$$\zeta_i(t) = k_r \zeta_i(t - 1) - \alpha x_i(t - 1) + a, \tag{10}$$

where $k_f(k_r)$: decay factor the feedback (refractoriness) ($0 \leq k_f < 1, 0 \leq k_r < 1$), α: refractory scaling parameter ($\alpha \geq 0$), w_{ij}: synaptic weight from neuron j to neuron i, a: threshold of neuron i.

We store two patterns:

$$\{\xi_i^1\} = (1, \cdots, 1, -1, \cdots, -1)^T, \tag{11}$$

$$\{\xi_i^2\} = (-1, \cdots, -1, 1, \cdots, 1)^T. \tag{12}$$

to the chaotic neural network. For storing the patterns, the synaptic weight w_{ij} is formed by Hebb's rule as follows:

$$w_{ij} = \begin{cases} 0 & (i = j) \\ \frac{2}{N} & (1 \leq i \leq N/2 \text{ and } 1 \leq j \leq N/2) \\ \frac{2}{N} & (N/2 + 1 \leq i \leq N \text{ and } N/2 + 1 \leq j \leq N) \\ -\frac{2}{N} & (1 \leq i \leq N/2 \text{ and } N/2 + 1 \leq j \leq N) \\ -\frac{2}{N} & (N/2 + 1 \leq i \leq N \text{ and } 1 \leq j \leq N/2) \end{cases} \tag{13}$$

In our simulation, we set the parameters to $(k_f = 0.7, k_r = 0.8, \alpha = 0.85, \varepsilon = 0.015, N = 4)$. Corresponding to the stored patterns, we adopt the input signal $S_i(t) = A \sin \pi \Omega t \ (1 \leq i \leq N/2), -A \sin \pi \Omega t \ (N/2 + 1 \leq i \leq N)$.

To measure the network state $\{x_i\}$, we utilize the relative overlap with pattern $\{\xi_i^1\}$:

$$m(t) = \frac{1}{N} \sum_{i=1}^{N} \xi_i^1 x_i(t). \tag{14}$$

To evaluate chaos in the system, we use the maximum Lyapunov exponent [18]:

$$\lambda_1 = \frac{1}{\tau M} \sum_{k=1}^{M} \ln\left(\frac{|\mathbf{d}^k(t_l = \tau)|}{|\mathbf{d}^k(t_l = 0)|}\right). \tag{15}$$

Here, $\mathbf{d}^k(t_l = 0) \ (k = 1, 2, \cdots, M)$ are M perturbed initial conditions for $\{\eta_i(t), \zeta_i(t)\}$ applied at $t = t_0 + (k - 1)\tau$, which are given by $\mathbf{d}^{k+1}(t_l = 0) = \mathbf{d}^k(t_l = \tau)/|\mathbf{d}^k(t_l = \tau)| \ (\mathbf{d}^1(t_l = 0) = \mathbf{d}_0, \mathbf{d}_0$: an initial vector). Their time evolutions for $t_l \in [0 : \tau]$ are calculated by $\mathbf{d}^k(t_l = \tau) = (\{\eta_i(t), \zeta_i(t)\} - \{\eta_i'(t), \zeta_i'(t)\})|_{t=t_0+k\tau}$, where $\{\eta_i'(t), \zeta_i'(t)\}$ indicates an orbit applied perturbation.

As shown in Fig. 3(a), the chaotic neural network embedded the patterns with the inversion relation has two symmetric chaotic attractors ($\lambda_1 > 0$ (Fig. 3(b)) against $m(t) = 0$ similar to cubic map under the signal-free condition ($A = 0$). In $0.26 \leq a \lesssim 0.27$, $m(t)$ is trapped inside one chaotic attractor depending on the initial condition. However, by merging these attractors in $a \gtrsim 0.27$, $m(t)$ hops between both attractors.

Furthermore, we investigate the signal response of this network by $\max_\tau C(\tau)$ and λ_1. Figure 4 shows the scatter plot of λ_1 and $\max_\tau C(\tau)$ under the additive weak signal ($A = 0.01, \Omega = 0.01$). In $0 \leq \lambda_1 \lesssim 0.3$, the signal response efficiency is low ($\max_\tau C(\tau) \lesssim 0.2$), but increases accompanied with λ_1, and $\max_\tau C(\tau)$ has the peak at the appropriate instability of chaotic orbit ($\lambda_1 \approx 0.35$), as is the case with cubic map.

(a)

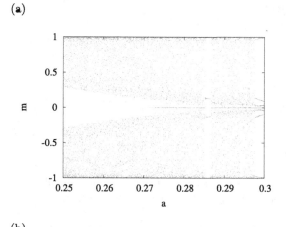

(b)

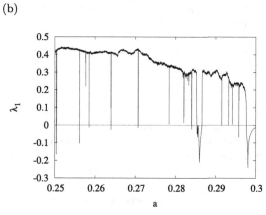

Fig. 3. Dependence of system behavior on parameter a in chaotic neural network. (a) Bifurcation diagram. (b) maximum Lyapunov exponent λ_1. ($k_f = 0.7, k_r = 0.8, \alpha = 1.12, \epsilon = 0.015$)

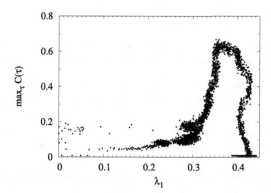

Fig. 4. Scatter plot of λ_1 and $\max_\tau C(\tau)$ in chaotic neural network. ($k_f = 0.7, k_r = 0.8, \alpha = 1.12, \epsilon = 0.015, A = 0.01, \Omega = 0.01$)

4   Conclusion

In this study, in order to reveal the relationship between the signal response of CR by chaos-chaos switching and the instability of chaotic orbit, we examined the signal response in the cubic map and the chaotic neural network embedded two symmetric patterns, using cross correlation $\max_\tau C(\tau)$ and Lyapunov exponent λ (or maximum Lyapunov exponent λ_1). As the results, it was confirmed that the efficiency of the signal response exhibits the unimodal maximum at the appropriate the values of λ (λ_1) in both systems. That is, the instability of chaotic orbits in CR can play a role of the noise strength in SR in not only spiking neural systems but also the systems with chaos-chaos intermittency.

Further research based on this study would be to evaluate CR in large scale chaotic neural networks, using the Lyapunov spectrum.

Acknowledgment. This work is supported by Grant-in-Aid for Young Scientists (B), grant number (15K21471).

References

1. Aihara, K., Takabe, T., Toyoda, M.: Chaotic neural networks. Phys. Lett. A **144**(6), 333–340 (1990)
2. Anishchenko, V.S., Astakhov, V., Neiman, A., Vadivasova, T., Schimansky-Geier, L.: Nonlinear Dynamics of Chaotic and Stochastic Systems: Tutorial and Modern Developments. Springer, Heidelberg (2007)
3. Benzi, R., Sutera, A., Vulpiani, A.: The mechanism of stochastic resonance. J. Phys. A Math. General **14**(11), L453 (1981)
4. Carroll, T., Pecora, L.: Stochastic resonance and crises. Phys. Rev. Lett. **70**(5), 576–579 (1993)
5. Carroll, T., Pecora, L.: Stochastic resonance as a crisis in a period-doubled circuit. Phys. Rev. E **47**(6), 3941–3949 (1993)
6. Crisanti, A., Falcioni, M., Paladin, G., Vulpiani, A.: Stochastic resonance in deterministic chaotic systems. J. Phys. A Math. General **27**(17), 597–603 (1994)
7. Gammaitoni, L., Hänggi, P., Jung, P., Marchesoni, F.: Stochastic resonance. Rev. Mod. Phys. **70**(1), 223–287 (1998)
8. Hänggi, P.: Stochastic resonance in biology how noise can enhance detection of weak signals and help improve biological information processing. ChemPhysChem **3**(3), 285–290 (2002)
9. McDonnell, M.D., Ward, L.M.: The benefits of noise in neural systems: bridging theory and experiment. Nat. Rev. Neurosci. **12**(7), 415–426 (2011)
10. Mori, T., Kai, S.: Noise-induced entrainment and stochastic resonance in human brain waves. Phys. Rev. Lett. **88**(21), 218101 (2002)
11. Moss, F., Wiesenfeld, K.: The benefits of background noise. Sci. Am. **273**, 66–69 (1995)
12. Nicolis, G., Nicolis, C., McKernan, D.: Stochastic resonance in chaotic dynamics. J. Stat. Phys. **70**(1–2), 125–139 (1993)
13. Nishimura, H., Katada, N., Aihara, K.: Coherent response in a chaotic neural network. Neural Process. Lett. **12**(1), 49–58 (2000)

14. Nobukawa, S., Nishimura, H., Katada, N.: Chaotic resonance by chaotic attractors merging in discrete cubic map and chaotic neural network. IEICE Trans. A **95**(4), 357–366 (2012). (in Japanese)
15. Nobukawa, S., Nishimura, H., Katada, N.: Evaluation of resonance phenomena in chaotic states through typical routes in Izhikevich neuron model. In: Proceedings of Nonlinear Theory and its Applications (NOLTA 2015), pp. 435–438. IEICE (2015)
16. Nobukawa, S., Nishimura, H.: Characteristic of signal response in coupled inferior olive neurons with Velarde-Llinás model. In: Proceedings of SICE Annual Conference (SICE 2013), pp. 1367–1374. IEEE (2013)
17. Nobukawa, S., Nishimura, H., Yamanishi, T., Liu, J.Q.: Analysis of chaotic resonance in Izhikevich neuron model. PloS One **10**(9), e0138919 (2015)
18. Parker, T.S., Chua, L.: Practical Numerical Algorithms for Chaotic Systems. Springer, New York (2012)
19. Schweighofer, N., Doya, K., Fukai, H., Chiron, J.V., Furukawa, T., Kawato, M.: Chaos may enhance information transmission in the inferior olive. Proc. Natl. Acad. Sci. U.S.A **101**(13), 4655–4660 (2004)
20. Schweighofer, N., Lang, E.J., Kawato, M.: Role of the olivo-cerebellar complex in motor learning and control. Front. Neural Circ. **7**(94), 10–3389 (2013)
21. Sinha, S., Chakrabarti, B.K.: Deterministic stochastic resonance in a piecewise linear chaotic map. Phys. Rev. E **58**(6), 8009–8012 (1998)
22. Tokuda, I.T., Han, C.E., Aihara, K., Kawato, M., Schweighofer, N.: The role of chaotic resonance in cerebellar learning. Neural Netw. **23**(7), 836–842 (2010)

Bioinformatics

Clustering-Based Weighted Extreme Learning Machine for Classification in Drug Discovery Process

Wasu Kudisthalert[✉] and Kitsuchart Pasupa

Faculty of Information Technology,
King Mongkuts Institute of Technology Ladkrabang, Bangkok 10520, Thailand
wasu.kudisthalert@gmail.com, kitsuchart@it.kmitl.ac.th

Abstract. Extreme Learning Machine (ELM) is a universal approximation method that is extremely fast and easy to implement, but the weights of the model are normally randomly selected so they can lead to poor prediction performance. In this work, we applied Weighted Similarity Extreme Learning Machine in combination with Jaccard/Tanimoto (WELM-JT) and cluster analysis (namely, k-means clustering and Support Vector Clustering) on similarity and distance measures (i.e., Jaccard/Tanimoto and Euclidean) in order to predict which compounds with not-so-different chemical structures have an activity for treating a certain symptom or disease. The proposed method was experimented on one of the most challenging datasets named Maximum Unbiased Validation (MUV) dataset with 4 different types of fingerprints (i.e. ECFP_4, ECFP_6, FCFP_4 and FCFP_6). The experimental results show that WELM-JT in combination with k-means-ED gave the best performance. It retrieved the highest number of active molecules and used the lowest number of nodes. Meanwhile, WELM-JT with k-means-JT and ECFP_6 encoding proved to be a robust contender for most of the activity classes.

1 Introduction

Drug discovery is a process of finding best new candidate compounds that have a certain targeted biological activity from a large library of compounds whose biological activity are not known. Chemoinformatics is a technique for improving the efficiency of drug discovery process. It utilizes computing and information theory to solve drug screening problems.

Recently, computers have played an increasingly importance role in medical and pharmaceutical research and development. Historically, chemists had to test many different compounds one at a time and find by trial and error which ones had the potential to be an effective drug for a certain symptom or disease. This process can take a decade to complete and waste a lot of money on testing yet-to-be-known ineffective compounds. The probability of success is also generally low. Therefore, a new technique, in silico drug screening (also known as virtual screening), has become popular. This technique improves the performance of

© Springer International Publishing AG 2016
A. Hirose et al. (Eds.): ICONIP 2016, Part I, LNCS 9947, pp. 441–450, 2016.
DOI: 10.1007/978-3-319-46687-3_49

discovery process in many aspects, e.g., speeding up the process and increasing the percentage of successful finds of new biologically active compounds.

The traditional virtual screening technique is similarity searching. It is a ligand-based method that measures the degree of similarity of the structure of a molecule in a database to a reference molecular structure. Chemists then only have to assay the similar molecules that are in the top rank. Recently, machine learning algorithms have been popular in the area of chemoinformatics. Proven successful techniques are Support Vector Machine (SVM) [1] and Binary Kernel Analysis (BKD) [2]. Currently, one efficient method is Extreme Learning Machine (ELM). It is very fast and easy to code. Its architecture is based on a single-layer feedforward neural network with no iterative process. Recently, Czarnecki (2015) has introduced a Tanimoto Extreme Learning Machine (TELM) [3]. It replaced the kernel function with Jaccard/Tanimoto (JT) coefficients in weight calculation process. His experimental results showed that the performance of TELM was much better than conventional ELM. Later, Kudisthalert and Pasupa (2016) introduced a Weighted Extreme Learning Machine (WELM) operating on 17 difference similarity measures and on one of the most challenging datasets named Maximum Unbiased Validation (MUV) [4]. The experimental results showed that WELM in combination with Jaccard/Tanimoto (WELM-JT) yielded the best performance.

However, the performance of ELM is not robust because its weights or hidden nodes are randomly selected from a set of training data. Hence, we aimed to improve the performance of ELM by methodically selecting the sample. This can be done by applying a clustering method. We proposed two approaches: (i) using k-means clustering to obtain a sample near the centroid of each cluster, and (ii) using Support Vector Clustering (SVC) to return a set of support vectors that bound each cluster. The proposed technique was investigated on the MUV dataset.

This paper is organized as follows: The next section briefly review methods used in this work namely, ELM, WELM, k-means clustering and Support Vector Clustering. We explain our experimental framework including fingerprint, dataset and the parameter settings in Sect. 3. Then, experimental results are shown in Sect. 4 followed by the conclusion in Sect. 5.

2   Methodology

2.1   Weighted Similarity Extreme Learning Machine (WELM)

WELM is originally based on Extreme Learning Machine (ELM), which was introduced by Huang et al. (2004) [5]. ELM was extended from single hidden-layer feedforward neural network (SLFN). It solved the bottleneck of SLFN. It was extremely simple and much faster than backpropagation neural network (BEP). Since the hidden layer does not need to iteratively tuned [6].

The number of hidden neurons and of input nodes are denoted as L and D, respectively, where $L \leq D$. The output function could describes as follow,

$$f_L(\boldsymbol{x}) = \sum_{i=1}^{L} \boldsymbol{\beta}_i h_i(\boldsymbol{x}) = h(\boldsymbol{x})\boldsymbol{\beta}, \tag{1}$$

where $h(\boldsymbol{x}) = [h_1(\boldsymbol{x}), \ldots, h_L(\boldsymbol{x})]$ are outputs of hidden layer for input $\mathbf{X}$ where $\mathbf{X} \in \mathbb{R}^{N \times D}$, N is a number of samples and $\boldsymbol{\beta} = [\beta_1, \ldots, \beta_L]^T$ is the output weights that are between the hidden layer and output layer. The design matrix corresponds to target labels $\mathbf{T} = [t_1, \ldots, t_N]^T$. The matrix can be calculated as follows,

$$\boldsymbol{\beta} = (\mathbf{H}^T \mathbf{H})^{-1} \mathbf{H}^T \mathbf{T}, \tag{2}$$

where $\mathbf{H}$ is the hidden-layer output matrix that is derived from a similarity function $\mathbf{S}$ from input $\mathbf{X}$, a set of weights $\mathbf{W}$ and biases $\boldsymbol{b}$,

$$\mathbf{H} = \begin{bmatrix} h(\boldsymbol{x}_1) \\ \vdots \\ h(\boldsymbol{x}_N) \end{bmatrix} = \begin{bmatrix} S(\boldsymbol{x}_1, \boldsymbol{w}_1, b_1) & \ldots & S(\boldsymbol{x}_1, \boldsymbol{w}_L, b_L) \\ \vdots & \ddots & \vdots \\ S(\boldsymbol{x}_N, \boldsymbol{w}_1, b_1) & \ldots & S(\boldsymbol{x}_N, \boldsymbol{w}_L, b_L) \end{bmatrix} \tag{3}$$

The weights of the model were not continuous probability distribution. They are randomly selected from a set of training data to ensure that the achieved weights are binary, sparse and has identical dimension span, thus $\mathbf{W} \subset \mathbf{X}$. WELM accommodate another similarity coefficient in weight calculation process. It was found that WELM with Jaccard/Tanimoto (WELM-JT) performance is the most robust for all biological activities [4]. The Jaccard/Tanimoto's equation can be described as follows,

$$\mathbf{S}_{JT}(\mathbf{X}, \mathbf{W}) = \frac{a}{(a + b + c)}, \tag{4}$$

where a is a number of bits that are set to 1 in both $\mathbf{X}$ and $\mathbf{W}$. b and c are number of bits that are set to 1 only in $\mathbf{X}$ and $\mathbf{W}$, respectively. The WELM algorithm is shown in Algorithm 1.

2.2 Clustering-Based WELM

Cluster analysis is an algorithm for grouping object. Its process is to partition a set of objects into k groups (or cluster) such that the objects in each group are more similar to the prototype (also known as centroid) of its group than prototypes of other groups; thus, objects in the same group are homogeneous.

We need automatic algorithms on high-dimensional data for finding characterized fingerprint of known molecules that are used as the weights of the model. The nodes of conventional WELM are randomly selected leading to non-robust performance, hence we methodically selected samples by utilizing clustering methods to organize and summarize data through group prototypes [7].

Algorithm 1. WELM

1: **function** WELM_TRAIN($\mathbf{X}, \mathbf{T}$)
2:    *Initialisation*: Randomly select $\mathbf{W}$ in $\mathbf{X}$. n_{-1} = Number of sample with label of -1, and n_1 = Number of sample with label of 1.
3:    $b = \sqrt{max(n_{-1}, n_{+1})}/n_{y_i}$ for $i = 1, \ldots, N$
4:    $\hat{\mathbf{H}} = b \cdot \mathbf{S}(\mathbf{X}, \mathbf{W})$
5:    $\beta = (\mathbf{I}/C + \hat{\mathbf{H}}^T \hat{\mathbf{H}})^{-1}(\hat{\mathbf{H}}^T b \cdot \mathbf{T})$
6:    **return** $\mathbf{W}, \beta$
7: **end function**

8: **function** WELM_PREDICT($\mathbf{W}, \beta, \mathbf{X}_{\text{test}}$)
9:    $\mathbf{H} = \mathbf{S}(\mathbf{X}_{\text{test}}, \mathbf{W})$
10:   $\hat{\mathbf{T}} = \mathbf{H}\beta$
11:   **return** $\hat{\mathbf{T}}$
12: **end function**

Many clustering algorithms have been introduced and re-introduced but, here, we used k-means clustering and SVC algorithms. The rationale behind the selection of both algorithms were the following: (i) k-means clustering is a conventional clustering algorithm; the algorithm will give a sample that represents the which is, in our framework, the sample that is closest to the centroid; and (ii) SVC returns a set of support vectors that represents a bound of the cluster. More details on both algorithms are described below:

k-means Clustering: Conventionally, this method aims to minimize the Euclidean distance between the centroid and each point in the cluster. However, we utilized Jaccard/Tanimoto coefficient as a similarity measure instead of Euclidean distance by maximizing a similarity between the centroid and each point in the cluster instead. The algorithm converges when all of centroids are stable. The procedures of k-means are the following [8]: (i) Randomly select k data points as initial centroids of k cluster. (ii) Assign data points to the cluster by considering the similarity/dissimilarity between each data point and the centroids; (iii) Update new cluster centroids. (iv) Stop if all of the centroids are not moved. Otherwise branch back to step (ii).

Incidentally, there are many concern issue regarding using k-means: first, performance depends on the initial cluster centroids chosen; second, the number of cluster k must be known; and finally, k-means is very sensitive to outlier data that can cause the poor results.

Support Vector Clustering (SVC): SVC is a clustering algorithm based on Support Vector Machine (SVM) [9]. It utilizes support vectors to characterize the support of a high dimensional distribution. The algorithm yields a set of contours that enclose the data points in each cluster as cluster boundaries and deal with outliers by using a soft margin constant that allows sphere not to enclose all data points in the feature space [10].

In our SVC implementation, data points are mapped from a data space to a high dimensional feature space by using Jaccard/Tanimoto distance to find a sphere where are smallest radius that encloses most of the data points. The Jaccard/Tanimoto distance is the inverse of Jaccard/Tanimoto similarity coefficient (It should be noted that we investigated using the conventional Euclidean distance as well). When this sphere is mapped back to the data space, these contours are interpreted as cluster boundaries.

3   Experimental Framework

3.1   Fingerprint

The most commonly used computer-aided molecular representation is molecular fingerprint. It involves transforming a molecule structure into a sequence of bit-string in which the value of each bit can be only either a "1" or a "0" representing the presence or absence of a substructure defined by encoding algorithm. There are several types of encoding algorithms of molecular fingerprints but circular fingerprints are the most widely used for full structure similarity searching [11]. Type of circular fingerprints are such as Molprint2D, Extended-Connectivity Fingerprints (ECFPs), Functional-Class Fingerprints (FCFPs).

In this paper, we used ECFPs and FCFPs with diameters of 4 and 6, which is referred as ECFP_4, ECFP_6, FCFP_4 and FCFP_6 respectively, since they give the best performance for similarity searching [12].

Several bioactivity datasets for virtual screening are analogue bias, low compound diversity, that leads to artificially high enrichment except this one, Maximum Unbiased Validation (MUV) dataset [13] that was derived from PubChem [14]. MUV is one of the most challenging datasets because its active molecules have been carefully chosen to be structurally diverse. This dataset consists of 17 activities. Each activity consists of 30 active and 15,000 decoy molecules, labeled as 1 and 0 respectively. A biological activity is the effect that a molecule has on humans or animals which inactive molecules do not have.

3.2   Experimental Settings

We constructed a training sets similar to the one reported in [12]. Ten representative reference structures from each activity class were randomly chosen to obtain 170 representative compounds as a training set.

In this experiment, each method WELM-JT, k-means-based WELM and SVC-based WELM needed to be tuned in order to get an optimal model. WELM-JT required 2 parameters to be tuned which are the number of hidden node h and regularization parameter C. The range of h was $[1, \ldots, 170]$ and of C was $[10^{-6}, 10^{-5}, \ldots, 10^{5}, 10^{6}]$. For clustering-based WELM, a regularization parameter for WELM needed to be tuned. In addition to a regularization parameter, the number of cluster for k-means-based WELM is considered to be tuned, while SVC-based WELM is a regularization constant for SVC. Regularization constant for SVC ranges from 0.1 to 1.0 [15]. We then used five-fold cross validation to find the best parameters for each training sets based on the area under the Receiver Operating Characteristic curve.

As for the input and output characteristics of WELM-JT, the input layer consists of 1024 nodes. Each node represented 1 bit of a fingerprint. The output layer consisted of 1 node which was a similarity score. In this work, we evaluated and compared the performance of the proposed methods to the WELM-JT.

To evaluate the performance, we sorted the output values in a descending order in order to compare the hit rates between methods (a hit rate is the number of active compounds retrieved in percentage to the total number of active compounds; only the top 1 % were used). For each method, we repeated the experiment 10 times with different random seeds and reported the average result of 10 runs.

4   Result Discussion

We compared our proposed methods on 4 different types of fingerprints i.e. ECFP_4, ECFP_6, FCFP_4 and FCFP_6. We also embedded and evaluated two different coefficients i.e. ED and JT on k-means clustering and SVC. The combinations of the algorithms and fingerprints were indexed as shown in Table 1. We reported the average performances across 17 activities of MUV dataset and across 10 runs.

Figure 1 shows the average of hit rates and the average number of nodes in each model in a descending order on hit rate. It can be seen that the first five methods i.e. A10, A2, A14, A18 and A6 perform equally well and were considering to yield best performance. They retrieved more active molecules than the other methods of which worst were A11, A3 and A19 in that order. However, the numbers of nodes used were not correlated to the hit rate performance.

Tables 2 and 3 show overall performance of each algorithm across four fingerprints and the overall performance of each fingerprint, respectively. The best results are in bold typeface. Table 2 shows that WELM-JT in a conjunction with k-means-ED was able to retrieve the highest number of active molecules and used smallest number of nodes. Meanwhile, WELM-JT with random weights (conventional) was the worst in accuracy. Considering the case of using the same similarity coefficient, utilizing centroids from k-mean clustering as node representatives achieve better performance than using support vectors that bounded the cluster from SVC. Table 3 shows that encoding fingerprint as ECFP_6 obviously yield the highest hit rates and used a highest number of nodes in the meantime. On the other hand, ECFP_4 only used only the smallest number of nodes. It is clearly seen that FCFP_4 was the worst of retrieving active molecules.

Table 4 shows ranks of the combinations given in Table 1 for each activity class. The ranks was based on the hit rates achieved by each contender on the MUV dataset. The sum of the ranks of each combination determined the overall rank of all activity classes.

Table 1. WELM-JT with different weight selection methods correspond to fingerprints.

Index	Weight selection methods	Fingerprints
A1	Random	ECFP_4
A2	Random	ECFP_6
A3	Random	FCFP_4
A4	Random	FCFP_6
A5	k-means (Euclidean)	ECFP_4
A6	k-means (Euclidean)	ECFP_6
A7	k-means (Euclidean)	FCFP_4
A8	k-means (Euclidean)	FCFP_6
A9	k-means (Jaccard/Tanimoto)	ECFP_4
A10	k-means (Jaccard/Tanimoto)	ECFP_6
A11	k-means (Jaccard/Tanimoto)	FCFP_4
A12	k-means (Jaccard/Tanimoto)	FCFP_6
A13	SVC (Euclidean)	ECFP_4
A14	SVC (Euclidean)	ECFP_6
A15	SVC (Euclidean)	FCFP_4
A16	SVC (Euclidean)	FCFP_6
A17	SVC (Jaccard/Tanimoto)	ECFP_4
A18	SVC (Jaccard/Tanimoto)	ECFP_6
A19	SVC (Jaccard/Tanimoto)	FCFP_4
A20	SVC (Jaccard/Tanimoto)	FCFP_6

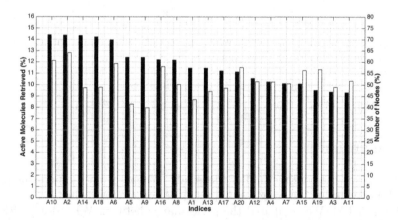

Fig. 1. Active molecules retrieved at top 1 % ranked databased (left y-axis) and percentage of number of nodes used in a model (right y-axis) with respect to indices given in Table 1.

Table 2. Performance of WELM-JT in each weight selection method.

Methods	Hit rates (%)	Number of nodes (%)
k-means-ED	**12.1691**	**50.3754**
SVC-ED	12.0294	52.545
k-means-JT	11.6838	50.9542
SVC-JT	11.5368	52.9862
Random	11.3824	52.0087

Table 3. Performance of WELM-JT on each fingerprint.

Fingerprints	Hit rates (%)	Number of nodes (%)
ECFP_6	**14.2794**	58.1981
ECFP_4	11.8824	**44.2967**
FCFP_6	11.3162	52.7768
FCFP_4	9.7647	53.8227

Furthermore, we evaluated the significance of the results by using the Kendall's Coefficient of Concordance (W) that expresses the degree of agreement among judges in ranking N objects [16]. We calculate the value of W for the data in Table 4 as follow,

$$W = \frac{12 \sum_{i=1}^{N} \bar{R}_i^2 - 3N(N+1)^2}{N(N^2 - 1)}, \tag{5}$$

where $\bar{R}$ is the average of the ranks assigned to the i-th object. In this work, N was 20 (the number of indices in Table 1.) and k was 17 (the number of activity classes in the MUV dataset), thus $W = 0.33$. However, if $N > 7$, the significance value has to be obtained from chi-square distribution. W can be transformed into χ^2 as follows,

$$\chi^2 = k(N-1)W \tag{6}$$

We then acquired $\chi^2 = 70.86$, which is statistically significant at a level of $p < 0.001$. Hence, the rank order is suggested by following:

$$A10 > A2 > A18 > A14 > A6 > A16 > A8 > A13 > A9 > A5 >$$
$$A12 > A17 > A20 > A4 > A15 > A1 > A7 > A3 > A19 > A11$$

5   Conclusion

WELM-JT was the core method of this study. We improved the performance of WELM-JT by using clustering techniques for selecting weights instead of random selection in ELM. We compared the performance between WELM-JT with our proposed clustering-based WELM. The clustering algorithms used in this work were k-means clustering and SVC. Usage of a similarity measure and a distance measure i.e. Jaccard/Tanimoto and Euclidean respectively, was investigated on one of the most challenging datasets for virtual screening, the MUV dataset.

Table 4. The rank of WELM with 20 indices on 17 activity classes

	A1	A2	A3	A4	A5	A6	A7	A8	A9	A10	A11	A12	A13	A14	A15	A16	A17	A18	A19	A20
FX1a	10	9	18	17	8	5	16	20	3	7	12.5	19	6	1	14.5	11	4	2	14.5	12.5
FXIIa	9	6	17	14	3	2	15	13	8	4.5	18	16	7	1	20	11	10	4.5	19	12
CG	17	5.5	11	9	18	3	12.5	8	15	2	15	5.5	20	12.5	5.5	1	19	15	10	5.5
EraP	20	18	4	12	18	15	12	4	18	12	15	8.5	4	1	8.5	8.5	6	2	15	8.5
M1	16	9	11.5	1.5	11.5	14	5.5	3.5	18.5	11.5	11.5	7.5	18.5	16	5.5	1.5	20	16	7.5	3.5
SF1	8	3	20	10.5	8	3	14	6	8	1	16	17	10.5	3	18.5	14	12	5	18.5	14
PKA	11	1	11	8	15	5	15	6	11	4	20	19	11	3	11	7	17.5	2	17.5	15
S1P1	11.5	3.5	16	13.5	10	1	18.5	16	6	2	18.5	11.5	8	3.5	16	8	8	5	20	13.5
RK2	11	2	17	16	6	14.5	14.5	10	12	4	8	5	19.5	18	8	3	19.5	13	8	1
ERaI	15	11.5	18	4.5	8.5	15	15	1	8.5	8.5	19	3	15	4.5	15	6	8.5	2	20	11.5
D1	15.5	5	18	12.5	9	6.5	20	18	15.5	1	18	8	6.5	2	12.5	10	3.5	3.5	12.5	12.5
ERbI	19	1.5	11.5	11.5	11.5	6.5	6.5	17	4	4	16	11.5	11.5	1.5	11.5	11.5	19	11.5	4	19
ERA4	5.5	2	14	19.5	10	2	16	17.5	8	4	14	19.5	9	2	11	12	7	5.5	17.5	14
SF1A	13	13	3	4	17	20	2	1	17	19	13	5	6.5	17	9	9	13	9	6.5	13
FAK	17	5.5	11	20	10	3	17	1	8	3	13	17	5.5	8	13	13	17	3	17	8
HSP	6	12	17.5	19.5	5	8.5	15.5	1.5	8.5	8.5	15.5	1.5	3	13	14	11	4	8.5	19.5	17.5
HIV	10	3	14.5	12.5	9	1.5	11	16	6	8	19	14.5	1.5	7	17.5	17.5	5	4	12.5	20
Sum Rank	214.5	110.5	233	205.5	177.5	125.5	226	159.5	175	104	262	189	163	114	211	155	193	111.5	239.5	201

We encoded molecules using several types of fingerprints i.e. ECFP_4, ECFP_6, FCFP_4, FCFP_6. The experimental results show that WELM-JT with k-means-JT in combination with ECFP_6 fingerprint was the best method for screening biologically active molecules that performed robustly on most of the activity classes.

References

1. Liu, X., Song, H., Zhang, J., Han, B., Wei, X., Ma, X., Cui, W., Chen, Y.: Identifying novel type ZBGs and nonhydroxamate HDAC inhibitors through a SVM based virtual screening approach. Mol. Inf. **29**(5), 407–420 (2010)
2. Chen, B., Harrison, R.F., Pasupa, K., Willett, P., Wilton, D.J., Wood, D.J., Lewell, X.Q.: Virtual screening using binary kernel discrimination: effect of noisy training data and the optimization of performance. J. Chem. Inf. Model. **46**(2), 478–486 (2006)
3. Czarnecki, W.M.: Weighted tanimoto extreme learning machine with case study in drug discovery. IEEE Comput. Intell. Mag. **10**(3), 19–29 (2015)
4. Kudisthalert, W., Pasupa, K.: A coefficient comparison of weighted similarity extreme learning machine for drug screening. In: 2016 8th International Conference on Knowledge and Smart Technology (KST), pp. 43–48. IEEE (2016)
5. Huang, G.B., Zhu, Q.Y., Siew, C.K.: Extreme learning machine: a new learning scheme of feedforward neural networks. In: Proceedings of the 2004 IEEE International Joint Conference on Neural Networks, vol. 2, pp. 985–990. IEEE (2004)
6. Huang, G.B., Zhu, Q.Y., Siew, C.K.: Extreme learning machine: theory and applications. Neurocomputing **70**(1), 489–501 (2006)
7. Jain, A.K.: Data clustering: 50 years beyond K-means. Pattern Recogn. Lett. **31**(8), 651–666 (2010)
8. Hartigan, J.A., Wong, M.A.: Algorithm as 136: a k-means clustering algorithm. J. Roy. Stat. Soc. Ser. C (Appl. Stat.) **28**(1), 100–108 (1979)
9. Vapnik, V.: The Nature of Statistical Learning Theory. Springer, New York (2013)
10. Ben-Hur, A., Horn, D., Siegelmann, H.T., Vapnik, V.: Support vector clustering. J. Mach. Learn. Res. **2**, 125–137 (2002)
11. Cereto-Massagué, A., Ojeda, M.J., Valls, C., Mulero, M., Garcia-Vallvé, S., Pujadas, G.: Molecular fingerprint similarity search in virtual screening. Methods **71**, 58–63 (2015)
12. Gardiner, E.J., Holliday, J.D., O'Dowd, C., Willett, P.: Effectiveness of 2D fingerprints for scaffold hopping. Future Med. Chem. **3**(4), 405–414 (2011)
13. Rohrer, S.G., Baumann, K.: Maximum unbiased validation (MUV) data sets for virtual screening based on PubChem bioactivity data. J. Chem. Inf. Model. **49**(2), 169–184 (2009)
14. Wang, Y., Xiao, J., Suzek, T.O., Zhang, J., Wang, J., Bryant, S.H.: PubChem: a public information system for analyzing bioactivities of small molecules. Nucleic Acids Res. **37**, W623–W633 (2009). gkp456
15. Lee, D., Lee, J.: Support vector clustering (SVC) toolbox
16. Siegel, S.: Nonparametric statistics for the behavioral sciences (1956)

Metabolite Named Entity Recognition: A Hybrid Approach

Wutthipong Kongburan$^{(\boxtimes)}$, Praisan Padungweang,
Worarat Krathu, and Jonathan H. Chan

School of Information Technology,
King Mongkut's University of Technology Thonburi, Bangkok, Thailand
wutthipong.k@mail.kmutt.ac.th,
{praisan.pad,worarat.kra,jonathan}@sit.kmutt.ac.th

Abstract. Since labor intensive and time consuming issue, manual curation in metabolic information extraction currently was replaced by text mining (TM). While TM in metabolic domain has been attempted previously, it is still challenging due to variety of specific terms and their meanings in different contexts. Named Entity Recognition (NER) generally used to identify interested keyword (protein and metabolite terms) in sentence, this preliminary task therefore highly influences the performance of metabolic TM framework. Conditional Random Fields (CRFs) NER has been actively used during a last decade, because it explicitly outperforms other approaches. However, an efficient CRFs-based NER depends purely on a quality of corpus which is a nontrivial task to produce. This paper introduced a hybrid solution which combines CRFs-based NER, dictionary usage, and complementary modules (constructed from existing corpus) in order to improve the performance of metabolic NER and another similar domain.

Keywords: Text Mining · Metabolic interaction · Named Entity Recognition · Hybrid NER

1 Introduction

Biologists need to perform gene and functional assignments towards metabolic interaction network using manual curation, which could be both a labor intensive and time consuming task [1,2]. Text Mining (TM) is currently used as an assisting tool for quickly scanning the entire documents with an essential goal to extract specific terms and concepts from free texts in biological literatures. This task is known as Named Entity Recognition (NER). Typically, most of NER tasks in biomedical domain mainly focused on protein or gene name, metabolic in contrast has received much less attention, especially the interaction between protein (enzyme) and metabolite that reside in cell. The challenges of NER in metabolic domain can be described as follows: the diversity of naming conventions, traditions for such entity, context or position of entity in sentence. These principal factors directly effect on wrong labeling between metabolite and other

© Springer International Publishing AG 2016
A. Hirose et al. (Eds.): ICONIP 2016, Part I, LNCS 9947, pp. 451–460, 2016.
DOI: 10.1007/978-3-319-46687-3_50

entity type. Moreover, there is a considerable amount of possible synonyms of entity, term sharing, cascaded entity, abbreviations as well as errors during tokenization or sentence splitting [3].

A Conditional Random Fields (CRFs)-based technique has been widely used in NER task, and provides several advantages. For example, it performs better in the case of spelling variations in entity names, thus it can recognize the new or unseen entity. CRFs-based NER outperforms other techniques for identifying long and variant entity such as *L-threo-dihydroneopterin*. However, CRFs-based often miss and provides wrong label for tagging *(vitamin)-B2* entity. This is the fact that CRFs-based performance depends on a quality of training corpus which very difficult to cover all possible entity in interested domain. In particular, as can be seen in [4] as well as from our previous work [5], CRFs-based usually provides a recall lower than the precision metric.

The previous work [4] proposed an integration of various TM tools in order to tackle above circumstances. They constructed a corpus of metabolic entity and metabolic events (i.e., events with a mechanical description of the enzyme and metabolite interaction). Firstly, a corpus used to train a classifier of CRFs-based NER in order to extract metabolic entity. After that, a metabolic event was automatically detected by Support Vector Machine-based event extraction tool. Finally, enzyme-metabolite interaction pairs are merged together, and then mapped into a metabolic interaction network. Although this TM framework and their corpus have been successfully used for simplified detection of metabolic events, some of limitations are NER step as well as a diversity of metabolic corpus. In other words, when they test the corpus with the real metabolic super pathways, the result of NER which detects enzyme and metabolite entity is not satisfactory; especially the recall of metabolite entity is quite low.

Typically, NER is one of the key factors in TM framework, because this preliminary task directly effects on overall accuracy of information extraction system. As [4] mentioned that their NER module has a weakness in detecting metabolite entity. In this paper, we concern with the text preprocessing and performance of metabolic NER. One way to improve the performance of CRFs-based NER module is to improve a quality of training corpus. However, this process is laborious task for domain expert. Moreover an appropriate size of a corpus is still controversial. We therefore try to against the performance of previous NER module with a new approach under existing resources.

This paper proposes a hybrid NER to improve the performance of stand-alone CRFs-based NER. Our framework firstly combines a CRF-based NER together with a dictionary matching technique. In addition, a small number of post-processing rules are also introduced to make the annotation results as consistence as possible. Both dictionary and post-processing rules are directly constructed from existing corpus. Finally, we apply Edit Distance algorithm [6] in order to conduct a variant matching. In general, CRFs-based NER provides the precision higher than recall, we consequently emphasis on an improvement of recall metric. The experiment with the metabolic super pathways articles will be used to demonstrate the effectiveness of our proposed method. Furthermore, we adapt this framework to work with another domain to represent a portability of our hybrid NER.

This paper is structured as follows. In the next section, we briefly explain the existing materials, together with the execution of proposed system. After that, we show and discuss the evaluation results on testing dataset. Lastly, in Sect. 4 we conclude the paper and reveal the future works.

2   Materials and Method

The main aim of this work is to improve a performance of NER step in metabolic event extraction module proposed in [4], especially the recall of identifying metabolite entity. Additionally, after we construct a hybrid NER, we try to adapt it to work with another domain. In this section, we firstly introduce the existing materials used in this work i.e. (1) Metabolic Entity (ME) corpus used for training a CRFs-based classifier, and (2) testing corpus of metabolic super pathways. (3) Training and testing corpus of thyroid cancer intervention entity [5]. (4) List of metabolite terms used for dictionary matching. We then explain the step by step of proposed method to construct a hybrid NER system in two domains, illustrated in detail of Fig. 1. The evaluation plan will be mentioned in the last subsection to support the hypothesis that a combination between CRFs-based NER, dictionary matching, as well as complementary modules should provide better performance than the individual CRFs-based system.

2.1   Materials

The ME training corpus[1] is developed by [4]. It is a collection of 271 abstracts and titles (2,288 sentences) from different databases that have been manually annotated for metabolic events by two domain experts who are biologists with different backgrounds. The original version of ME corpus consists of 2,513 gene and protein entities and 1,898 metabolite entities as well as 480 metabolic events (i.e., metabolic production, metabolic consumption, metabolic reaction, and positive regulation). In addition, the ME testing corpus was used for the experiment purpose publicly available at the same source. The ME testing corpus was constructed from a collection of 24 introduction and 27 abstract sections within two super pathways articles in the EcoCyc and PubMed databases, i.e., the super pathway of leucine, valine, and isoleucine biosynthesis (ME testing corpus 1) and the super pathway of pyridoxal 5'-phosphate biosynthesis and salvage (ME testing corpus 2). These articles contain 747 metabolite entities and 675 protein entities.

Another material in this work is the thyroid cancer intervention corpus [5]. This is because we tried to adapt our proposed method to work with another domain. This corpus consists of 143 abstracts of thyroid cancer articles. There are three label classes i.e. intervention, disease, and other class which not refer to

---

[1] www.sbi.kmutt.ac.th/~preecha/metrecon.

454   W. Kongburan et al.

both interested types. The intervention means a treatment and preventive care object or action that performed by doctor or other clinician targeted at a thyroid cancer patient. The disease means a disorder in which the cells of the thyroid gland become abnormal, grow uncontrollably, and form a mass of cells called a tumor. Disease entity includes common name and tumors name, group names, sub types or other forms of thyroid cancer. The training corpus contains 2,473 intervention (564 unique tokens), and 1,996 disease (94 unique tokens). For the testing corpus, we selected 50 abstracts of thyroid cancer article proposed in [5]. It contains 1,079 intervention tokens and 727 disease tokens.

As mentioned earlier, a hybrid NER utilizes a dictionary matching, we therefore need list of metabolite name. In this work, the list of metabolite name from [7] was used as a metabolite dictionary. It consists of 1,853 metabolite entities. Two domain experts (annotator A and B) annotated metabolite expressions in the MEDLINE (2007) abstracts. The target documents are 296 MEDLINE abstracts included in the yeast metabolic network reconstruction [8]. The annotations were restricted to only those names that appear in the context of metabolic super pathways. In contrast to the metabolic domain, adaptation experiment with thyroid cancer intervention domain did not use external dictionary. This is because we not only to demonstrate adaptability of our proposed method but also we aim to indicate that the hybrid NER can work well under existing corpus.

2.2   Proposed Method

This subsection, the step by step of proposed method will be explained according to a tag number that is illustrated in detail of Fig. 1. Note that this framework includes a preprocessing of materials and a testing process as well.

1. Entity extraction: ME corpus is in .ann annotation files format, which contain metabolic sentences. ME corpus includes entity, event, and relation. Each metabolic sentence composes of gene/protein entity and metabolite entity, connected by trigger word. We therefore need to extract only entity terms from the original version of corpus in order to construct a metabolic entity corpus.

2. Training a classifier: In this work, Stanford NER [9], a Java implementation of linear chain Conditional Random Field (CRF) sequence models, is used as a CRFs-based NER tool. We train a model or classifier by metabolic entity corpus obtained from step 1.

3-4. Sentence splitting and Tokenization: Stanford NER provides a set of natural language analysis tools; it relies on period and space for identifying sentence boundary. After that, tokenizer will split a sentence into the constituent meaningful units, called tokens.

5. Automatic annotation: We then use the NER classifier obtained from step 2 to label the testing documents automatically. Each token in article can be labeled as protein, metabolite, and other class.

6. Generate metabolite dictionary: This step is similar to step 1, however only metabolite entity is used to generate metabolite dictionary. This metabolite

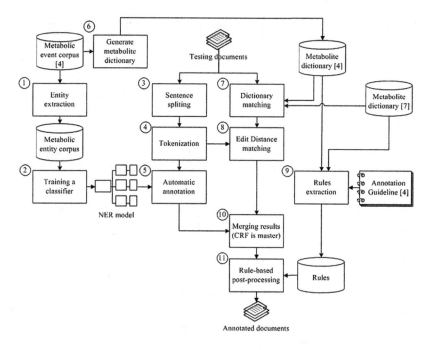

Fig. 1. The overview of proposed framework

list then was combined with metabolite dictionary from [7]. In particular, a number of conditions from annotation guideline [4] and sharing patterns of metabolite entities were used for preparing the dictionary. (1) Keep only character (abbreviation) length more than two. (2) Token which consists of only real number will be removed. (3) Special character should not be at the end of annotation. (4) Pronoun must be excluded. (5) Preposition and determiner are excluded from annotation, if it is at the beginning and at the end of annotation. (6) The interested entity has to be able to resolve itself without any additional or external information.

7. Dictionary matching: We use variantMatcher, simple matching method of dictionary-matching tool, LINNAEUS [10]. This process works with our dictionary from previous step for tagging metabolite entity, any matching tool consequently provide the same results.

8. Edit distance matching: After dictionary-matching step, we obtain metabolite entities of each document which exactly match to metabolite in the dictionary. We then use these metabolites for tagging metabolite entity in the same document again. In this step, we set a threshold in range [0 to 1] of operations required to transform one string into the other in order to get other similar words. This method known as edit distance algorithm, a way of quantifying how dissimilar between two strings.

9. Rule extraction: Some of metabolite entities in dictionary have a sharing pattern or are similar context, thus it can be used to extract a rule.

10. Merging results: We keep the union of all entities extracted by the dictionary-based or the CRFs-based NER. However, in such cases, there are many conflicts between two approaches; we resolve these overlaps by favoring a result of the CRFs-based over one from the dictionary-based. We decided to use this solution, because we observed that in most cases of a conflict the CRFs-based entity tagging is higher accurate than dictionary-based. This is also similar to the resolve method of [11].

11. Rule-based post processing: In fact, there are many rules or pattern of metabolite entity. However, these post-processing rules are directly extracted from the ME corpus, which is used as a case study in the experiments.

(a) If current token is labeled as a protein and next token equal *enzyme* or *protein* or *gene* or *clusters* or *family* or *homodimer* or *transcripts* or *product*, then the next token will be labeled as a protein.

(b) If current token is labeled as a metabolite and next token equal to *ion* or *analogue* or *acid* or *acids* or *nucleotide* or *nucleotides* or *carbohydrate* or *carbohydrates* or *precursor*, then the next token will be labeled as a metabolite.

(c) If current token ends with *ase* or *ases* and it is labeled as a protein and its length more than ten character, then a previous token will be labeled as a protein.

(d) If current token contains *dine*, it will be labeled as a metabolite.

(e) If current token contains *-keto* or *-methyl* or *-dihydroxy*, then a current token and previous token will be labeled as a metabolite.

2.3   A Hybrid NER of Thyroid Cancer Intervention

The adaptation of our proposed method which works with thyroid cancer intervention domain was described as follows. The process starts from step 2 by replacing metabolic entity corpus by thyroid cancer intervention corpus. Step 3 to step 5 are similar to the metabolic domain. For step 6, metabolite dictionary was replaced by intervention dictionary, and metabolite dictionary [7] was excluded. Step 7 to step 11 are similar to the first domain, except the last step which we replaced metabolic rules by a number of intervention rules. In particular, we focus on only the intervention entity. A number of post-processing rules in thyroid cancer intervention domain are described below.

(f) If a current token equals to *inhibitor* or *inhibitors* or *inhibition* or *ablation*, and it is not labeled as a disease or intervention. Then the current token and previous token will be labeled as intervention.

(g) An intervention token which has less than three characters, must be consists of only the letter (without a number). Otherwise it will be labeled as other.

(h) If an intervention token, previous token, and next token begin with uppercase, they will then labeled as other. This is because it might be a proper noun.

2.4   Evaluation Process

As an assessment of the proposed integrated hybrid NER system, two testing corpus of metabolic super pathway and testing corpus of thyroid cancer intervention were used for performance evaluation. We conducted the experiment under the hypothesis that our proposed method is able to compare against the standalone CRF-based approach. In order to evaluate the performance of the systems, standard precision, recall, and F1-score metrics were calculated based on a token criterion by comparing automatic annotating results with the ones provided by manual curators. In the experiments, we focused on only metabolite class in first domain and intervention class in second domain. The experiments were designed as follows: First, we calculated the performance of CRFs-based approach by using standard configurations of Stanford NER (steps 1–5 in Fig. 1), *Approach A*. This is a base case of the experiment. We then evaluated the performance of CRFs-based NER and dictionary matching (steps 1–7), *Approach B*. In addition, we put a number of post-processing rules to demonstrate that it has a positive impact to the overall performance (steps 1–11, except step 8), *Approach C*. Finally, the edit distance module was integrated into the system (steps 1–11), *Approach D*.

3   Results and Discussions

First of all, we evaluated a metabolic entity corpus by using standard configurations of Stanford NER, CRFs-based approach. This is a base case of the experiment that we were trying to compete with. As shown in Table 1, the precision and F1-score of the entity tagger for metabolite was quite high, more than 80 %. This is not surprising because the CRFs-based technique is usually provides high precision. However, the recall of the metabolite entity tagger was lower than our expectations for the first ME testing corpus. The recall showed about 70 %, this result seems to indicate that the entity tagger (i.e., Stanford NER) has a weakness in detecting metabolite entities in an abbreviated form (e.g., Pyridoxine (PN), Pyridoxal (PL), and 4-hydroxy-l-threonine phosphate (HTP)). This is the fact that a corpus contains only a small number of metabolite entities which might not well to deal with an unseen entity.

We then evaluated the performance of a combination result of CRFs-based NER and dictionary matching. This technique provides higher recall than standalone CRFs-based NER, this means that it can be used to reduce false negative rate significantly. In particular, dictionary approach can detect many missing label in an abbreviated form. This demonstrates that the additional list of metabolite entity can fulfill the weakness of CRFs-based NER. However, when the number of detected instances increased, it is more likely to increase more mistakes. This is because not all of metabolite terms are tagged as an interested entity in this corpus. As a result, although list of metabolite can boost recall metric, some of these metabolites still provide precision lower than standalone CRFs-based NER. The conflict label between CRFs-based NER and dictionary-matching will be therefore managed by keeping a result of the first approach.

Table 1. Performance of a hybrid NER on ME testing corpus and thyroid cancer intervention corpus

ME testing corpus 1	Approach A	Approach B	Approach C	Approach D
Recall (%)	70.95	80.93	84.92	**87.80**
Precision (%)	**94.39**	87.32	87.24	85.34
F-score (%)	81.01	84.00	86.06	**86.55**
ME testing corpus 2				
Recall (%)	76.82	83.13	83.28	**83.89**
Precision (%)	**90.03**	87.24	87.12	86.79
F-score (%)	83.03	85.13	85.15	**85.31**
Intervention corpus				
Recall (%)	80.49	83.13	84.90	**86.47**
Precision (%)	**91.83**	87.60	87.65	86.21
F-score (%)	85.78	85.31	86.25	**86.34**

After that, we introduced a number of post-processing rules, as mentioned in Sect. 2. The statistics of post-processing rules is shown in the Table 2. The definitions of the two basic measurement metrics are described follows: *Support*: denotes the frequency of rule. A high value means that the rule involves a great part of dataset. *Confidence*: denotes the percentage of correctness of rule. It is an estimation of conditioned probability.

A performance improvement of rule-based post-processing module is slightly better than without it, look at Table 1. This is the fact that, in some documents, such rules are not meaningful in entity recognition. For example, rule (e): If current token contains *-keto*, *-methyl* or *-dihydroxy*, then a current token and previous token will be labeled as a metabolite. Rule (e) is very meaningful, because it provides a high number of support and confidence score at 62.5 % and 80 % respectively. This means that there are many metabolite entities which can be detected by using this rule in super pathway of leucine, valine, and isoleucine biosynthesis. On the other hand, rule (d) not intertwines with this testing corpus.

Table 2. Confidence and support of post-processing rules

Corpus	Rule	(a)	(b)	(c)	(d)	(e)	(f)	(g)	(h)
ME testing corpus 1	Support	0.042	0.29	0.042	0	0.625	-	-	-
	Confidence	1	0.85	1	0	0.8	-	-	-
ME testing corpus 2	Support	0.25	0	0.125	0.5	0.125	-	-	-
	Confidence	1	0	1	0.5	1	-	-	-
Intervention corpus	Support	-	-	-	-	-	0.45	0.4	0.15
	Confidence	-	-	-	-	-	0.69	1	1

The last experiment is a hybrid NER which works with a complementary module; the edit distance module was combined into the system. We use list of metabolite entities obtained from dictionary matching process to conduct a re-matching. We apply edit distance algorithm with threshold equal to 0 or 1, this refer to exactly match and differ only one character respectively. In last column of Table 1, re-matching module for dictionary entries is useful to expand variant matching of metabolite entities. For example, testing corpus contains *PM* token, nevertheless this abbreviation form not appears in the original version of corpus and dictionary. The above three experiments cannot be used to detect *PM* metabolite entity. Fortunately, the dictionary has *PL* and *PN* metabolite entity, which we can apply edit distance with threshold equal to 1 to capture this entity. In the same way, *Co2+* and *Co2* can take this advantage.

The experiment of thyroid cancer intervention domain is similar to the first domain. The highest precision is the result of standalone CRFs-based, while highest recall and F1-score are the performance of the full option hybrid NER. Typically, a tradeoff between recall and precision is unavoidable. Although the dictionary-matching can increase a number of instances that are relevant and fulfill the gap of CRFs-based approach, there are many mislabeled or error tokens. Based on these experiments, we might conclude that CRFs-based is the most suitable NER model for identifying entity in term of precision. Alternatively, if the purpose of NER is to cast a wider net in order to cover more possible specific keywords of interest, our hybrid method is a reasonable model which outstanding the CRFs-based in terms of recall and F1-score.

Further analysis on confusion matrix, false positive and false negative tokens of metabolite entity reveal that our NER framework classifies metabolite entity as other class more than protein class, in both testing corpus. From these mistakes, we found that most errors between metabolite and other class fall into a case of such metabolite entities are not in dictionary, (e.g., *Zn2+*, *4PHT*, and *isoleucine-valine-requiring*). A tokenization is another reason of mislabeling, for example with default configuration of tokenizer; it splits *Vitamin B* into *Vitamin* and *B* which could not be detected by NER classifier. While most errors between metabolite and protein class fall into a case of entity's boundary. For instance, *L-serine* and *L-threonine* should be labeled as metabolite, however when the next token is *deaminases*, they must be labeled as protein. Another cause of wrong labeled entities is the challenging of entity meaning in different contexts.

4 Conclusion and Future Work

A quality of corpus is one of the most significant factors in efficiency NER. A construction and improvement of a corpus is a time-consuming process, and require much effort of domain expert. We have introduced a hybrid NER for extracting metabolite entity from natural language texts by using existing corpus. Our NER system is a combination of well-known CRFs-based NER and dictionary-matching, together with complementary modules. The edit distance algorithm was used to detect similar words in the same document. In addition, a number of post-processing rules were extracted from sharing patterns of metabolite entities in order to encourage the consistency of NER system. Through this study, we conclude that our hybrid NER produces competitive performance and a broad coverage of metabolite entities compared to the outputs of those systems alone. The further experiments, we adapted our proposed framework to work with thyroid cancer intervention domain. The two steps were changed i.e. intervention dictionary usage and post-processing rules module, our framework can boost performance of the new domain NER as well.

For the future work, we plan to improve the performance of our framework in several ways. For example, applying different tokenization may be particularly appropriate for interested entity. We also might explore the impact of using dictionary terms as features for training CRF model. Furthermore, we may focus on the normalization of entities to know identifiers which facilitates extraction rules as well. Ultimately, we aim at integrating our proposed method with event extraction in TM framework which could be achieved a relationship between entities.

References

1. Baumgartner, W.A., Cohen, K.B., Fox, L.M., Acquaah-Mensah, G., Hunter, L.: Manual curation is not sufficient for annotation of genomic databases. Bioinformatics **23**(13), i41–i48 (2007)
2. Andersen, M.R., Nielsen, M.L., Nielsen, J.: Metabolic model integration of the bibliome, genome, metabolome and reactome of Aspergillus niger. Mol. Syst. Biol. **4**(1), 178 (2008)
3. Hettne, K.M., Williams, A.J., van Mulligen, E.M., Kleinjans, J., Tkachenko, V., Kors, J.A.: Automatic vs. manual curation of a multi-source chemical dictionary: the impact on text mining. J. Cheminform. **2**(1), 3 (2010)
4. Patumcharoenpol, P., Doungpan, N., Meechai, A., Shen, B., Chan, J.H., Vongsangnak, W.: An integrated text mining framework for metabolic interaction network reconstruction. PeerJ **4**, e1811 (2016)
5. Kongburan, W., Padungweang, P., Krathu, W., Chan, J.H.: Semi-automatic construction of thyroid cancer intervention corpus from biomedical abstracts. In: 8th International Conference on Advanced Computational Intelligence (2016)
6. Ristad, E.S., Yianilos, P.N.: Learning string-edit distance. IEEE Trans. Pattern Anal. Mach. Intell. **20**(5), 522–532 (1998)
7. Nobata, C., Dobson, P.D., Iqbal, S.A., Mendes, P., Tsujii, J.I., Kell, D.B., Ananiadou, S.: Mining metabolites: extracting the yeast metabolome from the literature. Metabolomics **7**(1), 94–101 (2011)
8. Herrgard, M.J., Swainston, N., Dobson, P., Dunn, W.B., Arga, K.Y., Arvas, M., Hucka, M.: A consensus yeast metabolic network reconstruction obtained from a community approach to systems biology. Nat. Biotechnol. **26**(10), 1155–1160 (2008)
9. Finkel, J.R., Grenager, T., Manning, C.: Incorporating non-local information into information extraction systems by Gibbs sampling. In: 43rd Annual Meeting on Association for Computational Linguistics, pp. 363–370 (2005)
10. Gerner, M., Nenadic, G., Bergman, C.M.: LINNAEUS: a species name identification system for biomedical literature. BMC Bioinform. **11**(1), 1 (2010)
11. Rocktschel, T., Weidlich, M., Leser, U.: ChemSpot: a hybrid system for chemical named entity recognition. Bioinformatics **28**(12), 1633–1640 (2012)

Improving Strategy for Discovering Interacting Genetic Variants in Association Studies

Suneetha Uppu$^{(\boxtimes)}$ and Aneesh Krishna

Department of Computing, Curtin University, Perth, Australia
Suneetha.uppu@postgrad.curtin.edu.au, A.Krishna@curtin.edu.au

Abstract. Revealing the underlying complex architecture of human diseases has received considerable attention since the exploration of genotype-phenotype relationships in genetic epidemiology. Identification of these relationships becomes more challenging due to multiple factors acting together or independently. A deep neural network was trained in the previous work to identify two-locus interacting single nucleotide polymorphisms (SNPs) related to a complex disease. The model was assessed for all two-locus combinations under various simulated scenarios. The results showed significant improvements in predicting SNP-SNP interactions over the existing conventional machine learning techniques. Furthermore, the findings are confirmed on a published dataset. However, the performance of the proposed method in the higher-order interactions was unknown. The objective of this study is to validate the model for the higher-order interactions in high-dimensional data. The proposed method is further extended for unsupervised learning. A number of experiments were performed on the simulated datasets under same scenarios as well as a real dataset to show the performance of the extended model. On an average, the results illustrate improved performance over the previous methods. The model is further evaluated on a sporadic breast cancer dataset to identify higher-order interactions between SNPs. The results rank top 20 higher-order SNP interactions responsible for sporadic breast cancer.

Keywords: Deep feedforward neural network · Higher-order interactions · High-dimensional genome data · SNP-interactions · And multi-locus analysis

1 Introduction

Genome-wide association studies (GWAS) focuses on locating single-locus SNPs that may be responsible for complex diseases. In reality, the underlying cause of disease susceptibility is influenced by a number of factors acting together or independently. Identifying these interacting factors can provide insights into biological mechanisms and pathways of complex diseases. As a step forward in GWAS, a number of interaction studies brought great hope to identify multi-locus SNP interactions responsible for complex diseases [1]. However, detecting these interactions remains a biggest challenge to be considered in GWAS due to curse-of-dimensionality, and missing heritability [1–3]. Some of these critical issues have been partially addressed by a number of statistical and computational techniques.

© Springer International Publishing AG 2016
A. Hirose et al. (Eds.): ICONIP 2016, Part I, LNCS 9947, pp. 461–469, 2016.
DOI: 10.1007/978-3-319-46687-3_51

Conventionally, regression based approaches are most widely used parametric approaches, which search exhaustively for all the combinations of SNPs in genome-wide data [4]. PLINK [5], Logic Feature Selection (LogicFS) [6], and Park [7] are some of the logistic regression based models used to detect SNP-SNP interactions. Smoothly Clipped Absolute Deviation (SCAD), and Least Absolute Shrinkage and Selection Operator (LASSO) gained some popularity in penalized regression models [8]. Multi-factor Dimensionality Reduction (MDR) [9], Combinatorial Partitioning Method (CPM) [10] and Restricted Partitioning Method (RPM) [11] are some of the data reduction approaches. Some of the pioneering works in tree based approaches are Stratified sampling RF (SRF) [12], EpiForest [13], Random Jungle (RJ) [14], and SNPInterForest [15]. Pattern recognition methods such as Support Vector Machines (SVM) and Neural Networks (NNs) have also gained some popularity in identifying interactions in GWAS [3]. Bayesian Epistasis Association Mapping (BEAM) [16], and Bayesian Network Based Epistatic Association studies (bNEAT) [17] are some of the Bayesian methods used to discover interacting SNPs.

However, identifying these interacting SNPs in high-dimensional genome-wide data is still a challenging problem (search space increases as number of combinations increases) [18]. Flexible optimal methods with efficient computational algorithms can play an important role in revealing the complex architecture behind human diseases [3]. Deep learning is an emerging field that allows systems to learn representations of data at multiple levels of abstraction [19]. They allow the computational models to be fed with raw data and discover the representations needed for the classification automatically using general-purpose learning procedures [19]. Their applications in bioinformatics includes biomedical imaging and biomedical signal processing [20]. However, none of the studies progressed towards predicting multi-locus SNP-SNP interactions are responsible for complex diseases. In the previous research [21, 22], a deep learning model was trained to detect two-locus interactions between SNPs. The proposed model was trained and validated to predict the performance of the model in terms of accuracy and execution time. A number of experiments are performed on simulated datasets under different scenarios and a real dataset. The results showed remarkable improvements in predicting SNP interactions over some of the existing methods, such as MDR, RF, SVM, NN, Naïve Bayes', classification based on predictive association rules (CPAR), logistic regression (LR), and Gradient Boosted Machines (GBM). However, the performance of the trained model to identify three or more higher-order interactions is unknown. Hence, in this paper, the method is evaluated for three to ten loci SNP interactions in high-dimensional data. It reduces the high-dimensional genome-wide data to low-dimensional data using principal components analysis (PCA). The method is further extended for unsupervised learning. The extended model is evaluated on a sporadic breast cancer dataset to identify higher-order interactions between SNPs. The experimental results demonstrated improved performance over the previous methods. The results ranked top 20 higher-order interactions among 10 SNPs, which are included in 5 different estrogen-metabolism genes.

Rest of this paper is organised in the following sections as below. The workflow of the extended deep learning approach is introduced in Sect. 2. Further, the general steps of the core algorithm are summarised in this section. Section 3 illustrates the results with discussion. Finally, Sect. 4 presents conclusion and future works.

2   Methods

2.1   A Deep Learning Network

Figure 1 illustrates the extended deep learning method. This method is studied in detail in the previous work [21, 22]. The trained model is based on multi-layered feedforward neural networks [23].The general steps of the deep learning method (proposed in the previous work) are summarised briefly in this section. The trained deep learning network comprises of an input layer, multiple hidden layers and an output layer [19]. The basic computational units of the network are neurons which are biologically inspired from the human brain. The neurons in the input layer serves as the input to the hidden layers; whereas, the output of the hidden layers serves as input to the output layer. The computation in each layer transforms the representation of data into more abstract manner. In each computational unit, the weighted combinations of the inputs are combined together with a bias. The generalised parametric linear equation is expressed as: $\varphi = b + wx$, where b is bias, and w is the weight vector of the input x. The weighted sum transfer function of ith neuron in a layer is φ_i, and is represented as follows [19]:

$$\varphi_i = \sum_{i=1}^{n} w_i x_i + b_i \tag{1}$$

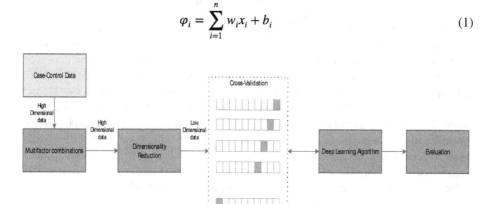

Fig. 1. Overview of the deep learning method (extension of [21, 22])

The non-linear activation function such as binary, rectifier, hyperbolic tangent, or sigmoid is applied to the weighted input sum to compute the output values of the layer. In the proposed model, the hyperbolic tangent activation function (a re-scaled and shifted logistic function whose range is $f(\cdot) \in [-1, 1]$) is used to transmit the classification output by the connected neurons. This function provides the algorithm to converge faster, and is proposed by [23]:

$$f(\varphi) = \tanh(\varphi) = \frac{sinh\varphi}{cosh\varphi} = \frac{e^{\varphi} - e^{-\varphi}}{e^{\varphi} + e^{-\varphi}} = \frac{e^{2\varphi} - 1}{e^{2\varphi} + 1} \tag{2}$$

More suitable representation of the data is learnt to adapt the weights by minimising the loss by using backpropagation. The output error is calculated by using cross entropy cost function and weights are trained by backpropagation. For training sample j, the cross entropy objective function is provided by [24]:

$$E = -\sum_{x} (\log y_j * t_j + \log(1 - y_j) * (1 - t_j)) \tag{3}$$

$$\frac{\partial E}{\partial y_j} = \frac{t_j - y_j}{y_j(1 - y_j)} \tag{4}$$

where, t and y represents the predicted and actual outputs respectively. The sum is the overall training input x. The weights are trained by backpropagation along with stochastic gradient decent (SGD) optimisation algorithm. This reduces the network error by adjusting the weights between two neurons. In addition, a lock-free approach is used to parallelise SGD to improve the efficiency of memory and execution time of the algorithm [24, 25].

2.2   Multifactor Combinations and Dimensionality Reduction

A SNP is the variation of a nucleotide at a specific location of DNA in the chromosome. SNPs are biallelic markers that contain two alleles (one with majority allele (A) and other with minority allele (a)). Due to duplication of DNA in each cell, the genotype combinations of biallelic SNPs are common homozygous (AA), heterozygous (Aa/aA), and variant homozygous (aa). Numerically, AA represented by '0', Aa/aA by '1', and aa represented by '2'. The input data are case-control based data with genotype combinations of SNPs along with their class labels. The class labels are represented by '0' for controls and '1' for cases. Consider k genetic factors are selected from k-dimensional space, where, n = k levels, and r = 1, 2, ... (k − 1), number of locus chosen from N factors in total. Within the current training set, all the factors are combined together in k-dimensional space, whose combination function is represented by:

$$f(N) = \sum_{n=k,r=1}^{r=(k-1)} nc_r = \sum_{n=k,r=1}^{r=(k-1)} \frac{n!}{(n-r)!r!} \tag{5}$$

The multifactor genotype combinations at different loci are combined together to improve the prediction accuracy of the model.

Evaluating multi-locus combinations of SNPs in genome-wide data increase exponentially. Finding an optimal combination among an unusually large number of combinations is not feasible within the existing computational techniques. That is, for a study of 300,000 SNPs in GWA, there will be $4.5*10^{10}$ two-way interactions and $4.5*10^{15}$ three-way interactions to be examined [18, 26]. This computational challenge has been addressed in this method by using Principal Component Analysis (PCA) [27]. PCA is used to reduce high-dimensional genome-wide data into a low-dimensional data, and applied to the proposed deep learning network [22] to detect higher-order SNP interactions associated with a disease. Further, the extended model is trained for unsupervised feature learning, and to detect anomalies in the data by using deep autoencoder [24]. It learns nonlinearly from the reduced representation of the actual data. The model is trained on a training data by ignoring class labels. Reconstruction error is computed between the output and input layers with anomaly detection to determine the outliers for higher-order interacting SNP test data.

3   Experimental Results

Multiple experiments were conducted over the proposed model using simulated datasets under two scenarios and breast cancer dataset. The detailed study on simulated datasets and real dataset for two-locus interactions are represented in the previous work [21, 22]. The objective of this study is to evaluate the model for detecting higher-order interactions (three-locus to ten-locus interactions) responsible for complex diseases. The deep feedforward neural network is trained and analysed in R using H2o package [24]. The model comprises of an input layer, three hidden layers and an output layer. Each hidden layer is trained with 50 computational units. The model processes 1000 epochs per 1000 iterations on 10 compute nodes. By default, entire data is processed on every node locally by shuffling the training samples in individual iteration.

The model is validated and analysed by passing various non-linear activation functions such as rectifier, tanh, maxout, rectifier with dropout, tanh with dropout, and maxout with dropout [22]. Among all, tanh with dropout has high prediction accuracy and low classification error. Hence, tanh with dropout is chosen as an appropriate activation function to achieve better approximation. The input drop out ratio is set to 0.2 and hidden dropout ratios for three hidden layers are set to 0.5 respectively [22]. The extended model is evaluated for one-locus to ten-locus SNP interactions individually on

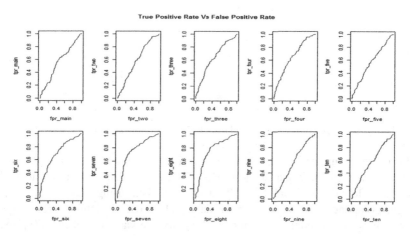

Fig. 2. True positives and false positives for one-locus to ten-locus interactions

sporadic breast cancer data [9]. Figure 2 plots true positives vs. false positives from one-locus to ten-locus SNP interactions. True positive rate constantly rises as false positive rate increases. Furthermore, the method is evaluated by binding all higher-order combinations together and predicted on test data for identifying interacting SNPs responsible for breast cancer. Top 20 highly ranked higher-order SNP interactions are shown in Fig. 3. Results show that two-locus SNP interaction (CypIAIm2_Cyp1B1.119) being highly associated with the breast cancer. Figure 4(a) and 4(b) shows the performance of the model on higher-order interactions during training and validation respectively.

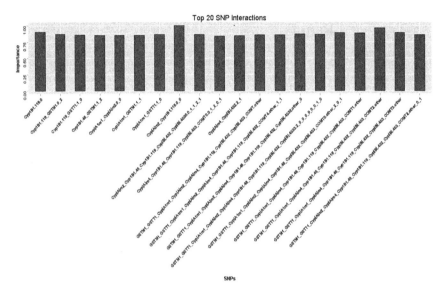

Fig. 3. Top 20 higher-order SNP-interactions for sporadic breast cancer data

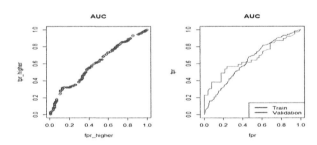

Fig. 4. Performance of higher-order model (a) AUC for training data (b) AUC for validation data

The model is further evaluated for unsupervised learning tasks such as dimensionality reduction and feature learning [24]. High dimensional data (888-dimensional data) is reduced to low dimensional data (50-dimensional data) using PCA. PCA computes principal components of breast cancer data and whose deviations are computed. Figure 5 shows the standard deviations and cumulative variances of first 50 principal components. It is observed that first 10 or 20 components cover the majority of variance of the breast cancer dataset. Deep autoencoder is used for unsupervised feature learning by discovering the anomalies in the reduced representation of the original data. Categorical offsets and reconstruction error between output layer and input layer are plotted in Fig. 6.

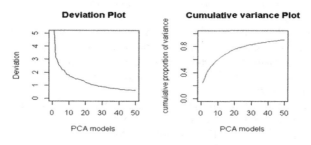

Fig. 5. (a) Deviation plot and (b) Cumulative variance plot of PCA

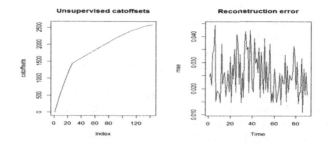

Fig. 6. (a) Catoffset plot (b) Reconstruction error plot for unsupervised learning

4   Conclusion and Future Work

In this paper, previously trained deep learning network is further extended and trained for detecting higher-order SNP interactions responsible for complex diseases. The approach is validated and analysed on a sporadic breast cancer dataset. The experimental results showed remarkable improvements in predicting three or more loci interactions over some of the existing machine learning approaches. Top 20 highly ranked interacting SNPs are illustrated for future studies and their role in the disease manifestation. The method is further evaluated for unsupervised learning tasks such as dimensionality reduction and feature learning. High dimensional data is reduced to low dimensional data using PCA. Deep autoencoder is used for unsupervised feature learning by discovering the anomalies in the reduced representation of the original data. Future studies will investigate the performance of the model in the presence of noise and family-based datasets. Parallel computational techniques and current optimisation algorithms will be explored and incorporated into the model to reduce the execution time.

References

1. Cordell, H.J.: Detecting gene–gene interactions that underlie human diseases. Nat. Rev. Genet. **10**(6), 392–404 (2009)
2. Van Steen, K.: Travelling the world of gene–gene interactions. Briefings Bioinform. **13**(1), 1–19 (2012)
3. Upstill-Goddard, R., et al.: Machine learning approaches for the discovery of gene–gene interactions in disease data. Briefings Bioinform. **14**(2), 251–260 (2013)
4. Chen, C.C., et al.: Methods for identifying SNP interactions: a review on variations of Logic regression, random forest and Bayesian logistic regression. IEEE/ACM Trans. Comput. Biol. Bioinform. **8**(6), 1580–1591 (2011)
5. Purcell, S., et al.: PLINK: a tool set for whole-genome association and population-based linkage analyses. Am. J. Hum. Genet. **81**(3), 559–575 (2007)
6. Schwender, H., Ickstadt, K.: Identification of SNP interactions using logic regression. Biostatistics **9**(1), 187–198 (2008)
7. Park, M.Y., Hastie, T.: Penalized logistic regression for detecting gene interactions. Biostatistics **9**(1), 30–50 (2008)
8. Niel, C., et al.: A survey about methods dedicated to epistasis detection. Front. Genet. **6**, 285 (2015). doi:10.3389/fgene.2015.00285
9. Ritchie, M.D., et al.: Multifactor-dimensionality reduction reveals high-order interactions among estrogen-metabolism genes in sporadic breast cancer. Am. J. Hum. Genet. **69**(1), 138–147 (2001)
10. Nelson, M., et al.: A combinatorial partitioning method to identify multilocus genotypic partitions that predict quantitative trait variation. Genome Res. **11**(3), 458–470 (2001)
11. Culverhouse, R., Klein, T., Shannon, W.: Detecting epistatic interactions contributing to quantitative traits. Genetic Epidemiol. **27**(2), 141–152 (2004)
12. Wu, Q., et al.: SNP selection and classification of genome-wide SNP data using stratified sampling random forests. IEEE Trans. Nanobiosci. **11**(3), 216–227 (2012)
13. Jiang, R., et al.: A random forest approach to the detection of epistatic interactions in case-control studies. BMC Bioinform. **10**(Suppl. 1), S65 (2009)
14. Schwarz, D.F., König, I.R., Ziegler, A.: On safari to random jungle: a fast implementation of random forests for high-dimensional data. Bioinformatics **26**(14), 1752–1758 (2010)
15. Yoshida, M., Koike, A.: SNPInterForest: a new method for detecting epistatic interactions. BMC Bioinform. **12**(1), 469 (2011)
16. Zhang, Y., Liu, J.S.: Bayesian inference of epistatic interactions in case-control studies. Nat. Genet. **39**(9), 1167–1173 (2007)
17. Han, B., Chen, X.-W.: bNEAT: a Bayesian network method for detecting epistatic interactions in genome-wide association studies. BMC Genom. **12**(Suppl. 2), S9 (2011)
18. Padyukov, L.: Between the Lines of Genetic Code: Genetic Interactions in Understanding Disease and Complex Phenotypes. Academic Press, Waltham (2013)
19. LeCun, Y., Bengio, Y., Hinton, G.: Deep learning. Nature **521**(7553), 436–444 (2015)
20. Min, S., Lee, B., Yoon, S.: Deep learning in bioinformatics. arXiv preprint arXiv:1603.06430 (2016)
21. Uppu, S., Krishna, A., and Gopalan, P.R., Towards deep learning in genome-wide association interaction studies. In: Pacific Asia Conference on Information System, Taiwan (2016). ISBN 9789860491029
22. Uppu, S., Krishna, A., Gopalan, P.R.: A deep learning appraoch to detect SNP interactions. J. Softw. (accepted), Will be published in vol. 11, no. 10, October 2016

23. Bengio, Y., I.J. Goodfellow, and A. Courville, Deep Learning. An MIT Press electronic book, version 10–18 (2015). http://www.deeplearningbook.org/
24. Candel, A., et al.: Deep Learning with H2O (2016). http://h2o.ai/resources/
25. Recht, B., et al.: Hogwild: a lock-free approach to parallelizing stochastic gradient descent. In: Advances in Neural Information Processing Systems (2011)
26. Uppu, S., Krishna, A., Gopalan, P.R.: Detecting SNP interactions in balanced and imbalanced datasets using associative classification. Aust. J. Intell. Inf. Process. Syst. **14**(1), 7–18 (2014)
27. Jolliffe, I.: Principal Component Analysis. Wiley Online Library (2002)

Improving Dependency Parsing on Clinical Text with Syntactic Clusters from Web Text

Xiuming Qiao, Hailong Cao$^{(\boxtimes)}$, Tiejun Zhao, and Kehai Chen

School of Computer Science and Technology,
Harbin Institute of Technology, Harbin, China
{xmqiao,hailong,tjzhao,khchen}@mtlab.hit.edu.cn

Abstract. Treebanks for clinical text are not enough for supervised dependency parsing no matter in their scale or diversity, leading to still unsatisfactory performance. Many unlabeled text from web can make up for the scarceness of treebanks in some extent. In this paper, we propose to gain syntactic knowledge from web text as syntactic cluster features to improve dependency parsing on clinical text. We parse the web text and compute the distributed representation of each words base on their contexts in dependency trees. Then we cluster words according to their distributed representation, and use these syntactic cluster features to solve the data sparseness problem. Experiments on Genia show that syntactic cluster features improve the LAS (Labled Attachment Score) of dependency parser on clinical text by 1.62 %. And when we use syntactic clusters combining with brown clusters, the performance gains by 1.93 % on LAS.

1 Introduction

Dependency parsing is a main NLP (Natural Language Processing) task of analyzing dependency relations ($head \rightarrow dependent$) between words in one sentence, as shown in Fig. 1. It is widely applied in entity disambiguation, information extraction, question answering, negation detection and so on.

The state-of-the-art dependency parsers [11,16], take English dependency parsing as an example, perform well on news data. With the large annotated treebanks, it is easy to train a dependency parser with high performance via supervised learning method. But the treebanks for clinical text are rather rare. And due to the particularity and diversity of words in clinical text, containing many long sequence of digits, complex noun phrases and appositives [2], the lexical sparseness problem in dependency parsing for clinical text is very serious.

Many researchers have devoted to consider clusters which is at a coarser level than word themselves. Koo et al. [7] demonstrates the effectiveness of brown clusters derived from large unannotated corpus in dependency parsing. Hogenhout et al. [5] studies the word clusters on the basis of their syntactic behaviour using Treebank data. Sagae and Gordon [18] derive syntactic word clusters from phrase-structured parsing trees. The scale and domain of Treebank data are limited, but the scale and diversity of unlabeled web text are very huge. More important, web text is free.

© Springer International Publishing AG 2016
A. Hirose et al. (Eds.): ICONIP 2016, Part I, LNCS 9947, pp. 470–478, 2016.
DOI: 10.1007/978-3-319-46687-3_52

Recently, word embedding has been very popular and applied in many NLP tasks [14,20]. It is a kind of distributed representation for word which represent a word by a dense low-dimensional and continuous vector according to the co-occurrence of words [13]. Word embedding can also be computed according to the context in dependency trees [1,8]. Dependency-based word embedding exhibit more functional similarity than original word embedding. We derive syntactic clusters according to the dependency-based word embedding. If two words always have the similar head or children in dependency trees, they belong to the same syntactic cluster.

Fig. 1. An example of dependency tree.

First, we parse web sentences and extract dependency contexts for each word according to the dependency trees. According to the dependency contexts, we represent the words by a vector. Then we derive the syntactic cluster of each word according to its distributed representation. Finally, syntactic cluster features are added to the dependency parser. After adding the syntactic cluster feature, the performance of MSTParser on Genia test set is improved by 1.62 %, showing that the syntactic clusters can tackle the data sparseness in bio dependency parsing. We make three contributions in this paper:

- We derive syntactic clusters from dependency-based word embeddings gained from web data.
- We augment the bio dependency parser with syntactic clusters as external knowledge.
- We also combine syntactic clusters with brown clusters to tackle the data sparseness in bio parsing more fully.

2 Graph-Based Parsing Model

Given a sentence x, dependency parsing is to produce its dependency tree y, assigning head-dependent relations between all the words in x. The score of a tree y is the sum of its subgraph scores [11]. The score of a subgraph is computed according to its high-dimensional feature vector $\mathbf{f}(x, g)$ and the feature weights in the weight vector $\mathbf{w}$.

$$s(x,y) = \sum_{g \in y} s(x,g) = \sum_{g \in y} \mathbf{w} \cdot \mathbf{f}(x,g) \qquad (1)$$

The decoding process is to find the max spanning tree y^*, that maximize the score $s(x, y)$ in the set of all the possible trees $Y(x)$.

$$y^* = \arg\max_{y \in Y(x)} \sum s(x, y) \tag{2}$$

The feature weight vector $\mathbf{w}$ is learned during training using the Margin Infused Relaxed Algorithm (MIRA) [11]. MIRA attempts to keep the norm of the change to the parameter vector as small as possible, subject to correctly classifying the instance under consideration with a margin at least as large as the loss of the incorrect classifications.

3 Syntactic Clusters

In this section, we will introduce how to gain syntactic clusters from unlabeled text and apply in dependency parsing.

3.1 Definition of Syntactic Clusters

According to the Distributional Hypothesis, words that occur in the same contexts tend to have similar semantics [4]. The underlying idea that *a word is characterized by the company it keeps* was popularized by Firth [3]. Then we believe that if two words have similar syntactic contexts in dependency trees, then they have functional similarities.

The syntactic cluster represents that a series of words have similar syntactic functions. For example, in two phrases of "drink a bottle of juice" and "drink a bottle of potion", *juice* and *potion* have the same syntactic role. But they are dissimilar in semantics.

If *potion* does not occur in the training data and *juice* occurs, but they share the same syntactic cluster, then we can derive the similar dependency structure for *potion* as *juice*. For dependency parsing, a syntactic analysis task, syntactic clusters are more meaningful than semantic clusters and experiments show that it is true.

3.2 Dependency Based Word Embedding

Recently, distributed representation of words has been applied in many NLP tasks [14,20]. Google released word2vec[1], a toolkit that represent a word by a vector according to its context. During training, word2vec has two models of CBOW and Skip-gram, which all use a simple neural network [13]. The difference is that Skip-gram uses current word to predict its contexts and its performance is better for low-frequency words, and CBOW uses contexts to predict current words and its speed is faster. There are many low-frequency words in biomedical

---

[1] http://code.google.com/p/word2vec/.

text, so we use Skip-gram model. Figure 2 shows the structure of skip-gram model.

Given a word sequence $w_1, w_2, .., w_T$, the goal of Skip-gram model is to maximize the objective function:

$$\frac{1}{T} \sum_{t=1}^{T} \sum_{-c \leq j \leq c, j \neq 0} log P(w_{t+j}|w_t), \tag{3}$$

in which c is the size of window, the inside sum is the log probability of predicting the word w_{t+j} correctly, and the outside sum represents traversing all words of training data. Each word w has two parameterized vectors u_w and v_w. u_w is the input vector for w and v_w is the output vector for w. Given w_i, the probability of predicting w_j correctly is the softmax function:

$$P(w_i|w_j) = \frac{exp(v_{w_i}^T u_{w_j})}{\sum_{t=1}^{W} exp(v_t^T u_{w_i})}, \tag{4}$$

where W is the length of vocabulary list.

Given a dependency parsing tree T, if the word w that has k modifiers $m_1, m_2, ..., m_k$ and the head h, then the contexts of w are $(m_1, lable_1)$,..., $(m_k, lable_k)$ and $(h, lable_h^I)$. $lable_k$ is the label of dependency relation for w and m_k. And $lable_h^I$ represents that w is the modifier and h is the head in this dependency relation.

3.3 Clustering Method

After representing each word as a dense, low-dimensional and real-value vector, we derive syntactic clusters according to the value of each dimension. The goal of clustering is to find the set C of cluster centers $c \in R^m$ and $|C| = k$. The objective is to minimize the following function over all examples x in the word set of X:

$$min \sum_{x \in X} \|f(C, x) - x\|^2 \tag{5}$$

According to the vectors of cluster center c and x, we can compute their Euclidean distance. Then $f(C, x)$ returns the nearest cluster center $c \in C$ to the word x.

The classic k-means algorithm is very expensive for large web data, so we use mini batch k-means as our clustering method [19]. Mini-batch k-means uses stochastic gradient descent (SGD) which converges quickly on large data sets. The input of mini-batch k-means algorithm are the number of clusters k, the batch size b, the iteration number t and the data set X.

3.4 Application of Syntactic Clusters

When we parse a sentence and there is a word that never occurs in the treebank, we cannot judge which word it will depend on. But if there are many words

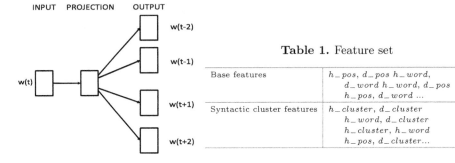

INPUT   PROJECTION   OUTPUT

w(t-2)

w(t-1)

w(t)

w(t+1)

w(t+2)

Table 1. Feature set

Base features	$h\_pos, d\_pos\ h\_word,$ $d\_word\ h\_word, d\_pos$ $h\_pos, d\_word\ ...$
Syntactic cluster features	$h\_cluster, d\_cluster$ $h\_word, d\_cluster$ $h\_cluster, h\_word$ $h\_pos, d\_cluster...$

Fig. 2. The Skip-gram model.

that belong to the same syntactic cluster with this word, we can easily decide which word it will depend on. For example, in the sentences of "He drinks a bottle of juice" and "She drinks a bottle of potion", *He* and *She* are in same syntactic cluster, *juice* and *potion* are dissimilar in semantics but in the same syntactic cluster. If the second sentence does not occur in the training set, then we can derive its dependency structure according to that of the first sentence. We integrate syntactic clusters as external features into graph-based dependency model.

The feature templates of baseline parser [12] and ours parser are shown in Table 1. h is the head of a dependency relation and d is the child. h_pos is the POS tag of h and h_word is the lexical of h. $h\_cluster$ is the syntactic cluster of h and these symbols also apply to d. We combine syntactic cluster features with word and POS tag features.

4   Experiment and Results

4.1   Experimental Settings

In order to show the effectiveness of syntactic cluster features, we conduct dependency parsing experiments on Genia [6], which is a semantically annotated corpus for bio-textmining. We follow the division of McClosky and Charniak [10]. The data we use to gain word syntactic clusters is the word embedding data provided by Levy and Goldberg [8], containing about 175000 words. These word embeddings are trained on English Wikipedia. Each word embedding contains 300 dimensions. According to the vectors, we derive 200 syntactic clusters. The discriminative dependency parser we use is MSTParser [11] which uses first order features.

4.2   Results

We evaluate the parser by UAS, UCAS, LAS and LCAS. As the standard practice, we exclude punctuation tokens of each sentence. UAS (unlabeled attachment score) is the ratio of words that have right heads. UCAS (unlabeled complete attachment score) is the ratio of sentences whose words all have the right heads. LAS (labeled attachment score) and LCAS (labeled complete attachment score) are based on UAS and UCAS respectively, and they look at both heads and dependency relation labels.

Table 2 shows the performance of different configurations. Base Parser represents the MSTParser with first order features [12]. We also conduct the experiment that uses the Genia training set and WSJ training set to train the MSTParser and test on Genia test set and it is denoted by Base+WSJ. Base+Brown is the MSTParser augmented with brown cluster features and the clusters are drawn from BLLIP corpus using Brown clustering algorithm [7]. Base+Syn is our system that using base features and syntactic cluster features. Finally, we also combine base features, brown cluster and syntactic cluster features and this configuration is denoted as Base+Syn+Brown.

From Table 2, we can see that our parser wins the Base Parser by 0.51 % in UCAS, 1.62 % in LAS and 6.62 % in LCAS. Base+Brown performs better than base parser in UAS, LAS and LCAS. But our system performs better than Base+Brown in UCAS and LCAS. The parser combined brown clusters and syntactic clusters performs the best. We can conclude that knowledge can be mined from different view and we should use different kinds of knowledge to improve dependency parsing for bio text.

The performance of Base+WSJ shows that it does not help if we add the newswire treebanks into the training data directly because of the domain difference between news and clinical text. Instead, we can use these newswire treebank to train a parser and get useful knowledge from unlabeled web data. Web data is very rich in scale and diversity, so it contains much useful knowledge to be mined.

We also compare the performance of our systems with other domain adaptation works. Plank et al. [17] uses topic models (Plank[2011] topic) and words (Plank[2011] words) as the similarity measures to select the relevant data for a target article from unknown domains. Ma[2013] is a kind of feature augmentation approach in which the features are acquired from subtrees of auto-parsed target domain data [9] (Table 3).

In more detail, we compare the result of the baseline parser Base Parser and our parser Base+Syn in each kind of dependency relation. Table 4 shows the number of errors and error reduction ratio for each dependency relation. We can see that the syntactic clusters notably works on *vc*, *obj*, *vmod* and *amod*.

4.3   Analysis

As observed in the results of syntactic clusters, the performance is rather satisfactory. It is much more distinguishable than POS tags. For one example, *their, her, his, our, its* are in the same cluster and their POS tags are same too. *merit* and *interval* have the same POS tag $NN(noun)$, but their syntactic clusters are different. *interval* is likely to be a child of a preposition such as *during*, *for* and *at*, while *merit* seems to depend on present tense verb such as *is, are*.

Table 2. Results of different systems.

System	UAS	UCAS	LAS	LCAS
Base Parser	88.08	25.37	86.05	18.82
Base+WSJ	87.80	25.44	83.64	12.06
Base+Brown	88.26	25.37	87.88	24.93
Base+Syn	88.06	25.88	87.67	25.44
Base+Syn+Brown	88.38	**26.4**	**87.98**	**25.59**
Plank[2011] topic	-	-	86.77	-
Plank[2011] words	-	-	86.44	-
Ma[2013]	**88.4**	-	87.1	-

Table 3. Examples of some words in the same syntactic clusters.

Word	Similar words
Interval	contact, line, precursor
Merit	note, thought
Measured	detected, assayed,observed
Co-operative	autonomous,provisional
Bloodstream	groove,axis,nucleus
Lipase	silica, serum,sucrose

Table 4. Error reductions of two systems in different dependency relations.

Relation	Explanation	Count	Baseline	Ours	Reduction
nmod	noun modifier	16509	2198	2176	1 %
vmod	verb modifier	5625	941	726	22.85
pmod	preposition modifier	4611	408	388	4.9 %
obj	object	1239	225	131	41.78 %
root	root of the sentence	1360	82	76	7.32 %
vc	auxiliary verb - main verb	904	44	9	79.55 %
amod	adjective modifier	854	417	314	24.7 %
sbar	complementizer - verb	655	83	67	19.28 %
dep	default classification	123	48	43	10.42 %

What's more, *measured* and *observed* are very different in semantic, but they are in the same syntactic cluster. Above all, the syntactic clusters carry different information from POS tags.

5   Related Work

Many researchers have focused on dependency parsing for biomedical text. McClosky and Charniak [10] use self-training to adapt the dependency parser from news domain to bio domain. Plank et al. [17] selects the most relevant data from Treebank for the article of target domain according to topic models or words frequency. Ma and Xia [9] acquire features from subtrees extracted in the auto-parsed dependency trees and retrain the target parser with the union of training data in the source and target domain.

In our parser for bio, the training data is clinical Treebank combining with web raw text. Clinical data has many various proper words, so the parser suffers from data sparseness. Additional knowledge has been explored to make up data sparseness in dependency parsing, such as brown clusters [7]. The brown cluster is computed according to the contextual words within one sentence. Our syntactic clusters are gained according to corresponding dependency trees of a sentence. Hogenhout et al. [5] studies the word clusters on the basis of their syntactic behaviour using Treebank data. Sagae and Gordon [18] derive syntactic word clusters from phrase-structured parsing trees, while we use dependency trees. Bansal and Curran [15] get the continuous representation of dependency links from dependency trees and apply them in dependency parsing for web text. Bansal et al. [1] and Levy and Goldberg [8] get the distributed representation of each word according to the dependency context. These studies inspire us to acquire syntactic knowledge from unlabeled web text to improve bio dependency parsing.

6 Conclusions

In this paper, we presented a simple yet effective method to improve dependency parsing for clinical text. We mined syntactic cluster knowledge from dependency-based word embedding to ease data sparseness in discriminative dependency parsing. We presented significant improvements on the dependency parsing for clinical text and also analyzed the performance of each kind of dependency relation.

In the future, we will try to derive soft syntactic clusters, because a word may belong to different syntactic roles. And we will gain more syntactic knowledge from web data to help dependency parsing for biomedical text.

Acknowledgments. This work is supported by the by the project of National Natural Science Foundation of China (No. 91520204, No. 61572154) and the project of National High Technology Research and Development Program of China (863 Program) (No. 2015AA015405).

References

1. Bansal, M., Gimpel, K., Livescu, K.: Tailoring continuous word representations for dependency parsing. In: Proceedings of the ACL 2014, pp. 809–815 (2014)
2. Dredze, M., Blitzer, J., Talukdar, P.P., Ganchev, K., Graca, J.a., Pereira, F.: Frustratingly hard domain adaptation for dependency parsing. In: Proceedings of the CoNLL Shared Task Session of EMNLP-CoNLL 2007, pp. 1051–1055 (2007)
3. Firth, J.: A Synopsis of Linguistic Theory 1930–1955. Studies in Linguistic Analysis, pp. 1–32 (1957)
4. Harris, Z.: Distributional structure. Word **10**(23), 146–162 (1954)
5. Hogenhout, W.R., Matsumoto, Y., Fast, J.S.W.: A preliminary study of word clustering based on syntactic behavior. In: Proceedings of the Computational Natural Language Learning, pp. 16–24 (1997)
6. Kim, J.D., Ohta, T., Tateisi, Y., Tsujii, J.: Genia corpus-semantically annotated corpus for bio-textmining. Bioinformatics **19**(Suppl 1), i180–i182 (2003)
7. Koo, T., Carreras, X., Collins, M.: Simple semi-supervised dependency parsing. In: Proceedings of ACL-08: HLT, pp. 595–603 (2008)

8. Levy, O., Goldberg, Y.: Dependency-based word embeddings. In: Proceedings of the 52nd Annual Meeting of the Association for Computational Linguistics, pp. 302–308 (2014)

9. Ma, X., Xia, F.: Dependency parser adaptation with subtrees from auto-parsed target domain data. In: Proceedings of the 51st Annual Meeting of the ACL, pp. 585–590 (2013)

10. McClosky, D., Charniak, E.: Self-training for biomedical parsing. In: Proceedings of ACL-08: HLT, Short Papers, pp. 101–104 (2008)

11. McDonald, R., Crammer, K., Pereira, F.: Online large-margin training of dependency parsers. In: Proceedings of the 43rd Annual Meeting on ACL 2005, pp. 91–98 (2005)

12. Mcdonald, R., Pereira, F.: Online learning of approximate dependency parsing algorithms. In: Proceedings of EACL, pp. 81–88 (2006)

13. Mikolov, T., Chen, K., Corrado, G., Dean, J.: Efficient estimation of word representations in vector space. CoRR abs/1301.3781 (2013)

14. Mikolov, T., Le, Q.V., Sutskever, I.: Exploiting similarities among languages for machine translation. CoRR abs/1309.4168 (2013)

15. Ng, D., Bansal, M., Curran, J.R.: Web-scale surface and syntactic n-gram features for dependency parsing. CoRR abs/1502.07038 (2015)

16. Nivre, J., Hall, J., Nilsson, J., Chanev, A., Eryiğit, G., Kübler, S., Marinov, S., Marsi, E.: MaltParser: a language-independent system for data-driven dependency parsing. Nat. Lang. Eng. **13**(2), 95–135 (2007)

17. Plank, B., van Noord, G.: Effective measures of domain similarity for parsing. In: Proceedings of the 49th Annual Meeting of the Association for Computational Linguistics: Human Language Technologies, pp. 1566–1576 (2011)

18. Sagae, K., Gordon, A.S.: Clustering words by syntactic similarity improves dependency parsing of predicate-argument structures. In: Proceedings of the 11th International Conference on Parsing Technologies, pp. 192–201 (2009)

19. Sculley, D.: Web-scale k-means clustering. In: Proceedings of the 19th International Conference on World Wide Web WWW 2010, pp. 1177–1178. ACM (2010)

20. Zhang, C., Zhao, T.: Bilingual lexicon extraction with forced correlation from comparable corpora. In: Arik, S., Huang, T., Lai, W.K., Liu, Q. (eds.) ICONIP 2015. LNCS, vol. 9490, pp. 528–535. Springer, Heidelberg (2015). doi:10.1007/978-3-319-26535-3_60

Exploiting Temporal Genetic Correlations for Enhancing Regulatory Network Optimization

Ahammed Sherief Kizhakkethil Youseph[1]([✉]), Madhu Chetty[2],
and Gour Karmakar[2]

[1] Faculty of Information Technology, Monash University, Clayton, Australia
ahammed.youseph@monash.edu
[2] Faculty of Science and Technology,
Federation University Australia, Gippsland, Australia
{madhu.chetty,gour.karmakar}@federation.edu.au

Abstract. Inferring gene regulatory networks (GRN) from microarray gene expression data is a highly challenging problem in computational and systems biology. To make GRN reconstruction process more accurate and faster, in this paper, we develop a technique to identify the gene having maximum in-degree in the network using the temporal correlation of gene expression profiles. The in-degree of the identified gene is estimated applying evolutionary optimization algorithm on a decoupled S-system GRN model. The value of in-degree thus obtained is set as the maximum in-degree for inference of the regulations in other genes. The simulations are carried out on *in silico* networks of small and medium sizes. The results show that both the prediction accuracy in terms of well known performance metrics and the computational time of the optimization process have been improved when compared with the traditional S-system model based inference.

Keywords: Gene regulatory network · S-system · Temporal correlation · Discrete cosine transform · Differential evolution

1 Introduction

Microarray data sets have not only been used extensively in multiclass gene classification [1], but also applied for reconstruction of Gene Regulatory Network (GRN) which represents the genes and their regulatory interactions. Identifying the regulatory network existing at the genetic level is significant in understanding the cellular processes in a biological system.

Various models have been proposed for the reconstruction of GRNs from microarray data. These can be widely classified into linear and nonlinear models. The dynamic Bayesian network (DBN) model, as a linear model, is often used in GRN inference [2] and utilizes the temporal information to be able to model feedback loops. Linear state space models have also been extensively used in GRN modeling [3,4]. Since the biological systems are nonlinear in nature, it limits the biological relevance of these linear models.

© Springer International Publishing AG 2016
A. Hirose et al. (Eds.): ICONIP 2016, Part I, LNCS 9947, pp. 479–487, 2016.
DOI: 10.1007/978-3-319-46687-3_53

Nonlinear state space models have also been proposed for modeling GRNs [5]. Among the other commonly used nonlinear models, S-system approach is widely accepted and considered to be a biologically relevant and generalized in nature. S-system is represented by a set of coupled nonlinear first-order ordinary differential equations. The reconstruction of a GRN of N genes using S-system model is essentially estimation of $2N(N+1)$ parameters with several approaches having been proposed for an accurate and fast estimation of these parameters. The idea of decoupling [6] for parameter estimation has been a big step towards reducing the computational cost. However, the number of parameters increases significantly with increase in number of genes and the modeling using S-system remains challenging for real-life networks.

In this paper, we have implemented a strategy of extracting information from a different perspective to complement an existing modeling approach. We have identified the genes having the higher in-degrees by exploiting the temporal correlations of expression profiles of genes. This information is then added as a prior knowledge to the existing traditional S-system model for inference. We have implemented the proposed method on two *in silico* data sets of small and medium sizes. The results show that this method is more accurate and faster than a traditional algorithm of S-system based inference of GRNs.

The rest of the paper is organized as follows: Sect. 2 details the background of the S-system model. The proposed method and the parameter estimation process are described in Sect. 3. In Sect. 4, we discuss the experimental results. The conclusions and future scope of the work are presented in Sect. 5.

2   Background

The traditional S-system model for the inference of GRNs is described next.

2.1   S-System Model

A set of coupled non-linear ordinary differential equations capable of modeling biochemical networks represents S-system Model.

$$\dot{x}_i = \frac{dx_i}{dt} = \alpha_i \prod_{j=1}^{N} x_j^{g_{ij}} - \beta_i \prod_{j=1}^{N} x_j^{h_{ij}} \quad i = 1, 2, \ldots, N \tag{1}$$

In the above model, N represents the total number of genes, x_i represents the expression of gene-i, α_i and β_i are the rate constants of synthesis and degradation of mRNA, respectively and $g_{i,j}$ and $h_{i,j}$ gives the direction and weight of regulation from gene-j to gene-i in formation and degradation phases, respectively.

The estimation of parameters from the given microarray data is usually achieved by minimizing the deviation of the data from the model predictions using evolutionary optimization algorithms. One of the most computationally expensive step in the process is finding the solutions of equations (e.g., Eq. (1)).

The decoupling or decomposition method of solving one equation at a time, instead of solving all the coupled equations together, reduces the computational cost to a great extent. The data for the time points missing in the original data are estimated using an interpolation technique. The decoupled set of equations can be expressed as follows:

$$\frac{dx_i}{dt} = \alpha_i x_i^{g_{ii}} \prod_{\substack{j=1 \\ j \neq i}}^{N} \widehat{x}_j^{g_{ij}} - \beta_i x_i^{h_{ii}} \prod_{\substack{j=1 \\ j \neq i}}^{N} \widehat{x}_j^{h_{ij}} \quad i = 1, 2, \ldots, N \tag{2}$$

where $\widehat{x}_j$ is the expressions value of gene-j, given in the data or estimated by the interpolation.

2.2   Fitness Function

A commonly employed fitness function during parameter optimization is squared relative error (SRE). In addition to the estimation error which corresponds to the deviation of the model predictions from the expected values, a complexity error has been included in the fitness function to account for the scale-free topology of the network. This penalization restricts the algorithms from adapting highly complex structures. One of the recently proposed penalty terms is included in adaptive squared relative error (ASRE) [7] as follows:

$$ASRE = \sum_{t=1}^{T} \left(\frac{x_i^{cal}(t) - x_i^{exp}(t)}{x_i^{exp}(t)} \right)^2 + B_i C_i \frac{2N}{2N - r_i} \tag{3}$$

where, r_i is the number of regulations for gene-i, B_i is the balancing factor which balances the two terms, SRE and penalty, on the R.H.S. of (3) and C_i is the penalty factor.

3   The Method

3.1   Preprocessing

The idea of this preprocessing was inspired by the temporal correlation among electroencephalogram (EEG) signals which have been effective in feature approximation and hence used in the classification of different medical conditions, e.g., the classification of EEG signals into ictal and interictal groups. The features of EEG signals derived by the two-dimensional discrete cosine transform (2D-DCT) have been shown to be more effective than those derived by 1D-DCT [8]. DCT coefficients can represent the instantaneous variations of EEG signals well. For this reason, to capture the temporal variations among the expression data of a gene profile due to activations or inhibitions from its regulatory genes, for this research, we have used the gene expression temporal correlation. The superimposition of the signals from regulatory genes enforces the expression value of a gene to vary proportionately to the number of its regulatory genes. This enables us to make an assumption "the gene which is regulated by the highest number of regulatory genes should have maximum energy value derived based on the temporal correlation among its expression data.

To validate this, we used a synthetic network of 10 genes, shown in Fig. 1 having 15 inter-regulatory arcs and 10 self-regulatory arcs. Data was generated using (1) assigning random weights and types for the arcs. The time-series data of a gene expression profile obtained after a perturbation, being non-stationary, was divided into a number of segments as in [8], before applying DCT. The samples in each segment of the profile were arranged in a two-dimensional matrix. Then, we applied the 2D-DCT and the coefficient matrix (F) was transformed into 1D using the zig-zg form. Since the high-frequency DCT coefficients represent the higher variations of a gene expression data, as with [8], the energies were calculated for each segment considering the 25 % of the high-frequency coefficients using the following equation:

$$E_i = \sum_{i=1}^{n} |F_i(t)|^2 \tag{4}$$

where, n indicates the number of the 25 % of the high-frequency coefficients and F_i is their i^{th} frequency coefficient. The average energies are calculated over all segments for all genes and are shown in Fig. 2.

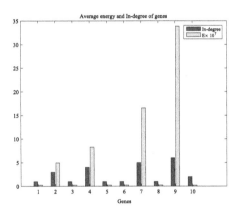

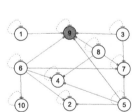

Fig. 1. Ten-gene regulatory network. Solid lines: regulations in the production phase. Dashed lines: the regulations in the degradation phase. Bar ended line indicates inhibition and circular end indicates activation or inhibition.

Fig. 2. Average energy and in-degree of genes

As expected, the gene having the highest in-degree showed the maximum energy, shown in Fig. 2. It can be noted that the gene-9 which has the maximum in-degree in the network shows the maximum energy. This observation has vindicated our earlier assumption "the gene having maximum in-degree can be identified as the gene having maximum average energy. This maximum in-degree based on temporal correlation (MITC) method can be briefly outlined as follows:

1. The time-series data of one gene is divided into segments of specific number of samples.
2. The samples in each segment are arranged in a 2-D matrix.
3. The discrete cosine transform (DCT) is applied on the matrix of all samples of each segment and a 1D vector is formed using zigzag manner.
4. A fixed percentage (in our case 25 %) of the high-frequency DCT coefficients are considered to calculate the energy using the equation defined in (4).
5. For each segment, the above three steps are repeated and the average energy is calculated.
6. For each gene, all the above steps are performed.
7. The gene with highest energy is identified as the gene with maximum in-degree.

3.2 Parameter Estimation

We used the S-system model for GRN inference. The decoupled equation described in (2) is used as they are computationally faster. In the process of parameter estimation, we first estimate the parameters involved in the equation for the gene identified as the one with maximum in-degree. The maximum and minimum in-degrees, I_i and J_i, respectively, are to be provided to the optimization. A random fixing could be twice the total number of genes for I_i and zero for J_i. Chowdhury *et al.* [7] made these parameters adaptive during the optimization. After every l iterations, the smallest and largest in-degrees of the population are examined and set as the new values for J_i and I_i, respectively.

The fitness function, ASRE defined in (3) is minimized during the optimization. The penalty factor C_i is fixed as in [9], according to the maximum and minimum in-degrees set during the optimization. As the same transcription factor usually does not regulate expression of a gene in both formation and degradation phases, a maximum in-degree less than N is set prior to the optimization. Once the equation (2) is solved, the parameters g and h provide the in-degree of gene-m which is then set as the maximum in-degree for parameter estimation of all other genes. Trigonometric differential evolution (TDE) is used for optimization. The hill-climbing local search (HCLS) [9] is also included in the algorithm. For the medium-scale network, the multi-refinement algorithm (MRA) used in [9] is applied for post-processing after completion of the given number of iterations.

4 Results and Discussion

We coded the inference algorithm in Matlab and the simulations were carried out for two *in silico* networks of small and medium sizes using the data sets generated by the S-system model. The parameters of evolutionary optimization were set as: the mutation factor, $M_f = 0.5$, the cross over ratio, $CR = 0.8$ and the trigonometric mutation ratio, $M_t = 0.05$. The population size was set to 50 for the small scale network and 100 for the medium scale network. The in-degree of the gene whose expression profile showed maximum temporal correlation is estimated by running the evolutionary optimization for 850 iterations which is then considered as the maximum-in-degree of the network. To compare the performance of the optimization algorithm using the temporal correlation information with that not using any correlation information, the simulations are carried out for 400 iterations using the same settings in both the algorithms. The inferred networks are assessed by calculating the four commonly used performance metrics: sensitivity (S_n), specificity (S_p), precision (P_r) and F-score (F). The computational complexity is assessed by estimating the number of function evaluations required per gene (N_f).

4.1   Small-Scale-Network

A small network of five genes and twelve regulations, including self-regulations, is considered. The gene which has maximum in-degree in the network is gene-4, having an in-degree of four. Gene-2 has an in-degree of three, gene-3 and gene-5 have an in-degree of two and gene-1 has an in-degree of one. Ten datasets having 21 samples in each are generated using S-system model. The MITC algorithm is performed first on the datasets. The average energies over various segments of the time-series data are calculated for different data sets. Gene-5, which has the maximum in-degree showed the maximum average energy.

The optimization algorithm is run for gene-5, setting the initial values of I_5 and J_5 to be 6 and 1, respectively. Once the optimization is completed, we get the in-degree as 4 including the self degradation. The network being small, the post-processing is excluded. For all other genes, I_i values are set as 4 at the beginning of the optimization which is adaptively changed during the optimization process.

Another set of simulations were run with the same optimization settings with the initial values of I and J set as 6 and 1, respectively, for all genes, i.e. without MITC. Five sets of simulations were carried out for both methods. The average values of the results obtained are shown in Table 1.

Table 1. The average values of the experimental results for small-scale *in silico* network

	S_n	S_p	P_r	F	N_f
Without MITC	0.8167	0.8842	0.7032	0.7521	3.651×10^4
With MITC	0.8667	0.9158	0.7718	0.8150	3.615×10^4

Table 1 shows that the method using MITC (our proposed method) produces better values for all performance metrics and computational complexity comparing with their corresponding counterparts of the method without using MITC.

The less average number of function evaluations required per gene for our proposed method indicates our method is computationally faster than that without MITC.

The values of the parameters inferred for the proposed method were more closer to the target values than for the method without MITC. The paired t-test is carried out for comparing the estimated parameters in both the methods with the original parameters expected. The p-values obtained for our proposed method and the method without MITC are 0.70 and 0.66, respectively. The larger p-value for our proposed method confirms the higher similitude to the target values.

4.2   Medium-Scale-Network

A twenty-gene network [9], widely been reconstructed using S-system model, is considered. The network consists of 20 self regulations in degradation phase and 26 inter-regulatory arcs in the production phase.

The energies based on temporal correlation were calculated for all genes using all data sets. The highest energy was shown by gene-15, one of the two genes that have the maximum in-degree. This gene is considered to be the gene with maximum in-degree. The optimization algorithm was run for gene-15, setting the initial values of I_{15} and J_{15} to be 8 and 1, respectively. Once the optimization and post-processing was completed, we obtained the in-degree as 4 including the self degradation. For all other genes, I_i values are set as 4 at the beginning of the optimization which is adaptively changed during the optimization process. For comparison purposes, another set of simulations were run with the same optimization settings with the initial values of I and J set as 8 and 1, respectively, for all genes, i.e. without MITC.

As the network is bigger in size, a post-processing is also necessary to achieve better optimal solutions. To study the influence of the proposed method at various stages of the inference algorithm, we have analyzed the results of simulations before as well as after applying the post-processing technique, multiple refinement algorithm (MRA). The average of values of the results over five sets of simulations before and after applying MRA are shown in Table 2 for the proposed method and those without applying MITC.

Similar to the previous dataset, for this 20-gene network dataset, all the performance metrics for our proposed method are better than those for the method without MITC. Also, the average number of evaluations required per gene for our proposed method is less than that for the method without MITC at both the stages of inference. It may be noted that the proposed method is computationally faster and concomitantly, has improved the prediction accuracy in terms of all performance metrics.

5   Conclusions

Reconstruction of GRNs remains a challenging problem in computational biology. In this paper, we have proposed a method of retrieving a useful information,

Table 2. The average values of the experimental results for medium-scale *in silico* network.

	Before applying MRA		After applying MRA	
	Without MITC	With MITC	Without MITC	With MITC
S_n	0.5088	0.5826	0.6255	0.6359
S_p	0.9340	0.9732	0.9120	0.9732
P_r	0.3220	0.5784	0.3032	0.5944
F	0.3934	0.5777	0.4074	0.6125
N_f	9.300×10^4	7.584×10^4	1.944×10^5	1.368×10^5

maximum in-degree of the given network, from the time-series data by comparing the temporal correlations of expression profiles of different genes. Input of this information to the inference algorithm makes the algorithm perform better in accuracy and speed. The results of simulations run on *in silico* networks confirm the superiority of the method.

Acknowledgment. This work is partly supported by Australian Federal and Victoria State Governments and the Australian Research Council through the ICT Centre of Excellence program, National ICT Australia (NICTA).

References

1. Ooi, C.H., Chetty, M., Teng, S.W.: Differential prioritization in feature selection and classifier aggregation for multiclass microarray datasets. Data Min. Knowl. Disc. **14**(3), 329–366 (2007)
2. Zou, M., Conzen, S.D.: A new dynamic Bayesian network (DBN) approach for identifying gene regulatory networks from time course microarray data. Bioinformatics **21**(1), 71–79 (2005)
3. Hirose, O., Yoshida, R., Imoto, S., Yamaguchi, R., Higuchi, T., Charnock-Jones, D.S., Print, C., Miyano, S.: Statistical inference of transcriptional module-based gene networks from time course gene expression profiles by using state space models. Bioinformatics **24**(7), 932–942 (2008)
4. Tamada, Y., Yamaguchi, R., Imoto, S., Hirose, O., Yoshida, R., Nagasaki, M., Miyano, S.: Sign-ssm: open source parallel software for estimating gene networks with state space models. Bioinformatics **27**(8), 1172–1173 (2011)
5. Noor, A., Serpedin, E., Nounou, M., Nounou, H.: Inferring gene regulatory networks via nonlinear state-space models and exploiting sparsity. IEEE/ACM Trans. Comput. Biol. Bioinf. **9**(4), 1203–1211 (2012)
6. Maki, Y., Ueda, T., Okamoto, M., Uematsu, N., Inamura, K., Uchida, K., Takahashi, Y., Eguchi, Y.: Inference of genetic network using the expression profile time course data of mouse p19 cells. Genome Inform. **13**, 382–383 (2002)
7. Chowdhury, A., Chetty, M., Vinh, N.: Incorporating time-delays in S-system model for reverse engineering genetic networks. BMC Bioinf. **14**(1), 196 (2013)
8. Parvez, M.Z., Paul, M.: Epileptic seizure detection by exploiting temporal correlation of electroencephalogram signals. IET Sig. Proc. **9**(6), 467–475 (2015)

9. Chowdhury, A.R.: Gene regulatory network reconstruction using time-delayed s-system model. Ph.D. thesis, Gippsland School of Information Technology, Monash University, Australia (2014)

Biomedical Engineering

Sleep Stage Prediction Using Respiration and Body-Movement Based on Probabilistic Classifier

Hirotaka Kaji[1]($^{(\boxtimes)}$), Hisashi Iizuka[1], and Mitsuo Hayashi[2]

[1] Toyota Motor Corporation, Susono, Shizuoka, Japan
{hirotaka_kaji,hisashi_iizuka}@mail.toyota.co.jp
[2] Hiroshima University, Higashi-hiroshima, Hiroshima, Japan
mhayashi@hiroshima-u.ac.jp

Abstract. In this paper, a sleep stage prediction method using respiration and body-movement based on probabilistic classifier is proposed. A pressure sensor is employed to capture respiratory signal. We propose to use least-squares probabilistic classifier (LSPC), a computationally effective probabilistic classifier, for four-class sleep stage classification (wakefulness, rapid-eye movement sleep, light sleep, deep sleep). Thanks to output of posterior probability of each class by LSPC, we can directly handle the confidence of predicted sleep stages. In addition, we introduce a method to handle imbalanced data problem which arises in sleep data collection. The experimental results demonstrate the effectiveness of sleep stage prediction by LSPC.

Keywords: Sleep stage prediction · Respiration · Body-movement · Probabilistic classifier

1 Introduction

Sleep is a fundamental and important state in daily life. Thanks to wide and rapid spread of sensor technologies, the collection of personal sleep data has entered the spotlight. These days, people who have health awareness use various sensors such as wristband sensors to measure their long-term sleep data and to check the trends. However, these data are predicted values, since the golden standard of sleep stage is determined by polysomnography (PSG) and manual scoring by experts. Sleep stage is basically categorized into six stages, that is, wakefulness (wake), rapid-eye-movement (REM) sleep, and non-REM (NREM) sleep classified to stage 1 to 4, based on Rechtschaffen and Kales' method [1]. These stages are mainly defined by brain wave measured by PSG. Most of previous studies using accelerometer and physiological sensors such as heart rate and respiration focused on two-stage (wake/sleep(REM+NREM)) or three-stage classification (wake/REM/NREM). However, deep sleep is an important factor of sleep quality. Therefore, we attempt to predict four-stage (wake/REM/light(stage $1 + 2$)/deep(stage $3 + 4$)) by using respiration and body-movement information.

© Springer International Publishing AG 2016
A. Hirose et al. (Eds.): ICONIP 2016, Part I, LNCS 9947, pp. 491–500, 2016.
DOI: 10.1007/978-3-319-46687-3_54

When a predictor using physiological information is developed, the following two problems have to be considered. One is the uncertainty of sensor signals. The measurement performance of low cost sensors is generally less than PSG, because the electrodes of PSG are securely attached with tape and paste. Sleep stages might be misclassified due to noisy physiological signals. Hence, addressing the reliability of prediction is required. The other is the imbalance of training samples. This problem naturally occurs in sleep data collection and deteriorates the performance of predictor. For instance, light sleep occupies the majority of total sleep time. To cope with aforementioned difficulties, we employ a probabilistic classifier, which can handle the confidence of prediction as the posterior probability.

This paper is organized as follows: Previous studies are introduced at first. Next, the detail of sleep data collection and preprocessing are described. Moreover, we introduce brief summary of least-squares probabilistic classifier (LSPC) [2] as a probabilistic classifier and indicate the effectiveness of LSPC for sleep stage prediction through experiments.

2   Related Work

Accelerometer based method which captures body movement during sleep is a typical approach to classify wake and sleep (REM+NREM) state [3]. Kawamoto et al. have demonstrated that accelerometer signal contains the fragments of respiratory information and have attempted to detect REM sleep by using support vector machine (SVM) [4]. Since cardiovascular system strongly correlates with autonomic nervous system (ASN), heart beat is a promising information to predict sleep stages. Various heart rate variability (HRV) based methods have been reported [5,6]. Watanabe and Watanabe used an air-filled cushion that can capture heart beat and body-movement, and proposed a predictor using the combination of those features [7].

On the other hand, respiration is also a promising physiological signal for the prediction. Chung et al. have proposed REM sleep estimation using respiration rate and variability, and they evaluated the discriminant performance between REM and NREM sleep [8]. Tataraidze et al. have proposed a respiratory based method using respiratory inductive plethysmography (RIP) [9]. They used features extracted from RIP of thorax and indicated good performance in three-stage classification (Wake/REM/NREM) by using bagging classifier.

As classification methods, various machine learning methods such as SVM, bagging, etc. have been adopted. However, the handling of imbalanced data and the confidence of sleep stage based on probabilistic classifier such as LSPC was scarcely mentioned in previous studies.

3   Data Collection

3.1   Experimental Setting

Ten healthy males in their 20 s participated in this experiment. The design of the experiment was approved and conducted according to the "Ethical Guidelines for

Research Involving with Human Subject" of Toyota Motor Corporation. Before the experiment, we explained the detail of the experiment and obtained their informed consents. We employed our prototype module which includes a pressure sensor and a three-axis accelerometer to measure the subjects' respiration and body-movement, respectively. The sensor module, of which size is about $5 \times 5 \times 2$ cm, was inserted between the front side of abdomen and bottoms.

Because the pressure sensor can detect the periodic movement of abdomen caused by breathing, we treated this signal as the surrogate of respiratory waveform. The sampling rate was 20 Hz. In addition, a physiological amplifier (Polymate AP1132, TEAC Corp.) with the sampling rate of 1 kHz was used to record their electroencephalogram (EEG) at central and occipital areas of the scalp, electrooculogram (EOG) and electromyogram (EMG) to determine the ground truth of sleep stages. Figure 1 shows the schematic diagram of the experiment system.

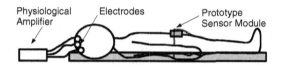

Fig. 1. The schematic diagram of experiment system

The measurement was executed at a quiet experiment room. In order to avoid first-night effect [10], which is caused by an unfamiliar environment, we measured the sleep data for two consecutive nights from all the subjects. All the subjects went to bed at 11:00 PM and waked up at 7:00 AM. The first night was treated as an adaptation session. We employed the data of second night for the analysis described below. Sleep scoring based on Rechtschaffen and Kales' method was executed each 30 s epoch by experts. As a result, the collected data was totally about 9,600 epochs (4,800 min).

3.2 Data Processing

In this paper, we used five respiratory features and a body-movement feature described below. All the features were calculated at 30-s intervals with 60-s sliding windows and were subtracted by the median to reduce the influence of individual difference.

Respiratory Feature. First of all, to extract respiratory waveform, the signal of pressure sensor was filtered with fourth-order Butterworth band-pass filter of which passband is from 0.5 to 1.0 Hz. Then following five features were computed from the respiratory waveform.

Mean respiration rate: Mean of respiration rate derived from the peak-to-peak interval of respiratory waveform.

Standard deviation of respiration rate: Standard deviation of respiration rate mentioned above.

Coefficient of variation of respiration rate: Mean respiration rate divided by standard deviation of respiration rate.

Coefficient of variation of respiration amplitude: Coefficient of variation of amplitude derived from the difference between a peak and a bottom of respiratory waveform.

Normalized autocorrelation coefficient: The first peak of normalized autocorrelation coefficient.

Body-Movement Feature. Following one feature was calculated by using raw accelerometer signal.

Difference norm of acceleration: Maximum value of $\log(\|a_t - a_{t-1}\|)$, where $a_t = (a_t^x, a_t^y, a_t^z)^\top$ is tth output of three-axis accelerometer with the sampling rate of 20 Hz.

Figure 2 depicts a conceptual diagram of peak-to-peak interval and amplitude of a respiratory waveform. Through the experiment, there were epochs that respiratory waveform did not indicate sufficient amplitude because of lack of pressure. Thus, we discarded such epochs from our training samples and finally obtained about 8,800 epochs. The rate of sleep stages was as follows: wake:8.0 %, REM:20.6 %, light(stage $1 + 2$): 65.3 %, and deep(stage $3 + 4$): 6.1 %.

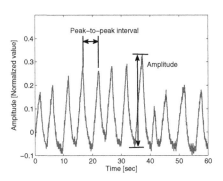

Fig. 2. The transition of respiratory waveform.

4   Sleep Stage Prediction

4.1   Least Square Probabilistic Classifier

Thanks to learning posterior probability directly, probabilistic classifiers can naturally handle multi-class classification problems. In addition, since the posterior probability represents the confidence of class prediction, the prediction output with low confidence can be rejected unlike deterministic classifiers such as SVM which only learn the decision boundary.

In this paper, we employed a computationally effective probabilistic classifier called least-squares probabilistic classifier (LSPC) proposed by Sugiyama et al. [2]. Let us consider training samples $\{(\boldsymbol{x}_i, y_i)\}_{i=1}^{n}$. In this study, $\boldsymbol{x}_i$ is a feature vector, and $y_i \in \{\text{wake}, \text{REM}, \text{light}, \text{deep}\}$ is a sleep stage of the ground truth. In LSPC, the posterior probability $p(y|\boldsymbol{x})$ of class $y = 1, \dots, c$ is modeled by linear model. In this paper, we use Gaussian kernel model:

$$q(y|\boldsymbol{x}; \boldsymbol{\theta}^y) = \boldsymbol{\theta}^{y\top}\boldsymbol{\phi}(\boldsymbol{x}) = \sum_{j:y_j=y} \theta_j^y K(\boldsymbol{x}, \boldsymbol{x}_j), \tag{1}$$

$$K(\boldsymbol{x}, \boldsymbol{x}') = \exp\left(-\frac{\|\boldsymbol{x} - \boldsymbol{x}'\|^2}{2h^2}\right), \tag{2}$$

where $\boldsymbol{\theta}^y$ and $\boldsymbol{\phi}(\boldsymbol{x})$ are the parameter and basis function vector, respectively. $\sum_{j:y_j=y}$ is the sum of jth parameter which satisfies $y_j = y$. LSPC minimizes the following criterion to learn the parameters:

$$
\begin{aligned}
\widehat{J}(\boldsymbol{\theta}^y) &= \frac{1}{2n}\sum_{i=1}^{n} q(y|\boldsymbol{x}_i; \boldsymbol{\theta}^y)^2 - \frac{1}{n}\sum_{i:y_i=y} q(y|\boldsymbol{x}_i; \boldsymbol{\theta}^y) + \frac{\lambda}{2n}\|\boldsymbol{\theta}^y\|^2 \\
&= \frac{1}{2n}\boldsymbol{\theta}^{y\top}\boldsymbol{\Phi}^\top\boldsymbol{\Phi}\boldsymbol{\theta}^y - \frac{1}{n}\boldsymbol{\theta}^{y\top}\boldsymbol{\Phi}^\top\boldsymbol{\pi}^y + \frac{\lambda}{2n}\|\boldsymbol{\theta}^y\|^2,
\end{aligned}
\tag{3}
$$

where $\boldsymbol{\Phi} = (\boldsymbol{\phi}(\boldsymbol{x}_1), \dots, \boldsymbol{\phi}(\boldsymbol{x}_n))^\top$, $\lambda \geq 0$ is the regularization parameter, and

$$\boldsymbol{\pi}^y = (\pi_1^y, \dots, \pi_n^y)^\top, \pi_i^y = \begin{cases} 1 & (y_i = y) \\ 0 & (y_i \neq y). \end{cases}$$

Then the minimizer $\widehat{\boldsymbol{\theta}}^y$ of the learning criterion $\widehat{J}$ can be analytically computed as

$$\widehat{\boldsymbol{\theta}}^y = \left(\boldsymbol{\Phi}^\top\boldsymbol{\Phi} + \lambda\boldsymbol{I}\right)^{-1}\boldsymbol{\Phi}^\top\boldsymbol{\pi}^y. \tag{4}$$

Finally, the posterior probability is calculated as follows:

$$\widehat{p}(y|\boldsymbol{x}) = \frac{\max(0, \widehat{\boldsymbol{\theta}}^{y\top}\boldsymbol{\phi}(\boldsymbol{x}))}{\sum_{y'=1}^{c} \max(0, \widehat{\boldsymbol{\theta}}^{y'\top}\boldsymbol{\phi}(\boldsymbol{x}))}. \tag{5}$$

Because the sleep measurement is executed in the course of nature, the rate of obtained sleep stages is biased. For example, the number of deep sleep epochs is less than one tenth of that of light sleep epoch in the training samples. Such a dataset is called imbalanced data, and it leads a difficulty to proper estimation of model parameters [11]. To cope with imbalanced data problem, we apply random selection to equate the number of training samples of each class. Here, the number is determined by the number of deep sleep epochs. The impact of imbalanced data problem is lightened by this procedure, but deep sleep might be overpredicted because the prior probability becomes equivalent to the others. Therefore, we add weight parameter w^y to Eq. (6) as follows and optimize it through cross-validation in order to correct the probabilities.

$$\widehat{p}(y|\boldsymbol{x}) = \frac{\max(0, w^y \widehat{\boldsymbol{\theta}}^{y\top} \boldsymbol{\phi}(\boldsymbol{x}))}{\sum_{y'=1}^{c} \max(0, w^{y'} \widehat{\boldsymbol{\theta}}^{(y')\top} \boldsymbol{\phi}(\boldsymbol{x}))}. \tag{6}$$

4.2   Performance Evaluation

We compare the performance of classification methods as follows:

SVM: Normal SVM with Gaussian kernel. We used LIBSVM [12] as the implementation of SVM. Optimized hyper parameters are C and γ.
SVM-RS: SVM with random selection procedure. Optimized hyper parameters are C and γ.
LSPC: Normal LSPC. Optimized hyper parameters are λ in Eq. (4) and h in Gaussian kernel.
LSPC-RS: LSPC with random selection procedure. Optimized hyper parameters are λ and h.
LSPC-WO: LSPC-RS with weight parameter optimization. Weight parameters w^{REM} and w^{light} in Eq. (6) are optimized. To avoid the huge number of combinations, we fix $w^{\text{wake}} = w^{\text{deep}} = 1$. In addition, we use $\lambda = 1$ and $h = 0.5$, which were the most commonly chosen in LSPC-RS.

All the training samples are used in the normal condition. On the other hand, the training samples are thinned out in the random selection and the weight optimization conditions. In this experiment, we employ median filter to the output of all the classification methods to suppress the fluctuation of prediction outputs. We use nine epochs as the window size of median filter. The following performance criteria are investigated [13].

Accuracy: The rate of test samples that are correctly predicted by a classifier.
Mean F-measure: The mean of F-measure in each class.

$$\text{mean F-measure} = \frac{1}{c} \sum_{i=1}^{c} \frac{2 \cdot \text{Precision}_i \cdot \text{Recall}_i}{\text{Precision}_i + \text{Recall}_i}$$

Because of the imbalance of sleep stages, mean F-measure is the significant criterion.

To evaluate the generalized performances of the classification methods, we employ leave one subject out cross-validation (LOSO-CV). In LOSO-CV, the training set is constructed from nine subjects, and the samples of the remaining subject are used as the test set. For hyper-parameter optimization and prediction performance evaluation, we employed the nested LOSO-CV procedure, where the outer loop computes prediction performance and the inner loop executes hyper-parameter optimization by using grid search.

In addition, we confirm the relation between the performance and the confidence. The rejection of outputs which are below a confidence threshold is a straightforward approach to deal with the uncertainty of physiological measurement, though the number of accepted outputs will be decreased when the confidence threshold is high. In this experiment, the confidence threshold T_C of LSPC-WO is changed in the range of 0.25–0.6 at 0.05 intervals.

4.3   Result and Discussion

Table 1 shows the evaluation results of five classification methods. LSPC-WO outperformed the others in terms of mean F-measure. In both of normal and random selection conditions, LSPCs indicated better performances than SVMs. Moreover, both LSPC-RS and SVM-RS exceeded normal LSPC and SVM due to the resolution of imbalanced data problem.

Table 1. Comparison of the performance of classification methods.

Subject	Accuracy					Mean F-measure				
	SVM	SVM-RS	LSPC	LSPC-RS	LSPC-WO	SVM	SVM-RS	LSPC	LSPC-RS	LSPC-WO
1	0.716	0.435	0.743	0.465	0.624	0.391	0.442	0.448	0.444	0.518
2	0.673	0.420	0.691	0.452	0.615	0.378	0.323	0.423	0.340	0.453
3	0.824	0.388	0.819	0.480	0.665	0.531	0.360	0.524	0.395	0.477
4	0.726	0.477	0.750	0.471	0.586	0.323	0.455	0.374	0.460	0.487
5	0.660	0.574	0.716	0.631	0.717	0.411	0.583	0.514	0.623	0.664
6	0.659	0.371	0.675	0.377	0.514	0.219	0.383	0.270	0.361	0.335
7	0.528	0.435	0.530	0.558	0.622	0.252	0.357	0.285	0.432	0.437
8	0.644	0.340	0.644	0.413	0.506	0.251	0.305	0.278	0.320	0.298
9	0.600	0.240	0.579	0.360	0.447	0.264	0.223	0.256	0.298	0.324
10	0.668	0.346	0.654	0.451	0.519	0.330	0.325	0.318	0.433	0.416
Mean	0.670	0.403	0.680	0.466	0.581	0.335	0.376	0.369	0.411	0.441

Figures 3, 4 and 5 show the transitions of sleep stages of subject 5. Solid and dotted line indicate the ground truth and prediction output, respectively. As indicated in the top of Fig. 5, we can confirm that the prediction of LSPC-WO captured the transition of ground truth. Even though deep sleep was not predicted, the prediction result of normal LSPC in the top of Fig. 4 seems adequately in the viewpoint of three-stage classification (wake/REM/NREM). This fact suggests that the influence of imbalanced data problem might be unremarkable in the three-stage classification scenario, because NREM sleep does not distinguish light and deep sleep. On the other hand, as shown in the bottom of Fig. 4, the deep sleep of LSPC-RS was overpredicted at 500–600 epochs. These results indicate that tuning weight parameters via cross-validation can lead to the intermediate performance of LSPC and LSPC-RS, and suppress the overprediction of deep sleep.

The transition of posterior probabilities is shown in the bottom of Fig. 5.

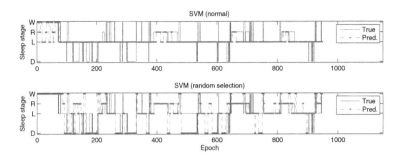

Fig. 3. The transitions of sleep stage of SVM and SVM-RS.

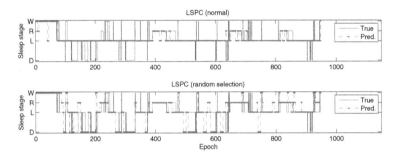

Fig. 4. The transitions of sleep stage of LSPC and LSPC-RS.

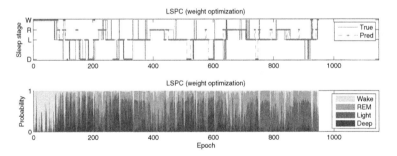

Fig. 5. The transitions of sleep stage and posterior probability of LSPC-WO.

The probability of wake was quite high in the early stage of sleep, and was suddenly diminished around 100 epochs. Then, the probabilities of REM and NREM (light+deep) were periodically fluctuated at intervals of about 200 epochs (100 min), and the light sleep gradually occupied the NREM sleep in the late stage of sleep.

Although the tuning procedure is able to apply to other classification methods such as SVM, the evaluation of confidence is a feature of probabilistic classifiers. Figure 6 presents the obvious trade-off between the mean F-measure and the number of accepted outputs. When $T_C = 0.25$, all the outputs were accepted due to the chance rate of four-class classification. On the other hand, when the threshold became high, the mean F-measure was made better though the accepted outputs were decreased. Through the experiments, we conclude that LSPC-WO is a promising classification method to handle imbalanced data problem and can address the uncertainty of physiological measurement.

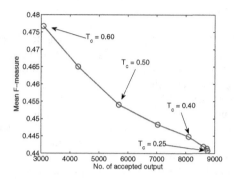

Fig. 6. The trade-off between mean F-measure and the number of accepted outputs.

5   Conclusion

We introduced a probabilistic classifier based sleep stage prediction using respiration and body-movement information. We indicated a method for least-squares probabilistic classifier (LSPC) to handle imbalanced data problem, which commonly arises in sleep data collection. Moreover, we confirmed the relation between the performance and the confidence. Through the experiments, the effectiveness of LSPC for sleep stage prediction was demonstrated. It can be expected that the proposed method is widely applied to various type of respiration sensors and accelerometer. In future works, in order to improve the limitations of this paper, we have plans to add the number of subjects including females, and to confirm the performance via long term evaluation in daily living.

References

1. Rechtschaffen, T., Kales, A.: A manual of standardized terminology, techniques and scoring system for sleep stage of human subjects. Public Health Service U.S. Goverment Printing Office (1968)
2. Sugiyama, M., Hachiya, H., Yamada, M., Simm, J., Nam, H.: Least-squares probabilistic classifier: a computationally efficient alternative to kernel logistic regression. Proc. IWSML **2012**, 1–10 (2012)
3. Cole, R.J., Kripke, F., Gruen, W., Mullaney, D.J., Gillin, J.C.: Automatic sleep/wake identification from wrist activity. Sleep **15**(5), 461–469 (1992)

4. Kawamoto, K., Kuriyama, H., Tajima, S.: Actigraphic detection of REM sleep based on respiratory rate estimation. J. Med. Bioeng. **2**(1), 20–25 (2013)
5. Adnane, M., Jiang, Z., Yan, Z.: Sleep-wake stages classification and sleep efficiency estimation using single-lead electrocardiogram. Expert Syst. Appl. **39**, 1401–1413 (2012)
6. Takeda, T., Mizuno, O., Tanaka, T.: Time-dependent sleep stage transition model based on heart rate variability. Proc. IEEE EMBC **2015**, 2343–2346 (2015)
7. Watanabe, T., Watanabe, K.: Noncontact method for sleep stage estimation. IEEE Trans. Biomed. Eng. **51**(10), 1735–1748 (2004)
8. Chung, G.S., Choi, B.H., Lee, J.S., Jeong, D.U., Park, K.S.: REM sleep estimation only using respiratory dynamics. Physiol. Meas. **30**, 1327–1340 (2009)
9. Tataraidze, A., Anishchenko, L., Korostovtseva, L., Kooij, B.J., Bochkarev, M., Syiryaev, Y.: Sleep stage classification based on respiratory signal. Proc. IEEE EMBC **2015**, 358–361 (2015)
10. Tamaki, M., Nittono, H., Hayashi, M., Hori, T.: Examination of the first-night effect during the sleep-onset period. Sleep **28**(2), 195–202 (2005)
11. Japkowicz, N.: The class imbalance problem: significance and strategies. In: Proceedings of IC-AI 2000 (2000)
12. Chang, C.C., Lin, C.J.: LIBSVM: a library for support vector machines. ACM Trans. Intell. Syst. Technol. **2**(3), 27:1–27:27 (2011)
13. Han, J., Kamber, M., Pei, J.: Data Mining: Concepts and Techniques, 3rd edn. Morgan Kaufmann (2011)

Removing Ring Artifacts in CBCT Images Using Smoothing Based on Relative Total Variation

Qirun Huo[1,2], Jianwu Li[1], Yao Lu[1($\boxtimes$)], and Ziye Yan[1,3]

[1] Beijing Key Laboratory of Intelligent Information Technology,
School of Computer Science and Technology,
Beijing Institute of Technology, Beijing 100081, China
vis_yl@bit.edu.cn
[2] College of Information Engineering,
Capital Normal University, Beijing 100048, China
[3] China Resource Wandong Medical Equipment Co., Ltd., Beijing 100015, China

Abstract. Removing ring artifacts in Cone Beam Computed Tomography (CBCT) images without impairing the image quality is critical for the application of CBCT. In this paper, we propose a novel method for the removal of ring artifacts in CBCT Images using an image smoothing based on relative total variation (RTV). After transforming the CBCT image into polar coordinates, we introduce a single-direction smoothing to separate the small scale textures, which include the artifacts, from the image structures. Then the artifact template is generated by median value extraction. Finally, the artifact template is transformed back into Cartesian coordinates and is subtracted from the original CBCT image. Experiments on different CBCT images show that the proposed method can obtain satisfactory results.

Keywords: CBCT image · Ring artifacts · Image smoothing · Relative total variation

1 Introduction

In recent years, Cone Beam Computed Tomography (CBCT) has been widely used in the medical fields, such as clinical diagnosis, 3D implants. However, due to the limitations of imaging system, CBCT images often have ring artifacts, which have the same center with the reconstructed image and different gray levels with the surrounding pixels [1]. Ring artifacts have great effect on the quality and authenticity of CBCT images. Many methods based on image processing have

Y. Lu—This work was supported in part by National Natural Science Foundation of China under Grant No. 61273273,61271374, the Specialized Fund for Joint Building Program of Beijing Municipal Education Commission.

© Springer International Publishing AG 2016
A. Hirose et al. (Eds.): ICONIP 2016, Part I, LNCS 9947, pp. 501–509, 2016.
DOI: 10.1007/978-3-319-46687-3_55

been proposed to remove ring artifacts. They can be roughly divided into two categories [2]: pre-processing approaches [3–7] based on the projection sinogram, and post-processing approaches [8–11] based on the reconstructed image.

Pre-processing approaches are performed on the projection sinogram [12], in which artifacts appear as parallel vertical lines. C. Raven et al. [3] adopted Fourier transform on the projection sinogram and used low-pass filter to remove artifacts. M. Boin et al. [4] used the average filter to eliminate artifacts during the reconstruction. B. Münch et al. [5] combined wavelet decomposition and Fourier low-pass filter to remove artifacts. Post-processing approaches are directly applied to the reconstructed images to reduce artifacts. J. Sijbers et al. [8] corrected ring artifacts based on the morphological operations. In [10], independent component analysis (ICA) was used to decompose the image into multiple independent components and the components containing streak artifact were selected for further filtering. In [11], a polar wavelet Gaussian filtering algorithm was applied to remove artifacts from the reconstructed image.

Although the exiting methods have certain correction effects, they still have some disadvantages. Most of them remove artifacts by different filters, which cause blurred edges and loss of details in the image. Furthermore, in pre-processing methods, the large memory space is required by the projection sinograms. While in most post-processing methods, the reconstructed image is corrected in polar coordinates. During the coordinate transformation, interpolation is often used to compensate the transformed image, and it is inevitable to produce the loss of edges and details. To avoid these problems, we propose a new artifact removing method in this paper. Inspired by the structure extraction algorithm [13], an effective image smoothing algorithm is introduced into our proposed method. Because the artifacts can be regarded as regular texture, we perform a single-direction smoothing to separate low-amplitude textures with artifacts, from main structures of the CBCT image. After that, artifacts are further extracted from these textures, and are subtracted from the original reconstructed image. Like most post-processing methods, we separate and extract artifacts in polar coordinates. However, different from them, we only convert the obtained artifacts back into Cartesian coordinates and directly subtract artifacts from the original CBCT image in Cartesian coordinates. Therefore, the meaningful information of the original image is ensured to be preserved largely. Experimental results show that this method can effectively remove ring artifacts and well preserve the details and edges of CBCT images.

The paper is organized as follows. The structure extraction algorithm is reviewed in Sect. 2. The proposed method is shown in Sect. 3. Experimental results are given in Sect. 4 and the last section concludes this paper.

2   Smoothing via Structure Extraction

Let S denote the smoothed result of an input image I. A pixel-wise *windowed total variation* measure is written as

$$D_x(p) = \sum_{q \in R(p)} g_{p,q} |(\partial_x S)_q|,$$

$$D_y(p) = \sum_{q \in R(p)} g_{p,q} |(\partial_y S)_q|, \tag{1}$$

where p and q index 2D pixels, and $R(p)$ is the rectangular region centered at pixel p. $D_x(p)$ and $D_y(p)$ are *windowed total variations* in the x and y directions, respectively, for pixel p, and they count the absolute spatial differences within the window $R(p)$. $g_{p,q}$ is a weight defined by a Gaussian function according to spatial affinity, expressed as

$$g_{p,q} \propto \exp(-\frac{(x_p - x_q)^2 + (y_p - y_q)^2}{2\sigma^2}), \tag{2}$$

where σ controls the spatial scale of the window. D is responsive for visual saliency. In an image with salient textures, both the detail and structure pixels yield large D.

To distinguish prominent structures from the texture details, *windowed inherent variation* is defined as

$$L_x(p) = \left| \sum_{q \in R(p)} g_{p,q} (\partial_x S)_q \right|,$$

$$L_y(p) = \left| \sum_{q \in R(p)} g_{p,q} (\partial_y S)_q \right|. \tag{3}$$

Different from D, L does not incorporate the modulus for each difference. Because spatial difference for a given pixel is possibly either positive or negative, the sum depends on whether or not the gradients in a window are coincident in terms of their directions. Usually, a major edge in a local window contributes more similar-direction gradients than near-regular textures. Thus, L in a window that only contains textures is generally smaller than that in a window that also includes structural edges.

Combining L with D, an even more effective regularizer $D_x(p)/(L_x(p) + \varepsilon) + D_y(p)/(L_y(p) + \varepsilon)$ is formed to further enhance the contrast between texture and structure. It is called *relative total variation* (RTV), where ε is a small positive real number to avoid divided by zero. After the regularizer RTV is added, the objective function of smoothing for structure-texture decomposition becomes

$$\min_s \sum_p \left((S_p - I_p)^2 + \lambda(\frac{D_x(p)}{L_x(p) + \varepsilon} + \frac{D_y(p)}{L_y(p) + \varepsilon}) \right), \tag{4}$$

where the term $(S_p - I_p)^2$ makes the smoothed image S similar to the original image I as much as possible. The regularizer RTV is used to remove image textures, and the smoothing parameter λ controls the smoothness of the result.

Texture and main structure exhibit completely different properties on RTV and thus they can be easily distinguished. Regarding the artifacts as the part of regular texture details, we can separate the texture details including artifacts from the main image structures.

3   Removing Ring Artifacts by Single-Direction Smoothing

The whole procedure of removing ring artifacts is described as follows.

3.1   Transformation into Polar Coordinates

In most existing post-processing methods, the reconstructed image is processed in polar coordinates, in which the artifacts appear as vertical stripes. Compared with concentric rings in Cartesian coordinates, stripe artifacts in polar coordinates are easier to be detected and eliminated [14].

Similarly, we transform the CBCT image into polar coordinates as described by [8] firstly. Figure 1 shows an example of the polar coordinate transformation. To better observe the details, we magnified a portion of image and expanded its image contrast, as shown in bottom right corner of figures.

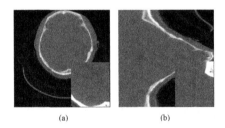

(a)   (b)

Fig. 1. Polar coordinate transformation. (a) An original brain CBCT image in Cartesian coordinates; (b) the transformed image in polar coordinates. A magnified portion is shown on bottom right corner.

3.2   Image Decomposition by Smoothing

As shown in Fig. 1(b), the artifacts appear as vertical stripes in polar coordinates. To preserve the original image information as much as possible, we only take the smoothing in horizontal direction into consideration for the tranformed image. By Eq. (4), the objective function is modified as

$$\min_{s} \sum_{p} \left((S_p - I_p)^2 + \lambda(\frac{D_x(p)}{L_x(p) + \varepsilon}) \right). \tag{5}$$

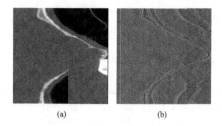

(a)   (b)

Fig. 2. Image smoothing. (a) A smoothed image, a magnified portion is shown on bottom right corner; (b) Texture image including artifacts and other details.

where the regularization term only includes RTV in the x directon. This is non-convex problem and the reference [13] provides an efficient solving strategy.

After smoothing, image textures that include artifacts and some original image details can be obtained by subtracting the smoothed result from the original image. The smoothed image and texture image of the example are shown in Fig. 2.

3.3   Artifact Template Extraction

In this subsection, we further extract artifacts from the texture image. The artifact pixels caused by the same detector basically have the same gray level in the image. Thus, we extract median values as artifacts. On the texture image, the median of each vertical column is computed as the value of corresponding column of the artifact template. Each column of the artifact template has the same value and the whole artifact template has the same size as the texture image.

Figure 3(a) shows the generated artifact template of the example. By subtracting the artifact template from the texture image, we can see that the remains are the original image details and need to be preserved, as shown in Fig. 3(b).

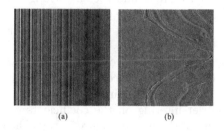

(a)   (b)

Fig. 3. Artifacts extraction. (a) Artifact template; (b) Other texture details.

3.4   Artifacts Subtraction

Finally, we transform the artifact template back into Cartesian coordinates and subtract it from the original CBCT image. Thus, the CBCT image without ring artifacts is obtained.

Figure 4(a) shows the artifact template in Cartesian coordinates, which has the same size as the original CBCT image. Figure 4(b) shows the result image, in which ring artifacts are removed effectively while the original information is well preserved.

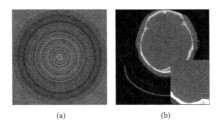

(a)   (b)

Fig. 4. Artifacts subtraction. (a) The artifact template in Cartesian coordinates; (b) The result without artifacts, a magnified portion is shown on bottom right corner.

4   Experiments

The proposed method was implemented in Matlab on real CBCT images of human tissues. We normalized all pixel values of the original CBCT image to the interval [0,1]. The size of image in Cartesian coordinates and in polar coordinates is set as 512 * 512 and 360 * 360, respectively. In the coordinate transformation, the bilinear interpolation method [15] was used to compensate the transformed image. ε in Eq. 5 was fixed to 1e-3, and the smoothing parameter λ was set to 0.005 and can be adjusted within (0, 0.05] according to the smoothing effect. The spatial parameter σ in Eq. 2 was set to 3 and is tunable for different images within a range (0,6].

To verify the effectiveness of the proposed method, we compared the proposed method with an existing post-processing method in [11],which applied the wavelet-Fourier filter [5] to the reconstructed images in polar coordinates. This technique performs Fourier filtering on the coefficients of 2-D wavelet decomposed vertical detail band of the image. Figure 5 shows the experimental results on three CBCT images including a brain CBCT image, a neck CBCT image and a skull CBCT image. Each subfigure is related to a CBCT image. From the left of each subfigure, the first one is the original CBCT image whose contrast was expanded properly to make ring artifacts distinct. The second one shows the whole result image processed by the proposed method. It can be seen that the ring artifacts are almost perfectly removed. The third one depicts a magnified portion of the original image, in which some ring artifacts can be found clearly. The last two images show the corresponding magnified portion of the corrected results. The former is the result processed by the wavelet-Fourier filter, in which ring artifacts at the center of images are corrected well, but some residual ring artifacts exist away from the central location. In the latter, the results obtained by our proposed method are presented. There is no significant residual ring artifacts, and the edges and details are remained.

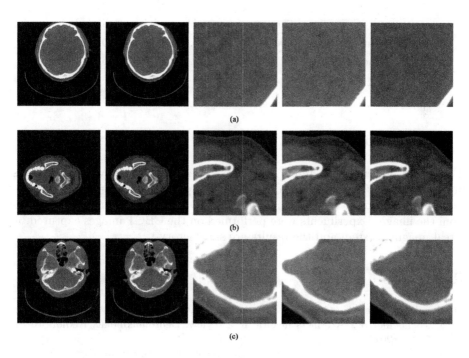

Fig. 5. Experimental results. (a) The brain CBCT image; (b) The neck CBCT image; (c) The skull CBCT image. From left to right, they are the original CBCT image, the result image by the proposed method, the magnified region of the original image, the magnified region of result by the wavelet-Fourier filter and the magnified region of result by the proposed method, respectively.

Additionally, the total energy of a signal is a well-known objective measure in signal processing for quantitative evaluation [5]. The loss of the energy can be expressed by the energy of the difference between the original image s_o and the corrected image s_r, relative to the original one, resulting in the relative mean square error (RMSE):

$$\text{RMSE} = \frac{\sum |s_o - s_r|^2}{\sum |s_o|^2}. \tag{6}$$

The combination of visual rating with a minimal relative change between the original and the processed image is supposed to be a robust technique for the assessment of the corrected performance. The smaller the value of RMSE is, the better the quality of the result is. The resulting RMSE of the two methods are listed in Table 1. Compared with the wavelet-Fourier filtering, the proposed method produces much smaller energy changes.

Table 1. Quantitative comparison on two methods based on the resulting RMSE.

	The brain image	The neck image	The skull image
The wavelet-Fourier filtering	0.0016	0.0040	0.0022
The proposed method	3.2893E–05	3.6032E–05	6.0362E–05

5 Conclusions

This paper presents is a simple but effective method to remove ring artifacts in CBCT image. The core of the method is the proposed single-direction image smoothing via an effective optimization, which separates the regular artifacts from the image. Experiments were performed on the CBCT images captured by a real imaging system and the results are satisfactory.

References

1. Schulze, R., Heil, U., Gross, D., Bruellmann, D.D., Dranischnikow, E., Schwanecke, U., Schoemer, E.: Artefacts in CBCT: a review. Dentomaxillofac. Radiol. **40**(5), 265–273 (2011)
2. Hasan, M.K., Sadi, F., Lee, S.Y.: Removal of ring artifacts in Micro-CT imaging using iterative morphological filters. Signal Image Video Process. **6**(1), 41–53 (2012)
3. Raven, C.: Numerical removal of ring artifacts in microtomography. Rev. Sci. Instrum. **69**(8), 2978–2980 (1998)
4. Boin, M., Haibel, A.: Compensation of ring artefacts in synchrotron tomographic images. Opt. Express **14**(25), 12071–12075 (2006)
5. Münch, B., Trtik, P., Marone, F., Stampanoni, M.: Stripe and ring artifact removal with combined waveletfourier filtering. Opt. Express **17**(10), 8567–8591 (2009)
6. Zeng, D., Ma, J., Zhang, Y., Bian, Z., Huang, J., Chen, W.: An Improved ring artifact removal approach for flat-panel detector based computed tomography images. In: Nuclear Science Symposium and Medical Imaging Conference, pp. 1–4. IEEE (2013)
7. Kim, Y., Baek, J., Hwang, D.: Ring artifact correction using detector line-ratios in computed tomography. Opt. Express **22**(11), 13380–13392 (2014)
8. Sijbers, J., Postnov, A.: Reduction of ring artefacts in high resolution Micro-CT reconstructions. Phys. Med. Biol. **49**(14), N247 (2004)

9. Brun, F., Kourousias, G., Dreossi, D., Mancini, L.: An improved method for ring artifacts removing in reconstructed tomographic images. In: Dössel, O., Schlegel, W.C. (eds.) World Congress on Medical Physics and Biomedical Engineering, September 7–12, 2009, Munich, Germany. IFMBE Proceedings, vol. 25/4, pp. 926–929. Springer, Munich (2009)
10. Chen, Y.W., Duan, G.: Independent component analysis based ring artifact reduction in cone-beam CT images. In: 16th IEEE International Conference on Image Processing, pp. 4189–4192. IEEE Press, Cairo (2009)
11. Wei, Z., Wiebe, S., Chapman, D.: Ring artifacts removal from synchrotron CT image slices. J. Instrum. **8**(06), C06006 (2013)
12. Kak, A.C., Slaney, M.: Principles of Computerized Tomographic Imaging. Society for Industrial and Applied Mathematics, Philadelphia (2001)
13. Xu, L., Yan, Q., Xia, Y., Jia, J.: Structure extraction from texture via relative total variation. ACM Trans. Graph. (TOG) **31**(6), 139 (2012)
14. Anas, E.M.A., Lee, S.Y., Hasan, M.K.: Removal of ring artifacts in CT imaging through detection and correction of stripes in the sinogram. Phys. Med. Biol. **55**(22), 6911 (2010)
15. Gribbon, K.T., Bailey, D.G.: A novel approach to real-time bilinear interpolation. In: Proceedings of Second IEEE International Workshop on Electronic Design, Test and Applications, DELTA 2004, pp. 126–131. IEEE (2004)

Proposal of a Human Heartbeat Detection/Monitoring System Employing Chirp Z-Transform and Time-Sequential Neural Prediction

Ayse Ecem Bezer and Akira Hirose[✉]

Department of Bioengineering, The University of Tokyo, 7-3-1 Hongo, Bunkyo-ku,
Tokyo 113-8656, Japan
bezerae@eis.t.u-tokyo.ac.jp, ahirose@ee.t.u-tokyo.ac.jp

Abstract. Heartbeat signal detection and/or monitoring is very important in the rescue of human beings existing under debris after disasters such as earthquakes as well as in the monitoring of patients in hospital. In this paper, we propose a human heartbeat detection/monitoring system employing chirp Z-transform and a time-sequential prediction neural network. The system is an adaptive radar using 2.5 GHz continuous microwave. The CZT realizes high resolution peak search in the frequency domain. We use a neural network to track adaptively the heartbeat signal which often has frequency fluctuation. The network learns the time-sequential peak frequency online in parallel to the detection and tracking. Even when the heartbeat frequency drifts, the network finds and tracks the heartbeat. Experiments demonstrate that the proposed system has high effectiveness in distinction between person-exist and person-non-exist observations, resulting in successful detection of persons.

Keywords: Heartbeat · Respiration · Drifting signal tracking

1 Introduction

Heartbeat and/or respiration signals are very useful and critical in detection of human beings existing under debris after disasters such as earthquakes as well as in monitoring of patients in hospital. Microwave/millimeter-wave radar detection systems are just under development by various research groups [1–5]. Heartbeat detection and/or monitoring is relatively more difficult than respiration detection because of its smaller movement. In particular, when the target person is far from the detection/monitor system or the location is unknown, the decision whether the system catches a person or not is highly affected by noise and fluctuation of heartbeat frequency.

To increase the signal-to-noise ratio (SNR) in heartbeat detection, we may adopt longer observation time, namely, larger window size for Fourier transform to catch the power peak in the frequency domain. However, the heartbeat frequency often fluctuates, leading to ambiguity in its frequency peak. Shorter window size results in lower resolution in frequency. This is an intrinsic trade-off.

In this paper, we propose a human heartbeat detection/monitoring system employing chirp Z-transform and time-sequential neural prediction. The CZT realizes an apparent

© Springer International Publishing AG 2016
A. Hirose et al. (Eds.): ICONIP 2016, Part I, LNCS 9947, pp. 510–516, 2016.
DOI: 10.1007/978-3-319-46687-3_56

enhancement of frequency resolution even for a small-size transform window in comparison with conventional Fourier transform by reducing the harmful discreteness in the frequency domain. It is very effective in locating the spectral peak precisely by utilizing this pseudo high resolution [6]. Then we run a neural network to track adaptively the heartbeat signal having fluctuating frequency, which is a modification of adaptive channel prediction network [7–10]. At every time point, the neural network predicts the heartbeat frequency in the next discrete time step. If the actual CZT peak frequency falls within a predicted frequency range, we decide that the network detects and tracks the heartbeat, indicating a detection of a human being with the information of heartbeat frequency. The network learns the time-sequential peak frequency online in parallel to the detection and tracking. Even when the heartbeat frequency changes, the network finds and tracks the signal. Contrarily, it cannot track noise data having white-noise like distribution. We also report the front-end and total construction of the microwave continuous-wave (CW) detection system that we constructed. Since we do not use pulses, we can realize a high SNR also in electronics.

2 Heartbeat Tracker Based on Chirp Z-Transform and Time-Sequential Neural Prediction

2.1 Microwave System Construction

Figure 1 illustrates the microwave measurement system. It transmits CW 2.5-GHz frequency electromagnetic wave and receives scattered and/or reflected wave to obtain time-sequential amplitude and phase information by using a vector network analyzer (VNA). The VNA may work as a so-called stepped-frequency transmission mode for ranging purpose. However, here we use the CW mode to focus on the slow movement of human body for detection of victims under debris and/or monitoring of his/her heartbeat.

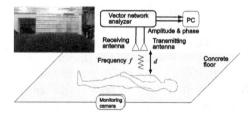
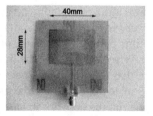

Fig. 1. Measurement system with an experiment photo.

Fig. 2. Patch antenna

Figure 2 shows the compact patch antenna that we fabricated for the use in this system. It works at 2.5 GHz. We chose the frequency to realize a sufficient phase sensitivity to the movement related to heartbeat as well to avoid excessively sharp beam forming.

2.2 Chirp Z-Transform and Neural Prediction for Tracking

Figure 3 is the flowchart showing the signal processing employing CZT and time-sequential neural prediction for detection of heartbeat. To obtain a low-discreteness frequency spectrum, we use the chirp Z-transform, instead of conventional fast Fourier transform (FFT), to zoom into the heartbeat frequency range. We obtain a short-time spectrum time-sequentially. Then we pick up an instantaneous maximum power frequency in each time bin. When the VNA yields only noise output, the peak frequency will show a random sequence or unstable changes. When it receives a heartbeat signal, we expect that the peak frequency point appears at around 1.3 Hz. However, in general, the heartbeat frequency changes gradually depending on the body-state variation. Simple averaging does not work effectively but, instead, smears the peak information. It also depends on individuals.

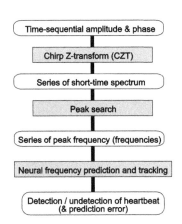

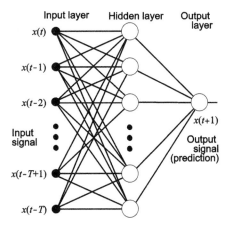

Fig. 3. Signal processing flowchart   **Fig. 4.** Construction of the neural network

Then we utilize the prediction ability of neural networks. The network construction is shown in Fig. 4. The inputs $x(t)$, $x(t - 1)$, ... are the peak-power frequency in the recent T time period, whereas the output $x(t + 1)$ is the predicted frequency in the next time step. When we examine the frequency spectrum data, which is to be shown below, we notice that we, the human beings, can catch the peak continuity in the frequency-time diagram to notice the heartbeat signal. We expect the neural network to do the same. That is, we run the network in an online learning mode and, if the prediction output is near to the actual peak frequency observed in the next time slot, then we decide that we find the continuity, meaning a heartbeat signal.

The periodicity learning will also result in meaningful and stable connections for the input terminals within an averaged coherence time length of the heartbeat after the learning. For noise-only input data, the network connections will get unstable because of the abrupt changes and low correlations between the desired output signal and the input past signals. This characteristic will also be an important index for decision whether a heartbeat is detected or not. Here in this paper, however, we focus on the prediction error mentioned above.

3   Experiments and Results

In the present experiment, the observation time step is $T = 0.1$ s. The input terminal number is 10, resulting in a total input data of 1 s, and the hidden neuron number is 10.

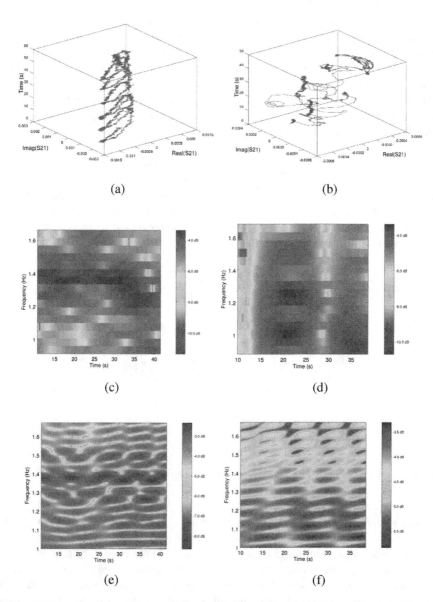

(a)   (b)

(c)   (d)

(e)   (f)

Fig. 5.   (a) Raw waveform for person observation, (b) raw waveform for noise-only observation, (c) person FFT spectrum, (d) noise-only FFT spectrum, (e) person CZT spectrum, and (f) noise-only CZT spectrum.

The learning process employs error backpropagation algorithm. For each set of input-output teachers (=0.1 s time step), the weights are updated 10 times.

Figure 5 shows the raw data example obtained for person measurement (distance $d \sim 40$ cm) as well as a measurement without a person (noise only). Figures 5(a) and (b) present the waveform shown as the real and imaginary parts in time domain. The person observation in (a) represents somewhat periodic evolution. Contrarily, the noise-only observation in (b) shows very small amplitude with less periodicity. The FFT spectra, calculated for a moving window of 20 s, reveals this characteristic. That is, (c) person measurement shows a curved ridge of high-power frequency pints, while in (d) noise-only data does not have such a continuous high-power curve. The FFT frequency resolution is not so high, which is limited by the window size.

By using the CZT, however, we can mitigate the discreteness in the frequency domain to generate a pseudo high-resolution spectrum. As shown in Fig. 6, this property is significantly effective for precise peak search both in Fig. 6. (a) person data and (b) noise-only data. Figure 5(e) presents the CZT spectrum obtained for heartbeat. We can find a fine structure in the heartbeat frequency even though the peak ridge is clearly observed. In contrast, in the noise-only case shown in Fig. 5(f), there is no continuous peak, suggesting no heartbeat included. By observing such textural difference, we can think intuitively that the former data catches a person's heartbeat signal, but the latter does not. We aim to realize such decision in a manner robust to gradual frequency variation automatically by using the neural network.

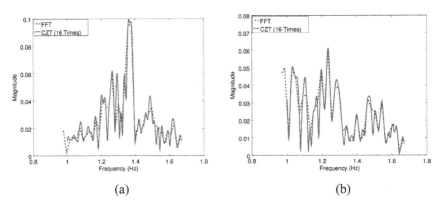

(a)   (b)

Fig. 6. Comparisons of FFT and CZT spectra for (a) person and (b) noise-only observations.

Figure 7 presents the decision results obtained by using the neural network. In the proposed system, we decide that *"the system catches a person's heartbeat"* when the neural network prediction is very near to the actual peak frequency in the next future bin because such a result implies a continuous peak frequency even if the peak gradually changes. Figure 7(a) shows the raw peak frequency curve (dashed), neural prediction curve (solid) for person observation, while those in (b) shows those for noise-only observation. Figure 7(a) includes also an indication of a contact-type device measuring at a subject's finger in parallel to the proposed electromagnetic-wave observation.

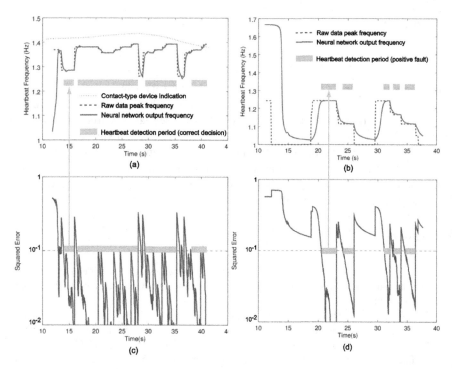

Fig. 7. (a) Raw peak frequency curve (dashed), neural prediction curve (solid) and "heartbeat found"-decision period for person observation (thick line), (b) those for non-person observation, (c) prediction error for person observation, and (d) prediction error for non-person observation where we have a threshold for "successful prediction" at relative error of 10^{-1}. Note that (a) includes also an indication of a contact-type device measuring at a subject's finger in parallel to the proposed electromagnetic-wave observation.

Figure 7(c) presents the prediction error and the automatic decision (light-green thick line segments) for person observation, and (d) those for noise-only observation. In the above decision in (a) and (b), we prepared a threshold for "successful prediction" at relative error of around 10^{-1}. We also observed the neural weight changes in time. We found for person observation that the weights often become stable quickly. However, for noise-only data, they are always varying unstably. It is consistent with the signal/noise characteristics as mentioned above.

Indication of the contact-type device included in a compact oximeter is also shown in Fig. 7(a) as a reference obtained by another equipment. In the experiment, the device touches the subject person at one of the fingers. Though the result shows a slight shift in the frequency value, we find that the trend is similar to the microwave measurement.

Accordingly, the system we proposed has found highly effective to distinguish heartbeat from noise-only observation. We are going to clarify the decision dependence on SNR as well as the usability in relation to how far a person can be away from the antennas in field experiments.

4 Summary

We proposed a human heartbeat detection/monitoring system with 2.5 GHz CW phase-sensitive microwave electronics. It employs chirp Z-transform and time-sequential neural prediction. The CZT realizes high resolution peak search in the frequency domain. We use a neural network to track adaptively the heartbeat signal which often has frequency fluctuation. The network learns the time-sequential peak frequency online in parallel to the detection and tracking. Even when the heartbeat frequency changes, the network finds and tracks the signal. The proposed system has high effectiveness in distinction between heartbeat and noise-only observations. It is useful in detection of human beings existing under debris after disasters such as earthquakes as well as in monitoring of patients in hospital.

References

1. Suzuki, A., Ikegami, T.: Biological information detection using ultra wideband signals. In: International Symposium on Medical Information and Communication Technology (ISMICT), pp. 73–77 (2015)
2. Chen, T.-C., Liu, J.-H., Chao, Pei-Yu., Li, P.-C.: Ultrawideband synthetic aperture radar for respiratory motion detection. IEEE Trans. Geosci. Remote Sens. **53**(7), 3749–3763 (2015)
3. Zhang, Y., Jiao, T., Lv, H., Li, S., Li, C., Guohua, L., Xiao, Yu., Li, Z., Wang, J.: An interference suppression technique for life detection using 5.75- and 35-GHz dual-frequency continuous-wave radar. IEEE Geosci. Remote Sens. Lett. **12**(3), 482–486 (2015)
4. Huang, M.-C., Liu, J.J., Xu, W., Gu, C., Li, C., Sarrafzadeh, M.: A self-calibrating radar sensor system for measuring vital signs. IEEE Trans. Biomed. Circ. Syst. (to be published)
5. Pino, E.J., De la Paz, A.D., Aqueveque, P.: Noninvasive monitoring device to evaluate sleep quality at mining facilities. IEEE Trans. Ind. Appl. **51**(1), 101–108 (2015)
6. Tan, S., Hirose, A.: Low-calculation-cost fading channel prediction using chirp Z-transform. Electron. Lett. **45**(8), 418–420 (2009)
7. Ozawa, S., Tan, S., Hirose, A.: Errors in channel prediction based on linear prediction in frequency domain - combination of time-domain and frequency-domain techniques. URSI Radio Sci. Bull. **337**, 25–29 (2011)
8. Ding, T., Hirose, A.: Fading channel prediction based on combination of complex-valued neural networks and chirp Z-transform. IEEE Trans. Neural Netw. Learn. Syst. **25**(9), 1686–1695 (2014)
9. Murata, T., Ding, T., Hirose, A.: Proposal of channel prediction by complex-valued neural networks that deals with polarization as a transverse wave entity. In: Arik, S., Huang, T., Lai, W.K., Liu, Q. (eds.) ICONIP 2015. LNCS, vol. 9491, pp. 541–549. Springer, Heidelberg (2015). doi:10.1007/978-3-319-26555-1_61
10. Hikosaka, M., Ding, T., Hirose, A.: Proposal of polarization state prediction using quaternion neural networks for fading channel prediction in mobile communications. In: International Joint Conference on Neural Networks (IJCNN), Vancouver (2016, to be presented)

Fast Dual-Tree Wavelet Composite Splitting Algorithms for Compressed Sensing MRI

Jianwu Li[1(✉)], Jinpeng Zhou[1], Qiang Tu[1], Javaria Ikram[1],
and Zhengchao Dong[2]

[1] Beijing Key Laboratory of Intelligent Information Technology,
School of Computer Science and Technology, Beijing Institute of Technology,
Beijing 100081, China
ljw@bit.edu.cn
[2] Department of Psychiatry, Columbia University, New York, NY 10032, USA

Abstract. We presented new reconstruction algorithms for compressed sensing magnetic resonance imaging (CS-MRI) based on the combination of the fast composite splitting algorithm (FCSA) and complex dual-tree wavelet transform (DT-CWT) and on the combination of FCSA and double density dual-tree wavelet transform (DDDT-DWT), respectively. We applied the bivariate thresholding to these two combinations. The proposed methods not only inherit the effectiveness and fast convergence of FCSA but also improve the sparse representation of both point-like and curve-like features. Experimental results validate the effectiveness and efficiency of the proposed methods.

Keywords: Compressed sensing · MRI · Composite splitting · Dual-tree wavelet · Double density dual-tree wavelet

1 Introduction

Magnetic resonance imaging (MRI) is an essential medical imaging modality with advantages of no harmful radiation and high soft tissue contrast. However, MRI is susceptible to motion artifacts because of its slow acquisition process or long data acquisition time. Recently developed compressed sensing (CS) [1, 2] technique shows that a signal with sparse representation can be reconstructed from significantly incomplete dataset sampled at rates or density lower than the Nyquist criterion. The emerging CS theory has been successfully applied to MRI and provided significant acceleration of acquisition process [3, 4].

Current research on compressed sensing MRI (CS-MRI) involves two attractive topics: One is the sparsifying transform, which aims at providing an efficient and sufficient sparse representation of MR images. Different sparse transforms have different characteristics in sparse representation. The wavelet can provide a good representation of point-like features but fail to represent curves because the conventional discrete wavelet transform (DWT) has shift sensitivity [5] and poor directionality [6]. Some efforts have been made to mitigate these disadvantages and the dual-tree complex wavelet (DT-CWT) was first used for compressed sensing in [7]. Though DT-CWT has approximate shift-invariance and better directional selectivity, it still may not be able to

© Springer International Publishing AG 2016
A. Hirose et al. (Eds.): ICONIP 2016, Part I, LNCS 9947, pp. 517–525, 2016.
DOI: 10.1007/978-3-319-46687-3_57

represent geometric regularity along the singularities well because it only has wavelets oriented in six directions which are not enough for clinical MR images [8]. Thus, the contourlet, an effective representation of curve-like features with low redundancy, has been proposed for CS-MRI [9]. However, the contourlet cannot represent point-like features well [10]. Thus, double-density dual-tree wavelet transform [11], which can provide effective sparse representation of both points and curves in MR images, was developed as a sparsifying transform for CS-MRI [8].

The other topic is the reconstruction algorithm aiming at providing an efficient method to solve the optimization problems in CS-MRI with expected reconstruction accuracy. In the pioneering study of CS-MRI [3], the conjugate gradient (CG) method was employed to solve an L_1 optimization. However, the CG has its bottleneck in computation. An improved CG with iterative algorithm was investigated for CS-MRI and better reconstruction quality was obtained [8]. Besides, other iteration method such as the iterative shrinkage-thresholding algorithm (ISTA) and its modified versions, such as the two-step ISTA [12] and the fast ISTA (FISTA) [13], were also applied to CS-MRI in succession. Though FISTA can provide faster convergence speed and better results, it can only work for simple optimization problem [14]. Since noise suppression tools such as total variation (TV) are usually used in CS-MRI [2], some methods are designed to solve the composite optimization problem that involves both L_1 norm term and TV term. In [14], a fast composite splitting algorithm (FCSA) for CS-MRI was proposed to solve the composite problem with powerful capability.

Many studies on these two topics have developed new approaches to the reconstruction of CS-MRI and have demonstrated their superiority in reconstruction quality and speed. Recently, a fast iterative contourlet thresholding algorithm (FICOTA) combining FISTA and contourlet transform, was proposed to improve the curve representation of MR images with fast computation [15]. However, FICOTA inherits the shortcomings of contourlet and FISTA mentioned above. Additionally, a method based on a 2D complex double-density dual-tree wavelet (CDDDT-DWT) has been proposed for CS-MRI to remedy the defects of wavelet [8], but its accuracy and efficiency are limited by the reconstruction algorithm it used.

Herein, we propose a new approach for the reconstruction of CS-MRI, using DT-CWT and DDDT-DWT, respectively, as sparsifying transforms of FCSA. We anticipate that the combinations, dubbed as DT-FCSA and DDDT-FCSA, may improve the sparse representation and reconstruction quality and speed the reconstruction.

2   CS-MRI with Dual-Tree Wavelet

We first briefly review basic CS-MRI model and then introduce the sparse transforms and reconstruction algorithm.

Suppose that x is an MR image, and F_u is an undersampled Fourier transform, then the undersampled measurement y of x can be defined as $y = F_u x$. Since MR images are naturally compressible, x can be reconstructed from the measurement y based on the compressed sensing theory. Let Ψ denote a transform operator that can represent x in a sparse domain. The reconstruction model can be formulated as an unconstrained problem [3]:

$$argmin_x \|\mathcal{F}_u x - y\|_2^2 + \lambda_1 \|x\|_{TV} + \lambda_2 \|\Psi_x\|_1 \tag{1}$$

where λ_1 and λ_2 are positive parameters for the total variation (TV) and L_1 regularization, respectively.

In this paper, we used DT-CWT and CDDDT-DWT as the sparse transform to provide better representation of both contours and points. We then implemented FCSA to solve the model in (1).

2.1 Sparse Transform

The conventional 2D discrete wavelet transform (DWT) has wavelets oriented at only 3 directions and its diagonal direction has no orientation, thus 2D DWT fails to sparsely represent curves and contours. The 2D dual-tree complex wavelet (DT-CWT) can provide 6 directions that support approximate shift-invariance and good orientation. Similarly, the double-density DWT (DD-DWT) improves shift-invariance by increasing redundancy. We hypothesize that a combination transform of DD-DWT and DT-CWT (CDDDT-DWT) would inherit the superiority of both DT-CWT and DD-DWT [8].

The CDDDT-DWT, implemented by parallel using four 2D DD-DWT, has 32 wavelets oriented at 16 directions and is capable of representing curves, contours and points simultaneously. Therefore, the CDDDT-DWT has higher redundancy.

We combined DT-CWT and CDDDT-DWT, respectively, with FCSA reconstruction algorithm to evaluate their performances.

2.2 FCSA Reconstruction

The FCSA was implemented by combining composite splitting denoising (CSD) method and FISTA [14]. The CSD method splits the reconstruction model into two sub-problems: L_1 term and TV term, solves each sub-problem, and then combines the sub-solutions linearly to obtain the final reconstruction. Besides, FCSA inherits the fast convergence and effective reconstruction of FISTA and thus can effectively solve the model (1).

3 Algorithm

We combined the aforementioned sparse representations with FCSA for the CS reconstruction and modified the shrink-thresholding process to fit the structure of the chosen sparse transform.

3.1   CS Reconstruction

The proposed combination was implemented in Algorithm 1 to solve the CS-MRI model (1). In Algorithm 1, $f(r^k) = \frac{1}{2}\|\mathcal{F}_u r^k - y\|_2^2$ is the data consistency function, $\nabla f(r^k) = \mathcal{F}_u^*(\mathcal{F}_u r^k - y)$ is the gradient of f at r^k, ρ and L_f are two positive parameters defined in [14], x has the same size as y and is initialized as zero. The L_1 and TV regularization problem was solved by a proximal map, which is defined as follows [13]:

$$prox_\rho(g)(x) = argmin_u\left\{g(u) + \frac{1}{2\rho}\|u - x\|^2\right\} \qquad (2)$$

where $g(x)$ a continuous convex function. Specifically, in solving L_1 term, we adopted the $prox_\rho(g)(x)$ process to the chosen sparse transform and implement $Thresholding(\cdot)$ with different shrink-thresholding methods.

3.2   Soft Thresholding

There are two frequently used thresholding methods: One is the hard thresholding, which is defined as:

$$\mathrm{hard}Thre(x) = x \cdot (|x| > T) \qquad (3)$$

where T is the threshold parameter. The other is the soft thresholding (HT), which is defined as:

$$softThre(x) = sgn(x) \cdot max(|x| - T, 0) \qquad (4)$$

The soft thresholding (ST) has better performance than the hard thresholding because the latter is discontinuous and yields abrupt artifacts in reconstruction [16, 17]. In [14], FCSA also implemented soft thresholding rather than the hard one. Therefore, we applied soft thresholding to all methods in this paper.

3.3   Bivariate Thresholding

The bivariate thresholding (BiT) is an effective and low-complexity denoising method and has better performance than the soft thresholding because it uses the joint statistics of the wavelet coefficients [18]. Hence, we further adopted the bivariate thresholding in our proposed method. The thresholding function can be described as follows:

$$biThre(x) = w \cdot \frac{max\left(\sqrt{w^2 + w_p^2} - \frac{\sqrt{3}\sigma^2}{\sigma_\epsilon}, 0\right)}{\sqrt{w^2 + w_p^2}} \qquad (5)$$

where w is a sparse transform coefficient and w_p is its parent coefficient in the next coarser scale. σ^2 is the estimation of noise variance calculated by coefficients of the finest scale using robust median estimator:

$$\sigma = \frac{Median(|w_{HH}|)}{06745}, \quad w_{HH} \in subband\ HH \tag{6}$$

$\sigma_{\in}^2$ is the marginal variance of coefficient w in a $M \times M$ neighborhood surrounding window [18]:

$$\sigma_{\in} = \sqrt{max\left(\frac{1}{M^2}\sum_{i,j}^{n} w_{ij}^2 - \sigma^2, 0\right)} \tag{7}$$

Algorithm 1 Reconstruction with FCSA based on DT-CWT

INPUTS: $\rho = {}^1/_{L_f}, \lambda_1, \lambda_2; r^1 = x^0; t^1 = 1;$

for $k = 1 : maxIter$

$\quad x_g = r^k - \rho\nabla f(r^k);$

$\quad x_{TV} = prox_\rho(2\lambda_1\|x\|_{TV})(x_g);$

$\quad \%\ x_{L_1} = prox_\rho(2\lambda_2\|\Phi x\|_1)(x_g)$

$\quad Coef_{DT} = \Psi_{DT}x;$

$\quad$**for** $j = 1 : scale_{DT}$

$\quad\quad realCoef = Coef_{DT}\{j\}\{1\};$

$\quad\quad imagCoef = Coef_{DT}\{j\}\{2\};$

$\quad\quad Coef_j = realCoef + i * imagCoef;$

$\quad\quad Coef_j = Thresholding(Coef_j, 2\lambda_2);$

$\quad\quad Coef_{DT}\{j\}\{1\} = Real(Coef_j);$

$\quad\quad Coef_{DT}\{j\}\{2\} = Imag(Coef_j);$

$\quad$**end**

$\quad x_{L_1} = \Psi_{DT}^* Coef_{DT};$

$\quad x^k = \dfrac{x_{TV} + x_{L_1}}{2};$

$\quad x^k = project(x^k, [l, u]);$

$\quad t^{k+1} = \dfrac{1 + \sqrt{1 + 4(t^k)^2}}{2};$

$\quad r^{k+1} = x^k + \dfrac{t^k - 1}{t^{k+1}}(x^k - x^{k-1});$

end

---

4 Experiments

We performed experiments based on two 256×256 MR images (Fig. 1). The proposed methods, DT-FCSA and DDDT-FCSA, were compared to the original FCSA [9] and FICOTA [15]. A 2D DT-CWT [19] with 6 scales and a 2D CDDDT-DWT [20] with 2 scales were used as sparse transforms.

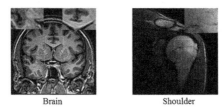

Fig. 1. Original MR Images

We implemented DT-FCSA and DDDT-FCSA with bivariate thresholding (BiT). For FCSA and FICOTA, we set the same regularization parameters as the original values [10]. For DT-FCSA and DDDT-FCSA, The regularization parameters (λ_1, λ_2) were set as $(0.01, 0.2)$ for DT-FCSA, and $(0.006, 0.1)$ for DDDT-FCSA. Besides, we chose the 'hi-lo' subband for dual-tree structure and 'hi1-hi1' subband for double density dual-tree to estimate σ. For each method, Gaussian white noise with standard deviation 0.01 was added to the k-space measurement y. Experiments were run on a 3.4-GHz Multi-core processor.

We used signal-to-noise ratio (SNR), peak SNR (PSNR), transferred edge information (TEI) [10] and L_2 norm error as criteria to assess the performance of the proposed methods.

We did experiments using different sampling ratios. Figures 2 and 3 show the brain and shoulder images reconstructed from 20 % sampling using different methods.

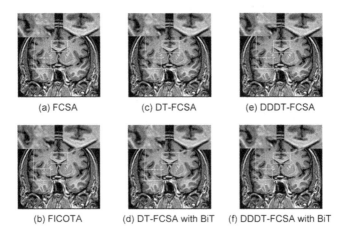

Fig. 2. Reconstruction of the **brain** image (sampling ratio = 20 %)

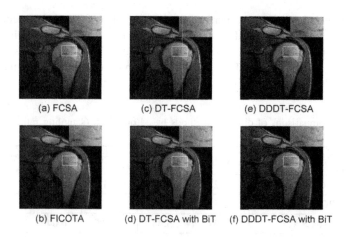

| (a) FCSA | (c) DT-FCSA | (e) DDDT-FCSA |
| (b) FICOTA | (d) DT-FCSA with BiT | (f) DDDT-FCSA with BiT |

Fig. 3. Reconstruction of the **shoulder** image (sampling ratio = 20 %)

Figure 4 shows the plots of PSNR and TEI versus sampling ratios. Tables 1 and 2 summarize the comparisons of different methods.

According to Figs. 2 and 3, both DT-FCSA and DDDT-FCSA improve the representation of curves. Meanwhile, they have better performance on denoising in reconstructed images than the FCSA method.

According to Fig. 4, both DT-FCSA and DDDT-FCSA show better results in TEI, which means better performance on edge strength and orientation preservation. When sampling ratio is lower than about 15 % for brain or 20 % for shoulder, the proposed

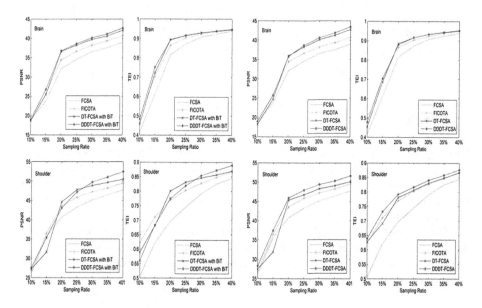

Fig. 4. PSNR and TEI versus sampling ratio for brain and shoulder

methods may not meet expected performance for denoising. The way of choosing the subband for σ cannot work well when the sampling rate is low because the lower the sampling ratio is, the larger the variance of the DT or DDDT coefficients is.

Furthermore, according to Tables 1 and 2, the proposed methods outperform the FCSA and FICOTA in term of five criteria with fast convergence speed.

Table 1. Comparisons of different methods based on **brain** (sampling ratio = 20 %)

Criterion / Algorithm	SNR	PSNR	TEI	L2 Error	Time(s)
FCSA	20.22	32.15	0.816	0.067	2.09
FICOTA	22.52	34.45	0.866	0.051	13.23
DT-FCSA	**24.01**	**35.94**	**0.885**	**0.043**	8.50
DDDT-FCSA	**23.87**	**35.80**	**0.879**	**0.044**	15.21
DT-FCSA with BiT	**24.69**	**36.62**	**0.895**	**0.040**	9.12
DDDT-FCSA with BiT	**24.86**	**36.80**	**0.895**	**0.039**	16.54

Table 2. Comparisons of different methods based on **shoulder** (sampling ratio = 20 %)

Criterion / Algorithm	SNR	PSNR	TEI	L2 Error	Time(s)
FCSA	21.91	40.71	0.6965	0.062	2.04
FICOTA	24.75	43.56	0.769	0.044	13.56
DT-FCSA	**26.54**	**45.37**	**0.781**	**0.036**	8.33
DDDT-FCSA	**27.07**	**45.88**	**0.792**	**0.034**	15.86
DT-FCSA with BiT	**25.82**	**44.63**	**0.801**	**0.039**	8.95
DDDT-FCSA with BiT	**24.27**	**43.08**	**0.775**	**0.047**	16.72

5   Conclusion

This paper proposes to combine FCSA with the complex dual-tree DWT or double density dual-tree DWT to enforce both the points and curves sparse representation with fast computation. Experimental results show our proposed methods can improve both precision and efficiency. Further significant improvement can be obtained when the proposed methods use bivariate thresholding.

The subbands in this work were chosen simply by experimental experience. Since the dual-tree or double density dual-tree wavelet structure has different variance in each subband, further effort is needed to investigate a more advanced thresholding to improve reconstruction effectiveness.

Acknowledgement. This work is supported by the National Natural Science Foundation of China (No. 61271374).

References

1. Donoho, D.: Compressed sensing. IEEE Trans. Inf. Theory **52**(4), 1289–1306 (2006)
2. Tsaig, Y., Donoho, D.L.: Extensions of compressed sensing. Sig. Process. **86**(3), 549–571 (2006)
3. Lustig, M., Donoho, D., Pauly, J.M.: Sparse MRI: the application of compressed sensing for rapid MR imaging. Magn. Reson. Med. **58**(6), 1182–1195 (2007)
4. Lustig, M., Donoho, D.L., Santos, J.M., Pauly, J.M.: Compressed sensing MRI. IEEE Signal Process. Mag. **25**(2), 72–82 (2008)
5. Kingsbury, N.: Complex wavelets for shift invariant analysis and filtering of signals. Appl. Comput. Harmonic Anal. **10**(3), 234–253 (2001)
6. Kingsbury, N.: Image processing with complex wavelets. Philos. Trans. R. Soc. A **357**, 2543–2560 (1999)
7. Kim, Y., Altbach, M., Trouard, T., Bilgin, A.: Compressed sensing using dual-tree complex wavelet transform. In: Proceedings of the International Society for Magnetic Resonance in Medicine, vol. 17, p. 2814 (2009)
8. Zhu, Z., Wahid, K., Babyn, P., Yang, R.: Compressed sensing-based MRI reconstruction using complex double-density dual-tree DWT. Int. J. Biomed. Imaging **2013**, 1–12 (2013)
9. Do, M.N., Vetterli, M.: The contourlet transform: an efficient directional multiresolution image representation. IEEE Trans. Image Process. **14**(12), 2091–2106 (2005)
10. Qu, X., Zhang, W., Guo, D., Cai, C., Cai, S., Chen, Z.: Iterative thresholding compressed sensing MRI based on contourlet transform. Inverse Prob. Sci. Eng. **18**, 737–758 (2010)
11. Selesnick, I.W.: The double-density dual-tree DWT. IEEE Trans. Sig. Process. **52**(5), 1304–1314 (2004)
12. Bioucas-Dias, J., Figueiredo, M.: A new TwIST: two step iterative shrinkage/thresholding algorithms for image restoration. IEEE Trans. Image Process. **16**(12), 2992–3004 (2007)
13. Beck, A., Teboulle, M.: A fast iterative shrinkage-thresholding algorithm for linear inverse problems. SIAM J. Imaging Sci. **2**(1), 183–202 (2009)
14. Huang, J., Zhang, S., Metaxas, D.: Efficient MR image reconstruction for compressed MR imaging. Med. Image Anal. **15**(5), 670–679 (2011)
15. Hao, W., Li, J., Qu, X., Dong, Z.: Fast iterative contourlet thresholding for compressed sensing MRI. Electron. Lett. **49**(19), 1206–1208 (2013)
16. Donoho, D.L.: De-noising by soft-thresholding. IEEE Trans. Inf. Theory **41**(3), 613–627 (1995)
17. Chang, S.G., Yu, B., Vetterli, M.: Adaptive wavelet thresholding for image denoising and compression. IEEE Trans. Image Process. **9**(9), 1532–1546 (2000)
18. Sendur, L., Selesnick, I.W.: Bivariate shrinkage functions for wavelet-based denoising exploiting interscale dependency. IEEE Trans. Sig. Process. **50**(11), 2744–2756 (2002)
19. http://eeweb.poly.edu/iselesni/WaveletSoftware/index.html
20. http://eeweb.poly.edu/iselesni/DoubleSoftware/index.html

Implementation of a Modular Growing When Required Neural Gas Architecture for Recognition of Falls

Frederico B. Klein[1]([✉]), Karla Štěpánová[2], and Angelo Cangelosi[1]

[1] School of Computing, Electronics and Mathematics,
Plymouth University, Plymouth, UK
{frederico.klein,a.cangelosi}@plymouth.ac.uk
[2] Department of Cybernetics, Czech Technical University, Prague, Czech Republic
stepakar@fel.cvut.cz
http://www.plymouth.ac.uk, http://www.fel.cvut.cz

Abstract. In this paper we aim for the replication of a state of the art architecture for recognition of human actions using skeleton poses obtained from a depth sensor. We review the usefulness of accurate human action recognition in the field of robotic elderly care, focusing on fall detection. We attempt fall recognition using a chained Growing When Required neural gas classifier that is fed only skeleton joints data. We test this architecture against Recurrent SOMs (RSOMs) to classify the TST Fall detection database ver. 2, a specialised dataset for fall sequences. We also introduce a simplified mathematical model of falls for easier and faster bench-testing of classification algorithms for fall detection.

The outcome of classifying falls from our mathematical model was successful with an accuracy of $97.12 \pm 1.65\,\%$ and from the TST Fall detection database ver. 2 with an accuracy of $90.2 \pm 2.68\,\%$ when a filter was added.

Keywords: Action recognition · Falls · Neural networks · Neural gas · Topological classifiers · Socially assistive robotics

1 Introduction

In the field of robotics, activity detection [11] is a fundamental concept if the robots are used in a setting where they are expected to cooperate with humans. The initial approaches to the detection of human actions involved the processing of RGB images, and as such were proven to be a hard problem due to the difficulty in segmenting the human body from the background and accurately processing pose information. Recently however this task was made considerably easier with the introduction of skeleton tracking based on depth-sensing cameras as implemented by the Microsoft Kinect and as it steadily improves, it also allows for more serious tasks that depend on activity recognition to be tackled, such as activity detection. We will focus on its use in the context of socially assistive robotics for social elderly care.

© Springer International Publishing AG 2016
A. Hirose et al. (Eds.): ICONIP 2016, Part I, LNCS 9947, pp. 526–534, 2016.
DOI: 10.1007/978-3-319-46687-3_58

1.1 Ageing Population

With the ageing of populations around the world, elderly care is a field of growing concern. Many different technological aids [3] are being developed specifically for this population and robotics has emerged as a possible solution as the mobilisation of human caretakers for such a large amount of persons seems infeasible. While robots in regards to human-robot-interaction are yet to find a particular field in which it is undeniably useful, an interesting approach [18] to their use is finding newer areas in which they can nothing but excel, simply because there are no persons nor other technology available to perform that task. One of such tasks is around the clock health monitoring for independent living.

1.2 Our Task of Interest: Fall Detection

It is medical fact [9] that diseases that can present themselves as a loss of consciousness, such as strokes and heart infarctions - those two, the leading causes of death world wide - can have excellent prognosis if treated within 3 h. Particularly cardiac arrests present a survival rate of about one in each three subjects, if CPR and defibrillation are initiated in less than 5 min, whereas the probability of survival without any help is virtually zero [19]. Also other diseases such as pneumonia or COPD exacerbations do tend to have better prognosis [20] if treated promptly. One must take care as to not make bold assumptions, even more under the light that major reviews [5] are yet to reveal clear benefits of telemedicine, but some interesting recent results [4,8] demonstrate COPD as a likely candidate to benefit from remote monitoring.

The specific task of fall detection has recently attracted a lot of research, with a primary focus on smart home environments. In fact most fall detection systems [21] involve wearing special sensors device with accelerometers or detectors built on the floor or a combination of video and wearable devices with a sensor fusion approach in order to increase the accuracy of detection even further. These approaches although more simple (and therefore robust) have however the limitation of needing either a sensor to be worn at all times, or that the person's house to be adapted for this, which in practice will vastly limit its adherence. We see coding fall detection into some sort of a multi-functional robotic companion- that could have as one of its many functionalities: fall detection-as a reasonable solution to this problem. A robot can follow the user in different environment, position itself in order to prevent image occlusion and we avoid the need to renovate someone's house or remember always to wear a sensor.

1.3 Our Approach: Use the Parisi's Multilayer GWR Classifier

Using an unsupervised method for topological description [6] of tasks is not a new idea, since these methods have many possible advantages such as the ability to "operate autonomously, on-line or life-long, and in a non-stationary environment". We chose to replicate the infrastructure implemented by [15] for it's overall performance in the CAD60 database and the theoretical generality of the method. In this paper we describe our implementation of a classifier based on an unsupervised Growing When Required Neural Gas with a sliding window scheme for time integration and chained in multiple layers to implement noise removal.

Source code (available in www.github.com/frederico-klein/ICONIP2016) of the Growing When Required Neural Gas implementation (in Matlab and Julia) is provided, as well as the full classification architecture and the inverted pendulum model (only in Matlab).

2   Materials and Methods

2.1   Justification for the Chosen Architecture: A Chained GWR Sliding Window Topological Classifier

A detailed discussion of different types of neural gases is outside of the scope of this text. For a more in depth understanding one should probably first refer to Martinetz's paper [12] that implemented the first neural gas and later to Marsland's paper [13], that implemented the Growing When Required neural gas (GWR). The justification of using multiple chained gases (as opposed to one) is, first the biological plausibility reviewed extensively by Parisi but secondly probably due to necessity regarding the way too long execution time of a gas with a high number of dimensions. Finally one must add that, although neural gases, due to their nature, adjust to data that changes over time, this feature does not seem useful in tracking movement. For this function a sliding window scheme was used.

The dataset we chose to test our implementation was the TST v2 dataset contains skeleton positions Microsoft Kinect v2 and IMU data for 11 subjects performing either ADLs (activities daily living) and simulated falls. The subjects were between 22 and 39 years old, with different height (1.62–1.97 m) and build. Each of the two main groups (ADLs or Falls) contains 4 activities that are repeated three times by each subject [7]. For our present study only the skeleton joints in depth and skeleton space and time information were used. The accelerometer, as well as other data, were not used for our algorithm.

In addition to the real dataset a simplified stick model was developed to test the ability of a Growing When Required multilayer with a sliding window classifier to discriminate between action sequences that included a fall. We modelled 2 different activities, a fall and a walk as the movement of a stick in a 3D space and then simply substituted the stick for a typical skeleton.

Fall. We simulated fall of a person by a free falling inverted pendulum rod with a random initial pitch angular velocity θ and perfect slippage. It can be shown [16] that the kinematics differential equations that describe angle and position changes for a rod are:

$$-mg\frac{L}{2}cos\theta = \left(I_c + \frac{mL^2}{4}cos^2\theta\right)\ddot{\theta} - \frac{mL^2}{4}cos\theta sin\dot{\theta}2 \tag{1}$$

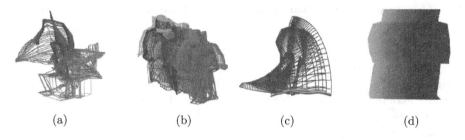

(a)   (b)   (c)   (d)

Fig. 1. (a) A typical fall from TST v2 dataset. (b) one of the ADLs from the TST v2 dataset, a walk. (c) a typical fall from our model. (d) a "walk" from our model.

$$0 = m\ddot{x}_c \tag{2}$$

And, approximating a person by a slender rod, one has $I_c = \frac{mL^2}{3}$. The model also was given simulation parameters to add random noise in the variables of height (1.6–1.9 m), initial position (within a square area) and any initial yaw angle.

Walk. The simulation of a person's walk was done by simply doing a linear space of displacements inside the area that would be covered by the Kinect sensor, with random initial positions and walking angle (Fig. 1).

2.2 Skeleton Data

The algorithm presented uses skeleton data and not RGB-D raw images. A more thorough descriptions [17] of the data obtained from the depth sensor should be referenced, but in short it is a set of J points (where J is the number of joints) with x, y and z coordinates, each representing a landmark on the body in time [10] in a 3D space. We represent thus a particular pose as the concatenation of these J points, such as that for each time frame k we have a pose p represented by the matrix:

$$p(k) = \begin{bmatrix} j_{1x}(k) & j_{1y}(k) & j_{1z}(k) \\ j_{2x}(k) & j_{2y}(k) & j_{2z}(k) \\ \cdots \\ j_{Jx}(k) & j_{Jy}(k) & j_{Jz}(k) \end{bmatrix} \tag{3}$$

An action sequence represented on discrete time steps $1...K$ could therefore represented as the multidimensional array resulting of the sequential concatenation of the k-th pose matrices. To use the pose information with a gas we change the representation of the pose matrix $p(k)$ into a vector size $3 * J$ and the action sequence is the horizontal concatenation of the all the k-th, $p(k)$ matrices. One may thus understand the pose vector as a single point in a high dimensional space and an action sequence as a necessarily continuous trajectory in that space.

2.3 Construction and Randomisation of Training and Validation Sets

The dataset was separated into training and validation sets containing 80 % and 20 % of data respectively, before each training similarly to a repeated learning-testing method [1,2]. They were separated by subject, so that each subject had all of its actions belonging exclusively to one set. This was done to describe a more realistic testing scenario, in which the subject performing the activities is completely new having no activity data of himself in the training set, preventing bias in accuracy estimation due to overfitting the training set.

2.4 Preconditioning

The GWR algorithm is not translation invariant, so the first action performed on the data was to select a joint - based on our reference algorithm we used the hips and subtracted the offset from the hips joint in both the z and x coordinates from all other joint vectors. Secondly, we normalised (scaled) the data so that after scaling variance of the data would be equal to 1. The final step was to implement a centroid generating function, so to generate a smaller dimensionality representation of the skeleton poses, in a similar fashion to the function tested by Parisi. We created a model of 3 centroids that were the average position of the skeleton points such that the upper centroid was composed by the joints: head, neck, left shoulder, right shoulder, left elbow, right elbow; middle centroid corresponded to torso and lower centroid: left knee, right knee, left hip, right hip. Many other preconditioning functions are available on the supplied code and maybe be tried by the interested reader.

2.5 Classifier Architecture

The classifier was implemented as a serial chaining of gas subunits. This was done to enable different structures to be tried with minimal effort. All classification attempts in this text were done using 5 gas subunits linked in manner as to implement the architecture in Parisi's [15] paper (see Fig. 2), that is, 2 parallel sets of 2 gas subunits in series, each stream dealing with either pose positions or pose velocities and a last gas that integrates both.

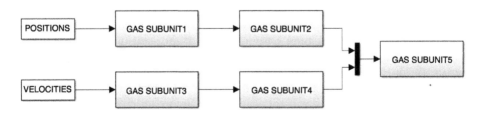

Fig. 2. Diagram of the classifier architecture.

For each gas subunit there are 5 main chained elements that are responsible for implementation estimation and classification:

- Sliding Window: implements the temporal concatenation of sample (also implements concatenation of multiple streams in case they exist).
- Gas Creator: receives data samples $p(k)$ or concatenated poses $W_l(k)$ and implements the learning algorithm for either the Growing When Required neural gas or the Growing Neural Gas.
- Mapping: finds the best matching pose from the nodes matrix A corresponding to each sample from the dataset.
- Labeller: simple labelling function that assigns the label of estimated concatenated pose as the same as the label of the pose to which it best matches.
- Activation checker: during training, checks to see if points are able to be well represented by the gas, and if not removes them from the sample.

3   Results and Discussion

For all the results here presented, the simulation parameters for the GWR neural gas are the same as in our reference paper [15].

3.1   Cornell CAD60 Dataset

As a means of comparing our implementation with that of Parisi, we also tested our architecture on the CAD60 dataset. Apparently our implementation does a lot of overfitting, as it reaches 99.6 % accuracy on the training set (average: 99.38 ± 0.2 % for 8 trials) but only reaches 71.7 % on the validation set. We noticed however that misclassifications were limited to some specific actions, with most having the same accuracy (greater than 90 %) in both sets. It is our conjecture that this difference reflects that the CAD60 dataset is too small to allow our stricter cross-validation method to produce generalization.

3.2   Falling Stick Model

Our algorithm, even with a much smaller network (100 nodes), seems to be quite consistently capable of classifying our faux fall/walk model. We simulated 20 subjects performing either a fall or a walk. The peak accuracy on the validation set of our implementation was 98.33 % (average: 97.12 ± 1.65 % for 8 trials).

3.3   TST Fall Detection ver.2 Dataset

Learning Across Layers. In order to understand how learning happens across layers we analyse the output classification from 5 gases with 1000 nodes run over 10 epochs (see Table 1) the results reflect what we would expect: there is a steady increase as we progress through the layers and there is a gain in accuracy.

Table 1. Progression of classification accuracy within different layers for chained GWR Neural Gas classifier with 1000 nodes and run over 10 epochs (for 8 trials).

Gas element	Validation set	Training set
GWR gas 1 Pos	67.69 ± 0.73 %	93.04 ± 0.12 %
GWR gas 2 Vel	64.80 ± 0.93 %	69.43 ± 0.66 %
GWR gas 3 Pos	66.93 ± 1.02 %	90.45 ± 0.18 %
GWR gas 4 Vel	65.14 ± 0.77 %	70.66 ± 0.65 %
GWR gas 5 STS	73.99 ± 1.16 %	88.35 ± 0.41 %

Mode Filter. With the intention of performing some sort of temporal filtering, we implemented a moving mode filter. The moving mode filter had an important positive effect on the classification results of the TST v2 database (see Table 2). Highest classification accuracy achieved (See Table 3) was 94.2 % on the validation set by 3rd gas with 1000 nodes and 10 epochs with mode filter length of 35 data samples. One must note that adding a moving mode filter of size 35 means a delay of 11.67 s (since we need $(9 + 1) * 35$ samples @30 Hz) much more than the 0.6 s Parisi reported.

Table 2. Accuracies (in %) after applying the moving mode filter on classification results of the final gas unit of the classifier (for 8, 8 and 1 trials respectively).

Epochs	10		20		30	
Filter length	Val	Train	Val	Train	Val	Train
5	81.25 ± 1.44	95.05 ± 0.69	79.72 ± 2.09	79.72 ± 2.09	82.7	94.9
10	85.95 ± 2.13	95.74 ± 0.54	83.87 ± 2.93	96.00 ± 0.38	86.9	95.5
15	88.57 ± 2.48	95.31 ± 0.36	85.91 ± 2.93	95.46 ± 0.30	88.6	97.0
20	89.95 ± 2.36	95.34 ± 0.23	86.70 ± 4.08	95.26 ± 0.31	88.6	95.0
25	90.16 ± 2.28	94.32 ± 0.27	87.37 ± 3.79	94.47 ± 0.25	89.2	95.4
35	90.20 ± 2.68	92.29 ± 0.37	88.49 ± 3.83	92.44 ± 0.38	91.5	93.3
40	89.28 ± 2.68	91.23 ± 0.32	87.75 ± 3.82	91.33 ± 0.35	91.5	92.2
50	87.39 ± 2.06	88.84 ± 0.33	85.35 ± 3.39	88.80 ± 0.49	91.3	89.9

Comparison with RSOM. As a means of comparing the performance of our implementation, we also classified the TST v2 dataset using an RSOM implementation. The RSOM used the same preconditioning as we did for the chained gas classifier and a set of 3 consecutive poses $(p(k), p(k-1), p(k-2))$. The simulation parameters were: 900 nodes, 30 epochs, method 'RSOMHebbV01'. The peak accuracy on the validation set of the RSOM with these parameters was 78.76 % (average: 77.67 ± 0.77 % for 5 trials).

Table 3. Confusion matrix for our most accurate gas classifier. The calculated accuracy for the validation set is 94.2 %, higher than the 92.0 % training set.

Validation set		Target		Training set		Target	
		1	2			1	2
Output	1	1532	36	Output	1	4006	258
	2	124	1054		2	328	2746

4   Conclusion

The resulting classification scheme does the task which we want, that is, discriminate falls within the TST v2 dataset, it does it better than the RSOM and it does it consistently with around 90.2 ± 2.68 % accuracy while using the mode filter. We believed we achieved our goal and we have now a classifier of falls with openly accessible code that will hopefully encourage persons into designing experiments using fall detection or using neural gases for classification of hard to classify data.

Acknowledgment. This work was partially supported by CNPq Brazil (scholarship 232590/2014-1) and by SGS grant No. 10/279/OHK3/3T/13, sponsored by the CTU in Prague, Czech Republic.

References

1. Arlot, S., Celisse, A.: A survey of cross-validation procedures for model selection. Stat. Surv. **4**, 40–79 (2010). doi:10.1214/09-SS054
2. Burman, P.: A comparative study of ordinary cross-validation, v-fold cross-validation and the repeated learning-testing methods. Biometrika **76**(3), 503–514 (1989)
3. Chan, M., Estéve, D., Escriba, C., Campo, E.: A review of smart homes-present state and future challenges. Comput. Methods Prog. Biomed. **91**, 55–81 (2008). doi:10.1016/j.cmpb.2008.02.001
4. Fernandez-Granero, M.A., Sanchez-Morillo, D., Leon-Jimenez, A.: Computerised analysis of telemonitored respiratory sounds for predicting acute exacerbations of copd. Sensors (Basel) **15**, 26978–26996 (2015). doi:10.3390/s151026978
5. Flodgren, G., Rachas, A., Farmer, A.J., Inzitari, M., Shepperd, S.: Interactive telemedicine: effects on professional practice and health care outcomes. In: Cochrane Database of Systematic Reviews. Wiley (2015)
6. Furao, S., Hasegawa, O.: An incremental network for on-line unsupervised classification and topology learning. Neural Netw. **19**, 90–106 (2006). doi:10.1016/j.neunet.2005.04.006
7. Gasparrini, S., Cippitelli, E., Gambi, E., Spinsante, S., Wåhslén, J., Orhan, I., Lindh, T.: Proposal and experimental evaluation of fall detection solution basedon wearable and depth data fusion. In: Loshkovska, S., Koceski, S. (eds.) ICTInnovations 2015, Advances in Intelligent Systems and Computing, pp. 99–108. Springer International Publishing, Switzerland (2016)

8. Ho, T.-W., Huang, C.-T., Chiu, H.-C., Ruan, S.-Y., Tsai, Y.-J., Yu, C.-J., Lai, F.: Effectiveness of telemonitoring in patients with chronic obstructive pulmonary disease in Taiwan-a randomized controlled Trial. Sci. Rep. **6** (2016). doi:10.1038/srep23797

9. Jauch, E.C., Saver, J.L., Demaerschalk, B.M., Khatri, P., McMullan Jr., P.W., Qureshi, A.I., Rosenfield, K., Scott, P.A., Summers, D.R., Wang, D.Z.: AHA/ASA Guideline. Stroke (2013)

10. JointType enumeration [WWW Document], n.d. https://msdn.microsoft.com/en-us/library/microsoft.kinect.jointtype.aspx. Accessed 14 May 16

11. Koppula, H.S., Saxena, A.: Anticipating human activities using object affordances for reactive robotic response. IEEE Trans. Pattern Anal. Mach. Intell. **38**, 14–29 (2016)

12. Martinetz, T.M., Schulten, K.J.: A "Neural Gas" networklearns topologies. In: Kohonen, T., Mäkisara, K., Simula, O., Kangas, J. (eds.) Proceedings of the International Conference on Artificial Neural Networks 1991, Espoo, Finland, pp. 397–402, Amsterdam, North-Holland, New York (1991)

13. Marsland, S., Shapiro, J., Nehmzow, U.: A self-organising network that grows when required. Neural Netw. **15**, 1041–1058 (2002)

14. Parisi, G., Wermter, S., others.: Hierarchical SOM-based detection of novel behavior for 3D human tracking. In: The 2013 International Joint Conference on Neural Networks (IJCNN), pp. 1–8. IEEE (2011)

15. Parisi, G.I., Weber, C., Wermter, S.: Self-organizing neural integration of pose-motion features for human action recognition. Front. Neurorobotics **9**, (2015). doi:10.3389/fnbot.2015.00003

16. Peacock, T., Hadjiconstantinou, N.: Course materials for 2.003J/1.053J dynamics and control I, Spring (2007). MIT OpenCourseWare (http://ocw.mit.edu), Massachusetts Institute of Technology. Accessed 13 May 2016

17. Prime sensor^{TM} NITE 1.3 framework programmer's guide - NITE.pdf. http://pr.cs.cornell.edu/humanactivities/data/NITE.pdf. Accessed 14 May 2016

18. Rabbitt, S.M., Kazdin, A.E., Scassellati, B.: Applications and recommendations for expanded use. Clin. Psychol. Rev. 35, 35–46. doi:10.1016/j.cpr.2014.07.001

19. Valenzuela, T.D., Roe, D.J., Cretin, S., Spaite, D.W., Larsen, M.P.: Estimating effectiveness of cardiac arrest interventions: a logistic regression survival model. Circulation **96**, 3308–3313 (1997). doi:10.1161/01.CIR.96.10.3308

20. Wilkinson, T.M.A., Donaldson, G.C., Hurst, J.R., Seemungal, T.A.R., Wedzicha, J.A.: Early therapy improves outcomes of exacerbations of chronic obstructive pulmonary disease. Am. J. Respir. Crit. Care Med. **169**, 1298–1303 (2004). doi:10.1164/rccm.200310-1443OC

21. Yongli, G., Yin, O.S., Han, P.Y.: State of the art: a study on fall detection. World Acad. Sci. Eng. Technol. **62**, 294–298 (2012)

Data Mining and Cybersecurity Workshop

Botnet Detection Using Graphical Lasso with Graph Density

Chansu Han[1]([✉]), Kento Kono[1], Shoma Tanaka[1],
Masanori Kawakita[2], and Jun'ichi Takeuchi[2]

[1] Graduate School of Information Science and Electrical Engineering,
Kyushu University, Fukuoka, Japan
tak@me.inf.kyushu-u.ac.jp
[2] Faculty of Information Science and Electrical Engineering,
Kyushu University, Fukuoka, Japan

Abstract. A botnet detection method using the graphical lasso is studied. Hamasaki et al. proposed a botnet detection method based on graphical lasso applied on darknet traffic, which captures change points of outputs of graphical lasso caused by a botnet activity. In their method, they estimate cooperative relationship of bots using graphical lasso. If the regularization coefficient of graphical lasso is appropriately tuned, it can remove false cooperative relationships to some extent. Though they represent the cooperative relationships of bots as a graph, they didn't use its graphical properties. We propose a new method of botnet detection based on 'graph density', for which we introduce a new method to set the regularization coefficient automatically. The effectiveness of the proposed method is illustrated by experiments on darknet data.

Keywords: Botnet · Darknet · Graphical lasso · Graph density

1 Introduction

In recent years, damage caused by cyber crime is a serious problem and technology of cyber security has become important. Anti-virus softwares using pattern matching and packet monitoring can detect only known attacks. In contrast, a large number of unknown attacks have been significantly increasing these days. According to this issue, there is a limit to detect botnet on usual method. Because of these facts, new detection methods based on machine learning which can automate detection and cope with unknown attacks have been attracting attention.

In this study, we focus on the problems caused by botnets in cyber crime. A group of infected computers with bots are called a botnet, that operates as part of a huge network. Bots are computer viruses for remotely controlling computers. In a botnet, infected hosts cooperate with other computers and attack other computers.

Hamasaki et al. [8] proposed a detection method based on the cooperative property of botnet traffic (the most hosts in botnet works cooperatively by synchronization [1]). In their method, they estimate the cooperative relationship

© Springer International Publishing AG 2016
A. Hirose et al. (Eds.): ICONIP 2016, Part I, LNCS 9947, pp. 537–545, 2016.
DOI: 10.1007/978-3-319-46687-3_59

of hosts and represent it as a Gaussian graphical model (GGM) by using the graphical lasso [3,7]. A remarkable property of graphical lasso is that it uses a sparse regularization. If the degree of sparse regularization is appropriately tuned, then the graphical lasso can detect and remove false cooperative relationships to some extent. Moreover, it scored a degree of changes of cooperative relationships using the Kullback-Leibler divergence between the obtained GGMs. If the score exceeds a threshold, it issues an alert. Although this method can detect botnets, it also shows false alerts a lot. To overcome this problem, Mukai et al. [9] modified the score calculation by using moving average of scores, which stabilizes behavior of score and reduces false alerts [4].

However, in the method of Hamasaki et al., though they represent the estimated cooperative relationships as a graph, they didn't exploit any graphical property of GGMs. In this paper, we propose a new method for botnet detection based on graph density of GGM. The graph density is an indicator to measure the ratio of the number of edges to the number of possible edges. Regularization coefficient must be tuned manually in the graphical lasso, but we propose a new method to select it automatically and appropriately by exploiting the graph density. Finally, we illustrate effectiveness of the proposed method via numerical experiments with real darknet data which are provided by PRACTICE [11]. We think our method is successful in automating adjustment of the value of regularization coefficient.

2   Darknet

A darknet is an accessible and unused IP address space on the Internet. Hence, received packets by darknet are almost caused by malwares, it is expected that the signal-noise ratio is very high. In this study, we used darknet data that are provided by the international collaboration darknet data. Which are provided by Proactive Response Against Cyber-Attacks Through International Collaborative Exchange (PRACTICE) [11]. PRACTICE is a project by Ministry of Internal Affairs and Communications (MIC) of Japan. Each record of darknet data contains information about received packets such as packet transmission time, source IP address, source port, destination IP address, destination port, protocol, etc. PRACTICE's darknets are installed in 10 countries. Among these, we used 3 countries' darknet data for experiments, sensorA, sensorB and sensorC (in country A, B and C). The number of IP addresses of sensorA, sensorB and sensorC is 128, 124 and 125, respectively.

3   Sparse Structure Learning Using the Graphical Lasso

In the method of Hamasaki et al. [8], botnets are detected by capturing changes in cooperative relationships between source hosts of darknet data. These cooperative relationships are expressed as a graph representing the dependencies between random variables (hosts) by using the graphical lasso [3,7], which learns dependencies between Gaussian random variables using ℓ_1-regularization and represents it as a Graphical Gaussian model (GGM) [2].

Let x denote an N-dimensional random variable according to a multivariate Gaussian distribution

$$N(x|\mu, \Sigma) = \frac{|\Sigma^{-1}|^{1/2}}{(2\pi)^{n/2}} \exp\left(-\frac{1}{2}(x - \mu)^T \Sigma^{-1}(x - \mu)\right), \tag{1}$$

where $x = (x_1, x_2, ..., x_N) \in \mathbb{R}^N$, $\mu \in \mathbb{R}^N$, and $\Sigma \in \mathbb{R}^{N \times N}$. The μ is a vector which represents the expectation of the random variable x, and Σ is a covariance matrix, which is non-negative definite. Under the assumption of the multivariate normal distribution, if $(\Sigma^{-1})_{i,j} = 0$, then x_i and x_j are conditionally independent given the other variables [5]. The matrix Σ^{-1} is referred to as the precision matrix.

To estimate Σ^{-1}, we employ the graphical lasso algorithm to learn the cooperative relationship between variables from the given data. Let $x^{(m)} \in \mathbb{R}^N$ denote the observed data at the mth time period. For each $i \in \{1, 2, ..., N\}$, let $x_i^{(m)}(m = 1, ..., M)$, let $x_i^{(1)}, x_i^{(2)}, ..., x_i^{(M)}$ a time series of the number of packets sent by the ith source host in a unit time interval. Let

$$D = \{x^{(m)} | m = 1, ..., M\}, \tag{2}$$

which is a data set of a time period. The following is the log likelihood function for the multivariate Gaussian model (1) given data D :

$$\ln \prod_{m=1}^{M} N(x^{(m)}|\hat{\mu}, \Sigma) = \text{const.} + \frac{M}{2}\left\{\ln|\Sigma^{-1}| - \text{Tr}(\hat{S}\Sigma^{-1})\right\}, \tag{3}$$

where $\hat{\mu}$ is the sample mean of $x^{(m)}$s that is the maximum likelihood estimation of μ and $|A|$ denotes the determinant of a square matrix A. Here $\hat{S}$ is the sample covariance matrix defined by

$$\hat{S}_{ij} = \frac{1}{M} \sum_{m=1}^{M} (x_i^{(m)} - \hat{\mu}_i)(x_j^{(m)} - \hat{\mu}_j). \tag{4}$$

As for the graphical lasso, it maximizes log likelihood function with ℓ_1-regularization term in the form

$$\ln|\Sigma^{-1}| - \text{Tr}(\hat{S}\Sigma^{-1}) - r||\Sigma^{-1}||_1, \tag{5}$$

where $||\Sigma^{-1}||_1$ is a ℓ_1-norm of Σ^{-1} defined as

$$||\Sigma^{-1}||_1 = \sum_{ij}^{N} |(\Sigma^{-1})_{ij}|. \tag{6}$$

The graphical lasso is an algorithm to solve the equation (5) in high accuracy and high speed. By the effect of ℓ_1-regularization term, Σ^{-1} is likely to be sparse, where the sparsity of the estimate depends on the value of r.

The method of Hamasaki et al. is a procedure as follows. It divides input data to time periods of a certain length, and obtains a time series of GGMs by the graphical lasso, each of which corresponds to a time period. Then it scores a degree of change in the time series of GGMs using the Kullback-Leibler divergence between them. If the score exceeds a threshold, it issues an alert. Although this method can detect botnets, it also shows false alerts a lot. To overcome this problem, Mukai et al. [9] modified the score calculation by introducing moving average, which stabilizes behavior of scores and reduces false alerts [4].

4 Proposed Method

4.1 Issue of the Alert Using the Graph Density

In the graphical lasso algorithm, by adjusting the value of regularization coefficient r of (4) appropriately, it is possible to estimate the precision matrix to be sparse. The appropriate value of r depends on properties of the data. Mukai et al. [10] proposed a method to issue an alert by exploiting the graph density. Let the number of nodes of a graph represented by precision matrix be N, and let the number of edges of the graph be E, the graph density defined as

$$\frac{E}{N^2}, \tag{7}$$

which is almost the ratio of the number of edges and the number of possible edges. The usual definition is $E/N(N-1)$, but we employ (7) for simplicity. Accordingly, when the graph density is high, it is estimated that there are cooperative relationships between a large number of hosts. However, in the method of Mukai et al., there was no way to adjust the value of regularization coefficient appropriately.

To illustrate this problem, we show line graphs of time series of graph densities with different values of r in Figs. 1, 2 and 3. The line graphs are obtained for international collaboration darknet data of October 23 to 25, 2015 of sensorA. The values of r in Figs. 1, 2 and 3 are 0.1, 0.8, and 2, respectively. When $r = 0.1$, overall graph densities turn out to be high (Fig. 1). In contrast, when $r = 2$ overall graph densities turn out to be low (Fig. 3). Compared to the cases of Figs. 1 and 3, in Fig. 2 it is easy to know the time period for which the graph density is significantly high. Hence, if we establish a method to select $r = 0.8$ automatically, then we can use graph density as a measure to detect the time periods when many hosts have stronger connection to each other than other time, which can be used for detection of botnet activity.

Below, we show our proposal to solve this issue. Assume that we have n time periods in a given darknet data for a certain time interval, and we calculate a graph density for each time period by using the graphical lasso algorithm with a certain fixed r. Let $D = (d_1, d_2, ..., d_n)$ be a sequence of graph densities observed in the time interval, arranged in descending order. Take a natural number i, let $D_i = (d_i, d_{i+1}, ..., d_n)$, and let σ_i^2 denote the variance of the elements of D_i.

Also, let $D'_i = (d_{i+1}, d_{i+2}, ..., d_n)$, and let σ'^2_i denote the variance of the elements of D'_i. Let the i_{max} be the maximum of i which satisfies $\sigma'^2_i / \sigma^2_i < 0.95$, issue an alert when the time period corresponds to $d_1, d_2, ..., d_{i_{max}}$. Also, regard as botnet only host group which send packets to a port 30 % or more of the host of the time period.

Moreover, we attempt to automate the adjustment of the value of regularization coefficient r. Let σ denote the standard deviation of $d_{i_{max}+1}, d_{i_{max}+2}, ..., d_n$, and let μ denote the mean of $d_{i_{max}+1}, d_{i_{max}+2}, ..., d_n$. We illustrate it in Fig. 2. Then, we perform this procedure for various values of r and employ the value of r which maximizes $(d_1 - \mu)/\sigma$.

4.2 Application of Proposed Method to Recent Real Data

In this section, we illustrate a part of our experiments on the real darknet data, which shows the effectiveness of the proposed method. We used sensorA, B and C which are provided by PRACTICE. At the same time, the method of Mukai et al. which used a moving average also be applied.

We set the time period be 10 min and the time interval be 1 min, i.e. we set $M = 10$ in Eq. (2). As pre-processing, we removed the packets with RST flag or SYN/ACK. In addition, concentrated high frequent packets were removed by

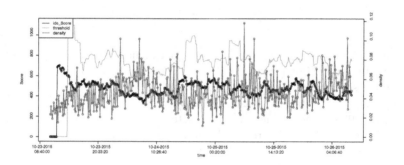

Fig. 1. sensorA, October 23 to 25, 2015, $r = 0.1$

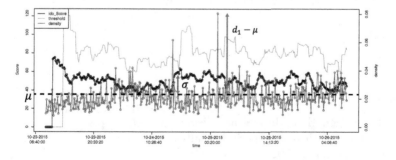

Fig. 2. sensorA, October 23 to 25, 2015, $r = 0.8$

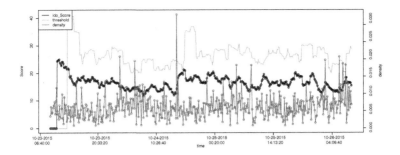

Fig. 3. sensorA, October 23 to 25, 2015, $r = 2$

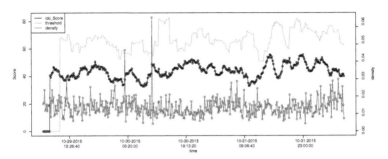

Fig. 4. sensorA, October 29 to 31, 2015

screening [6]. All the time shown in the experimental results in this paper is Japanese Standard Time (JST).

Figure 4 shows the line graph of graph densities for the data of sensorA October 29 to 31, 2015. The value of r is determined to be 0.9 by the proposed method. Figure 5 shows $(d_1 - \mu)/\sigma$ versus r, and we can confirm it is maximized at $r = 0.9$. We have alerts at [04:40–04:50], 30 Oct. and [11:10–11:20], 30 Oct., where we found host groups that are thought to be botnets which have the following features.

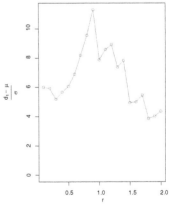

Fig. 5. Relationship of r and $(d_1 - \mu)/\sigma$

[04:40–04:50, 30 Oct.]

– The destination IP addresses are dispersed, 57 % of host of this time period, sends the packets to TCP port 502.
– The source IP addresses are dispersed, it consists of 55 hosts.

[11:10–11:20, 30 Oct.]

– The destination IP addresses are dispersed, 57 % of host of this time period, sends the packets to TCP port 80.
– The source IP addresses are dispersed, it consists of 59 hosts.

Also, Fig. 2 shows the result for the data of sensorA October 23 to 25, 2015. The value of r is determined to be 0.8 by the proposed method. We have alerts at [02:00–02:10], 25 Oct., where we found a host group that is thought to be a botnet which has the following features.

[02:00–02:10, 25 Oct.]

– The destination IP addresses are dispersed, 49 % of host of this time period, sends the packets to TCP port 631.
– The source IP addresses are dispersed, it consists of 54 hosts.

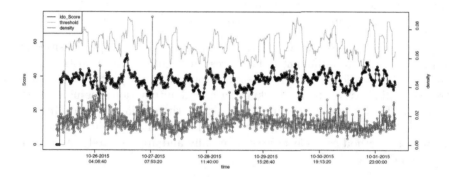

Fig. 6. sensorB, October 25 to November 1, 2015

Otherwise, Fig. 6 shows the result for the data of sensorB October 25 to November 1, 2015. The value of r is determined to be 1.0 by the proposed method. We have alerts at [08:40–08:50], 27 Oct., where we found a host group that is thought to be a botnet which has the following features.

[08:40–08:50, 27 Oct.]

– The destination IP addresses are dispersed, 59 % of host of this time period, sends the packets to TCP port 161.
– The source IP addresses are dispersed, it consists of 55 hosts.

Another result, Fig. 7 shows the result for the data of sensorC October 21 to 26, 2015. The value of r is determined to be 0.9 by the proposed method. We have alerts at [01:50–02:00], 25 Oct., where we found a host group that is thought to be a botnet which has the following features.

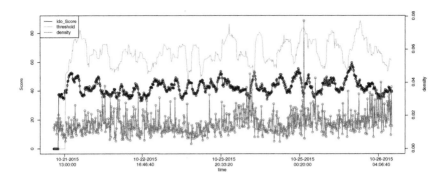

Fig. 7. sensorC, October 21 to 26, 2015

[01:50–02:00, 25 Oct.]

– The destination IP addresses are dispersed, 60 % of host of this time period, sends the packets to TCP port 102.
– The source IP addresses are dispersed, it consists of 53 hosts.

Since we can clearly know the time period with high graph densities from Figs. 2, 4, 6 and 7, it is thought to be successful in automating adjustment of the value of regularization coefficient. Also, it is found that the proposed method was effective in different countries' darknet data, and automation of issuing alerts by the graph density was successful. In addition, though in the preceding work, there was no way to adjust the value of regularization coefficient appropriately, the propose method can select it automatically and appropriately by exploiting the graph density.

5   Conclusion

In this paper, we propose a new method for botnet detection using the graphical lasso and 'graph density', for which we establish a solution to determine the value of regularization coefficient, and show its effectiveness by experiments using darknet data.

However, the proposed method cannot work in an online manner, since we have to determine the value of regularization coefficient using the given data in advance. To solve this issue is our future work.

Acknowledgments. We appreciate the feedback offered by everyone of Cybersecurity Laboratory, NICT. Our work was partially supported by "PRACTICE: Proactive Response Against Cyber-Attacks Through International Collaborative Exchange" administered by the Ministry of Internal Affairs and Communications.

References

1. Akiyama, M., Kawamoto, T., Shimamura, M., Yokoyama, T., Kadobayashi, Y., Yamaguchi, S.: A proposal of metrics for botnet detection based on its cooperative behavior. In: Proceedings of the SAINT Internet Measurement Technology and its Applications to Building Next Generation Internet Workshop, p. 82 (2007)
2. Bishop, C.M.: Pattern Recognition and Machine Learning. Springer, New York (2006)
3. Friedman, J., Hastie, T., Tibshirani, R.: Sparse inverse covariance estimation with the graphical lasso. Biostatistics 9(3), 432–441 (2008)
4. Falt, J., Blostein, S.D.: A Bayesian approach to two-sided quickest change detection. In: IEEE International Symoposium on Information Theory, pp. 736–740 (2014)
5. Ide, T., Lozano, A.C., Abe, N., Liu, Y.: Proximity-based anomaly detection using sparse structure learning. In: SDM, pp. 97–108 (2009)
6. Okamoto, A., Shoudai, T.: Mining first-come-first-served frequent time sequence patterns in streaming data. In: IADIS International Conference e-Society 2013 (ES2013) (2013)
7. Witten, D., Friedman, J., Simon, N.: New insights and faster computations for the graphical lasso. J. Comput. Graph. Stat. 20(4), 892–900 (2011)
8. Hamasaki, H., Kawakita, M., Takeuchi, J., Yoshioka, K., Inoue, D., Etoh, M., Nakao, K.: Proposal of botnet detection based on structure learning and its application to darknet data. In: Symposium on Cryptography and Information Security (2011). (in Japanese)
9. Mukai, S., Kawamura, Y., Kawakita, M., Takeuchi, J.: Performance evaluation of botnet detection method using a sparse structure learning. In: Symposium on Cryptography and Information Security (2015). (in Japanese)
10. Mukai, S., Kawamura, Y., Kawakita, M., Takeuchi, J.: Botnet detection using anomaly detection based on sparse structure learning. In: IEICE Technical report, ISEC2014-105, vol. 114, no. 471, pp. 193–198 (2015). (in Japanese)
11. Research and Development of Technologies to Analyze the Similarity, the Locality and the Time Series Properties of Cyber Attack Information. http://itslab.inf.kyushu-u.ac.jp/cyber/en/index.html. Accessed 15 Dec 2015

The Usability of Metadata for Android Application Analysis

Takeshi Takahashi[1](✉), Tao Ban[1], Chin-Wei Tien[2], Chih-Hung Lin[2], Daisuke Inoue[1], and Koji Nakao[1]

[1] National Institute of Information and Communications Technology, Tokyo, Japan
takeshi_takahashi@nict.go.jp
[2] Institute for Information Industry, Taipei, Taiwan

Abstract. The number of security incidents faced by Android users is growing, along with the surge in malware targeting Android terminals. Such malware arrives at the Android terminals in the form of Android Packages (APKs). Assorted techniques for protecting Android users from such malware have been reported, but most of them focus on the APK files themselves. Different from these approaches, we use metadata, such as web information obtained from the online APK markets, to improve the accuracy of malware identification. In this paper, we introduce malware detection schemes using metadata, which includes categories and descriptions of APKs. We introduce two types of schemes: statistical scheme and support vector machine-based scheme. Finally, we analyze and discuss the performance and usability of the schemes, and confirm the usability of web information for the purpose of identifying malware.

Keywords: Android · Malware · APK · Risk analysis · Machine learning

1 Introduction

The number of Android malware is increasing. Malware arrives at the Android terminals in the form of Android Packages (APKs). APK analysis techniques are needed to identify malware. Especially, automated ones are needed to facilitate the malware analysis operations. DroidRisk [15] determines whether an APK file contains malware based on the permissions requested by an APK. It calculates the numerical value of the risk associated with a permission request by multiplying the percentage of APKs that use that permission maliciously and the impact of the abuse of the permission. DroidRisk then sums up the values for all the permissions required by an APK to produce the final numerical value of APK risk. Sarma et al. [9] also proposed a scheme that determines whether an APK file is malware by introducing Category-based Rare Critical Permission (CRCP). A CRCP is a permission that should not usually be required for the category of application. When any of the CRCPs are called, the APK is regarded as malware. Various related studies on APK analysis have been reported, many of which need optimization and refinement before putting into practical use.

© Springer International Publishing AG 2016
A. Hirose et al. (Eds.): ICONIP 2016, Part I, LNCS 9947, pp. 546–554, 2016.
DOI: 10.1007/978-3-319-46687-3_60

Contribution. This paper analyzes the usability of metadata for APK analysis. Unlike conventional schemes, we use an APK's metadata obtained from the web, in addition to the characteristics derived from the APK file. Two types of APK analysis approaches are examined: statistical and machine learning-based. DroidRisk, which quantifies the risk of an APK file, is an example of the statistical approach. It uses permission information as its input, but we extend the scheme by using metadata, i.e., category and application description obtained from the web to improve the quality of risk quantification. Regarding the machine learning-based approach, we examine the use of support vector machine (SVM) [10,13] by using the same metadata. We also analyze the usability of API calls instead of permission requests. Through these analyses, we clarify the usability of metadata and API calls for APK analysis.

Differences from Our Earlier Work. This paper advances our preliminary work [11], which addressed the usability of metadata for risk quantification schemes. Section 3 newly introduces the use of metadata for malware detection schemes using SVM, and Sect. 4 evaluates the usability of metadata for that purpose. Different from our earlier work, this paper evaluates performance of schemes in terms of accuracy instead of area under the curve (AUC). Moreover, this paper compares the (un)usefulness of API calls and permission requests.

2 Risk Quantification (Statistical Analysis)

This section introduces risk quantification schemes. It begins with reviewing DroidRisk and proposes two extended schemes using APKs' metadata. Note that supplementary information is available in our earlier work [11].

DroidRisk. DroidRisk quantifies the risk of an APK file based on its permission request patterns. It first quantifies the security risk of each permission, and then quantifies the security risk of an APK file by summing up the quantified security risk of all permissions used by the APK file. The quantified risk value, denoted as r is derived from Eq. (1), where $L(p)$ denotes the likelihood of permission p being used by malware, and $I(p)$ denotes the impact of permission p being misused by malware. The risk value can be used to determine whether an APK file is malware by setting an appropriate threshold value.

$$r = \sum_i \{L(p_i) \times I(p_i)\} \tag{1}$$

Category-Based DroidRisk. We introduce a new scheme, referred to as $DR(p, ct)$. It extends DroidRisk to quantify security risks based on application categories of APK files. DroidRisk determines $L(p)$ and $I(p)$ by studying the whole dataset, but the optimal values should be different depending on the type of application. For instance, many applications within the category of "Travel&Maps" request permission for using GPS, whereas few within the "Multimedia" category request this permission. Thus, DR(p, ct) sets different values for $L(p)$ and $I(p)$ for each application category by studying the data in each category, as in Eq. (2), where ct denotes the application category.

$$r = \sum_i \{L(p_i, ct) \times I(p_i, ct)\} \tag{2}$$

Cluster-Based DroidRisk. We introduce another scheme, referred to as DR(p, cl), by extending DroidRisk. It uses automatically-generated application clusters instead of application categories. The clusters are generated from the application descriptions available on the web. DR(p, cl) runs the following four processes to quantify the risk of APKs: data preprocessing, topic model generation, cluster generation, and risk quantification. Note that the first three processes are based on CHABADA [3], which was adjusted for our purposes. The *data preprocessing process* uses the application descriptions and produces words usable for Latent Dirichlet Allocation (LDA) [1]. First, we check the description language and discard non-English descriptions. We also discard non-text items, i.e., numbers, HTML tags, web links, and email addresses. Second, we extract stems from the descriptions and truncate stop words. Third, we count the number of words in the resultant descriptions. We discard a description if it contains less than 10 words[1]. The *topic model generation process* processes the words in the descriptions with LDA. We import the words in the descriptions and train a number of topics. We consider a total of 300 topics, with 0.05 as the threshold value of topic proportion and a maximum of four topics per entry. As a result, this process outputs several pairs of {topic number, proportion value} (maximum number of four pairs)[2]. The *cluster generation process* clusters the APKs according to the {topic number, proportion value} pairs for each description. The number of categories was set to 12, which is the same as the number of categories used in the Opera Mobile Store[3]. The *risk quantification process*, as with the category-based scheme, calculates the risk value r, following the Eq. (3), where cl denotes application cluster.

$$r = \sum_i \{L(p_i, cl) \times I(p_i, cl)\} \tag{3}$$

---

[1] We used the language-detection library [2] to detect the language, stemmify [7] for the stemming operation, and stoplist/en.txt of MALLET [5] as the list of stop words.
[2] We used MALLET for running LDA and considered 300 topics because the MALLET documentation states that "The number of topics should depend to some degree on the size of the collection, but 200 to 400 will produce reasonably fine-grained results."
[3] We used the "kmeans" [4] function of Ruby gem [8].

3   Malware Detection (Machine Learning-Based Analysis)

This section introduces malware detection schemes based on machine learning. SVM is known to be effective for detecting malware, thus this section begins with reviewing SVM and proposes its extended uses with APK's metadata.

3.1   Support Vector Machine

The idea of a two-class SVM is described as follows. From a set of training samples $D = \{(x_i, y_i)|x_i \in R^d, y_i \in \{-1, +1\}, i = 1, ..., l\}$, SVM learns a norm 1 linear function, suppose its existence,

$$f(x) = \langle w, x \rangle + b, \tag{4}$$

determined by a weight vector w and threshold b that realizes the maximum margin, where margin is defined as the distance from the hyperplane to the nearest training data point of either class. According to Vapnik's VC dimension theory, margin maximization in the training set is equivalent to minimization of the generalization error of the classifier. In case that the training set is non-separable by a linear hyperplane, the so-called kernel trick and slack variables are often employed to produce the following equation:

$$f(x) = \sum_{i=1}^{l} \alpha_i^* y_i K(x, x_i) + b \tag{5}$$

This equation is a kernelized version of Eq. (4). If $f(x) > 0$, then x is assigned to the positive class, otherwise it is assigned to the negative class.

3.2   SVM-Based APK Analysis

We use SVM to detect malware among APK files. We also use APK's metadata, i.e., application categories and clusters, for improving the performance of the malware detection. Different from the risk quantification approaches introduced in Sect. 2, SVM does not output the quantified value of risk level but the binary value on whether an APK is a malware. Note that we use application clusters instead of application descriptions. More rich information, such as the stems of application descriptions can be used, but it increases dimensionality of characteristics. For simplicity, we use the same cluster as DR(p, cl) does.

We define four types of schemes that run SVM with different parameters, i.e., SVM(p), SVM(p, ct), SVM(p, cl), and SVM(p, ct, cl). SVM(p) runs SVM using only permission requests, SVM(p, ct) using permission requests and application categories, SVM(p, cl) using permission requests and application clusters, and SVM(p, ct, cl) using permission requests, application categories, and application clusters. Each of the schemes calculates the value $f(x)$ of Eq. (5) and judges positive if $f(x) > 0$.

<table>
<thead>
<tr><th colspan="4">Table 1. Dataset by category</th></tr>
</thead>
</table>

Category	Benign	Malicious	Total
Business&Finance	2784	479	3263
Communication	3779	268	4047
eBooks	327	132	459
Entertainment	14138	2453	16591
Games	2545	445	2990
Health	2114	323	2437
Languages translators	1300	87	1387
Multimedia	2422	567	2989
Organizers	1536	228	1764
Ringtone	734	41	775
Theme Skins	12090	2603	14693
Travel&Maps	5276	5059	10335
Total	49045	12685	61730

Table 2. Dataset by cluster

Cluster	Benign	Malicious	Total
Cluster1	3574	934	4508
Cluster2	3883	889	4772
Cluster3	3945	976	4921
Cluster4	5247	1206	6453
Cluster5	4317	1174	5491
Cluster6	3820	1077	4897
Cluster7	3474	919	4393
Cluster8	5337	2091	7428
Cluster9	4104	811	4915
Cluster10	4346	832	5178
Cluster11	3496	818	4314
Cluster12	3502	958	4460
Total	49045	12685	61730

4 Numerical Results

This section evaluates the usability of metadata from the standpoint of risk quantification and malware detection by using our own dataset.

4.1 Dataset

We collected 87,182 APK files from the Opera Mobile Store [6] over the period of January–September 2014. Files from which we could not extract permission requests were excluded, as permission requests are a necessary input for our risk quantification schemes. The files were then checked by VirusTotal [14] to determine whether they were malware. VirusTotal analyzes the risk of an APK file using multiple evaluation engines from different vendors. If one or more of the results indicated that the file was malicious, we considered the APK file as malware. Note that adware was not counted as malware, and the APK files that VirusTotal could not handle were excluded from the dataset in advance. As a result, we obtained a dataset of 78,649 APK files, consisting of 52,251 benign files and 26,398 malicious ones.

We also collected metadata of the APK files over the same period from the Opera Mobile Store. The metadata includes the application category, description, and the number of downloads, though we use only the application category and description in this paper. We stored the metadata in XML, following the data structure mentioned in [12]. The breakdowns of the dataset by the categories and clusters are shown in Tables 1 and 2, respectively.

4.2   Usability of Metadata for Risk Quantification Techniques

Table 3 shows the performance comparison of the risk quantification schemes introduced in Sect. 2 in terms of accuracy, precision, recall, and false positive rate (FPR). Note that DroidRisk is represented as DR. The values are the averages of 10 times 10-cross validation results. As can be seen, the performance of DR(p) is improved by the use of application categories and clusters. Compared to application clusters, application categories improves the performance more, though the contribution of application clusters could be improved further by fine-tuning the cluster generation algorithms. However, the performance improvement is not significant. We believe this is because both DR(p, ct) and DR(p, cl) suffer from overfitting caused by the dataset division for contextual-based analysis.

Table 3. Performance of DroidRisk-based schemes

	Accuracy[%]	Precision[%]	Recall[%]	FPR[%]
DR(p)	83.59±0.14	67.02±0.61	39.65±1.29	5.05±0.25
DR(p, ct)	85.63±0.20	59.68±2.19	29.85±1.69	4.93±0.48
DR(p, cl)	83.88±0.17	65.78±0.88	42.35±1.75	6.05±0.40
DR(api)	79.50±0.04	51.77±18.47	0.93±0.34	0.18±0.06
DR(api, ct)	82.41±0.12	45.82±9.01	14.07±0.70	6.75±0.86
DR(api, cl)	79.53±0.04	53.84±9.17	0.85±0.26	0.13±0.04

One may argue that API calls should be used instead of permission requests for more accurate analysis. To verify that, we have measured the performance of these schemes using API calls instead of permission requests. More than 30,000 types of API calls, including Android Framework APIs, JAVA APIs, and third party APIs, are analyzed to calculate parameters L and I. For simplicity, the value of the L was optimized for the top 10 % of API calls used by malware, while the value was set to 1 for the rest. Table 3 shows the performances of DR(api), DR(api, ct), and DR(api, cl), where api denotes api calls. As can be seen, the performance of these api-based schemes are largely behind the permission-based schemes. We believe this is because the api-based schemes suffer from more overfitting problem than the permission-based ones. The degree of overfitting problem is determined by the number of dataset and the dimensionality of characteristics. Since the dataset is the same, the api-based schemes that use larger dimensionality of characteristics suffer from overfitting more than the permission-based schemes. When using API-calls and metadata, we need to consider the degree of overfitting problem.

4.3   Usability of Metadata for Malware Detection Techniques

Table 4 shows the performances of SVM(p), SVM(p, ct), SVM(p, cl), and SVM(p, ct, cl) in terms of accuracy, precision, recall, and false positive rate

(FPR). The values are the averages of 10 times 10-cross validation results. As with Sect. 4.2, the use of metadata improves the performance of SVM(p), and application categories improves the performance more than application clusters, though the contribution of application clusters could be improved further by fine-tuning the cluster generation algorithms.

We have also measured the performance of theses schemes using API calls instead of permission requests, as with Sect. 4.2. The performance of SVM(api), SVM(api, ct), SVM(api, cl), and SVM(api, ct, cl) are also shown in Table 4. Contrary to the DroidRisk-based schemes, the use of API calls improves the performance of SVM-based schemes. We believe this is because the use of API calls does not incur overfitting issues by using metadata. Moreover, as with SVM(p), the use of metadata improves the performance of SVM(p), though the performance contribution of application clusters is rather inferior to that of application categories.

The use of metadata improves the performance, but it incurs non-trivial amount of extra calculation costs. Moreover, the improvement is insignificant. Therefore, the use of metadata is not always recommended. Meanwhile, the use of API call significantly improves performance, as can be seen by comparing SVM(p) to SVM(api). The performance improvement gained by the use of metadata is small compared to the improvement gained by the use of API calls.

Table 4. Performance of SVM-based schemes

	Accuracy[%]	Precision[%]	Recall[%]	FPR[%]
SVM(p)	88.87±0.12	81.19±0.65	59.67±0.41	3.58±0.16
SVM(p, ct)	89.45±0.15	83.87±0.60	60.27±0.80	3.00±0.15
SVM(p, cl)	89.38±0.10	83.48±0.67	60.25±0.38	3.08±0.16
SVM(p, ct, cl)	89.45±0.09	83.76±0.53	60.35±0.59	3.03±0.14
SVM(api)	94.07±0.16	87.23±0.45	83.34±1.00	3.16±0.15
SVM(api, ct)	94.09±0.17	87.37±0.49	83.26±0.73	3.11±0.14
SVM(api, cl)	94.08±0.17	87.37±0.53	83.40±0.79	3.16±0.16
SVM(api, ct, cl)	94.07±0.15	87.20±0.38	83.36±0.89	3.16±0.12

One may be interested in the effectiveness of the use of DroidRisk for the purpose of malware detection. By comparing Tables 3 and 4, we can see that SVM-based schemes provide better performance than DroidRisk-based ones. All of the DroidRisk-based schemes provide better performance than a random classifier, but they could be regarded as weak classifiers. It is natural since DroidRisk is designed for quantifying risks and not for a classifier, while SVM is designed to be an efficient classifier.

When applying these schemes for practical environment, some fine-tuning per individual use cases should be considered. For instance, applications that provide security alerts to the users may wish to implement a scheme that minimizes false negative rate (FNR), which is equivalent to $(1 - recall)$, while applications that provide automated countermeasures may wish to implement a scheme that minimizes FPR. When measuring the performance shown in Table 4, each of the schemes are tuned so that they can maximize accuracy, but it can be tuned so that the other parameters, such as recall or FPR, can be optimized.

5 Conclusion

This paper showed the usefulness of APK metadata for APK analysis. The use of APK metadata improves the performance of risk level quantification schemes, but it also causes overfitting and cancel some of the improvement. Likewise, it improves the performance of malware detection using SVM. It does not suffer from overfitting, but the improvement is rather marginal. Apart from that, this paper also showed the usefulness of API calls for APK analysis. The use of API calls worsened the performance of risk level quantification since it causes significant overfitting. On the contrary, it was very effective for malware detection using SVM. Though the use of metadata can improve performance of risk analysis, the improvement is not that large, and non-trivial amount of calculation costs will be imposed. Indeed, the performance could be improved further by taking other means, such as the use of API calls. Therefore, the use of metadata is not always the best approach for APK analysis, and we need to decide whether to use metadata, API calls and the other parameters depending on individual use cases.

References

1. Blei, D.M., Ng, A.Y., Jordan, M.I.: Latent Dirichlet allocation. J. Mach. Learn. Res. **3**, 993–1022 (2003)
2. Cybozu Labs: Language Detection Library for Java, December 2014. https://code.google.com/p/language-detection/
3. Gorla, A., Tavecchia, I., Gross, F., Zeller, A.: Checking app behavior against app descriptions. In: ICSE 2014, Proceedings of the 36th International Conference on Software Engineering (2014)
4. MacQueen, J.: Some methods for classification and analysis of multivariate observations. In: Proceedings of the Fifth Berkeley Symposium on Mathematical Statistics and Probability, vol. 1: Statistics, pp. 281–297 (1967)
5. McCallum, A.K.: MALLET: a machine learning for language toolkit, December 2014. http://mallet.cs.umass.edu
6. OPERA SOFTWARE ASA: Opera Mobile Store, January 2015. http://apps.opera.com/
7. Ray Pereda: stemmify, December 2014. https://rubygems.org/gems/stemmify
8. RubyGems.org: kmeans, December 2014. https://rubygems.org/gems/kmeans/
9. Sarma, B.P., Li, N., Gates, C., Potharaju, R., Nita-Rotaru, C., Molloy, I.: Android permissions: a perspective combining risks and benefits. In: Proceedings of the 17th ACM Symposium on Access Control Models and Technologies, SACMAT 2012, pp. 13–22. ACM, New York (2012). http://doi.acm.org/10.1145/2295136.2295141

10. Schölkopf, B., Smola, A.J.: Learning with Kernels: Support Vector Machines, Regularization, Optimization, and Beyond. MIT Press, Cambridge (2001)
11. Takahashi, T., Ban, T., Mimura, T., Nakao, K.: Fine-grained risk level quantication schemes based on APK metadata. In: Arik, S., Huang, T., Lai, W.K., Liu, Q. (eds.) ICONIP 2015. LNCS, vol. 9491, pp. 663–673. Springer, Heidelberg (2015). doi:10.1007/978-3-319-26555-1_75
12. Takahashi, T., Nakao, K., Kanaoka, A.: Data model for android package information and its application to risk analysis system. In: First ACM Workshop on Information Sharing and Collaborative Security. ACM, November 2014
13. Vapnik, V.: Statistical Learning Theory. Wiley, New York (1998)
14. VirusTotal: virustotal for android, January 2015. http://www.virustotal.com/ja
15. Wang, Y., Zheng, J., Sun, C., Mukkamala, S.: Quantitative security risk assessment of android permissions and applications. In: Wang, L., Shafiq, B. (eds.) DBSec 2013. LNCS, vol. 7964, pp. 226–241. Springer, Heidelberg (2013). doi:10.1007/978-3-642-39256-6_15

Preserving Privacy of Agents in Reinforcement Learning for Distributed Cognitive Radio Networks

Geong Sen Poh[1]([✉]) and Kok-Lim Alvin Yau[2]

[1] MIMOS Berhad, Technology Park Malaysia, 57000 Kuala Lumpur, Malaysia
gspoh@mimos.my
[2] Sunway University, Jalan Universiti,
Bandar Sunway, 47500 Petaling Jaya, Selangor, Malaysia
koklimy@sunway.edu.my

Abstract. Reinforcement learning (RL) is one of the artificial intelligence approaches that has been deployed effectively to improve performance of distributed cognitive radio networks (DCRNs). However, in existing proposals that involve multi-agents, perceptions of the agents are shared in plain in order to calculate optimal actions. This raises privacy concern where an agent learns private information (e.g. Q-values) of the others, which can then be used to infer, for instance, the actions of these other agents. In this paper, we provide a preliminary investigation and a privacy-preserving protocol on multi-agent RL in DCRNs. The proposed protocol provides RL computations without revealing agents' private information. We also discuss the security and performance of the protocol.

1 Introduction

Cognitive Radio Networks (CRNs) [1,10] are next generation wireless networks that exploit underutilized spectrum (or white spaces) in licensed spectrum whilst minimizing interference to licensed users (or Primary Users, PUs). It allows unlicensed users (or Secondary Users, SUs) to use this spectrum in an opportunistic manner. Distributed Cognitive Radio Networks (DCRNs) are distributed version of CRNs, in which a number of SUs interact with one another without a fixed infrastructure, such as a base station. One of the approaches that was applied to enhance performance of CRNs and DCRNs is Reinforcement Learning (RL) [16]. It has been used in a wide range of schemes, such as for network performance enhancement [19], dynamic channel selection [18] and routing [12].

However in existing proposals for DCRNs, information are shared between agents in plain (e.g. Q-values). This raises privacy concern, where a corrupt agent may use it to learn actions of a certain agent(s) or to skew the actions of the agent(s) to its advantage.

Contributions. We propose a privacy-preserving protocol that performs multi-agent RL operations in DCRNs in a private manner. As far as we know, this

A. Hirose et al. (Eds.): ICONIP 2016, Part I, LNCS 9947, pp. 555–562, 2016.
DOI: 10.1007/978-3-319-46687-3_61

is the first instance of privacy-preserving RL for CRNs. We utilise the concept of privacy-preserving RL [15] and homomorphic encryption scheme for privately computing the sum of Q-values. We also suggest the use of efficient multi-party computation (MPC) schemes [17] as an alternative approach. We further discuss security and performance of our protocol.

2   Related Works

As was previously stated, RL was used to provide various operational enhancements on CRNs in [12,18,19]. Ling et al. [9] further discussed how RL can be applied on CRNs for security enhancements, but did not examine agents' privacy. On the other hand, privacy-preserving mechanisms were well-established and have been deployed for various applications, including RL [15] and facial expression recognition [14]. More recent proposals include schemes on smart grids [7] and extreme learning machine algorithm [5]. Most schemes, however, are confined to two-party computation. We require mechanisms, as in Çatak's scheme [5], for multiple parties since we anticipate more than two agents in our setting. Also, formalism for privacy-preserving data mining using MPC is provided by Lindell and Pinkas in [8]. Practical multi-party version includes FairPlayMP [2] and ShareMind [3]. Real-world applications using these schemes were described in [4]. In terms of privacy-preserving works on CRNs, Qin et al. [13] proposed a method to preserve SU privacy, in the scenario where PU requires information from SU in order to calculate payments. The application and goal of Qin et al. are different from ours, as Qin et al.'s scheme does not involve RL and focus on secure two-party computation between a SU and a PU.

3   Multi-agent RL in DCRNs

In this section we define informally what is RL and describe how multi-agent RL (MARL) are applied in DCRNs. According to Sutton and Barto [16], RL involves learning what and how to map situations to actions so as to gain maximum reward. It is an unsupervised technique where the learner (or agent) is not given directive on what action to take but instead must explore and discover through online learning. Our MARL model and description are based on [19], which allows payoff message exchanges between agents as shown in Fig. 1.

The important representations in the model are *state*, *action* and *reward*. A *state* represents the factors affecting how an agent makes decisions, and it is observed through the operating environment. *Action* is taken by an agent to maximise its reward. The *reward* is a performance metric that is either to be maximized or minimized. Underlying the model is an online learning algorithm for RL known as Q-learning. It enables an agent to learn in an interactive manner in the operating environment, through estimating the Q-values of the state-action pairs $Q_t(s_t, a_t)$, where t denotes a specific decision epoch. For every $Q_t(s_t, a_t)$, an agent calculates its short-term rewards and subsequently its future rewards as time progress.

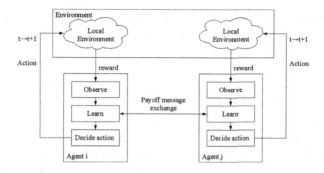

Fig. 1. A RL model with payoff message exchange [19]

A MARL setting involves embedding the above RL model in every SU. The main idea then is to define a payoff message mechanism in which an agent in a SU shares its perceptions with others. Each agent i maintains a Q-table of size $|A|$, where A is the set of actions. Since Q-values are used to estimate the level of local reward of an action, changes in Q-values result in changes in an agent's action. It is updated as follow:

$$Q^i_{t+1}(a^i_t) \leftarrow (1 - \alpha)Q^i_t(a^i_t) + \alpha r^i_{t+1}(a^i_t), \qquad (1)$$

where $0 \le \alpha \le 1$ is the learning rate. A higher value of α means recent local reward is of more importance compared to the past knowledge. Over-time, an optimal value can be achieved by searching for the maximum value as $\max_{a \in A}(Q^i_t(a))$.

Each agent i also additionally maintains a μ-table of size $|\Gamma(i)| \times |A|$ to store the payoff messages from neighbouring agents, where $\Gamma(i)$ represents all neighbours of i. Given the two tables, agent i computes its action $a^i_t \in A$ using its local Q-value $Q^i_t(a^i_t)$ from Q-table and its neighbours' Q-values from the μ-table. Each agent exchanges payoff message $\mu_{i\psi}(a^\psi_t)$ constantly among the agents until a fixed optimal point is obtained. In this case $\mu_{i\psi}(a^\psi_t)$ denotes the local Q-values of the agent's current action as follow:

$$\mu_{i\psi}(a^\psi_t) = [Q^i_t(a^i_t)]. \qquad (2)$$

When an agent j receives $\mu_{i\psi}(a^\psi_t)$ while it is taking action a^j_t, we denote $\mu_{i\psi}(a^\psi_t) = \mu_{ij}(a^j_t)$ as the local reward of agent i. Given the local reward, every agent calculates its optimal action to maximise local payoff as follow:

$$g^i_t(a^i_t) = \max_{a \in A} \left[Q^i_t(a) + \sum_{j \in \Gamma(i)} \mu_{ji}(a) \right], \qquad (3)$$

and agent i then determines its optimal action as follow:

$$a^i_t = \arg \max_{a \in A} g^i_t(a). \qquad (4)$$

The above processes compute the global payoff of an agent. The MARL steps are executed until the agent converges to an optimal local action. This occurs when changes in Q-values and payoff values between epoches are insignificant.

4   Privacy Preserving RL Computation in DCRNs

We first examine operations that are to be executed in a private manner. Based on the processes described in the previous section, Q-value update (Eq. 1) is calculated by the agent itself. This means Eq. 1 can be computed in plain. However, these Q-values (in the form of $\mu_{ji}(a)$) are shared as payoff messages among agents and stored in μ-tables, hence it must be encrypted before sharing in order to preserve privacy. For every agent i, calculation of Eq. 3 involves the encrypted $\mu_{ji}(a)$ in the μ-tables. It means the summation part of Eq. 3 requires *addition in the encrypted domain*, which is the main focus of our protocol. Once an agent i obtained the result of the summation, max and arg max (Eqs. 3 and 4) can be computed in plain since these are performed by agent i locally.

In summary, our scenario involves an agent i collecting payoff messages from all its neighbouring agents, where these messages should be in private and not known by agent i, yet allowing agent i to compute its global payoff value based on summing the values in these messages. This means we require an encryption scheme that has the additive homomorphic property, caters for multi-agents and, decryption can be performed on the resulting encrypted sum value by authorised agent(s). We note that this scenario is somewhat similar to the problem of electronic voting and thus we adopt the generalised Paillier encryption scheme [6] that has been used for this purpose. Also, we assume all agents are honest-but-curious, in that they will not provide invalid encryption of their messages.

Generalised Paillier Encryption Scheme [6]. We require the threshold variant of this scheme. It defines a common public key pk and a set of private key shares $(sk_1, \ldots, sk_v)$. It is based on computations of modulo n^{s+1} where n is an RSA modulus and s a natural number. Given its additive homomorphic property, there is an operation $\cdot$ whereby for two messages m_1 and m_2, $\mathrm{E}_{pk}(m_1 + m_2) = \mathrm{E}_{pk}(m_1) \cdot \mathrm{E}_{pk}(m_2)$, where $\mathrm{E}_{pk}(x) = (n+1)^x r^{n^s}$ denotes the encryption function, $pk = (n, s)$ and $r \in Z_n^*$ a random value. Threshold decryption can then be performed, in which a (u, v)-threshold setting means that at least u out of v agents are required for decrypting a ciphertext. Share decryption D is performed using a polynomial-based recovery algorithm together with the public key pk and the private keys shares $sk_1, \ldots, sk_u$. The scheme is provably secure under the decisional composite residuosity assumption introduced by Paillier [11].

Scaling. The cryptographic primitives for our constructions required values to be represented in integers. As such all values involved in the calculations must be scaled with an integer scaling factor S, which can be performed through the approach proposed in [14, Sect. 3.2.1].

4.1   A Basic Approach with a Trusted Third Party

We remark that a straightforward protocol can be constructed by introducing a fully trusted third party (TTP). Firstly, the TTP initiates by generating its public-private key pair (pk_{ttp}, sk_{ttp}) based on an additive homomorphic encryption scheme (e.g. Paillier [11]). The public key pk_{ttp} is shared to all participating agents. Each agent i for every decision epoch t encrypts its Q-values as $E_{pk_{ttp}}(S\mu_{ji}(a))$, using the public key of the TTP, where S is the scaling factor. This value is sent to all the other neighbouring agents. Once an agent i collected all encrypted Q-values from its neighbours, it is able to compute the encrypted sum of the collected values using the $\cdot$ operator where $E_{pk_{ttp}}(S\mu_{ji}(a)) \cdot E_{pk_{ttp}}(S\mu_{ji}(a)) \cdot \ldots \cdot E_{pk_{ttp}}(S\mu_{ji}(a)) = E_{pk_{ttp}}\left(\sum_j (S\mu_{ji}(a))\right)$ for all $j \in \Gamma(i)$. Agent i then sends the encrypted sum to the TTP, who decrypts using its private key sk_{ttp} and returns the sum value to agent i. Based on the sum value (divided with the scaling factor S), computations for optimised action of agent i (Eqs. 3 and 4) can be performed straightforwardly in plain.

While this approach is simple, a drawback is that the introduction of a TTP defeats the main purpose of the multi-agent DCRN setting, which is to do away with any centralised entity. Next, we propose a construction without TTP.

4.2   A Privacy-Preserving Protocol for RL in DCRNs

We propose and describe our protocol to compute an agent i optimal action in a private manner, without a TTP, using a threshold-based additive homomorphic encryption scheme (i.e. generalised Paillier scheme). Figure 2 gives the steps of the protocols. It contains three phases, which are *initial setup*, *private computation* and *calculating optiaml action*.

During setup, we assume there is a trusted controller who generates the public key pk and a set of private key shares for agent i and its neighbouring agents $(sk_1, \ldots, sk_v)$. This means using the generalised Paillier encryption scheme based on (u, v)-threshold, where if all agents must be involved to decrypt a value then $u = v$. While requiring a controller might be a limitation, it is required only during setup to generate the key pairs, which differs from the basic approach where the TTP must always be online. The public key pk is broadcast to all agents and every agent receives its respective key share.

In the private computation phase, each agent computes and encrypts its Q-values using pk that its received. The encrypted Q-values is shared among all neighbouring agents. Once an agent collected all its neighbours' encrypted Q-values, it sums them in the private manner, as shown in Fig. 2 (this step is similar to the previous approach that utilises the TTP). After that the agent requests the assistance from u neighbouring agents to decrypt the encrypted sum based on their private key share. Once the agent receives the decrypted sum value, it proceeds with the calculation of Eq. 3 and chooses its optimal action based on Eq. 4. We remark that alternatively secure multi-party computation (MPC) scheme such as FairPlayMP [2], a system based on garbled circuit with constant number of communication rounds or Sharemind [3] based

Initial Setup:

1. Trusted controller: Generates a common public key pk and a set of private key shares $(sk_1, \ldots, sk_v)$ for a (u, v)-threshold additive homomorphic scheme (i.e. generalised Paillier scheme).
2. Trusted controller: Broadcasts pk to all participating agents.
3. Trusted controller: Sends sk_i to agent i under a secure channel, for $i = 1$ to v.

Private Computation:

1. For every decision epoch t and $i = 1$ to v:
   (a) Agent i: Encrypts all Q-values $Q_t^i(a_t^i)$ as $\mathsf{E}_{pk}(S\mu_{ij}(a))$.
   (b) Agent i: Broadcasts $\mathsf{E}_{pk}(S\mu_{ij}(a))$ to all other $v - 1$ neighbouring agents.
   (c) Agent i: Receives $\mathsf{E}_{pk}(S\mu_{ji}(a))$ for $j \in \Gamma(i)$ of all neighbouring agents, where $|\Gamma(i)| = v - 1$.
   (d) Agent i: Calculates sum in the encrypted domain for all $j \in \Gamma(i)$:

$$\mathsf{E}_{pk}\left(\sum_j (S\mu_{ji}(a))\right) = \mathsf{E}_{pk}(S\mu_{ji}(a)) \cdot \mathsf{E}_{pk}(S\mu_{ji}(a)) \cdot \ldots \cdot \mathsf{E}_{pk}(S\mu_{ji}(a))$$

   (e) Agent i: Requests u neighbouring agents to assist in decrypting the encrypted sum using their private key share:

$$\sum_j (S\mu_{ji}(a)) = \mathsf{D}_{(sk_j)_{j \in \Gamma(i)}^u}\left(\mathsf{E}_{pk}\left(\sum_j (S\mu_{ji}(a))\right)\right)$$

Calculating Optimal Action:

1. For every decision epoch t and $i = 1$ to v:
   (a) Agent i: Given $\sum_j (S\mu_{ji}(a))$, divides it with S and finds maximum $g_t^i(a_t^i)$ based on eq. 3. Then given $g_t^i(a_t^i)$, finds optimal action based on eq. 4.

Fig. 2. A privacy-preserving protocol for RL in DCRNs

on secret sharing schemes can also be utilised. These systems involve more intensive computations but provide operations beyond private addition. For example, if finding the maximum must be private, these systems can be deployed.

5 Discussions

Security. The main objective of our proposed protocol is to preserve privacy of agents' Q-values. The use of the provably secure threshold-based generalised Paillier scheme to encrypt these values achieves this objective. This is because any of the agents is not able to learn the content the encrypted values provided by other neighbouring agents. Similarly, the addition in the encrypted domain does not leak the content of these values as well. However, the agents who participate in assisting an agent i for decryption does learn the sum value, as can be observed from Step 1(c) in Fig. 2. We note that learning the sum does not allow an agent j involved in decryption to calculate the payoff and optimal action of the agent i since agent j does not know agent i's Q-value.

Performance. The main trade-off of our proposed protocol is the performance overhead compared to a plain RL computations, which is inevitable in any privacy-preserving setup. However, the privacy-preserving computation is kept to the minimum on computing the sum only and all other computations remain as they are. In terms of bandwidth there is message expansion since the encrypted values are of the size of the encryption scheme's modulus (i.e. n).

6 Conclusions and Future Works

We introduced privacy-preserving RL on DCRNs for addressing potential privacy threats against shared agents' information in a multi-agent environment. A high-level privacy-preserving protocol was proposed based on provably secure cryptographic primitives that cater for secure multi-party computation. As future works, we intend to provide concrete constructions by simulating the environment with implementation of our protocol, in order to study the performance (especially the issue of timing constraints of DCRNs that may limit the usage of privacy preserving mechanisms) using three different approaches: (1) basic construction with a TTP, (2) using threshold cryptosystems such as generalised Paillier and (3) using MPC systems.

Acknowledgement. We would like to thank the anonymous referees for their comments. Kok-Lim Alvin Yau is supported by the Ministry of Education Malaysia (MOE) Fundamental Research Grant Scheme (FRGS) FRGS/1/2014/ ICT03/SYUC/02/2.

References

1. Akyildiz, I.F., Lee, W.-Y., Vuran, M.C., Mohanty, S.: Next generation/dynamic spectrum access/cognitive radio wireless networks: a survey. Comput. Netw. **50**(13), 2127–2159 (2006)
2. Ben-David, A., Nisan, N., Pinkas, B.: FairplayMP: a system for secure multi-party computation. In: Ning, P., Syverson, P.F., Jha, S. (eds.) CCS, pp. 257–266. ACM (2008)
3. Bogdanov, D., Laur, S., Willemson, J.: Sharemind: a framework for fast privacy-preserving computations. In: Jajodia, S., Lopez, J. (eds.) ESORICS 2008. LNCS, vol. 5283, pp. 192–206. Springer, Heidelberg (2008)
4. Bogetoft, P., Christensen, D.L., Damgård, I., Geisler, M., Jakobsen, T., Krøigaard, M., Nielsen, J.D., Nielsen, J.B., Nielsen, K., Pagter, J., Schwartzbach, M., Toft, T.: Secure multiparty computation goes live. In: Dingledine, R., Golle, P. (eds.) FC 2009. LNCS, vol. 5628, pp. 325–343. Springer, Heidelberg (2009)

5. Çatak, F.Ö.: Secure multi-party computation based privacy preserving extreme learning machine algorithm over vertically distributed data. In: Arik, S., Hunag, T., Lai, W.K., Liu, Q. (eds.) ICONIP 2015. LNCS, vol. 9490, pp. 337–345. Springer, Heidelberg (2015). doi:10.1007/978-3-319-26535-3_39

6. Damgård, I., Jurik, M., Nielsen, J.B.: A generalization of Paillier's public-key system with applications to electronic voting. Int. J. Inf. Sec. **9**(6), 371–385 (2010)

7. Erkin, Z.: Private data aggregation with groups for smart grids in a dynamic setting using CRT. In: WIFS, pp. 1–6. IEEE (2015)

8. Lindell, Y., Pinkas, B.: Secure multiparty computation for privacy-preserving data mining. IACR Cryptology ePrint Archive 2008:197 (2008)

9. Ling, M.H., Yau, K.-L.A., Qadir, J., Poh, G.S., Ni, Q.: Application of reinforcement learning for security enhancement in cognitive radio networks. Appl. Soft Comput. **37**, 809–829 (2015)

10. Mitola, J., Maguire, G.Q.: Cognitive radio: making software radios more personal. IEEE Pers. Commun. **6**(4), 13–18 (1999)

11. Paillier, P.: Public-key cryptosystems based on composite degree residuosity classes. In: Stern, J. (ed.) EUROCRYPT 1999. LNCS, vol. 1592, pp. 223–238. Springer, Heidelberg (1999). doi:10.1007/3-540-48910-X_16

12. Peng, J., Li, J., Li, S., Li, J.: Multi-relay cooperative mechanism with q-learning in cognitive radio multimedia sensor networks. In: IEEE 10th International Conference on Trust, Security and Privacy in Computing and Communications, pp. 1624–1629 (2011)

13. Qin, Z., Yi, S., Li, Q., Zamkov, D.: Preserving secondary users' privacy in cognitive radio networks. In: INFOCOM, pp. 772–780. IEEE (2014)

14. Rahulamathavan, Y., Phan, R.C.-W., Chambers, J.A., Parish, D.J.: Facial expression recognition in the encrypted domain based on local fisher discriminant analysis. IEEE Trans. Affect. Comput. **4**(1), 83–92 (2013)

15. Sakuma, J., Kobayashi, S., Wright, R.N.: Privacy-preserving reinforcement learning. In: Cohen, W.W., McCallum, A., Roweis, S.T., (eds.) ICML, vol. 307. ACM International Conference Proceeding Series, pp. 864–871. ACM (2008)

16. Sutton, R.S., Barto, A.G.: Reinforcement Learning: An Introduction. The MIT Press, Cambridge (1998)

17. Talviste, R.: Applying Secure Multi-party Computation in Practice. Ph.D. thesis, University of Tartu (2016)

18. Tang, Y., Grace, D., Clarke, T., Wei, J.: Multichannel non-persistent CSMA MAC schemes with reinforcement learning for cognitive radio networks. In: ISCIT 2011, pp. 502–506 (2011)

19. Yau, K.L.A., Komisarczuk, P., Paul, D.T.: Enhancing network performance in distributed cognitive radio networks using single-agent and multi-agent reinforcement learning. In: LCN 2010, pp. 152–159 (2010)

Campus Wireless LAN Usage Analysis and Its Applications

Kensuke Miyashita$^{(\boxtimes)}$ and Yuki Maruno

Faculty for the Study of Contemporary Society, Kyoto Women's University,
Higashiyama-ku, Kyoto 605-8501, Japan
{miyasita,maruno}@kyoto-wu.ac.jp

Abstract. Wireless LAN (WLAN) service has been provided in many companies, universities, hotels, coffee shops and even on the street, which supports the growing number of users with mobile devices. Kyoto Women's University has offered WLAN service with many access points in centralized control style since 2011, which allows mobile users to access the network at any location covered by its access points while on campus. It is useful for evacuation planning to figure out when and where people gather, and therefore, it is worth understanding the trends of WLAN usage in each organization at all times. In this paper, we analysed the trends of WLAN usage in the university and described some applications.

Keywords: Information and communication infrastructure · Campus network · Network usage trends · Wireless LAN

1 Introduction

Nowadays, mobile devices including smart phones and laptop computers have become a common part of our daily lives. Wireless LAN (WLAN) access points (APs) are installed to various buildings such as companies, universities, hotels, coffee shops and even on the street, which allows mobile users to access the network at any location covered by its APs [1].

The information and communication systems (ICS) of Kyoto Women's University (KWU) was formed in 2000 and restructured several times. The WLAN service of KWU has begun in 2001. Ever since then the information systems at KWU have been considered to play a fundamental role as the infrastructure for ICS in research and education while keeping up with the revolution in computer and network technologies [2].

There is a lot of information about the connected devices/users in the WLAN system log. This information used to be preserved in each AP of WLAN system and the administrator can gather a lot of information about WLAN devices. In recent years it is in centralized controller, thus the administrator can gather it more easily.

Since the user always carries her/his WLAN devices, the information of WLAN devices is closely linked up with the information of user who carries

© Springer International Publishing AG 2016
A. Hirose et al. (Eds.): ICONIP 2016, Part I, LNCS 9947, pp. 563–569, 2016.
DOI: 10.1007/978-3-319-46687-3_62

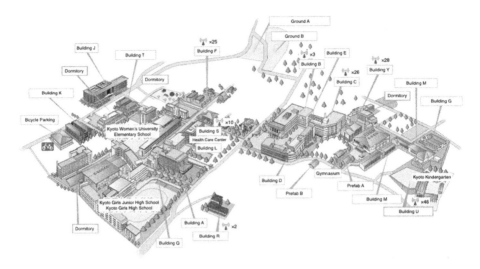

Fig. 1. Campus Map

them. It is worth analysing the trends of WLAN usage on the university campus to grasp where staffs and students move around and arrange the APs in the campus [3–7].

In this paper, we report the WLAN usage trends in KWU, and describe some applications. This work is based on the continuous investigation from [8] to [10]. In [8–10] we had WLAN logs for only a few month. We have kept the logs since July 2013 and we analyse the logs for approximately two years and 10 months in this paper.

2 The WLAN System

KWU has its main campus which consists of about 30 buildings basically marked with letters A through Y. WLAN is available in the buildings B, C, F, R, S, U and Y (Fig. 1)[1]. The buildings C and S have both seminar room and computer room, and the buildings B, F and Y have seminar room. The buildings R and U are a counseling and nutrition center and a preschool education center, respectively.

Aruba Networks' APs[2], which support IEEE 802.11a/b/g/n WLAN, are installed in each building (Table 1). These APs are controlled by the central controllers in the building S and have the same SSID in this system, which enables users to move around the campus without losing their connections. The topology of KWU network system is shown in Fig. 2.

Access authentication ensures network security. WLAN supports IEEE 802.1X authentication, pre-shared key (PSK) authentication, captive portal

[1] http://www.kyoto-wu.ac.jp/student/campus/map/ (translated by the authors).
[2] http://www.arubanetworks.com/.

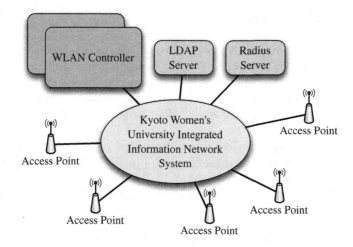

Fig. 2. Network Topology

Table 1. Number of access points

Building	# of APs
B	7
C	28
F	25
R	2
S	10
U	47
Y	28
Total	147

authentication, and MAC address authentication. Almost all users of KWU connect to the wireless network with IEEE 802.1X authentication. A user who wants to connect her/his device to this WLAN must be authenticated by the authentication server with a user name and password. Manual authentication prompts users for a user name and password the first time they access the Internet through a browser. Her/his device keeps the digital certificate of 802.1X, the user name and password, thus the connection to WLAN is automatically done after the second time. People in campus always carry WLAN devices such as smart phone, tablet PC or laptop PC and thus the WLAN usage is closely related with the users' behaviour.

3   WLAN Usage Trends

The central controller of WLAN system always collects the detailed information about the devices which connect to WLAN, including IP address of each device, MAC address, AP's name where the device connects, the duration of each connection and user name. The information has been collected every 10 min from the central controllers by SNMP query. We clarify the trends of WLAN usage by the information.

The number of unique devices which connected to WLAN and the number of unique users have almost doubled from 2014 to 2015. There were 3490 unique devices and 2061 unique users connected to WLAN in 2014, and the numbers were 6710 and 3818 respectively in 2015. The numbers of users who connects a device, two devices and three devices are about 73 %, 20 % and 4 % respectively in 2014 (the left-hand of Fig. 3), and they are 66 %, 23 % and 8 % respectively in 2015 (the right-hand of Fig. 3). It is evident that the proportion of users who connects more than one device to ones who connects only a device has grown.

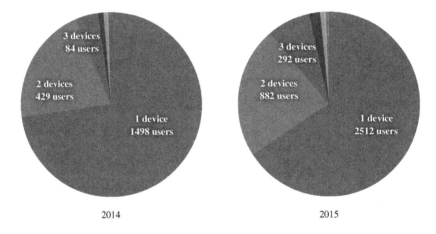

Fig. 3. Number of devices per user

Figure 4 shows hourly distribution of users and Fig. 5 shows it by building. Both figures describe a growth in users. Between 8 a.m. and 8 p.m., there are more users than the average in the university. This is a reasonable result because the classes begin at 8:50 and end at 19:40, and it confirms most WLAN devices automatically connect to WLAN.

From Fig. 5, there are most of the users in the buildings C, S and Y. It is due to these buildings consist of lecture rooms, computer rooms and laboratories while the other buildings consist of mainly laboratories and a few lecture rooms. Therefore, most students are in the buildings C, S and Y in the daytime to participate in classes.

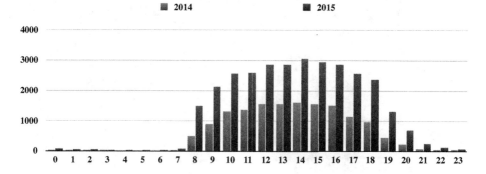

Fig. 4. Number of unique users (hourly, cumulative)

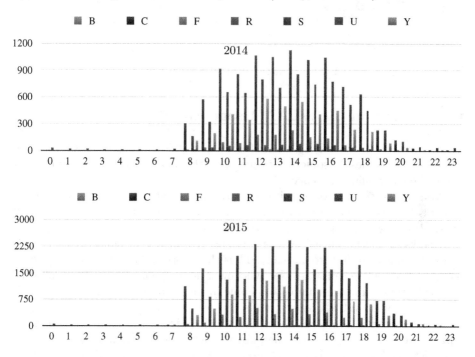

Fig. 5. Number of unique users by building (hourly, cumulative)

It is interesting to know user's movement in campus. In Fig. 6, the number of users who connects to WLAN in two buildings is the most large proportion in 2014 and the sum of the number of users who connects in one and two buildings is in a majority. In 2015, the number of users who connects to WLAN in three buildings is the most large proportion and the sum of the number of users who connects in three and four buildings is in a majority. This shows that WLAN become popular among university students in recent years.

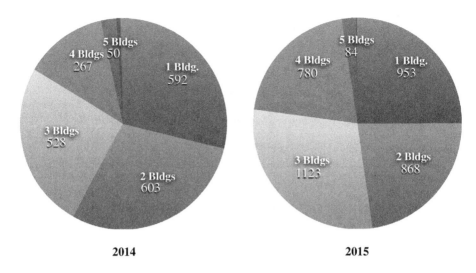

2014   2015

Fig. 6. Number of buildings per user

4   Applications of WLAN Usage Trends

Kyoto Women's University is at the foot of Higashiyama mountains which are formed by activities of some faults including Hanaori Fault, as a result, it is feared the earthquake may occur.

When an earthquake occurs, it is important to notify the people in campus of the latest and correct information. They must decide whether they really should stay in the building or get out immediately.

Thanks to 802.1X authentication, the device which has connected to WLAN once can connect to it automatically. Thus the information that a device is connecting to an AP means the user who owns it is around there. The network administrators know the number of people in each building and who is in specified building with WLAN usage trends we have described above.

Of course the information about the user location is sensitive and should be kept secure. It is required that the information has some anonymities to utilize as statistical data. At this time, a user must be distinguished from each other but not be specified. WLAN logs can be converted into the information which has such anonymities [10], and it is useful information for even users.

5   Conclusion

In this paper, the authors analyse WLAN logs to clarify usage trends in KWU. It is worth knowing such as the population distribution of WLAN users and hourly distribution. These information about WLAN devices includes physical presence of the users, thus it is very useful for daily life in campus even for evacuation plan.

The authors continue the survey of WLAN usage and analyze it in more detail. There are some related works such as [3–7] that survey WLAN usage on the university campus to understand its trends. When the standardized method to investigate WLAN trends is established, the knowledge of each university can be integrated and thus such survey is applicable to evacuation plan and/or crisis management.

References

1. Ministry of Internal Affairs and Communications, 2015 WHITE PAPER Information and Communications in Japan (2015). http://www.soumu.go.jp/johotsusintokei/whitepaper/eng/WP2015/2015-index.html
2. Miyashita, K., Mizuno, Y.: A study of development in information and communication technologies for the university-wide education and information system at Kyoto Women's University. IPSJ J. **53**(3), 997–1004 (2012)
3. Satoh, M., Murakami, T., Isogami, S., Kidokoro, H., Kuboyama, T.: Analysis of wireless LAN usage trends in campus, IPSJ SIG Technical report, vol. 2013-IOT-22, no. 3 (2013)
4. Sugiki, A., Satoh, A., Wada, K.: Analyses and problems of usage statistics on campus-wide wireless LAN system, ISPJ SIG Technical report, vol. 2013-IOT-23, no. 7 (2013)
5. Yanagita, N., Hatsukade, I., Aoki, K., Sonoda, M., Kawabata, K.: Usage analysis of wireless LAN of Miyazaki University, IPSJ SIG Technical report, vol. 2015-IOT-31, no. 9 (2015)
6. Hatono, I.: Analysis and application of university-wide wireless LAN usage log information, IPSJ SIG Technical report, vol. 2015-IOT-31, no. 10 (2015)
7. Fukuda, Y., Nakamura, Y.: Performance measurement of campus WiFi at Kyushu Institute of Technology, IPSJ SIG Technical report, vol. 2016-IOT-32, no. 1 (2016)
8. Miyashita, K.: A survey on trends of WiFi use in Kyoto Women's University, IPSJ SIG Technical report, vol. 2013-IOT-23, no. 6 (2013)
9. Miyashita, K.: A wireless LAN usage trends survey on campus for evacuation planning, workshop on Resilient Internet based Systems (REIS). In: Proceedings of International Conference on Signal-Image Technology and Internet-Based Systems (SITIS 2013), pp. 865–869 (2013)
10. Miyashita, K.: A study of usage logs in campus networks, IPSJ SIG Technical report, vol. 2014-IOT-27, no. 15 (2014)

MDL Criterion for NMF with Application to Botnet Detection

Shoma Tanaka[1]([✉]), Yuki Kawamura[1,3], Masanori Kawakita[1], Noboru Murata[2], and Jun'ichi Takeuchi[1]

[1] Graduate School of Information Science and Electrical Engineering,
Kyushu University, Fukuoka, Japan
`takeuchi@inf.kyushu-u.ac.jp`
[2] Faculty of Science and Engineering, Waseda University, Tokyo, Japan
[3] Nihon Unisys, Ltd., Tokyo, Japan

Abstract. A method for botnet detection from traffic data of the Internet by the Non-negative Matrix Factorization (NMF) was proposed by (Yamauchi et al. 2012). This method assumes that traffic data is composed by several types of communications, and estimates the number of types in the data by the minimum description length (MDL) criterion. However, consideration on the MDL criterion was not sufficient and validity has not been guaranteed. In this paper, we refine the MDL criterion for NMF and report results of experiments for the new MDL criterion on synthetic and real data.

Keywords: Botnet · NMF · MDL principle

1 Introduction

We detect unknown botnets by focusing on the cooperative relationship on hosts. We discuss the method for botnet detection from traffic data of the Internet proposed by Yamauchi et al. [7], which is based on the Non-negative Matrix Factorization (NMF) [1]. NMF is a method to decompose an $n \times m$ matrix into product of $n \times r$ matrix and $r \times m$ matrix approximately, where r is an input parameter. It is known to be useful for many applications including pattern recognition, text mining, document clustering, signal processing, and cyber security.

Yamauchi et al. treated determination of the optimal r in NMF as a statistical model selection problem, and proposed a method based on the minimum description length (MDL) criterion [3]. They derived the MDL criterion for NMF by assuming a statistical model corresponding to NMF and evaluating the number of free parameters. However, consideration on the assumed statistical model was insufficient and the proposed criterion is not sound in view of description length.

In this paper, we discuss properties of the assumed statistical model based on the notion of non-negative rank of non-negative matrices and derive a new

---

Y. Kawamura—Moved to Nihon Unisys, Ltd. in April 2015.

A. Hirose et al. (Eds.): ICONIP 2016, Part I, LNCS 9947, pp. 570–578, 2016.
DOI: 10.1007/978-3-319-46687-3_63

form of MDL criterion for NMF. Finally, we report the results of experiments on the new MDL criterion using synthetic and recent real data.

2   A Method for Botnet Detection by NMF

We review the botnet detection method by Yamauchi et al. based on [7]. The problem setting of NMF is as follows:

[Problem Setting]
Given a data matrix $V \in \Re^{n \times m}$, solve the following optimization problem, where r is a positive integer not greater than $\min\{n, m\}$, $W \in \Re^{n \times r}$, and $H \in \Re^{r \times m}$.

$$\min_{W,H} ||V - WH||^2 \text{ (Frobenius norm)},$$

$$\text{subject to } W \geq 0, H \geq 0.$$

This problem is not tractable. Then, several algorithms to obtain local optimal solutions have been proposed. Among them, we employ the multiplicative update algorithm by Lee and Seung [6]. Using NMF, we analyze vector valued time series data as follows. For each μ ($\mu = 1, \cdots, m$), let $v^{1\mu}, v^{2\mu}, \cdots, v^{n\mu}$ a time series of the number of packets sent by the μth source host in a unit time interval and $\mathbf{v}^\mu$ be a vector $(v^{1\mu}, v^{2\mu}, \cdots, v^{n\mu})^T$. Let $V \in \Re^{n \times m}$ be a matrix such that $\mathbf{v}^\mu$ is the μth column vector. Given a data matrix V, we are to find two non-negative matrices $W \in \Re^{n \times r}$ and $H \in \Re^{r \times m}$ so that $V \approx WH$ using NMF. Here, r corresponds to the number of activity patterns which are contained in the data matrix V.

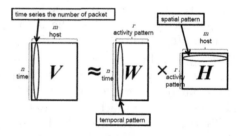

Fig. 1. The role of each matrix of NMF

The role of each matrix is shown in Fig. 1. As we observe the column vectors of W, we can see the temporal behavior of the number of packets of each potential pattern. We refer to it as a "temporal pattern". Similarly, as we observe the row vectors of H, we can see the distribution of the number of packets released by each host in the potential pattern. We refer to it as a "spatial pattern".

3   MDL Principle

We review the MDL criterion for NMF proposed in [7]. The MDL criterion is an information criterion introduced by Rissanen [3]. The MDL principle says that, when the data is compressed with the help from a statistical model, the model with shortest description length is optimal. For a data set $x^N = x_1 x_2 \ldots x_N$ and a k-dimensional model $q(x^N | \theta)$ ($\theta \in \Re^k$), the MDL criterion is generally known as $MDL = -\ln f + (k/2)\ln N$. Here, $f = \max_\theta q(x^N|\theta)$ is the maximum likelihood. To apply MDL to NMF, we introduce a statistical model corresponding to NMF as

$$V = WH + \epsilon. \tag{1}$$

Here, each element of ϵ is an independent normal random variable with mean 0. Let $p(V|\theta)$ $(\theta = (W, H))$ denote the density function of the model (1). Using this model, the formula of MDL criterion in [7] was given as

$$MDL_{NMF} = -\ln f + \frac{nr + rm - r^2}{2}\ln\frac{nm}{2\pi}, \tag{2}$$

where the maximum log likelihood $\ln f$ is

$$\ln f = -\frac{nm}{2}\ln 2\pi\sigma^2 - \frac{1}{2\sigma^2}\sum_{i=1}^{n}\sum_{\mu=1}^{m}(V_{i\mu} - (\hat{W}\hat{H})_{i\mu})^2.$$

Here, $(\hat{W}, \hat{H})$ is the maximum likelihood estimate, but it is not efficiently computable. Hence, we use the output of NMF instead, that is, we assume that $(\hat{W}, \hat{H})$ is an output of NMF algorithm, $\sigma^2 = (nm)^{-1}\sum_{i=1}^{n}\sum_{\mu=1}^{m}(V_{i\mu} - (\hat{W}\hat{H})_{i\mu})^2$, and the factor $nr + rm - r^2$ in (2) is the dimension (degree of freedom) of the model defined by (1). Note that the dimension is less than the sum of the numbers of entries of W and H, since they contain r^2 redundant entries. Let $\hat{r}$ denote the r which minimizes (2). We refer to $\hat{r}$ as the MDL estimate of r.

To derive (2), Yamauchi et al. make use of an asymptotic expression of stochastic complexity (SC) [4], which is the strict value of MDL. For a regular model, SC is evaluated as

$$SC = -\log q(x^N|\hat{\theta}) + \frac{k}{2}\log\frac{N}{2\pi} + \log\int\sqrt{|J(\theta)|}d\theta + o(1), \tag{3}$$

where $\hat{\theta}$ is the maximum likelihood estimate given x^N and $J(\theta)$ is the Fisher information matrix. Regard the data matrix V as a single nm-dimensional vector valued datum. Then, (3) can be transformed to

$$SC_{NMF} = -\log f + \frac{nr + rm - r^2}{2}\log\frac{1}{2\pi} + \log\int\sqrt{|J(\theta)|}d\theta + o(1).$$

The third term can be evaluated by noting that the following holds for each element of $J(\theta)$.

$$J_{ij}(\theta) = -E_\theta\frac{\partial^2\log p(V|\theta)}{\partial\theta_i\partial\theta_j}$$

$$= -E_\theta\left[\frac{\partial^2}{\partial\theta_i\partial\theta_j}\left(\frac{1}{2}(x^N - \mu)\Sigma^{-1}(x^N - \mu)^T + \frac{1}{2}\ln|\Sigma|\right)\right]$$

$$= -E_\theta \left[\frac{\partial^2}{\partial \theta_i \partial \theta_j} \left(\frac{1}{2}(x^N - \mu) \Sigma^{-1}(x^N - \mu)^T \right) \right]$$

$$= -E_\theta \left[\frac{\partial^2}{\partial \theta_i \partial \theta_j} \frac{1}{2} \mathrm{Tr}(\Sigma^{-1}(x^N - \mu)(x^N - \mu)^T) \right]$$

$$= nm A_{ij}(\theta) + O(1). \tag{4}$$

Here, $A_{ij}(\theta)$ is a quantity independent of nm. The factor nm in (4) appears because the dimension of the datum V is nm. Therefore, we have

$$\int \sqrt{|J(\theta)|} d\theta = (nm(1 + o(1)))^{\frac{nr+rm-r^2}{2}} C,$$

where C is a certain constant. Neglecting a term of $o(\frac{nr+rm-r^2}{2})$, we obtain (2).

From the above, Yamauchi et al. calculated MDL criterion for NMF. However, there were obscure points in the number of free parameters and the region of integral in (2), which influence the main term of the MDL criterion.

4   Non-negative Rank

Yamauchi et al. calculated the degree of freedom of the matrices whose rank is r in (2), but this only holds for few limited situations. We have to use the notion of non-negative rank, which we explain below.

We review the notion of non-negative rank following [2] and discuss the degree of freedom of the space of WH. Let $\Re_+$ stand for the set of non-negative real numbers. Non-negative rank $rank_+(A)$ for $\forall A \in \Re_+^{n \times m}$ is defined as

$$rank_+(A) = \min \left\{ q \, \middle| \, \sum_{j=1}^{q} R_j = A, rank(R_j) = 1, R_j \in \Re_+^{n \times m} \right\}.$$

Note that we can write

$$WH = \sum_{i=1}^{r} \begin{pmatrix} w_{i1} \\ \vdots \\ w_{in} \end{pmatrix} \left(h_{i1} \cdots h_{im} \right) = \sum_{i=1}^{r} \mathbf{w}_i^T \mathbf{h}_i.$$

Since $w_i^T h_i$ is a rank 1 non-negative matrix, we can think that the space of WH is the space of matrices of non-negative rank r. Therefore, we should consider the degree of freedom in the model of NMF as that of the space of fixed non-negative rank. Apparently, non-negative rank is not less than rank, and there is a possibility that previous MDL criterion for NMF is incorrect.

Define the scaling factor $\sigma(A)$ by $\sigma(A) := \mathrm{diag}\{\|\mathbf{a}_1\|_1, \cdots, \|\mathbf{a}_m\|_1\}$ where $A = [\mathbf{a}_1, \cdots, \mathbf{a}_m]$ and, $\|\cdot\|_1$ stands for the 1-norm of a vector. Define the pullback map $\mathcal{V}(A)$ by $\mathcal{V}(A) := A\sigma(A)^{-1}$ Each column of $\mathcal{V}(A)$ can be regarded as a point on the $(n-1)$-dimensional probability simplex $\mathcal{D}_n$ defined by $\mathcal{D}_n := \{\mathbf{a} \in \Re_+^n \mid \mathbf{1}_n^T \mathbf{a} = 1\}$ where $\mathbf{1}_n$ stands for the vector of all 1's in $\Re^n$

For a given matrix $A \in \Re_+^{n \times m}$, denote its non-negative matrix factorization $A = UV$, where $U \in \Re_+^{n \times p}$, $V \in \Re_+^{p \times m}$, $UV = (UD)(D^{-1}V)$ for any invertible diagonal matrix $D \in \Re_+^{p \times p}$. We may assume without loss of generality that U is already a pullback so that $\sigma(U) = I_m$. It follows that

$$A = \mathcal{V}(A)\sigma(A) = UV = \mathcal{V}(U)\mathcal{V}(V)\sigma(V).$$

Since the columns of the product $\mathcal{V}(U)\mathcal{V}(V)$ and the pullback map $\mathcal{V}(A)$ are all on the simplex $\mathcal{D}_n$, we have

$$\mathcal{V}(A) = \mathcal{V}(U)\mathcal{V}(V),$$
$$\sigma(A) = \sigma(V).$$

It thus suffices to consider the geometric meaning of $rank_+(\mathcal{V}(A))$ on the simplex $\mathcal{D}_n$. The following lemma holds [2].

Lemma 1. *Given a non-negative matrix* $A \in \Re_+^{n \times m}$, $rank_+(A) = rank_+(\mathcal{V}(A))$.

Proof. Assuming $rank_+(A) = r$, we can denote that $A = \sum_{i=1}^r w_i h_i^T$. Then,

$$\mathcal{V}(A) = A\sigma(A)^{-1} = \sum_{i=1}^r w_i h_i^T \sigma(A)^{-1} = \sum_{i=1}^r w_i h_i'^T$$

Thus, $rank_+(\mathcal{V}(A)) \leq r = rank_+(A)$. Similarly, assuming $rank_+(\mathcal{V}(A)) = r$, we can derive $rank_+(A) \leq r = rank_+(\mathcal{V}(A))$. Therefore, $A \in \Re_+^{n \times m}$, $rank_+(A) = rank_+(\mathcal{V}(A))$.

Note that the relationship $\mathcal{V}(A) = \mathcal{V}(U)\mathcal{V}(V)$ implies that the columns in the pullback $\mathcal{V}(A)$ are convex combination of columns of $\mathcal{V}(U)$. The following interesting geometrical interpretation of non-negative rank is shown in [2].

Lemma 2. *The non-negative rank* $rank_+(A)$ *stands for the minimal number of vertices on* $\mathcal{D}_n$ *so that the resulting convex polytope encloses all columns of the pullback* $\mathcal{V}(A)$.

In [2], the conditional probabilities $P(rank_+(A) = 3|rank(A) = 3)$ and $P(rank(A) = 3|rank_+(A) = 3)$ are evaluated, assuming that each column vector of A follows absolutely continuous probability density over the three-dimensional space in the unit tetrahedron. As noted in [2], the former depends on the assumed probability density. On the other hand, the following theorem holds for the case of general about the latter.

Theorem 1. *The following holds on the condition that* $r < min\{m, n\}$

$$P(rank(A) = r|rank_+(A) = r) = 1$$

That is, in the set of matrices whose non-negative rank is r, the set of matrices whose rank is less than $r - 1$ is measure zero. Therefore, it is concluded that the degree of freedom of the set of matrices whose non-negative rank is r is equal to the degree of freedom of the set of matrices whose rank is r. Then it follows that, the degree of freedom in (2) is correct.

5   Discussion on Main Term of MDL Criterion for NMF

Yamauchi et al. derived the MDL criterion ignoring following 2 conditions.
1. There exists non-negative value constraints in the set of data series
2. Integration that depends on degree of freedom of parameter
We reconsider the above second condition and calculate the upper bound of MDL criterion.

5.1   Modification About $\int |J(\theta)|^{1/2} d\theta$

In Chap. 3, Yamauchi et al. calculated $\ln \int |J(\theta)|^{1/2} d\theta$ as follows.

$$\int |J(\theta)|^{1/2} d\theta = \frac{k}{2} \ln nm + O(1).$$

They integrated in θ in the derivation process of this formula. However θ is a k-dimensional vector and the integration result should depend on k. Considering this point, we calculate the upper bound of $\int |J(\theta)|^{1/2} d\theta$ as

$$
\begin{aligned}
\ln \int |J(\theta)|^{1/2} d\theta &= \ln \left\{ \int |A(\theta)|^{1/2} (nm(1+o(1)))^{k/2} d\theta \right\} \\
&= \ln \left\{ (nm(1+o(1)))^{k/2} \int |A(\theta)|^{1/2} d\theta \right\} \\
&\leq \ln \left\{ (nm(1+o(1)))^{k/2} C \int d\theta \right\} \\
&\leq \ln \left\{ (nm(1+o(1)))^{k/2} C \cdot \prod_{i=1}^{r+1} \left(\frac{M_i^{n-1}}{(n-1)!} \right) \cdot \prod_{j=r+2}^{m} \left(\frac{M_j^{r-1}}{(r-1)!} \right) \right\}
\end{aligned}
\tag{5}
$$

where $M_1, M_2, ..., M_m$ are the 1-norm of column vectors, assuming they are in ascending order of 1-norm and $|A(\theta)|^{1/2} \leq C$. Then, we assume that the ith column vectors for $i \leq r+1$ are elements of $(n-1)$-dimensional simplex of size M_i and that the jth column vectors for $j > r+1$ is an element of the affine space defined by the first $r+1$ column vectors. Note that the volume of $(n-1)$-dimensional simplex of size M_i is $M_i^{n-1}/(n-1)!$, and that the volume of the above affine space restricted in positive region is bounded by $M_j^{r-1}/(r-1)!$. By this inference, we have obtained the evaluation (5).

6   Simulation

Here we show the results of numerical experiments with various sizes of input matrices.

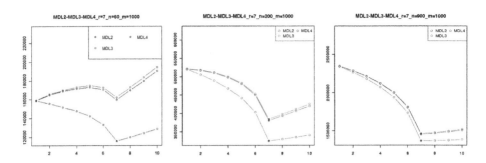

Fig. 2. Comparison by the difference of the main term

6.1   Numerical Experiments

We verify the difference between MDL4(new MDL) and MDL2(old MDL). We make $W \in \Re^{200 \times 7}, H \in \Re^{7 \times 1000}$ where true $r = 7$. Each element of the matrices takes the value from 1 to 5 which follows the uniform distribution. We make the data matrix $V = WH + \epsilon$, where the each elements of ϵ follows the normal distribution with mean 0 and variance 1. We decompose V by NMF, and estimate r by the above two MDL criteria.

6.2   Results

In the graphs in Fig. 2, horizontal axis shows r of NMF and vertical axis shows value of the main terms of MDL2 and MDL4.

6.3   Influence on MDL4 by the Size of Matrices

When the size of data matrix V is 200×1000 (the center of Fig. 2), both MDL2 and MDL4 select the true model. The left side of Fig. 2 is the graph on the condition that the size of data matrix V is 60×1000. MDL2 select the true model but MDL4 select $r = 1$. However, if the size of V is enlarged to 900×1000, the value of MDL4 becomes almost equal to MDL2 (The right side of Fig. 2).

7   Application to Botnet Detection

We apply our method to botnet detection using darknet data.

7.1   Detection Method for Botnet Activity Pattern

We assume that the hosts of botnets send packets synchronously. So, we issue an alert when some hosts in a spatial pattern of a cluster send packets synchronously. Specifically, we pay attention to a cluster in which the number of the synchronized hosts is more than 5.

Since the number of packets sent by each bot is small, it is usually difficult to detect anomaly from the Internet traffic. To overcome this problem, it is effective to use the data of darknets. A darknet is an accessible and unused IP address space. Since the packets which reach a darknet are due to infection activities by malwares or misconfiguration of networks, darknets are more useful to detect the behavior of botnets than real networks. We use several data sets observed by a part of the darknet managed by National Institute of Information and Communications Technology (NICT), which consists of over 30000 IP addresses.

7.2   Results

We apply our method to the 2014 and 2015 darknet data. The following is some results of our experiments. Based on the data of April 30, 2014 from February 1, 2014, we got alerts to 53/UDP, 123/UDP and 161/UDP that are DRDoS related ports. About this event, it reported that there is an increase of scan to DRDoS related port around April 2014 [5]. It would be highly possible that we detected scan activities to evaluate vulnerability to the target ports.

Figure 3 is a temporal pattern and a spatial pattern of one alert (the upper side cluster) of DRDoS related port. In the temporal pattern, there is a peak in the time series of the number of packets. In the spatial pattern, there is a synchronism in the number of packets of each host. Therefore, the hosts in the upper side cluster act synchronously, and we judged that this cluster corresponded to botnet.

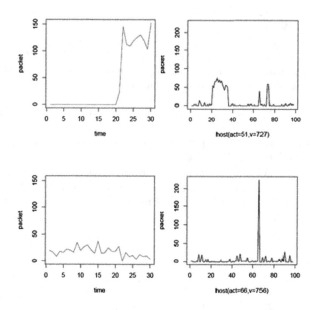

Fig. 3. Temporal pattern and spatial pattern of an alert (2014/04/13)

However, as a result of the investigation of the hosts of the cluster, they turned out to be hosts of research institutes that evaluate vulnerability to the target ports. It would be highly possible that we detected coordinated activities of their hosts.

We also evaluate the difference in the number of alert between MDL2 and MDL4 during April 2014. In MDL2, 179 alerts is detected and in MDL4, 84 alerts is detected. The DRDoS related alerts reported above were detected by both MDL2 and MDL4. Therefore, there is a possibility that MDL4 can reduce the number of false alerts.

8   Conclusion

We verified the MDL criterion for NMF introduced by Yamauchi et al. It was found that their derivation about degree of freedom is correct. We also consider the main term of MDL criterion and evaluated its upper bound. A future issue is that we need to experiment on more data and verify the validity of this technique.

Acknowledgment. We thank everyone of Cybersecurity Laboratory, NICT, who provides the darknet data, and the members of Proactive Response Against Cyber-attacks Through International Collaborative Exchange (PRACTICE).

References

1. Cichocki, A., Zdunek, R., Phan, A.H., Amari, S.: Nonnegative Matrix and Tensor Factorizations. Wiley, Chichester (2009)
2. Dong, B., Lin, M.M., Chu, M.T.: Nonnegative rank factorization - a heuristic approach via rank reduction. Numer. Algorithms **65**(2), 251–274 (2014)
3. Rissanen, J.: Modeling by shortest data description. Automatica **14**, 465–471 (1978)
4. Rissanen, J.: Fisher information and stochastic complexity. IEEE Trans. Inf. Theor. **42**(1), 40–47 (1996)
5. Kramer, L., et al.: AmpPot: monitoring and defending amplification DDoS attacks. In: Proceedings of RAID 2015 (2015)
6. Lee, D.D., Seung, H.S.: Algorithms for non-negative matrix factorization. Neural Inf. Process. Syst. **13**, 556–562 (2001)
7. Yamauchi, S., Kawakita, M., Takeuchi, J.: Botnet detection based on non-negative matrix factorization and the MDL principle. In: Proceedings of the 19th International Conference on Neural Information Processing, Doha, Qatar, 12–15 November, pp. 400–409 (2012)

A Brief Review of Spin-Glass Applications in Unsupervised and Semi-supervised Learning

Lei Zhu[1](✉), Kazushi Ikeda[2], Paul Pang[1],
Ruibin Zhang[1], and Abdolhossein Sarrafzadeh[1]

[1] Unitec Institute of Technology, Auckland, New Zealand
yiming.peng@ecs.vuw.ac.nz
[2] NARA Institute of Science and Technology, Ikoma, Japan

Abstract. Spin-glass theory developed in statistical mechanics has found its usage in various information science problems. In this study, we focus on the application of spin-glass models in unsupervised and semi-supervised learning. Several key papers in this field are reviewed, to answer the question that why and how spin-glass is adopted. The question can be answered from two aspects.

Firstly, adopting spin-glass models enables the vast knowledge base developed in statistical mechanics to be used, such as the self-organizing grains at the superparamagnetic phase has a natural connection to clustering. Secondly, spin-glass model can serve as a bridge for model development, i.e., one can map existing model into spin-glass manner, facilitate it with new features and finally map it back.

1 Introduction

Spin-glass is a family of theories studying the disordered magnetic system, using statistical mechanics tools, which interprets the macroscopic properties of many-body systems starting from the knowledge of interactions between microscopic elements [1]. Since late 1980's, spin-glass models and statistical mechanics find their usage in many information science fields, such as error-correcting code [2], image restoration [3], associative memory [4] etc.

This paper focuses on the usage of spin-glass in unsupervised and semi-supervised learning. Several key works in this field is reviewed, trying to sketch a big picture of why spin-glass is preferred and how it is adopted in unsupervised and semi-supervised learning.

In the next section, we will start by introducing the widely adopted Potts spin-glass model and its characteristics related to our topic.

2 Potts Spin-Glass Model

Potts spin-glass model [5] has a long history in statistical mechanics. The spin s can take one of q states in Potts model, $s = 1, 2, \ldots, q$. Spins on sites i and j are coupled by an interaction of strength $J_{ij} > 0$. Let $S = \{s_i\}_{i=1}^{N}$ be the

© Springer International Publishing AG 2016
A. Hirose et al. (Eds.): ICONIP 2016, Part I, LNCS 9947, pp. 579–586, 2016.
DOI: 10.1007/978-3-319-46687-3_64

configuration of a system with N sites, the total energy of such system is given
by the Hamiltonian

$$\mathcal{H}(S) = \sum_{\langle i,j \rangle} J_{ij}(1 - \delta(s_i, s_j)) \quad s_i = 1, 2, \ldots, q, \tag{1}$$

where $\langle i, j \rangle$ denotes neighboring sites i and j, $\delta(s_i, s_j)$ is the Kronecker delta
function which equals to 1 if $s_i = s_j$ and 0 otherwise.

Ground state of a system refers to the spin configuration S that minimizes
energy $\mathcal{H}(S)$. There are various methods available for finding the ground state of
a given Hamiltonian, from approximation to exact solution [1]. Thus modeling a
problem into a spin-glass system to find the ground state serves as alternatives
of direct optimization, which we will see later.

To calculate the thermodynamic average of a physical quantity A at a given
temperature T, one has to calculate

$$\langle A \rangle = \sum_S A(S)P(S), \quad P(S) = \frac{1}{Z}\exp(-\frac{\mathcal{H}(S)}{T}) \tag{2}$$

where the probability density $P(S)$ gives the statistical weight of each spin configuration in thermal equilibrium and $Z = \sum_S \exp(-\mathcal{H}(S)/T)$ is a normalization constant. Since the number of configurations increases exponentially with
the system size N, analytically computing (2) is simply impractical for large
system. In practice, $\langle A \rangle$ is always calculated through Monte Carlo simulation
methods [6].

Next we introduce some important quantities that will be used later. The
order parameter $\langle m \rangle$ of the system is the thermodynamic average of magnetization $m(S)$, $m(S)$ is defined as

$$m(S) = \frac{qN_{max}(S) - N}{(q-1)N}, \quad N_{max}(S) = \max\{N_1(S), N_2(S), \ldots, N_q(S)\}, \tag{3}$$

where $N_\mu(S)$ is the number of spins in state μ. The thermal average of $\delta(s_i, s_j)$ is
called spin-spin correlation $G_{ij} = \langle \delta(s_i, s_j) \rangle$, which denots the probability that
spins s_i and s_j being aligned.

3   Clustering and Semi-supervised Classification

The papers discussed in this section share a major idea in common: the natural
cluster structure in input data is revealed by the self-organization phenomena
at superparamagnetic phase. This self-organization is a physical property found
in inhomogeneous ferromagnetic materials.

3.1   From Phase Transition to Clustering

In the work of Blatt et al. [7], each data point is assigned as a Potts spin, and
the spin state denotes the cluster identity of that data point. The interaction

between neighboring points (i.e. J_{ij} in (1)) is defined as a decreasing function of the distance between the points, such as the localized interaction setting

$$J_{ij} = \begin{array}{l} \frac{1}{\bar{K}} \exp(-\frac{d_{ij}^2}{2a^2}) \text{ if i and j are neighbors} \\ 0 \qquad\qquad\quad \text{otherwise} \end{array} \tag{4}$$

where a is a scale to define high-density regions, d_{ij} is the distance, and $\bar{K}$ is the average number of neighbors. Using such interaction setup, there are strong interactions in high-density regions and weak interactions in low-density regions, and the system is modeled as strongly inhomogeneous granular magnet.

For strongly inhomogeneous Potts systems (in which the spins form magnetic grains, with very strong couplings between neighbors that belong to the same grain and very weak interactions between all other pairs), such as the ones build as above on data with natural cluster structure, there are three phases with respect of the temperature change [8]. At very low temperatures, it is completely ordered, one spin state dominates, namely ferromagnetic phase. At ferromagnetic phase, most data points belongs to the same cluster. At very high temperatures, the system does not exhibit any ordering, all spins are randomly oriented, namely paramagnetic phase. In between these two phase, there is a superparamagnetic phase. In this phase strongly coupled grains are aligned (that is, are in their respective ferromagnetic phases), while there is no relative ordering of different grains.

Clearly, data clusters can be identified by the internal alignment within grains at the superparamagnetic phase. In order to locate superparamagnetic phase, susceptibility $\mathcal{X}$, the variance of the magnetization, is computed at all temperatures

$$\mathcal{X} = \frac{N}{T}(\langle m^2 \rangle - \langle m \rangle^2). \tag{5}$$

At low temperatures, fluctuations of the magnetizations are negligible, thus $\mathcal{X}$ is small in the ferromagnetic phase. At the transition from the ferromagnetic to superparamagnetic phase a pronounced peak of $\mathcal{X}$ is observed [9], because in the superparamagnetic phase fluctuations of the state taken by grains acting as a whole (as giant superspins) produce large fluctuations in the magnetization. As the temperature is raised further, each grain disorders and $\mathcal{X}$ decreases abruptly. So by observing the change of $\mathcal{X}$ over temperatures, superparamagnetic phase can be located.

At the superparamagnetic phase, the spin-spin correlation G_{ij} is computed for any pair of spins (data points). If the correlation is higher than the expected value on a null model, the pair of points are clustered as the same group.

Similar to above approach, Clement et al. [10] also take advantage of self-organizing property of close related spins at superparamagnetic phase. After fist round clustering like [7], they treat each cluster as a Ising spin, then cteat a Hamiltonian to study the relation between clusters obtained. The characteristic of this scaled Hamiltonian reveals the structure among clusters.

Recently, Qin et al. [11] present a set of numerical methods to determine critical temperature at which phase transition occurs. With the help of their methods, one can directly target the superparamagnetic phase rather than locate superparamagnetic phase via costly simulation and observation.

3.2 Incorporating Labelled Data Points

The work of Getz et al. [12] can be seen as an extension of above work [7]. With the presence of labelled data, the Potts spin system can no longer be described as (1). Assume the first M data points are labelled, the system is represented as

$$\mathcal{H}(S) = \sum_{\langle i,j \rangle} J_{ij}(1 - \delta(s_i, s_j)) + \sum_{k=1}^{M} h_k(1 - \delta(s_k, c_k)), \qquad (6)$$

where c_k is the known class label for point k. The second term in (6) serves as penalty in case that s_k violates the assigned class c_k. In their work, h_k is set to be infinity, which means violation is not allowed.

Have the system defined, the rest of work in [12] is similar to that in [7]. Firstly locating the superparamagnetic phase by observing $\mathcal{X}$. And then clustering data points at superparamagnetic phase by evaluating spin-spin correlation G_{ij}. The cluster has points with given class label is classified into corresponding class. Clusters without labelled points are the hints for new classes. And the data points not associated with any cluster is considered as unknown points.

The major difference between [7,12] appears in the Monte Carlo simulation step. The introduction of labelled points changes the Hamiltonian meanwhile inherently changes the properties of the system. The energy landscape for (6) is more ragged, thus standard MCMC methods may be confined to certain energy 'valley' and give highly biased estimation. Getz et al. suggest using extended MCMC methods like [13] to solve the such problem by allowing the system to 'jump' between 'valleys'.

Next section we will review the family of community detection, which is closely related to clustering.

4 Unsupervised and Semi-supervised Community Detection

Given a relational network over a set of entities, community detection methods seek to identify 'Communities' which are sets of nodes such that nodes within each community are more densely connected than between communities. As an unsupervised learning task, fully automated community detection can be seen as clustering upon graphs.

The input for community detection is an undirected weighted graph $G = (V, A)$. The node set $V = \{v_1, v_2 \ldots, v_n\}$ represents entities, and the adjacency matrix A represents the relation between entities. $A_{ij} \in (0, 1])$ specifies that there is an edge e_{ij} between v_i and v_j with weight A_{ij} (v_i and v_j have a relation with strength A_{ij}), and $A_{ij} = 0$ otherwise (v_i and v_j have no relation). The degree of node v_i is defined as the summation of the weights it relates to, $d_i = \sum_j A_{ij}$. And the total weight on graph G is given by $m = \frac{1}{2}\sum_{i,j} A_{ij}$. The objective of community detection is to determine a community partition C such that nodes from the same community are densely interconnected and different communities are sparsely connected.

Newman-Girvan graph modularity [14] is arguably the most widely used model in community detection. Community structure of the graph is measured from a global perspective, as the difference of the graph's structure from an expected null model presumed to have no community structure. The modularity Q of a community partitioning C is given by

$$Q(C) = \frac{1}{2m}\sum_{i,j}(A_{ij} - P_{ij})\delta(C_i, C_j), \tag{7}$$

where P_{ij} represents the probability of an edge between v_i and v_j in the null model, C_k states the community that v_k belongs to, and $\delta(C_i, C_j)$ is the Kronecker delta function which equals to 1 if $C_i = C_j$ (v_i and v_j belongs to the same community) and 0 otherwise. Newman-Girvan graph modularity employs a null model that randomly rewrites the given graph while maintaining the total number of edges and the degree distribution of the nodes, which follows that

$$P_{ij} = \frac{d_i d_j}{2m}. \tag{8}$$

Thus high modularity Q value indicates strong community structure in the graph. Newman and Girvan present several methods for community identification using spectral clustering over the modularity matrix $(\boldsymbol{A} - \boldsymbol{P})$ [14]. Since then, large number of other community detection methods employ modularity as their foundation [15].

4.1   Potts Spin-Glass Modeling for Community Detection

Reichardt and Bornholdt [16] interpret the relational network and community using a Potts spin-glass system, in which any community partitioning is considered as a corresponding spin configuration and community detection problem falls into finding the ground state of the Potts spin-glass system. This work also shows that by certain parameter setting, their Potts Spin-glass model is equivalent to Newman-Girvan modularity. Thus it bridges various community detection methods to statistical mechanics tools.

Reichardt and Bornholdt's work starts from a Potts Hamiltonian to be minimized

$$\mathcal{H}(C) = -\sum_{i \neq j} a_{ij}A_{ij}\delta(C_i, C_j) + \sum_{i \neq j} b_{ij}(1 - A_{ij})\delta(C_i, C_j)$$
$$+ \sum_{i \neq j} c_{ij}A_{ij}(1 - \delta(C_i, C_j)) - \sum_{i \neq j} d_{ij}(1 - A_{ij})(1 - \delta(C_i, C_j)). \tag{9}$$

In (9), (i) internal links between nodes of the same community and (ii) non-links between different community are rewarded, meanwhile (iii) missing links between nodes of the same community and (iv) existing links between different community are penalized. These (i)–(iv) objectives are balanced by a_{ij}, d_{ij}, b_{ij} and c_{ij}, respectively. Then they consider weighting links and nonlinks equally, no matter they are external or internal, i.e. $a_{ij} = c_{ij}$ and $b_{ij} = d_{ij}$. Using γ to balance a_{ij} and b_{ij}, set $a_{ij} = 1 - \gamma p_{ij}$ and $b_{ij} = \gamma p_{ij}$, where p_{ij} represents the probability that there is a link between node v_i and v_j, when normalized, have $\sum_{i \neq j} p_{ij} = 2m$. Adopt all these settings, Hamiltonian (9) is simplified into

$$\mathcal{H}(C) = -\sum_{i \neq j}(A_{ij} - \gamma p_{ij})\delta(C_i, C_j) \tag{10}$$

As can see, (11) and (9) are closely related. Actually, having γ set to be 1 and define p_{ij} as P_{ij} following (8), minimizing $\mathcal{H}(C)$ in (11) is equivalent to maximizing $Q(C)$ in (9). Detailed proof of this equivalence can be found in [16]. And the $\gamma = 1$ here leads to natural situation that the total energy that can possibly be contributed by links and nonlinks is equal, having

$$\sum_{i \neq j} A_{ij} a_{ij} = \sum (1 - A_{ij})b_{ij}. \tag{11}$$

4.2   Incorporating Guidance into Community Detection

When extra guidance information rather than only the relational network is available, the community detection task changes from unsupervised into semi-supervised. Base on the work [16] introduced in last section, Eaton and Mansbach adopt Potts spin-glass model into semi-supervised community detection scenario [17]. Eaton and Mansbach's model is also closely connected to Newman-Girvan graph modularity model, thus various existing community detection methods can take advantage of.

   In context of community detection, the guidance can only be in form of two cases: (a) v_i and v_j come from the same community, or (b) v_i and v_j belong to different communities. To incorporate such external guidance, [17] constructs an energy function penalizing community structures that violate the guidance, the function is given by

$$U(C) = \sum_{i \neq j}(u_{ij}(1 - \delta(C_i, C_j)) + \bar{u}_{ij}\delta(C_i, C_j)), \tag{12}$$

where u_{ij} and $\bar{u}_{ij}$ are the penalty for violating above guidance (a) and (b) respectively. Rewrite(12) and incorporate it into the community detection Hamiltonian, have

$$\mathcal{H}'(C) = \mathcal{H}(C) + \mu \sum_{i \neq j}(u_{ij} - (u_{ij} - \bar{u}_{ij})\delta(C_i, C_j)), \tag{13}$$

where $\mu \geq 0$ balances the inherent community structure and the external guidance. μ is suggested to be set proportionally to the expected quality of the guidance. (13) can be further expressed as

$$
\begin{aligned}
\mathcal{H}'(C) &= -\sum_{i \neq j}(A_{ij} - \gamma p_{ij})\delta(C_i, C_j) + \mu\sum_{i \neq j}(u_{ij} - (u_{ij} - \bar{u}_{ij})\delta(C_i, C_j)) \\
&= -\sum_{i \neq j}((A_{ij} - \gamma p_{ij})\delta(C_i, C_j) + \mu(u_{ij} - \bar{u}_{ij})\delta(C_i, C_j)) + \mu\sum_{i \neq j}u_{ij} \quad (14) \\
&= -\sum_{i \neq j}(A_{ij} - \gamma(P_{ij} - \frac{\mu}{\gamma}(u_{ij} - \bar{u}_{ij})))\delta(C_i, C_j) + \mu\sum_{i \neq j}u_{ij},
\end{aligned}
$$

where P_{ij} is the probability of edge e_{ij} in the original null model (without external guidance). Since $\sum_{i \neq j} u_{ij}$ in (14) is a constant for any community partitioning C, it can be discarded in optimization of $\mathcal{H}'(C)$. Then modified (14) is of the same form as (11), and the modified null model is given by

$$
P'_{ij} = P_{ij} - \frac{\mu}{\gamma}(u_{ij} - \bar{u}_{ij}). \tag{15}
$$

Here, the guidance reduces the null probability of edges between nodes that should be in the same community and increases the null probability of edges for nodes pairs that should be in different communities.

Another major contribution of [17] is that it presents a new form of Newman-Girvan modularity which incorporates external guidance. From Hamiltonian (14), setting P_{ij} as (8), choosing γ to be 1 and then normalizing, just as [16] has done, the new modularity is given by

$$
Q'(C) = \frac{1}{2m}\sum_{i \neq j}(A_{ij} - (\frac{d_i d_j}{2m} - \mu(u_{ij} - \bar{u}_{ij})))\delta(C_i, C_j) - \frac{\mu}{2m}\sum_{i \neq j}u_{ij}. \tag{16}
$$

It worth noting that when external guidance is ignored (i.e. $\mu = 0$) or absent (i.e. $\forall i, j$ $u_{ij} = \bar{u}_{ij} = 0$), (16) falls back to (7). In the other words, Newman-Girvan graph modularity is a special case of (16).

5   Conclusion

In summary, we can see at least two reasons for spin-glass models being preferred. Firstly, adopting spin-glass models enables the vast knowledge base developed in statistical mechanics to be used, such as the self-organizing grains at the superparamagnetic phase for clustering, and ground state finding methods for optimization. Secondly, as seen in Sect. 4, spin-glass model can serve as a bridge for model development, i.e., one can map existing model into spin-glass manner, facilitate it with new features and then map it back.

References

1. Nishimori, H.: Statistical Physics of Spin Glasses, Information Processing: An Introduction. Clarendon Press, Oxford (2001)
2. Sourlas, N.: Spin-glass models as error-correcting codes. Nature **339**(6227), 693–695 (1989)
3. Inoue, J.-I., Carlucci, D.M.: Image restoration using the q-ising spin glass. Phys. Rev. E **64**(3), 036121 (2001)
4. Amit, D.J., Gutfreund, H., Sompolinsky, H.: Storing infinite numbers of patterns in a spin-glass model of neural networks. Phys. Rev. Lett. **55**(14), 1530 (1985)
5. Wu, F.-Y.: The potts model. Rev. Mod. Phys. **54**(1), 235 (1982)
6. Wang, J.-S., Swendsen, R.H.: Cluster monte carlo algorithms. Phys. A **167**(3), 565–579 (1990)
7. Blatt, M., Wiseman, S., Domany, E.: Data clustering using a model granular magnet. Neural Comput. **9**(8), 1805–1842 (1997)
8. Fortuin, C.M., Kasteleyn, P.W.: On the random-cluster model: I. introduction. and relation to other models. Physica **57**(4), 536–564 (1972)
9. Blatt, M., Wiseman, S., Domany, E.: Superparamagnetic clustering of data. Phys. Rev. Lett. **76**(18), 3251 (1996)
10. Clement, C., Liarte, D., Middleton, A., Sethna, J.: Effective hamiltonians of 2D spin glass clusters. In: APS Meeting Abstracts, vol. 1, p. 50004 (2015)
11. Qin, S.-M., Zeng, Y., Zhou, H.-J.: Spin glass phase transitions in the random feedback vertex set problem, arXiv preprint arXiv:1603.09032 (2016)
12. Getz, G., Shental, N., Domany, E.: Semi-supervised learning-a statistical physics approach, arXiv preprint cs/0604011 (2006)
13. Berg, B.A., Neuhaus, T.: Multicanonical ensemble: a new approach to simulate first-order phase transitions. Phys. Rev. Lett. **68**(1), 9 (1992)
14. Newman, M.E.: Finding community structure in networks using the eigenvectors of matrices. Phys. Rev. E **74**(3), 036104 (2006)
15. Fortunato, S.: Community detection in graphs. Phys. Rep. **486**(3), 75–174 (2010)
16. Reichardt, J., Bornholdt, S.: Statistical mechanics of community detection. Phys. Rev. E **74**(1), 016110 (2006)
17. Eaton, E., Mansbach, R.: A spin-glass model for semi-supervised community detection. In: AAAI, Citeseer (2012)

Learning Latent Features with Infinite Non-negative Binary Matrix Tri-factorization

Xi Yang[1], Kaizhu Huang[1(✉)], Rui Zhang[1], and Amir Hussain[2]

[1] Xi'an Jiaotong-Liverpool University, SIP, Suzhou, China
{Xi.Yang,Kaizhu.Huang,Rui.Zhang02}@xjtlu.edu.cn
[2] Division of Computing Science and Maths, School of Natural Sciences,
University of Stirling, Stirling FK9 4LA, UK
ahu@cs.stir.ac.uk

Abstract. Non-negative Matrix Factorization (NMF) has been widely exploited to learn latent features from data. However, previous NMF models often assume a fixed number of features, say p features, where p is simply searched by experiments. Moreover, it is even difficult to learn binary features, since binary matrix involves more challenging optimization problems. In this paper, we propose a new Bayesian model called infinite non-negative binary matrix tri-factorizations model (iNBMT), capable of learning automatically the latent binary features as well as feature number based on Indian Buffet Process (IBP). Moreover, iNBMT engages a tri-factorization process that decomposes a nonnegative matrix into the product of three components including two binary matrices and a non-negative real matrix. Compared with traditional bi-factorization, the tri-factorization can better reveal the latent structures among items (samples) and attributes (features). Specifically, we impose an IBP prior on the two infinite binary matrices while a truncated Gaussian distribution is assumed on the weight matrix. To optimize the model, we develop an efficient modified maximization-expectation algorithm (ME-algorithm), with the iteration complexity one order lower than another recently-proposed Maximization-Expectation-IBP model [9]. We present the model definition, detail the optimization, and finally conduct a series of experiments. Experimental results demonstrate that our proposed iNBMT model significantly outperforms the other comparison algorithms in both synthetic and real data.

Keywords: Infinite non-negative binary matrix tri-factorization · Infinite latent feature model · Indian Buffet Process prior

1 Introduction

Non-negative matrix factorization (NMF), a popular matrix decomposition technique, has been widely applied in data analysis and machine learning [8]. Typically, NMF can be exploited to reveal from observations the latent features and consequently be used in semantic recognition or clustering. However, previous

© Springer International Publishing AG 2016
A. Hirose et al. (Eds.): ICONIP 2016, Part I, LNCS 9947, pp. 587–596, 2016.
DOI: 10.1007/978-3-319-46687-3_65

NMF models usually assume the number of features as a constant parameter, which is generally tuned or searched by trial and error. Such algorithms include the methods proposed in [1,2,10]. Moreover, when the factor matrix is assumed as binary, NMF is even challenging, since binary matrices usually lead to more difficult optimization.

To tackle the above problems, we extend standard NMF to learn binary features with a novel Bayesian model called infinite non-negative binary matrix tri-factorization (iNBMT) in this paper. Different from traditional NMF, the novel iNBMT model can select automatically from infinite latent features an optimal set by applying Indian Buffet Process (IBP) prior to the factor matrices. In addition, we manage to decompose the input sample matrix $\mathbf{Y}$ into triple matrix factors i.e., $\mathbf{Y} = \mathbf{ZWX}^T$, where $\mathbf{Z}$ and $\mathbf{X}$ are two binary matrices, and non-negative matrix $\mathbf{W}$ can be considered as a weight matrix. Compared from bi-factorization typically involved in NMF, tri-factorization can better capture latent features and reveal hidden structures underlying the samples [2]. Importantly, although two binary matrices are involved, we further propose an efficient modified maximization-expectation algorithm (ME-algorithm), which can be even fast used in very large matrix decomposition. In particular, the time complexity of our proposed ME-algorithm proves one order lower than another competitive model called Maximization-Expectation-IBP (ME-IBP) [9].

In the literature, there have been several proposals of NMF for binary matrix decomposition. However, all of them have certain drawbacks. Binary Matrix Factorization (BMF) proposed in [10] limits the input data to be binary; this is however too strong in real cases. On the other hand, the correlated IBP-IBP model enforces a product of two binary matrices to be still binary; such assumption is in general invalid unfortunately. Despite of its good properties, the recently-proposed Maximization-Expectation-IBP (ME-IBP) model [9] is slow in optimization. In particular, the iteration complexity for the ME-IBP model is $O(\gamma ND)$, which is significantly higher than $O(\alpha N + \beta D)$, the iteration complexity of our iNBMT model. Here, N and D, usually two big numbers, denote respectively the number of observations and the dimensionality. α, β, and γ are three coefficients.

2   Notation and Background

2.1   Indian Buffet Process

IBP can be considered as a prior defined on models with infinite binary matrices. It is typically used to infer how many latent features each observation processes. Suppose $\mathbf{Y} \in \mathbb{R}^{N \times D}$ be generated by linear combination with K-dimensional vector of latent factors $\mathbf{W} \in \mathbb{R}^{K \times D}$ and the assignment matrix $\mathbf{Z} \in \mathbb{R}^{N \times K}$. The observed data $\mathbf{Y}$ is then modeled as $\mathbf{Y} = \mathbf{ZW} + \epsilon$. ϵ is noise term of distributed independently over $\mathcal{N}(0, \sigma\mathbf{I})$.

Let $\mathbf{Z}$ be a binary matrix where $z_{nk} = 1$ presents the latent feature k belongs to the observation n. The following IBP prior on binary feature matrix $\mathbf{Z}$ is derived by placing independent beta priors on Bernoulli. π_k's are generated independently for each column following a Beta prior. And then each object possessing feature k are generated independently from a Bernoulli with mean π_k.

$$\pi_{\mathbf{k}} \mid (\alpha) \sim Beta(\alpha/\mathbf{K}, 1), \qquad \mathbf{Z} \mid \pi_k \sim Bernoulli(\pi_k),$$

$$p([\mathbf{Z}]) = \frac{\alpha^K}{\prod_{h>0} K_h!} e^{\{-\alpha H_N\}} \prod_{k=1}^{K} \frac{(N - m_k)!(m_k - 1)!}{N!}, \qquad (1)$$

where K_h is the number of rows corresponding to the non-zero number h, $m_k = \sum_{i=1}^{N} z_{ik}$ is the number of objects possessing feature k, and $H_N = \sum_{j=1}^{N} \frac{1}{j}$ is the Nth harmonic number.

The IBP inspired several infinite-limit versions of classic matrix factorization models, e.g. infinite ICA models [6]. In infinite limit, Grinffiths et al. take the IBP prior into the infinite limit by defining equivalence classes on binary matrices [5]. The equivalence classes are matrices permutating the order of columns through eliminating all the null columns. Therefore, let K be unbounded and assume that we allow the number of active features K_+ to be learned from the data while remaining finite with probability one. By defining a scheme to re-order the non-zero columns of Z we can take $K \to \infty$ and find

$$p([\mathbf{Z}]) = \frac{\alpha^{K_+}}{\prod_{h>0} K_h!} e^{\{-\alpha H_N\}} \prod_{k=1}^{K_+} \frac{(N - m_k)!(m_k - 1)!}{N!}. \qquad (2)$$

2.2 Maximization-Expectation Algorithm

The ME algorithm just reverses the roles of two steps in the classical EM algorithm by maximization over hidden variables and marginalization over random parameters [7]. Given a dataset $\mathbf{Y}$, $p(\mathbf{Y}, \mathbf{Z}, \mathbf{W})$ is a probabilistic model where $\mathbf{Z}$ and $\mathbf{W}$ are all hidden random variables. To perform approximate MAP inference, it is necessary to compute posterior or marginal probabilities such as $p(\mathbf{Z}|\mathbf{Y}), p(\mathbf{W}|\mathbf{Y})$ or $p(Y)$. It can be viewed as a special case of a Mean-Field Variational Bayes (MFVB) approximation to a posterior that cannot be computed analytically. $p(\mathbf{Z}, \mathbf{W}|\mathbf{Y})$ is approximated by $q(\mathbf{Z})q(\mathbf{W})$ [4] if we assume independent variational distributions.

In MFVB, the variational Bayesian approximation alternatively estimates these distributions by minimizing the KL-divergence between the approximation and the exact distribution: $KL[q(\mathbf{Z})q(\mathbf{W})\|p(\mathbf{Z}, \mathbf{W}|\mathbf{Y})]$. The results are close-formed with the updates,

$$q(\mathbf{Z}) \propto exp(\mathbb{E}[\ln p(\mathbf{Y} \mid \mathbf{Z}, \mathbf{W})_{q(\mathbf{Z})}]), \quad q(\mathbf{W}) \propto exp(\mathbb{E}[\ln p(\mathbf{Y} \mid \mathbf{Z}, \mathbf{W})_{q(\mathbf{W})}]). \quad (3)$$

3 Infinite Non-negative Binary Matrix Tri-factorization

3.1 Model Description

The iNBMT model is applied on a real-valued observation data $\mathbf{Y} \in \mathbf{R}^{\mathbf{N} \times \mathbf{D}}$ where the rows and columns could be exchangeable. For a latent feature model, we use the matrix $\mathbf{F}$ to indicate the latent feature values. Then our focus will be on a distribution over observations conditioned on features $p(\mathbf{Y}|\mathbf{F})$, where $p(\mathbf{F})$ is the prior over features. F can be expressed as the element-wise product of these three components, $\mathbf{F} = \mathbf{Z} \otimes \mathbf{W} \otimes \mathbf{X}$, where a latent feature binary vector $\mathbf{x_j}$ is associated with each attribute, each item has a potential binary vector $\mathbf{z_i}$, and a matrix $\mathbf{W}$ represents the interaction weights parameter. Furthermore, the prior of the features is also defined by $p(\mathbf{F}) = p(\mathbf{Z})p(\mathbf{W})p(\mathbf{X})$.

In effect, we factorized $\mathbf{Y}$ into the linear inner product of the features and weight, $\mathbf{ZWX^T}$, generated by a fixed observation process $f(\cdot)$, as illustrated in Fig. 1. This process is equivalent to factorization or approximation of the data:

$$\mathbf{Y} \mid \mathbf{Z}, \mathbf{W}, \mathbf{X} \sim f(\mathbf{ZWX}^T, \theta),$$

where θ are hyperparameters specific to the model variant.

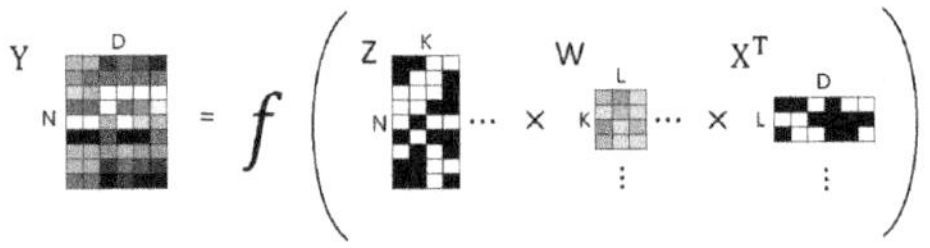

Fig. 1. Representation of the iNBMT model. The process $f(\cdot)$ applied to the linear inner product of the three components. Here $\mathbf{Z}, \mathbf{X}$ are infinite binary matrices, $\mathbf{W}$ present non-negative matrix.

We now develop our iNBMT model using Bayesian non-parametric priors. Specifically, IBP priors are imposed over binary matrices Z and X, while any non-negative prior $\mathcal{F}$ (e.g. exponential and truncated Gaussian) is assumed on the weight matrix W:

$$\mathbf{Z} \sim \mathcal{IBP}(\alpha), \qquad \mathbf{X} \sim \mathcal{IBP}(\lambda), \qquad \mathbf{W} \sim \mathcal{F}(\mathbf{W}; \mu, \sigma_W^2).$$

We assumed the hyperparameters were estimated from the data. By placing conjugate gamma hyperpriors on these parameters, we can have a straightforward extension to infer their values. Formally,

$$\mathbf{Y} \mid \mathbf{Z}, \mathbf{W}, \mathbf{X}, \theta \sim p(\mathbf{Y} \mid \theta), \qquad \theta = \{\alpha, \lambda, \sigma_Y \sigma_W\} \sim Gamma(a, b).$$

3.2 Linear-Gaussian iNBMT Model

To illustrate the iNBMT model for capturing the latent features, we set the linear-Gaussian model as the observation distribution with mean $\mathbf{ZWX}^T$ and covariance $(1/\theta)\mathbf{I}$ throughout this paper. This can be thought of a two-sided version of the linear-Gaussian model.

The marginal probabilities of the linear-Gaussian iNBMT model, is shown as below:

$$p(\mathbf{Y}|\mathbf{Z}, \mathbf{W}, \mathbf{X}, \sigma_X^2) = \frac{1}{(2\pi\sigma_Y^2)^{ND/2}} \exp -\frac{1}{2\sigma_Y^2} tr((\mathbf{Y} - \mathbf{ZWX}^T)^T(\mathbf{Y} - \mathbf{ZWX}^T)).$$

The weight matrix $\mathbf{W}$ uses the truncated Gaussian priors with a zero-mean i.i.d.

$$p(\mathbf{W}|0,\sigma_W^2) = \prod_{k=1}^{K}\prod_{l=1}^{L} TN(a_{kl};0,\sigma_W^2).$$

The marginal probabilities $p([\mathbf{Z}])$ and $p([\mathbf{X}])$ are specified with infinite IBP prior (given in Eq. (2)):

$$p(\mathbf{Z}|\alpha) = \frac{\alpha^{K_+}}{K_+!}\prod_{k=1}^{K}\frac{(N-m_k)!(m_k-1)!}{N!},$$

$$p(\mathbf{X}|\lambda) = \frac{\lambda^{L_+}}{L_+!}\prod_{l=1}^{L}[\frac{(D-m_l)!(m_l-1)!}{D!}].$$

From the Bayesian theorem, the posterior can be write as follows:

$$p(\mathbf{Y},\mathbf{Z},\mathbf{W},\mathbf{X}|\boldsymbol{\theta}) = p(\mathbf{Y}|\mathbf{Z},\mathbf{W},\mathbf{X},\sigma_Y^2)p(\mathbf{W}|0,\sigma_W^2)p(\mathbf{Z}|\alpha)p(\mathbf{X}|\lambda),$$

where the hyperparameters $\boldsymbol{\theta}$ conjugate gamma priors on inference parameters.

3.3   Evidence of iNBMT

In this part, we will present the approximate MAP inference, derived from the ME algorithm, for the linear-Gaussian iNBMT model.

Given the MFVB constraint, we determine the variational distributions by minimizing the KL-divergence, $\mathcal{D}(q\|p)$, between the variational distribution and the true posterior; this is equivalent to maximizing a lower bound on the evidence:

$$\ln p(\mathbf{Y}|\theta) = \mathbb{E}_q[\ln p(\mathbf{Y},\mathbf{Z},\mathbf{W},\mathbf{X}|\theta)] + \mathcal{H}[q] + \mathcal{D}(q\|p) \tag{4}$$

$$\geq \mathbb{E}_q[\ln p(\mathbf{Y},\mathbf{Z},\mathbf{W},\mathbf{X}|\theta)] + \mathcal{H}[q] \tag{5}$$

$$\equiv \mathcal{T}, \tag{6}$$

where $\mathcal{H}[q]$ is the entropy of q. The lower bound of evidence, $\mathcal{T}$, for the linear-Gaussian iNBMT model is:

$$\mathcal{T} \equiv \frac{1}{\sigma_Y^2}[-\frac{1}{2}(\mathbf{Z}\mathbb{E}[W]\mathbf{X}^T)(\mathbf{Z}\mathbb{E}[W]\mathbf{X}^T)^T + \mathbf{Z}(\mathbb{E}[W]\mathbf{Y}^T + \mathbf{Z}\gamma)\mathbf{X}^T]$$

$$+ \sum_{k=1}^{K}[\ln\frac{(N-m_k)!(m_k-1)!}{N!}] + \sum_{l=1}^{L}[\ln\frac{(D-m_l)!(m_l-1)!}{D!}]$$

$$- \ln K_+! - \ln L_+! + \sum_{k=1}^{K}\sum_{l=1}^{L}\varphi_{kl} + const; \tag{7}$$

$$\gamma = \frac{1}{2}\sum_{k=1}^{K}\sum_{l=1}^{L}[\mathbb{E}[w_{kl}]^2 - \mathbb{E}[w_{kl}^2]]^T,$$

$$\varphi_{kl} = -\frac{KL}{2}\ln(\frac{\pi\sigma_W^2}{2}) - \frac{\mathbb{E}[w_{kl}^2]}{2\sigma_W^2} + \mathcal{H}(q(w_{kl})).$$

Here $\mathbb{E}[\mathbf{W}]$ is a matrix with each element defined as $\mathbb{E}[w_{kl}]$.

3.4 Parameter Updates

The updates for the variational parameters of the non-negative $\mathbf{W}$ over the truncate Gaussian distribution are shown as follows:

$$q(\mathbf{W}) = \prod_{k=1}^{K} \prod_{l=1}^{L} TN(w_{kl}; \mu_{kl}, \sigma_{kl}^2) = \prod_{k=1}^{K} \prod_{l=1}^{L} \frac{N(\mu_{kl}, \sigma_{kl}^2)}{\mathbf{\Phi}(\infty) - \mathbf{\Phi}(0)},$$

where $t = -\frac{\mu_{kl}}{\sqrt{2}\sigma_{kl}}$, $\mathbf{\Phi}(\mathbf{a}) = \frac{1}{2}(1 + erf(\frac{a-\mu_{kl}}{\sqrt{2}\sigma_{kl}}))$, $\mathbf{\Phi}(\infty) = 1$, $erf(\cdot)$ is the Gaussian error function. According to the upper tail truncation, the parameters are updated as follows:

$$\mathbb{E}[w_{kl}] = \mu_{kl} + \sigma_{kl}\lambda(t), \qquad \mathbb{E}[w_{kl}^2] = \mu_{kl}\mathbb{E}[w_{kl}] + \sigma_{kl}^2,$$
$$\lambda(t) = \frac{\sqrt{2}}{\sqrt{\pi}e^{t^2}(1-erf(t))}.$$

Meanwhile, the mean and variance of truncated Gaussian distributions can be updated as follows:

$$\mu_{kl} = \begin{cases} \tau^2 \sum_{n=1}^{N} z_{nk}^T(y_{nd} - \sum_{k'/k} z_{nk'}\mathbb{E}[w_{k'l}x_{dl}^T])x_{dl}, K \to \infty; \\ \tau^2 \sum_{d=1}^{D} x_{dl}(y_{nd}^T - \sum_{l'/l} x_{dl}^T\mathbb{E}[w_{kl'}z_{nk'}])z_{nk}^T, L \to \infty. \end{cases} \tag{8}$$

$$\sigma_{kd} = \tau\sigma_Y, \tag{9}$$

where $\tau = (m_k^T m_l + \frac{\sigma_Y^2}{\sigma_W^2})^{-\frac{1}{2}}$. Then the entropy of truncated Gaussian distribution is given as

$$\mathcal{H}(q(w_{kl})) = \frac{1}{2\sigma_{kl}^2}\{\mathbb{E}[w_{kl}]^2 - \mathbb{E}[w_{kl}^2] - (\mathbb{E}[w_{kl}] - \mu_{kl})^2$$
$$-[\frac{1}{2}\ln\frac{2}{\pi\sigma_{kl}^2} - \ln(1 - erf(t))]\}.$$

The updates on $\mathbf{Z}$ and $\mathbf{X}$ are relatively straightforward by computing Eq. (3). Given $q(\mathbf{W})$, we compute MAP estimates of $\mathbf{X}, \mathbf{Z}$ by maximizing the evidence Eq. (7). Similar to variational IBP methods, we must split the expectation in Eq. (6) into terms depending on each of the latent variables [3], with the benefit that the binary variables updates are not affected by inactive features. Therefore, we decompose the relevant terms of $\mathbf{X}$ in Eq. (7). Similarly, we also decompose the terms depending on $\mathbf{Z}$ during updating. First, to decompose $\ln\frac{(D-m_l)!(m_l-1)!}{D!}$, we define a quadratic pseudo-Boolean function:

$$f(x_{dl}) = \begin{cases} 0, \; if \; m_{l\backslash d} = 0 \; and \; x_{dl} = 0; \\ \ln\frac{(D-m_{l\backslash d}-x_{dl})!(m_{l\backslash d}+x_{dl}-1)!}{D!}, otherwise. \end{cases}$$

Here the subscript "." indicates that the given variable is determined after removing the dth row from L. Therefore the terms $\sum^{K_+}[\ln \frac{(D-m_l)!(m_l-1)!}{D!}]$ is changed to: $\sum_{l=1}^{L_+} f(x_{dl}) = \sum_{l=1}^{L_+} x_{dl}(f(x_{dl}=1) - f(x_{dl}=0)) + f(x_{dl}=0)$. Moreover, $\ln L!$ becomes $\ln L_+! = ln(L_{+\backslash n} + \sum_{l=1}^{L_+}[\mathbf{1}_{\{m_{l\backslash d}=0\}}x_{dl}])!$, where $\mathbf{1}_{\{\cdot\}}$ is the indicator function. Here we show that the evidence lower bound Eq. (7) is well-defined in the limit $L \to \infty$.

$$
\begin{aligned}
\mathcal{T}(\mathbf{X}_{d\cdot}) = & -\frac{1}{2\sigma_{\mathbf{Y}}^2}(\mathbf{A}_{n\cdot}\mathbf{X}d\cdot^T)(\mathbf{A}_{n\cdot}\mathbf{X}d\cdot^T)^T + \boldsymbol{\omega}_{n\cdot}\mathbf{X}_{d\cdot} \\
& + \sum_{k=1}^{K}[\frac{(N-m_k)!(m_k-1)!}{N!} + \mathbf{1}_{\{m_{l\backslash d}=0\}}x_{dl}\boldsymbol{\varphi}_{kl}] \\
& + [x_{dl}(f(x_{dl}=1) - f(x_{dl}=0)) + f(x_{dl}=0)] \\
& - \ln\mathbf{K}! - ln(L_{+\backslash l} + \sum_{l=1}^{L_+}[\mathbf{1}_{\{m_{l\backslash d}=0\}}l_{dl}])! + const,
\end{aligned}
$$

where $\omega_{nk} = -\frac{1}{\sigma_{\mathbf{Y}}^2}(\mathbf{A}_{n\cdot}\mathbf{Y}_{nd}^T + \boldsymbol{\gamma})$ and $\mathbf{A}_{n\cdot} = \mathbf{Z}\mathbb{E}[W]$.

3.5   Complexity Analysis

In this part, we show that, under a linear-Gaussian likelihood model, the per-iteration complexity of our model outperforms another recently-proposed latent feature model via IBP [9]. The iNBMT model reduces many operations when updating the parameters of non-negative matrix. $q(\mathbf{W})$ is updated twice per iteration from Eq. (9). $O(K^2L)$ operations are involved when updating $\mathbf{ZW}$, while $O(L^2K)$ operations are needed in updating $\mathbf{WX}^T$. Hence it yields a per-iteration complexity of $O(N(K^2L) + D(L^2K))$ for the $p(\mathbf{W})$ updates. The latent feature model via IBP proposed in [9] uses similar ME inference over the latent factors. Its per-iteration complexity on $q(\mathbf{W})$ is easily checked as $O(NK^2D)$. Updating $p(Y|Z)$ and $p(Y|X)$ are independent of the remaining observations and only require the computation of $\mathcal{T}(\cdot)$. We can update $\mathcal{T}(\mathbf{Z})$ in $O(N(K^2 \ln K))$ operations and $O(D(L^2 \ln L))$ operations when updating $\mathbf{X}$. The total per-iteration complexity of iNBMF is then $O(NK^2(L + \ln K) + DL^2(K + \ln L))$. The traditional model just has an infinite variable $\mathbf{Z}$, therefore its total per-iteration complexity is $O(NK^2(D + \ln K))$. In practice, N and D are usually sufficiently larger than K and L, hence, the per-iteration complexity of iNBMF can be written as a simple form: $O(\alpha N + \beta D)$, while that of ME-IBP model is simplified as $O(\gamma ND)$, where α, β, and γ are small coefficients. Clearly, our proposed iNBMF has the per-iteration complexity one order lower than that of the ME-IBP model.

4   Experiments

In this section, we conduct experimental analysis of our proposed iNBMT. We study the latent features on a synthetic and a real digit dataset. We also compare the performance of iNBMT with two competitive algorithms: Maximization-Expectation-IBP (ME-IBP) and Correlated IBP-IBP (IBP-IBP).

4.1 Synthetic Dataset

The synthetic dataset has $4,500$ samples consisting of 6×6 grey images. Different from the dataset used in Griffiths [5], our dataset is added combination of three different luminance, illustrating as Fig. 3(b). Each row of the observations $\mathbf{Y}$ was a 36-dimension vector, which is generated by using $\mathbf{Z}$ to linearly combine a subset of the four binary factors $\mathbf{X}$. And $\mathbf{W}$ is loading different luminance combination (see Fig. 2(a)). The input datasets are shown in Fig. 3(a) by adding Gaussian noise $\sigma = 0.8$.

The first demonstration shown in Fig. 2 is used to evaluate various algorithms' ability to extract the latent features from the generated data. Figure 2(c) shows the inferred features are closely match the truth features, however, each feature is repeated twice and have some noise. Compared with ME-IBP, the learning features of IBP-IBP shown in Fig. 2(c) also repeated and learning more noise. It is obvious that iNBMT outperforms other competitors by perfectly matching the truth features as well as identifying the feature number automatically.

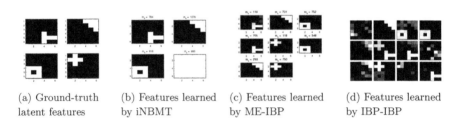

(a) Ground-truth latent features   (b) Features learned by iNBMT   (c) Features learned by ME-IBP   (d) Features learned by IBP-IBP

Fig. 2. Comparison of iNBMT, ME-IBP and IBP-IBP on synthetic dataset. iNBMT perfectly matches the truth features.

We also show the reconstruction power of our iNBMT model in Fig. 3.[1]

4.2 Digit Dataset

In this experiment, we further demonstrate the power of our iNBMT model on handwritten digit images. The digit dataset contains $2,000$ 64×64 samples which are randomly combined with the digits $0, 1, 2, 3$ from the USPS dataset. We then corrupted the images with Gaussian noise $\sigma = 0.8$. Some examples of the randomly generated images and their corrupted version are shown in Fig. 4(a) and (b).

---

[1] Since IBP-IBP is mainly for clustering, we do not show its (almost messy) reconstruction results for fairness.

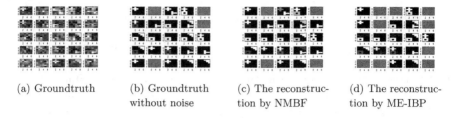

(a) Groundtruth   (b) Groundtruth without noise   (c) The reconstruction by NMBF   (d) The reconstruction by ME-IBP

Fig. 3. Comparison of sample reconstruction on synthetic data. iNBMT best matches the groundtruth than ME-IBP.

It is interesting to see from Fig. 4(e), our proposed iNBMT not only captures the latent features, i.e., each of the clear digits, but also their image contours. Moreover, from the framework of iNBMT, $\mathbf{W} \times \mathbf{X}^T$ can be thought of as a set of basis images which can be added together with binary coefficients $\mathbf{Z}$ to recover images. In particular, Fig. 4(g) shows the basis images which are captured by iNBMT. It is apparent that all digit combinations are detected. In terms of reconstruction, iNBMT almost perfectly recovers the images, as shown in Fig. 4(c). In comparison, ME-IBP extracts almost every different digit as the latent features, and their reconstruction results are also worse than our method.

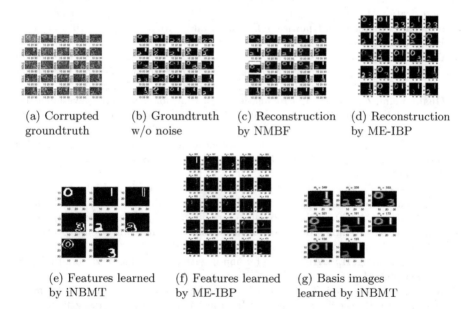

(a) Corrupted groundtruth   (b) Groundtruth w/o noise   (c) Reconstruction by NMBF   (d) Reconstruction by ME-IBP

(e) Features learned by iNBMT   (f) Features learned by ME-IBP   (g) Basis images learned by iNBMT

Fig. 4. Comparison of iNBMT and ME-IBP on Digits dataset. iNBMT clearly shows the best performance. We did not report IBP-IBP, since it is difficult to obtain reasonable results in this data set.

5 Conclusion

This paper proposes a new Bayesian model called infinite non-negative binary matrix tri-factorizations model (iNBMT), capable of learning automatically the latent binary features as well as feature number based on Indian Buffet Process (IBP). iNBMT engages a tri-factorization process that decomposes a nonnegative matrix into the product of three components including two binary matrices and a non-negative real matrix; this is also different from bi-factorization exploited by many other NMF models. A series of experiments show that our proposed model outperforms the other competitive algorithms.

Acknowledgement. The paper was supported by the National Basic Research Program of China (2012CB316301), National Science Foundation of China (NSFC 61473236), and Jiangsu University Natural Science Research Programme (14KJB520037).

References

1. Ding, C., He, X., Simon, H.D.: On the equivalence of nonnegative matrix factorization and spectral clustering. In: SIAM International Conference on Data Mining (2005)
2. Ding, C., Li, T., Peng, W., Park, H.: Orthogonal nonnegative matrix tri-factorizations for clustering. In: Proceedings of the Twelfth ACM SIGKDD International Conference on Knowledge Discovery and Data Mining, pp. 126–135. ACM Press (2006)
3. Doshi-velez, F., Miller, K.T., Van Gael, J., Teh, Y.W.: Variational inference for the Indian buffet process. In: Proceedings of the Twelfth International Conference on Artificial Intelligence and Statistics, AISTATS, pp. 137–144 (2009)
4. Ghahramani, Z., Beal, M.J.: Propagation algorithms for variational Bayesian learning. In: Advances in Neural Information Processing Systems, vol. 13, pp. 507–513 (2000)
5. Griffiths, T.L., Ghahramani, Z.: Infinite latent feature models and the Indian buffet process. In: Advances in Neural Information Processing Systems, vol. 18, pp. 475–482 (2005)
6. Knowles, D., Ghahramani, Z.: Infinite sparse factor analysis and infinite independent components analysis. In: Davies, M.E., James, C.J., Abdallah, S.A., Plumbley, M.D. (eds.) ICA 2007. LNCS, vol. 4666, pp. 381–388. Springer, Heidelberg (2007)
7. Kurihara, K., Welling, M.: Bayesian k-means as a "maximization-expectation" algorithm. Neural Comput. **21**(4), 1145–1172 (2009)
8. Lee, D.D., Seung, H.S.: Algorithms for non-negative matrix factorization. In: NIPS, pp. 556–562. MIT Press (2001)
9. Reed, C., Ghahramani, Z.: Scaling the Indian buffet process via submodular maximization. In: Proceedings of the 30th International Conference on Machine Learning, pp. 1013–1021(2013)
10. Zhang, Z., Li, T., Ding, C.H.Q., Ren, X.-W., Zhang, X.-S.: Binary matrix factorization for analyzing gene expression data. Data Min. Knowl. Discov. **20**(1), 28–52 (2010)

A Novel Manifold Regularized Online Semi-supervised Learning Algorithm

Shuguang Ding[1], Xuanyang Xi[2], Zhiyong Liu[2,3]($^{(\boxtimes)}$),
Hong Qiao[2,3], and Bo Zhang[1]

[1] LSEC, Institute of Applied Mathematics,
AMSS, Chinese Academy of Sciences, Beijing 100190, China
[2] State Key Lab of Management and Control for Complex Systems,
Institute of Automation, Chinese Academy of Sciences, Beijing 100190, China
zhiyong.liu@ia.ac.cn
[3] CAS Centre for Excellence in Brain Science and Intelligence Technology (CEBSIT),
Shanghai 200031, China

Abstract. In this paper, we propose a novel manifold regularized online semi-supervised learning (OS^2L) model in an Reproducing Kernel Hilbert Space (RK-HS). The proposed algorithm, named Model-Based Online Manifold Regularization (MOMR), is derived by solving a constrained optimization problem, which is different from the stochastic gradient algorithm used for solving the online version of the primal problem of Laplacian support vector machine (LapSVM). Taking advantage of the convex property of the proposed model, an exact solution can be obtained iteratively by solving its Lagrange dual problem. Furthermore, a buffering strategy is introduced to improve the computational efficiency of the algorithm. Finally, the proposed algorithm is experimentally shown to have a comparable performance to the standard batch manifold regularization algorithm.

Keywords: Manifold regularization · Online semi-supervised learning · Lagrange dual problem

1 Introduction

Cognitive science has drawn a lot of attentions for its significance in understanding human categorization in recent years [5]. In human learning, learners can incrementally learn the classes of various objects from the surrounding environment, where only a few objects are labeled by a knowledgeable source. This scenario can be regarded as online semi-supervised learning, that is, the label of a new arrived sample is unavailable or presented very sporadically in the online process.

In this paper, we focus on the online semi-supervised learning (OS^2L) problems. Several online semi-supervised learning algorithms have been proposed in the past several years. By using a heuristic method to greedily label the unlabeled examples, Babenko et al. [1] and Grabner et al. [9] tried to solve the OS^2L

© Springer International Publishing AG 2016
A. Hirose et al. (Eds.): ICONIP 2016, Part I, LNCS 9947, pp. 597–605, 2016.
DOI: 10.1007/978-3-319-46687-3_66

problems in an online supervised learning framework. Dyer et al. [3] presented a semi-supervised learning (SSL) framework called COMPOSE (COMPacted Object Sample Extraction), where a few labeled samples are given initially, and then a SSL problem is solved based on the currently labeled samples and new unlabeled samples, which are from a drift distribution. To reduce the computational complexity of manifold construction in the online training process, Kveton et al. [11] and Farajtabar et al. [4] proposed the harmonic solution for manifold regularization on an approximate graph. By using online convex programming, Goldberg et al. [6] proposed an online manifold learning framework for SSL in a kernel space with stochastic gradient descent. In addition, they extended their method to online active learning by adding an optional component to select which instances to label [8]. Sun et al. [7,14] exploited the property of Fenchel conjugate of hinge loss and gradient ascend method to solve the dual problem of their online manifold learning model. These algorithms in [6,7,14] are derived by using online gradient methods, implying that these methods can be regarded as solving the off-line semi-supervised learning models by stochastic gradient methods. However, none of these stochastic gradient methods can obtain exact solution because they do not directly solve the constrained optimization problem involved.

Note that an algorithm with an exact solution can obtain better performance. Therefore, to exploit the internal geometry information of the unlabeled data and take advantage of the kernel methods, in this paper we propose a novel online manifold regularization learning model in an Reproducing Kernel Hilbert Space (RKHS). In each iteration of online training, by considering the new arrived sample and the previous samples, an online model based on a constrained optimization problem is presented. Unlike the stochastic gradient method for solving the off-line model, the exact solution of the proposed model can be obtained by exploiting the Lagrange dual problem. In addition, the regularization parameter of the proposed model can be regarded as a forgetting factor, which can be used to control the number of support vectors by considering a buffering strategy in the online leaning process. By such merit, the proposed online predictor experimentally exhibits a high accuracy comparable to batch algorithm LapSVM.

The rest of this paper is organized as follows. Section 2 presents the proposed model and algorithm. Experimental results on several data sets are shown in Sect. 3. Some concluding remarks are given in Sect. 4.

2   Online Manifold Learning

In this section, the proposed model is presented in detail. In Sect. 2.1, a new model is proposed for online manifold regularization learning in an RKHS. In Sect. 2.2, the proposed model is solved by exploiting the property of Lagrange dual problem.

2.1   Online Model Based on Manifold Regularization

Assume that the current learning data for semi-supervised learning are $(x_1, y_1, \delta_1), (x_2, y_2, \delta_2), \ldots, (x_t, y_t, \delta_t)$ where $x_i \in \mathcal{X}$ is a point, $y_i \in \mathcal{Y} = \{-1, 1\}$ is its label and δ_i is a flag to determine whether the label y_i is available (y_i is available if and only if $\delta_i = 1$). At round t, the current predictor is $h_t(x) = \text{sign}(f_t(x))$ and f_0 is set as $f_0 = 0$ in our algorithm. In online semi-supervised learning, when a new sample $(x_{t+1}, y_{t+1}, \delta_{t+1})$ is available, the function f_{t+1} is updated based on the current decision function f_t and the implicit feedback, that is, the manifold structure of the samples.

Suppose that $K(\cdot, \cdot)$ is a chosen Kernel function over the training samples and $\mathcal{H}$ is the corresponding RKHS. Therefore, according to the Representer Theory [13], f_t and f_{t+1} can be written as:

$$f_t(\cdot) = \sum_{i=1}^{t} \alpha_i^t K(x_i, \cdot), f_{t+1}(\cdot) = \sum_{i=1}^{t} \alpha_i^{t+1} K(x_i, \cdot) + \alpha_{t+1}^{t+1} K(x_{t+1}, \cdot). \quad (1)$$

In the online learning process, our aim is to update $\{\alpha_i^{t+1}\}_{i=1}^{t+1}$ from $\{\alpha_i^t\}_{i=1}^{t}$ based on a proper algorithm. Considering the trade-off between the amount of progress made on each round and the amount of information retained from previous rounds, and compromise the classification error, the manifold constraint and the complexity of f as LapSVM, our online semi-supervised learning model with manifold regularization is presented as:

$$\min_{f, \xi_{t+1}} \frac{1}{2} \|f - f_t\|_{\mathcal{H}}^2 + \frac{\lambda_1}{2} \|f\|_{\mathcal{H}}^2 + C\delta_{t+1}\xi_{t+1} + \frac{1}{2}\lambda_2 \sum_{i=1}^{t} (f(x_i) - f(x_{t+1}))^2 w_{it+1} \quad (2)$$

$$\text{s.t.}\quad y_{t+1}f(x_{t+1}) \geq 1 - \xi_{t+1}, \xi_{t+1} \geq 0$$

where $\frac{1}{2}\|f - f_t\|_{\mathcal{H}}^2$ measures the difference between f and the previous f_t, the term $\|f\|_{\mathcal{H}}^2$ controls the complexity of the decision function f, $\sum(f(x_i) - f(x_{t+1}))^2 w_{it+1}$ is the manifold regularizer which depends on the edge weight w_{it+1}, f and x_i, and ξ_{t+1} is the slack variable denoting a possible error for the newly arrived data$(x_{t+1}, y_{t+1}, \delta_{t+1})$ after f is determined, λ_1, λ_2 and C are parameters reflecting the weights compromising complexity, the manifold regularizer and the classification error.

In the objective function of (2), the manifold structure of the samples is reflected in the term $\sum_{i=1}^{t}(f(x_i) - f(x_{t+1}))^2 w_{it+1}$, which can be regarded as an implicit feedback. This regularization term makes the new sample gain a similar decision value to its close sample in the manifold. Therefore, the proposed model can take advantage of the implicit feedback and the kernel methods. The solution of the proposed model is presented in the next section.

2.2   Online Algorithm of the Proposed Model

In this section, we give a detailed solution of the proposed model by exploiting the property of Lagrange dual problem. Assuming that $\delta_{t+1} = 1$ (if $\delta_{t+1} = 0$, the solution of (2) can be obtained by the similar process as below), the Lagrange dual problem of (2) is

$$\max_{\gamma_{t+1}} \min_{f,\xi_{t+1}} \quad L(f, \xi_{t+1}, \gamma_{t+1}, \beta_{t+1})$$
$$\text{s.t.} \quad \gamma_{t+1} \geq 0, \quad \beta_{t+1} \geq 0 \tag{3}$$

where γ_{t+1} and β_{t+1} are the Lagrange multipliers corresponding to the constraints $y_{t+1}f(x_{t+1}) \geq 1 - \xi_{t+1}$ and $\xi_{t+1} \geq 0$, respectively.

For simplicity, we define D and W as

$$D_{ij} = \begin{cases} w_{ij} & \text{if} & 0 < i = j < t+1 \\ \sum_{i=1}^{t} w_{it+1} & \text{if} & i = j = t+1 \\ 0 & \text{otherwise} \end{cases} \tag{4}$$

$$W_{ij} = \begin{cases} w_{ij} & \text{if} & 0 < i < t+1, j = t+1 \\ w_{ij} & \text{if} & i = t+1, 0 < j < t+1 \\ 0 & \text{otherwise} \end{cases} \tag{5}$$

Substituting (1), (4), (5) into (3) and let $L = D - W$, we have

$$\begin{aligned} L(\alpha, \xi_{t+1}, \gamma_{t+1}, \beta_{t+1}) = &\frac{1}{2}\alpha^T(K + \lambda_1 K + \lambda_2 KLK)\alpha \\ &- \gamma_{t+1}(y_{t+1}\alpha^T J - 1 + \xi_{t+1}) + c_0 \\ &- \alpha^T K\tilde{\alpha}^t - \beta_{t+1}\xi_{t+1} + C\xi_{t+1} \end{aligned} \tag{6}$$

where $\alpha = [\alpha_1, \ldots, \alpha_{t+1}]^T$, $\tilde{\alpha}^t = [\alpha_1^t, \ldots, \alpha_t^t, 0]^T$, K is a $(t+1) \times (t+1)$ Gram Matrix with $K_{ij} = K(x_i, x_j)$, $J = Ke$, $e = [0, \ldots, 0, 1]^T$ is a $(t+1)$-dimensional vector and c_0 is a constant.

Note that $L(\alpha, \xi_{t+1}, \gamma_{t+1}, \beta_{t+1})$ attains its minimum with respect to α and ξ_{t+1}, if and only if the following conditions are satisfied:

$$\nabla_\alpha L(\alpha, \xi_{t+1}, \gamma_{t+1}, \beta_{t+1}) = 0, \tag{7}$$

$$\nabla_{\xi_{t+1}} L(\alpha, \xi_{t+1}, \gamma_{t+1}, \beta_{t+1}) = 0. \tag{8}$$

Therefore, we formulate a reduced Lagrangian:

$$\begin{aligned} L^R(\alpha, \gamma_{t+1}) = &\frac{1}{2}\alpha^T(K + \lambda_1 K + \lambda_2 KLK)\alpha + c_0 \\ &- \gamma_{t+1}(y_{t+1}\alpha^T J - 1) - \alpha^T K\tilde{\alpha}^t. \end{aligned} \tag{9}$$

Taking derivative of the reduced Lagrangian with respect to α, we have:

$$\alpha = (K + \lambda_1 K + \lambda_2 KLK)^{-1}(K\tilde{\alpha}^t + Jy_{t+1}\gamma_{t+1}). \tag{10}$$

Substituting back in the reduced Lagrangian we get:

$$\max_{\gamma_{t+1}} -\frac{1}{2}(K\tilde{\alpha}^t + Jy_{t+1}\gamma_{t+1})^T A^{-1}(K\tilde{\alpha}^t + Jy_{t+1}\gamma_{t+1}) + \gamma_{t+1} \tag{11}$$

$$\text{s.t.} \qquad 0 \le \gamma_{t+1} \le C,$$

where $A = K + \lambda_1 K + \lambda_2 KLK$.

Let $\overline{\gamma}_{t+1}$ be the stationary point of the object function of (11). Therefore

$$\overline{\gamma}_{t+1} = \frac{1 - y_{t+1}J^T A^{-1} K\tilde{\alpha}^t}{J^T A^{-1} J}. \tag{12}$$

Assume that optimal solution of (11) is γ_{t+1}^*. Note that the object function (11) is quadratic, so the optimal solution γ_{t+1}^* in the interval $[0, C]$ is at either $0, C$ or $\overline{\gamma}_{t+1}$. Hence

$$\gamma_{t+1}^* = \begin{cases} 0, & \text{if } \overline{\gamma}_{t+1} \le 0 \\ C, & \text{if } \overline{\gamma}_{t+1} \ge 0 \\ \overline{\gamma}_{t+1}, & \text{otherwise} \end{cases} \tag{13}$$

Furthermore, if $\delta_{t+1} = 0$, we can obtain the solution of the proposed model by the similar process as above. Thus, the online manifold regularization for classification is presented as

$$f_{t+1}(x) = \sum_{i=1}^{t+1} \alpha_i^{t+1} K(x_{t+1}, x), \tag{14}$$

$$h_{t+1} = \text{sign}(f_{t+1}(x)),$$

where

$$\alpha^{t+1} = A^{-1}(K\tilde{\alpha}^t + \delta_{t+1}y_{t+1}\gamma_{t+1}^* J)$$

The above process of solving the proposed model is denoted by MOMR. In practice, the parameter λ_1 can be regard as a forgetting factor. Suppose λ_2 is very small and $\lambda_1 > 0$. According to (10), we have $\alpha \simeq (1+\lambda_1)^{-1}(\tilde{\alpha}^t + y_{t+1}e\gamma_{t+1})$, that is, $\alpha_i^{t+1} \simeq \alpha_i^t/(1 + \lambda_1)$ for $i = 1, \ldots, t$, which means that the absolute value of α_i^t will continually decrease in the online process. Thus, if the absolute value of a coefficient is small in the current decision function, the corresponding sample can be deleted safely from the current support vectors set. Based on this, a buffering strategy is introduced to make the online training feasible in the RKHS. The detailed process of the buffering strategy is presented in Sect. 3.

3   Experiments

To verify the effectiveness, we compare the proposed algorithm with two online manifold regularization algorithms and a batch algorithm on two data sets. In Sect. 3.1, the experimental setups are introduced in detail. In Sect. 3.2, several experiments are processed and the results are summarized and analyzed in detail.

3.1   Experimental Setup

Two data sets are used in our experiments. The first data set is the MNIST [12]. We focus on the binary classification task of separating '6' from '8' (MNIST6VS8) in our experiment. The sizes of the training set and test set are 11769 and 1932 respectively. The second data set is the FACEMIT [10] which contains 361-dimensional images of faces and non-faces. A balanced subset from FACEMIT (size 5000) is sampled and divided into two sets: the training set and the test set with a proportion 1:1 for our experiment. Similar to the experimental settings in [6,7], the labeled rate of the training samples is set to be 2% in all the experiments.

In our experiments, we focus on online manifold regularization algorithms derived from the dual problem. Therefore, We compare the performance of our algorithm MOMR with an online manifold regularization algorithm based on Example-Associate Update (denoted by OMR-EA), an online manifold regularization algorithm based on Overall Update (denoted by OMR-Overall) [7] and a batch manifold regularization algorithm LapSVM [2].

To reduce the storage for online learning in an RKHS, we use a buffering strategy for all the online algorithms: Let the buffer size be B. If the buffer is full, the sample with the smallest absolute coefficient in the buffer is replaced by the new arrived sample. We evaluate the three online algorithms separately with different buffer sizes ($B \in \{50, 200\}$) in our experiments.

In all the experiments, the RBF kernel $k(x_i, x_j) = \exp(-\|x_i - x_j\|^2/(2\sigma_K^2))$ is used for classification and the edge weights are Gaussian weights $k(x_i, x_j) = \exp(-\|x_i - x_j\|^2/(2\sigma_W^2))$ which define a fully connected graph. The parameter values $\sigma_K, \sigma_W, \lambda_1$ and λ_2 are selected by using 5-fold cross validation on the first 500 samples of the training data, where $\sigma_K, \sigma_W \in \{2^{-3}, 2^{-2}, 2^{-1}, 2^0, 2^1, 2^2, 2^3\}$ and $\lambda_1, \lambda_2 \in \{10^{-5}, 10^{-4}, 10^{-3}, 10^{-2}, 10^{-1}, 10^0, 10^1, 10^2\}$. In addition, as suggested in [7], the step sizes of the OMR-EA and OMR-Overall are set to be a small value 0.01. The value of parameter C is set to be 1 for the proposed algorithm MOMR. The computational efficiencies of all the algorithms are evaluated in terms of their CPU running time (in seconds). All the experiments are implemented in Matlab over a desktop PC with Inter(R) Core(TM) 3.2 GHz CPU, 4G RAM and Windows 7 operating system.

3.2   Online Processing and Performance Evaluation

In this subsection, we give out a detailed process of the experiments and evaluate the performance of the proposed algorithm for online manifold regularization learning.

All the three online algorithms are performed in the same way which are divided into two steps: (1) Online processing. Train a classifier with a new arrived sample; (2) Test. Test the final model on a test set. In each learning round, the batch algorithm LapSVM is trained with all the visible samples. We repeat all the experiments 10 times (each with an independent random permutation of the training samples) and the results presented below are all average over 10 trials.

Table 1. The accuracy of different algorithms on the data set MNIST6VS8 and FACEMIT with different buffer sizes. The best classification results are marked in boldface.

Date set	B	MOMR	OMR-EA	OMR-Overall	LapSVM
MNIST6VS8	50	98.012 ± 0.442	96.491 ± 1.775	97.495 ± 0.714	98.861 ± 0
MNIST6VS8	200	$\mathbf{99.048 \pm 0.078}$	98.954 ± 0.177	97.981 ± 0.543	98.861 ± 0
FACEMIT	50	$\mathbf{78.024 \pm 3.411}$	77.992 ± 3.390	78.000 ± 3.478	77.600 ± 0
FACEMIT	200	$\mathbf{78.552 \pm 3.360}$	77.948 ± 3.126	77.920 ± 3.237	77.600 ± 0

The test accuracies on the two data sets are summarized in Table 1. From the results, the test accuracy of MOMR is comparable with the off-line algorithm LapSVM on the two data sets and higher than those of the two online algorithms OMR-EA and OMR-Overall. These are reasonable since that: (a) in our algorithm, the exact solution is obtained from the proposed model; (b) in OMR-EA and OMR-Overall, the approximate solutions of their models are derived by online gradient method.

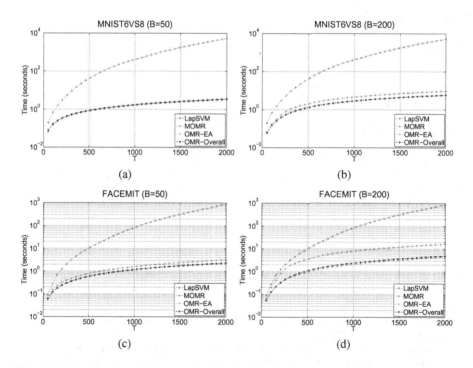

Fig. 1. Cumulative running time of online updating the classifiers with different buffer sizes on the data set MNIST6VS8 and FACEMIT.

The online updating time of the four algorithms are presented in Fig. 1. With respect to the running time, we can see that MOMR is comparable to the online algorithms OMR-EA and OMR-Overall when the buffer size is small and much faster than the off-line algorithm LapSVM. These can be explained by: (a) each sample is only trained once by the online algorithms; (b) a buffering strategy is used to reduce the repeated training process.

Considering above two results, it can be inferred that the proposed algorithm is in the first grade among the four algorithms both on the test accuracy aspect and on the running time aspect.

Additionally, in practice, the buffer size can be used to trade-off the accuracy and the time cost of online classifiers. The appropriate buffer size can be derived by using cross validation on the first N arrived samples, where N is a predefined number.

4   Conclusion

In this paper, the proposed model offers a new method to solve the OS^2L problem. Experiment results verify the effectiveness of the proposed algorithm. In addition, the proposed method enriches the research fields of cognitive computation and LapSVM.

Acknowledgment. This work is partly supported by NFSC grants 61375005 and MOST grants 2015BAK-35B01.

References

1. Babenko, B., Yang, M.H., Belongie, S.: Visual tracking with online multiple instance learning. In: IEEE Conference on Computer Vision and Pattern Recognition, pp. 983–990 (2009)
2. Belkin, M., Niyogi, P., Sindhwani, V.: Manifold regularization: a geometric framework for learning from labeled and unlabeled examples. J. Mach. Learn. Res. **7**, 2399–2434 (2006)
3. Dyer, K.B., Capo, R., Polikar, R.: Compose: a semisupervised learning framework for initially labeled nonstationary streaming data. IEEE Trans. Neural Netw. Learn. Syst. **25**(1), 12–26 (2014)
4. Farajtabar, M., Shaban, A., Rabiee, H.R., Rohban, M.H.: Manifold coarse graining for online semi-supervised learning. In: Gunopulos, D., Hofmann, T., Malerba, D., Vazirgiannis, M. (eds.) ECML PKDD 2011. LNCS (LNAI), vol. 6911, pp. 391–406. Springer, Heidelberg (2011). doi:10.1007/978-3-642-23780-5_35
5. Gibson, B.R., Rogers, T.T., Zhu, X.: Human semi-supervised learning. Top. Cogn. Sci. **5**(1), 132–172 (2013)
6. Goldberg, A.B., Li, M., Zhu, X.: Online manifold regularization: a new learning setting and empirical study. In: Daelemans, W., Goethals, B., Morik, K. (eds.) ECML PKDD 2008. LNCS (LNAI), vol. 5211, pp. 393–407. Springer, Heidelberg (2008). doi:10.1007/978-3-540-87479-9_44
7. Sun, B.L., Li, G.H., Jia, L., Zhang, H.: Online manifold regularization by dual ascending procedure. Math. Probl. Eng. **2013** (2013). doi:10.1155/2013/838439

8. Goldberg, A.B., Zhu, X.J., Furger, A., Xu, J.M.: Oasis: online active semi-supervised learning. In: Proceedings of the Twenty-Fifth AAAI Conference on Artificial Intelligence, pp. 1–6 (2011)
9. Grabner, H., Leistner, C., Bischof, H.: Semi-supervised on-line boosting for robust tracking. In: Forsyth, D., Torr, P., Zisserman, A. (eds.) ECCV 2008. LNCS, vol. 5302, pp. 234–247. Springer, Heidelberg (2008). doi:10.1007/978-3-540-88682-2_19
10. Heisele, B., Poggio, T., Pontil, M.: Face detection in still gray images. AI Memo 1697, Center for Biological and Computational Learning, MIT, Cambridge, MA (2000)
11. Kveton, B., Philipose, M., Valko, M., Huang, L.: Online semi-supervised perception: real-time learning without explicit feedback. In: IEEE Computer Society Conference on Computer Vision and Pattern Recognition Workshops (CVPRW), pp. 15–21 (2010)
12. LeCun, Y., Bottou, L., Bengio, Y., Haffner, P.: Gradient-based learning applied to document recognition. Proc. IEEE 86(11), 2278–2324 (1998)
13. Schölkopf, B., Herbrich, R., Smola, A.J.: A generalized representer theorem. In: Helmbold, D., Williamson, B. (eds.) COLT 2001. LNCS (LNAI), vol. 2111, pp. 416–426. Springer, Heidelberg (2001). doi:10.1007/3-540-44581-1_27
14. Sun, B.L., Li, G.H., Jia, L., Huang, K.H.: Online coregularization for multiview semisupervised learning. Sci. World J. **2013** (2013). doi:10.1155/2013/398146

Learning from Few Samples
with Memory Network

Shufei Zhang and Kaizhu Huang[✉]

Department of EEE, Xi'an Jiaotong-Liverpool University, SIP, Suzhou 215123, China
zsftesila@gmail.com, Kaizhu.Huang@xjtlu.edu.cn

Abstract. Neural Networks (NN) have achieved great success in pattern recognition and machine learning. However, the success of NNs usually relies on a sufficiently large number of samples. When fed with limited data, NN's performance may be degraded significantly. In this paper, we introduce a novel neural network called Memory Network, which can learn better from limited data. Taking advantages of the memory from previous samples, the new model could achieve remarkable performance improvement on limited data. We demonstrate the memory network in Multi-Layer Perceptron (MLP). However, it keeps straightforward to extend our idea to other neural networks, e.g., Convolutional Neural Networks (CNN). We detail the network structure, present the training algorithm, and conduct a series of experiments to validate the proposed framework. Experimental results show that our model outperforms the traditional MLP and other competitive algorithms in two real data sets.

Keywords: Memory · Multi-layer perceptron

1 Introduction

Conventional Neural Networks (NN), e.g., Multi-Layer Perceptrons (MLP), are widely used in pattern recognition, computer vision, and machine learning. To succeed, NN usually requires to be trained with a sufficiently large number of samples [4]. When only few data are available, NN's performance may however be significantly limited. Moreover, to facilitate the training of NN, input samples are usually assumed identically and independently distributed (i.i.d.). With the i.i.d. assumption, samples can be fed to NN sequentially; this hence enables a stochastic gradient descent algorithm for training a NN conveniently and efficiently. However, an i.i.d assumption may often be violated in practice; on the other hand, the learning procedure of human is not independent, but rather relies on previous knowledge. For example, if a child would like to learn running, his previous experience of walking can provide him some relevant knowledge which can help him learn running easier. Another example is that, if a British tries to learn French, previous memory about English study would benefit greatly the learning. Both examples above indicate that the memory and previous knowledge are very important and might be used to improve the present learning.

© Springer International Publishing AG 2016
A. Hirose et al. (Eds.): ICONIP 2016, Part I, LNCS 9947, pp. 606–614, 2016.
DOI: 10.1007/978-3-319-46687-3_67

Motived from these examples, we propose a novel neural network framework called Memory Network (MN). Enjoying the similar structure with traditional neural networks (including input, hidden, and output layers), MN introduces additional memory structures that can appropriately take advantages of previous knowledge learned from previous samples for the present learning. When only limited data are available, previous learned knowledge (stored in the memory network) could significantly benefit the training for the present learning. More specifically, we keep a memory of the network structures for previous N training samples. Depending on if the present training sample share or not the same category (class label), we enforce different constraints on two activations in each layer of the present network and previous network. For example, if the previous sample shares the same label with present sample, we then force similar the two activations of the same layers between the present network and previous network; otherwise, we try to enlarge their activations of the same layers between the present and previous network. One appealing feature of our proposed MN is that, despite a seemingly complicated network, an efficient stochastic gradient descent algorithm can be readily applied to make the network easily optimized.

2 Notation and Background

In this section, we present the notation used throughout the paper and also review the basic principles of conventional NN and Back Propagation (BP) algorithm. Essentially, NN is a stack of parametric non-linear and linear transformations [7]. Suppose an NN (with $L - 1$ hidden layers) is trained to perform predication in the scenario of classification. NN will map the M-dimension vector to the D-dimension label space. The matrix X_0 denotes the input data matrix where each row of X_0 represents a sample vector ($X_{0,i}$ is the ith sample vector with M dimensions). X_l indicates the activation of the lth layer of NN (where $l = 1, 2, \ldots, L-1$) and X_L denotes the output of the NN. Y represents the labels each row of which is the label for corresponding sample with D dimensions. The problem of NN can be formulated as the following optimization problem:

$$\min_{W_{1:L}, b_{1:L}} \frac{1}{2}\|X_L - Y\|^2 \quad \text{s. t.}$$
$$X_l = \sigma(X_{l-1}W_l + b_l), l = 1, \ldots, L - 1 \tag{1}$$
$$X_L = X_{L-1}W_L + b_L$$

where $\sigma(.)$ is the element-wise sigmoid function for a matrix. For each element x of matrix, the sigmoid function is defined as $\sigma(x) = \frac{1}{1+\exp(-x)}$.

In NN, the sigmoid function is used to perform the non-linear transformation and it can be also replaced by other functions such as $\max(0, x)$ and $\tanh(x)$.

We plot an illustrative example of a typical L-layer NN in Fig. 1, where X_l ($l = 1, 2, \ldots, L - 1$) represent the hidden layers. X_0 denotes the input for the NN, and X_L indicates the output of the NN. The aim is to learn the optimum parameters $W_{1:L}$ and $b_{1:L}$. The common approach is BP and stochastic gradient decent (SGD).

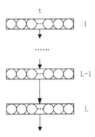

Fig. 1. The structure of conventional Neural Network

Back Propagation is an abbreviation for "backward propagation of errors" which is a common approach for training NNs with an optimization method such as gradient descent. The method calculates the gradient of a loss function with respect to all the parameters of the network. The gradient is used in the optimization method which in turn uses it to update the parameters in order to minimize the loss function.

BP requires the inputs with corresponding labels in order to calculate the loss function gradient. Therefore, it is considered as a supervised learning method, although it is also used in some unsupervised models such as auto encoders. It is a generalization of the delta rule to multi-layered feed-forward networks, made possible by using the chain rule to iteratively calculate the gradients for each layer. Assuming that the activation function be differentiable, the whole procedure is shown as below:

$$\frac{\mathrm{d}E}{\mathrm{d}X_L} = 2(X_L - Y) \tag{2}$$

$$\frac{\mathrm{d}E}{\mathrm{d}X_l} = (\frac{\mathrm{d}E}{\mathrm{d}X_{l+1}} \circ X_{l+1} \circ (1 - X_{l+1}))W_{l+1} \tag{3}$$

$$\frac{\mathrm{d}E}{\mathrm{d}W_l} = X_{l-1}^T(\frac{\mathrm{d}E}{\mathrm{d}X_l} \circ X_l \circ (1 - X_l)) \tag{4}$$

$$\frac{\mathrm{d}E}{\mathrm{d}b_l} = mean(\frac{\mathrm{d}E}{\mathrm{d}X_l} \circ X_l \circ (1 - X_l), 1) \tag{5}$$

where E is the value of the loss function and we can compute the gradients using the chain rule above. $\circ$ represents the element-wise product and $l = 1, 2, ...L$. $mean(., 1)$ denotes the average operation on matrices.

3   Memory Network

This section will introduce our proposed novel Memory Network (MN) in details. We will first present the structure of MN and then introduce the corresponding optimization algorithm.

3.1   Network Structure

The structure of MN is plotted in Fig. 2. As can be seen, the structure of MN consists of two parts, i.e., the present network and the memory part. We will detail these two parts one by one.

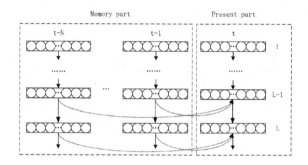

Fig. 2. Structure of Memory Network

Present Network. The structure of present network is the same as the traditional NN (consisting of the input layer, hidden layers and the output layer).

Memory Part. The memory part contains N copies of present network. They have totally the same parameters with present network. The difference is that the past N samples are fed into the memory part. There are also additional connections between each layer of memory part and present network. These connections indicate the minus operations which are used to calculate the difference of activations between memory and present part.

   The purpose of using the memory part is to exploit past knowledge (obtained from past samples) to help the present learning (present sample). There are two different cases: (1) if the present sample has the same class label as the past sample, we then try to make the activations of the same layers for present sample and past sample more similar; (2) if the present sample shares a different class from the past one, we should try to make the activations of top two layers for present sample and past sample more different. Motivated from these two cases, we then formulate the training of MN as follows.

3.2   Model Formulation

In order to exploit past knowledge (obtained from previous examples) for present learning, we design the model of our proposed MN as follows:

$$\min_{W_{1:L}, b_{1:L}} \frac{1}{2}\|X_L^t - Y^t\|^2 + \frac{1}{2}\sum_{j=1}^{p}\sum_{i=1}^{N} k_j^i \|X_{L-j+1}^t - X_{L-j+1}^{t-i}\|^2 \quad \text{s. t.} \quad (6)$$

$$X_l^t = \sigma(X_{l-1}^t W_l + b_l), \quad l = 1, ..., L-1,$$

$$X_L^t = X_{L-1}^t W_L + b_L$$

where X_l^t represents the activation of layer l for the present sample at time t, and Y^t is its corresponding class label; X_l^{t-i} represents activation of the previous i^{th} sample in layer l at time $t - i$, while Y^{t-i} describes the corresponding label for this specific sample. The matrix $\mathbf{k}$ (of the size $p \times N$) is a coefficient matrix. Its element k_j^i is defined as a positive value, if $Y^t = Y^{t-i}$ (i.e., the previous i^{th} sample X_0^{t-i} shares the same class label as the present sample X_0^t); otherwise it is a negative value. In more details, a positive k_j^i encourages more similarity between activations (in the $L-j+1$ layer) of the present learning (at t time) and the previous learning (at $t-i$ time); this is reasonable, since the present sample, X_0^t shares the same label as the previous sample X_0^{t-i}. Similarly, a negative k_j^i would enlarge the difference between the activations of the current learning and the previous learning, since the present sample and the previous sample have a different class label. We could also adapt the value of k_j^i, depending on if how deep the layer $L - j + 1$ is. Usually, a deeper or topper layer (i.e., smaller j) is more important, leading that k_j^i should be set to a bigger value.

In a short summary, on one hand, the optimization problem (6) would try to minimize the loss at the current time t (when a sample X_0^t is fed), i.e., the first term in (6); on the other hand, the proposed MN would also try to reduce (or enlarge) the difference of the activations up to the last p layers between the present network and the previous networks, i.e., the memory loss in the second term of (6), depending if the present sample shares the same class label as the previous sample. By this process, knowledge trained from previous samples can be transferred to the present learning, making the network possible to achieve remarkable performance even if the training samples are limited.

3.3 Optimization

For solving the modified optimization problem above, we can still rely on the BP algorithm, since the gradients with respect to the parameters can be easily computed from Eq. (6). For example, when p is set to 2, we could calculate the gradients for the output layer L as:

$$\frac{dE}{dX_L} = (X_L^t - Y^t) + \sum_{i=1}^{N} k_1^i (X_L^t - X_L^{t-i})$$

$$\frac{dE}{dX_{L-1}} = (\frac{dE}{dX_L} \circ X_L \circ (1 - X_L))W_L + \sum_{i=1}^{N} k_2^i (X_{L-1}^t - X_{L-1}^{t-i})$$

$$\frac{dE}{dX_l} = (\frac{dE}{dX_{l+1}} \circ X_{l+1} \circ (1 - X_{l+1}))W_{l+1}$$

$$\frac{\mathrm{d}E}{\mathrm{d}W_l} = X_{l-1}^T(\frac{\mathrm{d}E}{\mathrm{d}X_l} \circ X_l \circ (1 - X_l))$$

$$\frac{\mathrm{d}E}{\mathrm{d}b_l} = mean(\frac{\mathrm{d}E}{\mathrm{d}X_l} \circ X_l \circ (1 - X_l), 1)$$

It is straightforward to extend the above cases to bigger p's. With the above gradients, a BP can be easily conducted so that a local minimum can eventually obtained for the memory network.

4   Experiments

In this section, we conduct a series of experiments on two small-size data sets including face and handwriting data.

4.1   Experimental Setup

The face data set just contains 120 training samples [1] and the handwriting data set is a small portion of MNIST data set [3]. In the face data, a training and test set is respectively provided by following [8]. We hence train the different models on the training set and then report their performance on the test set. In the handwriting data, we randomly sampled 50, 100, and 500 digits from MNIST training set. We then report the performance on the test set. For fairness, we do the sampling five times and report the average classification accuracy. In order to compare the performance of the proposed Memory Network, we have implemented the conventional MLP, Linear and nonlinear Support Vector Machine with the rbf kernel function (in short, linear-SVM, and rbf-SVM) on these two data sets.

For these two data sets, the structure and parameters of the proposed network are set up differently. For different data sets, the network share the same depth with totally 5 layers, i.e., 1 input layer, 3 hidden layers, and 1 output layer. We exploit the deep structure, since deep networks are more flexible. For face data set, the input-hidden-output units are respectively set to $100-300-100-40-15$, and for handwriting data set, the input-hidden-output units are $100-200-300-100-10$. Both the structures are tuned in experiments. Again, p is set to 2, since the top layers are usually more stable. The memory weights k_j^i are tuned from the set $\{0.0001, 0.001, 0.01, 0.1\}$. For SVM, the trade-off parameter C and the width γ is tuned via cross validation.

4.2   Face Recognition with Different Pose

The face data set contains totally 195 images for 15 persons [1]. Each person has 13 horizontal poses from -90 to $90°$ with interval $15°$. We have done a series of preprocessing including resizing the images to 48×36 and then reducing the dimension to 100 with Principal Component Analysis (PCA). We divide this data set into two parts, number 1–8 poses are used as the training set and number 9–13 poses are used as the test set.

Table 1. Recognition rates of different models on face data. The proposed Memory Network and RBF-SVM significantly outperforms the other models. The other results (except Memory Network and conventional Neural Network) were copied from the associated papers due to the same setting.

Classifier	Accuracy (%)
Bilinear (Field) [6]	60.00
Style mixture (Singlet) [5]	70.00
Style mixture (Field) [5]	73.33
Nearest class mean [8]	60.00
FDA [8]	69.33
FBM [8]	74.67
linear-SVM	84.00
rbf-SVM	**85.33**
MLP	81.33
Memory Network	**85.33**

Table 1 reports the performance (recognition rate) of different models. It can be noted that the test set shares very different pose from the training set which makes the problem very challenging. As observed, our novel Memory Network and rbf-SVM achieves the best performance with 85.33 %. More specifically, the proposed MN significantly improves the performance of MLP from 81.33 to 85.33! On the other hand, Fisher Discriminant Analysis (FDA) is the state-of-the-art algorithm for face recognition, which only achieved the error rate of 69.33 % [8]. Moreover, other approaches such as the bilinear model, the style mixture model, the Field Bayesian Model and conventional Neural Network are obviously worse than our proposed Memory Network.

4.3 Handwriting Classification

We also test our proposed model on very famous handwriting digits data set, MNIST. MNIST is a large handwriting data set which has 60,000 training samples and 10,000 test samples. It is a portion of a larger data set NIST [2] and the samples have been size-normalized and centered in a fixed-size image (28×28). In this experiment, we focus on the small sample set. Therefore, we sample the small portions from MNIST. In particular, 50, 100 and 500 samples are chosen from 60000 samples of MNIST database randomly. Before training, for increasing training speed, we reduce the dimension of samples from 28×28 to 10×10. For testing, we use all test samples of MNIST database, totally 10,000 samples. We perform the experiments five times and then report the average accuracy.

We compare the performance of our proposed MN model with the conventional MLP, linear-SVM, and rbf-SVM. Table 2 shows the performance (recognition rate). Our proposed MN demonstrates a distinct performance improvement

Table 2. Recognition rates (%) of different models on hand-writing data.

# Training Samples	50	100	500
MLP	56.04 ± 1.60	67.76 ± 2.76	88.79 ± 0.87
Linear-SVM	63.10 ± 0.45	71.41 ± 2.96	86.69 ± 0.54
rbf-SVM	64.56 ± 2.30	73.78 ± 1.78	89.20 ± 0.84
Memory Network	**75.65 ± 1.02**	**81.60 ± 1.92**	**91.35 ± 0.78**

when the training samples are fewer. In particular, it can be noted that our proposed model achieves much better performance over MLP on 50-sample set (from 56.04 % to 75.65 %) and 100-sample set (from 67.76 % to 81.60 %). There is just a slight improvement on 500-sample set (from 88.79 % to 91.35 %). Our proposed MN also outperform both linear-SVM and rbf-SVM significantly. This experiment further validates the advantages of our proposed MN, especially when the training samples are limited.

5   Conclusion

In this paper, we proposed a novel Memory Network which can appropriately take advantages of past knowledge. Specifically, we built a novel network with two parts: memory part and present part both of which share the same structures. We proposed to connect the top p layers of memory part and present part, which are exploited to deliver the past knowledge. We developed a modified stochastic optimization algorithm, which can efficiently optimize the proposed MN model. We conducted experiments on two small-size databases including face and handwriting data. Experimental results showed that our proposed model achieves the best performance on both the data sets compared with the other competitive models.

Acknowledgement. The paper was supported by the National Basic Research Program of China (2012CB316301), National Science Foundation of China (NSFC 61473236), and Jiangsu University Natural Science Research Programme (14KJB520037).

References

1. Gourier, N., Hall, D., Crowley, J.: Estimating face orientation from robust detection of salient facial features. In: International Conference on Pattern Recognition (ICPR) (2004)
2. Grother, P.J.: NIST special database 19 handprinted forms, characters database. National Institute of Standards and Technology (1995)
3. Lecun, Y., et al.: Gradient-based learning applied to document recognition. In: Proceedings of the IEEE 86.11, pp. 2278–2324, November 1998. ISSN 0018-9219, doi:10.1109/5.726791

4. Rumelhart, D.E., Hinton, G.E., Williams, R.J.: Learning representations by back-propagating errors. In: Cognitive Modeling 5.3, p. 1 (1988)
5. Sarkar, P., Nagy, G.: Style consistent classification of isogenous patterns. IEEE Trans. Pattern Anal. Mach. Intell. **27**(1), 88–98 (2005)
6. Tenenbaum, J.B., Freeman, W.T.: Separating style and content with bilinear models. Neural Comput. **12**(6), 1247–1283 (2000)
7. Wang, H., Yeung, D.-Y.: Towards Bayesian deep learning: a survey. arXiv preprint arXiv:1604.01662 (2016)
8. Zhang, X.-Y., Huang, K., Liu, C.-L.: Pattern field classification with style normalized transformation. In: International Joint Conference on Artificial Intelligence (IJCAI), pp. 1621–1626 (2011)

Generalized Compatible Function Approximation for Policy Gradient Search

Yiming Peng[1]([✉]), Gang Chen[1], Mengjie Zhang[1], and Shaoning Pang[2]

[1] School of Engineering and Computer Science, Victoria University of Wellington,
Wellington, New Zealand
{yiming.peng,aaron.chen,mengjie.zhang}@ecs.vuw.ac.nz
[2] Department of Computing, Unitec Institute of Technology, Auckland, New Zealand
ppang@unitec.ac.nz
http://www.dmli.info, http://ecs.victoria.ac.nz/Groups/ECRG/

Abstract. Reinforcement learning aims at solving stochastic sequential decision making problems through direct trial-and-error interactions with the learning environment. In this paper, we will develop generalized compatible features to approximate value functions for reliable Reinforcement Learning. Further guided by an Actor-Critic Reinforcement Learning paradigm, we will also develop a generalized updating rule for policy gradient search in order to constantly improve learning performance. Our new updating rule has been examined on several benchmark learning problems. The experimental results on two problems will be reported specifically in this paper. Our results show that, under suitable generalization of the updating rule, the learning performance and reliability can be noticeably improved.

Keywords: Markov decision process · Reinforcement learning · Actor critic · Policy gradient search · Function approximation · Generalization · Compatible feature

1 Introduction

Stochastic sequential decision making problems frequently appear in diverse real-world applications and are increasingly gaining attentions [8]. These problems can be described through Markov Decision Processes (MDPs) [8]. To solve an MDP, traditional techniques such as dynamic programming can be applied but they often suffer from the issue of *curse of dimensionality* whenever the state space is large. In this situation, an alternative technique, known as Reinforcement Learning (RL), is widely considered to be more suitable for practical use [5].

In RL, to break the curse, a common approach is to adopt function approximation for value functions (aka. critic) via parameterization, the dimension of which is generally smaller than that of states [1,6]. Clearly the success of RL depends on appropriate and generalized function approximators [1,6,7]. Based on properly approximated value functions, we can further pursue the solutions to MDPs in the form of parametric policies (aka. actor). Under this *Actor-Critic*

© Springer International Publishing AG 2016
A. Hirose et al. (Eds.): ICONIP 2016, Part I, LNCS 9947, pp. 615–622, 2016.
DOI: 10.1007/978-3-319-46687-3_68

(AC) framework, various policy gradient search methods can be implemented [1,5]. In particular, through iterative updates to *policy parameters*, the learning performance is expected to be constantly improved [5].

Guided by the AC framework, Sutton proved the first time that the gradient of J with respect to policy parameters $\boldsymbol{\theta}$ follows the formula below [6],

$$\Delta\boldsymbol{\theta} \propto \nabla_{\boldsymbol{\theta}} J^{\pi}(\boldsymbol{\theta}) = \sum d^{\pi}(\boldsymbol{s}) \sum \nabla_{\boldsymbol{\theta}} \pi(\boldsymbol{s}, a) Q^{\pi}(\boldsymbol{s}, a), \tag{1}$$

where J^{π} (see (4)) stands for the *expected long term payoff* obtainable by an agent upon following policy π which is parameterized by $\boldsymbol{\theta}$ (i.e. the policy parameters). In (1), $\boldsymbol{s}$ and a represent the state and action of an MDP respectively (for more information on MDP, please refer to Sect. 2). $Q^{\pi}(\boldsymbol{s}, a)$ is the expected total reward while initially taking action a at state $\boldsymbol{s}$ following policy π. To use (1) for policy gradient search, a key issue is to approximate $Q^{\pi}(\boldsymbol{s}, a)$. In [6], Sutton proposed an important method to estimate $Q^{\pi}(\boldsymbol{s}, a)$, i.e.

$$Q^{\pi}(\boldsymbol{s}, a) \approx \hat{Q}^{\pi}(\boldsymbol{s}, a) = \boldsymbol{\omega}^{T} \Phi(\boldsymbol{s}, a) = \boldsymbol{\omega}^{T} \nabla_{\boldsymbol{\theta}} \ln \pi(\boldsymbol{s}, a), \tag{2}$$

where $\boldsymbol{\omega}$ is used to parameterize $\hat{Q}^{\pi}$, and $\Phi(\boldsymbol{s}, a) = \nabla_{\boldsymbol{\theta}} \ln \pi(\boldsymbol{s}, a)$ is the so-called *compatible feature vector*. Furthermore, Sutton proved that the approximation of Q^{π} by $\hat{Q}^{\pi}$ in (2) will still ensure precise evaluation of $\nabla_{\boldsymbol{\theta}} J^{\pi}$ in (1), i.e.

$$\begin{aligned} \nabla_{\boldsymbol{\theta}} J(\boldsymbol{\theta}) &\equiv \sum d^{\pi}(\boldsymbol{s}) \sum \nabla_{\boldsymbol{\theta}} \pi(\boldsymbol{s}, a) \hat{Q}^{\pi}(\boldsymbol{s}, a) \\ &= \sum d^{\pi}(\boldsymbol{s}) \sum \nabla_{\boldsymbol{\theta}} \pi(\boldsymbol{s}, a) \boldsymbol{\omega}^{T} \Phi(\boldsymbol{s}, a) \end{aligned} \tag{3}$$

Based on (3), many policy gradient search algorithms follow strictly (2) to approximate Q^{π} [1]. However, we found that, under proper conditions (to be detailed in a separate venue due to space limitation), when the compatible feature vector in (2) is represented as $\Phi(\boldsymbol{s}, a) = \nabla_{\boldsymbol{\theta}} \mathcal{G}(\pi(\boldsymbol{s}, a), \nu)$ where $\mathcal{G}(\cdot)$ is a generalization of the logarithm function and ν controls the level of generalization, the result in (3) will continue to hold. In the literature, $\mathcal{G}(\cdot)$ has been utilized to define *Tsallis entropy* with substantial practical applications in many disciplines [2]. Inspired by this understanding, we seek to take the first step towards answering an important research question: *when the generalized logarithm function $\mathcal{G}(\cdot)$ is used to produce the compatible feature Φ in (2), will policy gradient search algorithm become more effective and reliable for RL?* To the utmost of our knowledge, so far this research question hasn't been studied sufficiently.

In this paper, based on an incremental regular-gradient actor-critic algorithm (i.e. RAC) proposed by Bhatnagar et al. in [1], we will investigate experimentally the usefulness of $\mathcal{G}(\cdot)$ for building compatible features Φ and develop some empirical answer to our research question. Several benchmark problems, including Puddle World [5] and Cart-Pole [5], will be employed for this purpose. Our experiment results clearly show that, under suitable generalization, the learning performance can exhibit observable improvement.

2   Markov Decision Process

An MDP is described as an agent repetitively interacting with a stochastic environment at discrete time intervals [5]. At any time t, the agent can observe its state $s_t \in S$ in the environment and choose an action $a_t \in \mathcal{A}$ to execute, where S and $\mathcal{A}$ (i.e., state and action spaces) represent countable sets of all possible states and actions, respectively. Meanwhile, the environment responds with an arbitrary scalar reward $r(s_t, a_t, s_{t+1})$, depending partially on the next state s_{t+1} governed by the state transition probability $P(s_t, a_t, s_t)$ [5]. The learning objective in an MDP is to achieve the maximum *long term payoff*, which can be modeled in several different ways [5]. In this paper, we focus specifically on the discounted infinite horizon model. Accordingly, the *expected long term payoff* of policy π can be formulated as,

$$J^\pi = V^\pi(s_0) = \mathbf{E}_\pi[\sum_{k=0}^\infty \gamma^k r_{t+k+1}|s_t = s_0]. \tag{4}$$

where $\gamma \in [0, 1)$ is the *discount factor*. Meanwhile r_{t+k+1} is the immediate reward at time step $k + 1$ provided that the agent started from state s_t. Similarly, function Q^π introduced in Sect. 1 can be defined as below,

$$Q^\pi(s, a) = \mathbf{E}_\pi[\sum_{k=0}^\infty \gamma^k r_{t+k+1}|s_t = s, a_t = a]. \tag{5}$$

Based on (4) and (5), we can re-formulate the expected long term payoff starting from any state s as

$$V^\pi(s) = \sum_{a \in \mathcal{A}} \pi(s, a)Q^\pi(s, a). \tag{6}$$

Hence, the ultimate goal of RL is to identify the optimal policy π^* below,

$$\pi^* = \underset{\pi}{\operatorname{argmax}} V^\pi(s_0). \tag{7}$$

3   A Regular-Gradient Actor-Critic Algorithm

In this work, we decide to base our experimental study on an incremental regular-gradient actor-critic algorithm (i.e. RAC) proposed in [1]. We choose RAC due to its simplicity and proven effectiveness on many benchmark RL problems. According to [1,6], Q^π is approximated by $\hat{Q}_\pi$ in RAC by minimizing the *Mean Square Error* (MSE),

$$\epsilon^\pi(\omega) = \sum_{s \in S} d^\pi(s) \sum_{a \in \mathcal{A}} \pi(s, a)[Q^\pi(s, a) - \hat{Q}^\pi(s, a)]^2. \tag{8}$$

The minimization can be achieved in theory by solving the equation below

$$\nabla_\omega \epsilon^\pi(\omega) = \sum_{s \in S} d^\pi(s) \sum_{a \in \mathcal{A}} \pi(s, a)[Q^\pi(s, a) - \hat{Q}(s, a)]\nabla_\omega \hat{Q}^\pi(s, a) = 0, \tag{9}$$

Because of (2), $\nabla_\omega \hat{Q}^\pi(s, a) = \Phi(s, a)$ in (9). Following (9), the temporal difference (TD) error δ is

$$\delta_t^\pi = r(s_t, a_t, s_{t+1}) + \gamma^{t+1}V^\pi(s_{t+1}) - V^\pi(s_t) \tag{10}$$

By defining $v^{\pi T}\phi(s)$ as an unbiased estimation of the value function $V^{\pi}(s)$, where v^{π} is the so-called *value-function parameters* and $\phi(s)$ is another set of *state features* to be distinguished from $\Phi(s,a)$, the TD error can be further described as,

$$\delta_t^{\pi} = r(s_t, a_t, s_{t+1}) + \gamma^{t+1} v_{t+1}^{\pi}{}^T \phi(s_{t+1}) - v_t^{\pi}{}^T \phi(s_t) \tag{11}$$

Consequently, following (11) and (4), we can eventually determine the formula in (12) for incremental update of v^{π}.

$$v_{t+1}^{\pi} \leftarrow v_t^{\pi} + \alpha \cdot \delta_t^{\pi} \cdot \phi(s_t), \tag{12}$$

where α is a learning rate. Subsequently, the updating formula in (13) for incrementally updating the policy parameters θ can also be determined easily.

$$\theta_{t+1} \leftarrow \theta_t + \beta \cdot \delta_t^{\pi} \cdot \Phi(s_t, a_t), \tag{13}$$

Similar to (12), β in (13) refers to a separate learning rate. To summarize our discussion above, Algorithm 1 presents the pseudo-code for RAC. It is to be noted that, for RAC, the *compatible feature* is computed from $\Phi(s_t, a_t) = \frac{1}{\pi(s,a)}\nabla_{\theta}\pi(s,a)$. In the next section, we will generalize the calculation of $\Phi(s_t, a_t)$ by using the generalized logarithm function $\mathcal{G}(\cdot)$.

Algorithm 1. Regular-Gradient Actor-Critic Algorithm [1]

Input: an MDP $\langle \mathcal{S}, \mathcal{A}, P(s_t, a_t, s_{t+1}), r(s_t, a_t, s_{t+1}), \gamma \rangle$
Output: θ, v^{π}, $\phi(s)$
    Initialization:
1: $\theta \leftarrow \theta_0$
2: $v^{\pi} \leftarrow v_0^{\pi}$
3: $s \leftarrow s_0$
4: $a \leftarrow a_0$
    Learning Process:
5: **for** $t = 0, 1, 2, ...$ **do**
6:    $\delta_t^{\pi} \leftarrow r(s_t, a_t, s_{t+1}) + \gamma v_{t+1}^{\pi}{}^T \phi(s_{t+1}) - v_t^{\pi}{}^T \phi(s_t)$
7:    $v_{t+1}^{\pi} \leftarrow v_t^{\pi} + \alpha \delta_t^{\pi} \phi(s_t)$
8:    $\theta_{t+1} \leftarrow \theta_t + \beta \delta_t^{\pi} \Phi(s_t, a_t)$
9: **end for**
10: **return** θ, v^{π}

---

4   Generalized Compatible Function Approximation

In this research, we focus primarily on generalizing the computation of the *compatible feature* $\Phi(s,a)$. The learning procedure still follows closely Algorithm 1.

As we explained in Sect. 1, we propose to use a generalized logarithm function $\mathcal{G}(\cdot)$ as shown below to determine the compatible feature Φ in (2),

$$\mathcal{G}(\pi(s,a),\nu) = \frac{\pi(s,a)^{1-\nu} - 1}{1 - \nu}. \tag{14}$$

Clearly, the level of generalization in $\mathcal{G}(\cdot)$ is controlled by ν. Whenever $\nu = 1$, $\mathcal{G}(\cdot)$ degrades to the standard logarithm function, i.e.

$$\lim_{\nu \to 1} \mathcal{G}(\pi(s,a),\nu) = \ln \pi(s,a). \tag{15}$$

For simplicity, ν in (14) will be called the *compatible generalization factor*. In consequence, we can obtain a new way to determine the compatible feature below,

$$\begin{aligned}
\tilde{\Phi}(s,a) &= \nabla_{\boldsymbol{\theta}}\mathcal{G}(\pi(s,a),\nu) \\
&= \nabla_{\boldsymbol{\theta}}\pi(s,a)\pi(s,a)^{1-\nu},
\end{aligned} \tag{16}$$

Given a Gaussion Distribution for policy $\pi(s,a)$, we have,

$$\pi(s,a) = \frac{1}{\sigma\sqrt{2\pi}}e^{-\frac{(a-\mu)^2}{2\sigma^2}}, \tag{17}$$

where $\mu = \boldsymbol{\theta}^T\phi(s)$ is the mean action output from policy π in state s, which can be adjusted by changing policy parameters $\boldsymbol{\theta}$. On the other hand, the standard deviation σ in (17) is pre-determined. Note that π in RHS of (17) refers to the circumference ratio.

Based on the generalized compatible feature in (16), the updating rule shown in (13) is now modified to become

$$\begin{aligned}
\boldsymbol{\theta}_{t+1} &\leftarrow \boldsymbol{\theta}_t + \beta \cdot \eta \cdot \delta_t^\pi \cdot \tilde{\Phi}(s_t,a_t) \\
&= \boldsymbol{\theta}_t + \beta \cdot \eta \cdot \delta_t^\pi \nabla_{\boldsymbol{\theta}}\pi(s_t,a_t)\pi(s_t,a_t)^{1-\nu},
\end{aligned} \tag{18}$$

where a new constant factor η is introduced in (18) to ensure that $\boldsymbol{\theta}$ will be updated at the same scale in Algorithm 1 regardless of using either updating rule (13) or (18). For this purpose, we need to solve the following equation,

$$\mathbf{E}_a[|\nabla_{\boldsymbol{\theta}} \ln \pi(s,a)|] = \mathbf{E}_a[|\eta \nabla_{\boldsymbol{\theta}}\mathcal{G}(\pi(s,a),\nu))|]. \tag{19}$$

From (19), we can directly determine the value for η below,

$$\eta = \begin{cases}
\dfrac{(2\pi)^{\frac{1-\nu}{2}}(\nu-2)\sigma^{1-\nu}\left(1-e^{-\frac{(\mu+t)^2}{2\sigma^2}}\right)}{e^{\frac{(\nu-2)(t-\mu)^2}{2\sigma^2}} + e^{\frac{(\nu-2)(\mu+t)^2}{2\sigma^2}} - 2}, & 0 < \nu < 2 \\[4ex]
\dfrac{\sqrt{\frac{2}{\pi}}\sigma\left(1-e^{-\frac{(\mu+t)^2}{2\sigma^2}}\right)}{(\mu+t)^2}, & \nu = 2
\end{cases}. \tag{20}$$

Again π in (20) is the circumference ratio. Meanwhile t and $-t$ give the upper and lower bounds for the action output from any policy, respectively.

Table 1. Experiment general settings for one trial.

Problem	Training		Testing		Evaluating ν values
	Episodes	Steps	Episodes	Steps	
Puddle world	10000	1000	50	100	$\{0.5, 0.7, 0.9, \mathbf{1.0}, 1.1, 1.5, 2.0\}$
Cart pole	20000	50	50	1000	$\{0.5, 0.7, 0.9, \mathbf{1.0}, 1.1, 1.5, 2.0\}$

5   Experimental Results

To understand the efficacy of using generalized updating rule for learning policy parameters in (18), we will evaluate the performance and reliability of RAC on two benchmark problems. In order to obtain reliable results, 50 independent trials will be performed on each benchmark problem and with respect to different settings of the compatible generalization factor ν (note that when $\nu = 1$, the original updating rule in (13) is realized). A total of 8 different settings for ν have been examined in our experiments (see Table 1). To simplify our discussion, in this section, CASE-X will denote the experiments on RAC when $\nu = X$.

Common settings that we follow on every trial have been summarized in Table 1. As evidenced in this table, after every 20 training episodes during each trial, the learned policy will be further tested on 50 independent testing episodes to measure learning performance.

5.1   Experiments on the Puddle World Problem

To compare the performance differences among various settings of ν, we firstly evaluate RAC on the classic Puddle World problem [3,5]. Figure 1 presents the average steps to reach the goal region upon using the policies learned through RAC. As seen from Fig. 1, near-optimal policies can be learned successfully whenever $\nu \in [0.9, 2.0]$. On the other hand, CASE-0.5 and CASE-0.7 failed to solve this problem satisfactorily. Among all the results presented in Fig. 1, CASE-1.5 appears to achieve the best performance by observation (i.e. on average 19.2 steps to reach the goal region). In comparison, for CASE-1.0 (i.e. normal RAC), the average steps to reach the goal region is 46.4 after 10,000 learning episodes have been completed. A t-test is performed in between CASE-1.0 and CASE-1.5 and it produces a p-value of 0.10, insufficient to prove that CASE-1.5 is significantly better. However, we found that the problem is solved successfully by CASE-1.5 100 % of the time. For CASE-1.0, the problem is solved on only 94 % of the trials. This observation suggests that CASE-1.5 can solve the problem more reliably.

5.2   Experiments on the Cart-Pole Problem

Next, we have compared the learning performances on the Cart-Pole problem (aka. inverted pendulum problem) [5]. The performances in terms of the average

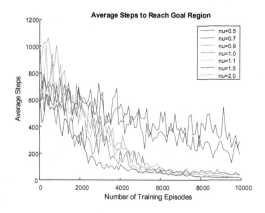

Fig. 1. Average steps to goal region.

ξ (i.e. the angle of the pole) and the average balancing steps (i.e. the duration for the pole to be balanced continually) are presented in Fig. 2. It can be observed in Fig. 2(a) that, in comparison to other cases, during a long learning period from 3000 training episodes to the end, CASE-1.5 can manage to bring the pole closer to the upright position on average. For example, at 3000 training episodes, the average ξ achieved by CASE-1.5 is -0.01. In comparison, CASE-1.0 can only manage to achieve on average of -0.07 for ξ. However, this observed performance difference is not verifiable through statistical tests (perhaps more repeated tests are to be performed in order to reveal significant differences in between CASE-1.0 and CASE-1.5).

On the other hand, by checking the average balancing steps in Fig. 2(b), we found that most of the cases can solve this problem reasonably well. The only

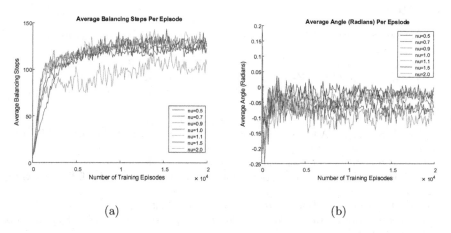

Fig. 2. The performance evaluation for the Cart-Pole problem. Part (a) shows the average ξ, and part (b) shows the average balancing steps.

case that falls apart is when $\nu = 2.0$, suggesting that the value for ν cannot significantly differ from 1.0.

5.3 Experiment Summary

In summary, our experimental evaluation clearly shows that there exists a strong correlation in between the degree of generalization in (18) and the learning performance as well as reliability. Whenever we set ν in (18) to a proper value, e.g. 1.5, observable improvement on learning performance and reliably can be witnessed. However, when ν deviates significantly from the common setting of 1.0, the learning performance may degrade abruptly.

6 Conclusions

In this paper we studied the possibility of using generalized compatible features for value function approximation. Guided by an Actor-Critic framework for RL, a generalized rule for policy gradient search in RL algorithms has been developed successfully. We have experimentally examined the usefulness of this rule on two benchmark RL problems. We found that, under suitable generalization (e.g. $\nu = 1.5$), the learning performance and reliability can be noticeably improved. We have therefore obtained some promising and empirical answer to the research question highlighted in Sect. 1. Based on the research results reported in this paper, we will further explore our research question in the future through in-depth theoretical analysis and extensive experimental assessments.

Acknowledgments. Authors appreciate all the supports from NeSI [4], who provides the High Performance Computing facility to ensure the success of our computationally heavy experiments.

References

1. Bhatnagar, S., Sutton, R.S., Ghavamzadeh, M., Lee, M.: Natural actor-critic algorithms. Automatica **45**(11), 2471–2482 (2009)
2. Cartwright, J.: Roll over, boltzmann. Phys. World **27**(5), 31–35 (2014)
3. Chen, G., Douch, C.I.J., Zhang, M.: Reinforcement learning in continuous spaces by using learning fuzzy classifier systems. IEEE Trans. Evol. Comput. **PP**(99), 1 (2016)
4. NeSI: New Zealand eScience Infrastructure (2016). https://www.nesi.org.nz/
5. Sutton, R.S., Barto, A.G.: Reinforcement Learning : An Introduction. MIT Press, Cambridge (1998)
6. Sutton, R.S., Mcallester, D., Singh, S., Mansour, Y.: Policy gradient methods for reinforcement learning with function approximation. Adv. Neural Inf. Process. Syst. **12**, 1057–1063 (1999)
7. Sutton, R.: Generalization in reinforcement learning: successful examples using sparse coarse coding. In: Advances in Neural Information Processing Systems, pp. 1038–1044 (1996)
8. White, D.J.: A survey of applications of Markov decision processes. J. Oper. Res. Soc. **44**, 1073–1096 (1993)

A Combo Object Model for Maritime Boat Ramps Traffic Monitoring

Jing Zhao[1], Shaoning Pang[1(✉)], Bruce Hartill[2], and Abdolhossein Sarrafzadeh[1]

[1] Department of Computing, Unitec Institute of Technology,
139 Carrington Road, Mount Albert, Auckland 1025, New Zealand
{jzhao,ppang,hsarrafzadeh}@unitec.ac.nz
[2] The National Institute of Water and Atmospheric Research,
41 Market Place, Viaduct Harbour, Auckland, New Zealand
Bruce.Hartill@niwa.co.nz

Abstract. Conventional tracking methods are incapable of tracking boats towed by vehicles on boat ramps because the relative geometry of these combined objects changes as they move up and down the ramp. In the context of maritime boat ramp surveillance, fishing trailer boat is the object of interest for monitoring the amount of recreational fishing activities over the time. Instead of tracking trailer boat as a single object, this paper proposes a novel boat-vehicle combo object model, by which each boat is tracked as a combination of a trailered boat and a towing vehicle, and the relationship between these two components is modelled in multi-feature space and traced across consecutive frames. Experimental results show that the proposed combo modelling tracks the object of interest accurately and reliably in real-world boat traffic videos.

1 Introduction

Traffic surveillance at maritime boat ramps is a highly challenging problem in the computer vision community, because object tracking in this scenario involves severe unpredictability of the object and dynamic land water composition scene. In particular, the object of interest is a combo object, which is a combination of a trailered boat and a towing vehicle. A combo object is even more unpredictable than that of a single object, because each component object of the combo (i.e., vehicle or boat) varies in scale, transformation, rotation and viewpoint. In this sense, tracking methods require more accurate object modelling, and should exploit more complex temporal matching approach and data association to deal with all possible unpredictable variations of two objects.

In the track of fishing trailer boats, a launching boat is normally pushed by a vehicle to approach toward the water area. After being put in the water, the boat and the vehicle are separated from each other and disappear individually. Similarly, a retrieving boat stays in the water to wait for a vehicle to pick it up. After being taken out of the water, the boat is dragged by the vehicle and they leave the ramp together. In general, we consider the following four scenarios for boat ramp surveillance:

ⓒ Springer International Publishing AG 2016
A. Hirose et al. (Eds.): ICONIP 2016, Part I, LNCS 9947, pp. 623–630, 2016.
DOI: 10.1007/978-3-319-46687-3_69

(1) A single boat in water: a launching boat or a retrieving boat;
(2) A single vehicle on ground: on the way to pick up a retrieving boat or leave the ramp after sending a launching boat;
(3) A boat-vehicle combo on ground: on the way to send a launching boat or leave the ramp after picking up a retrieving boat;
(4) A boat-vehicle combo at the intersection of land and water with the boat in water and the vehicle on ground.

Therefore, the object of interest is either in the status of single object (i.e., boat or vehicle) or a boat-vehicle combo. The philosophy of our method is to conduct boat-vehicle combo modelling to enhance tracking performance. By recognizing the interactions between pairs of object and estimating their collective activities, two correlated objects are tracked simultaneously under low frame rate conditions.

2   Proposed Method

2.1   Modeling Boat-Vehicle Combo

In the operation of fishing trailer boat, the vehicle is active driving and the boat is passive. The minimum distance between two objects is shown when the boat front faces the rear of vehicle. This distance increases when the vehicle turns to either left or right. This distance reaches a maximum when the passive object (i.e., the trailer boat) starts to move in following the active object motion. Figure 1 illustrates the scenario of combo object detection. Let us denote the passive and active object by o^i and o^j with their centers identified as A and B, respectively. As seen, when the o^i front faces the rear of o^j, the minimum distance between two objects gives as the length of line segment AB. When the active object o^j turns direction until point A, D and E lie on a straight line, the two objects distance reaches the maximum, which is the length of line segment AB'.

Therefore, o^i and o^j form a boat-vehicle combo if the following rule is satisfied:

$$\frac{1}{2}w_i + \frac{1}{2}w_j + \varepsilon \le d_{ij}$$
$$\le \frac{1}{2}\sqrt{(w_i{}^2 + h_i{}^2) + (w_j{}^2 + h_j{}^2) + 2w_j\sqrt{w_i{}^2 + h_i{}^2}} + \varepsilon, \tag{1}$$

where d_{ij} is the Euclidean distance between the center of o^i and o^j, (w_i, h_i) and (w_j, h_j) are the width and height of bounding box of o^i and o^j, respectively, and ε is the gap between two bounding boxes. Consider ε is negligible as compared to the size of boat and vehicle. Then (1) can be simplified as

$$\frac{1}{2}w_i + \frac{1}{2}w_j \le d_{ij}$$
$$\le \frac{1}{2}\sqrt{(w_i{}^2 + h_i{}^2) + (w_j{}^2 + h_j{}^2) + 2w_j\sqrt{w_i{}^2 + h_i{}^2}}. \tag{2}$$

A straightforward solution to combo object tracking is to treat a combo as a combined single object, which is represented as a bounding box containing

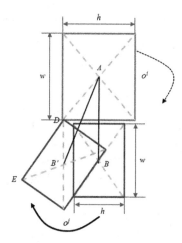

Fig. 1. Illustration of combo object detection in terms of distance estimation between two single objects in the combination.

both boat and vehicle, and conduct typical single object tracking. As we know, object tracking very much relies on robust object feature description. A combo object has normally bigger than single object bounding box, which more likely to cause invalidity of feature for object tracking across frames, and which causes eventually false negatives.

Consider the fact that the retrieving boat is always pulled by the vehicle when they leave the boat ramp, and the launching boat is always pushed by the vehicle and they move backwards to the shoreline. The boat should be always closer to water area than the vehicle in a combo. Thus, we can have a safe assumption that the object closer to the shoreline is the boat, and the remaining object in the combo is the vehicle. Then we have the second solution to combo object tracking: Two combos are considered to be matched only if their boats and their vehicles are matched respectively.

The disadvantage of this method is, under low frame rate conditions, one object may be more recognizable than the other. We can not treat them equally without the consideration of unbalanced variations among different objects. To cope with the unbalanced variations, we revise the above solution as follows: two combos are considered to be matched if their boats or their vehicles are matched. Apparently, the method does not make use of the distinctions provided by the combined single objects in a combo. In practice, we consider reducing false negatives by characterising both the combined single object and two single objects in the combination, we have the proposed combo object tracking strategy as: two combos are considered as the same if their combined single objects or any category of single objects are the same.

2.2 Combo Object Features

Let us denote boat-vehicle combos in frame t and $t-1$ by z_t and z_{t-1}, respectively. m_t and m_{t-1} refer to the corresponding combined single object (i.e., one bounding box containing boat and vehicle). Vehicle object v_t and boat object b_t are in combo z_t, and vehicle object v_{t-1} and boat object b_{t-1} are from combo z_{t-1}. As matching objects, distance calculation is performed on three pairs: $\{m_t, m_{t-1}\}$, $\{v_t, v_{t-1}\}$ and $\{b_t, b_{t-1}\}$. The dissimilarity of two combos is defined as the minimum of distance of these three pairs, according to the strategy described in Sect. 2.1. To measure the similarity between two combos, four features, namely histogram of intensity, average intensity of interest points, texture feature and average intensity ratio of vehicle to boat, are investigated as follows:

Let $D(z_t, z_{t-1})$ be the intensity distance of combo z_t and z_{t-1}. The intensity histogram of m_t, m_{t-1}, v_t, b_t, v_{t-1} and b_{t-1} is given as $\phi(m_t)$, $\phi(m_{t-1})$, $\phi(v_t)$, $\phi(b_t)$, $\phi(v_{t-1})$ and $\phi(b_{t-1})$, respectively.

The difference between the intensity distribution of three pairs, $\{m_t, m_{t-1}\}$, $\{v_t, v_{t-1}\}$ and $\{b_t, b_{t-1}\}$, can be computed by the vector cosine angle distance,

$$D(\phi(m_t), \phi(m_{t-1})) = \frac{\sum_{i=1}^{M} \phi(m_t)\phi(m_{t-1})}{\sqrt{\sum_{i=1}^{M} \phi(m_t)^2}\sqrt{\sum_{i=1}^{M} \phi(m_{t-1})^2}}, \tag{3}$$

$$D(\phi(v_t), \phi(v_{t-1})) = \frac{\sum_{i=1}^{M} \phi(v_t)\phi(v_{t-1})}{\sqrt{\sum_{i=1}^{M} \phi(v_t)^2}\sqrt{\sum_{i=1}^{M} \phi(v_{t-1})^2}}, \tag{4}$$

$$D(\phi(b_t), \phi(b_{t-1})) = \frac{\sum_{i=1}^{M} \phi(b_t)\phi(b_{t-1})}{\sqrt{\sum_{i=1}^{M} \phi(b_t)^2}\sqrt{\sum_{i=1}^{M} \phi(b_{t-1})^2}}, \tag{5}$$

where i is the histogram bin index, and M is the number of bins. The same as the case for single object matching, M is set as 16. Then the difference of intensity distribution between z_t and z_{t-1} is calculated as,

$$D(z_t, z_{t-1}) = min\{D(\phi(m_t), \phi(m_{t-1})), \tag{6}$$
$$D(\phi(v_t), \phi(v_{t-1})), D(\phi(b_t), \phi(b_{t-1}))\}.$$

Based on (6), the likelihood of intensity distribution for combo matching between z_t and z_{t-1} is defined as,

$$p_v(z_{t-1}|z_t) \propto exp\left(-\frac{D^2(z_t, z_{t-1})}{2\sigma^2}\right), \tag{7}$$

To characterize the relationship between two single objects (i.e., vehicle and boat) in the combo, the ratio of vehicle to boat in terms of average intensity is particularly considered for combo matching.

Let us denote the average intensity ratio of z_t and z_{t-1} by $\zeta(z_t)$ and $\zeta(z_{t-1})$, respectively. Then, the difference between $\zeta(z_t)$ and $\zeta(z_{t-1})$ is calculated as

$$D(\zeta(z_t), \zeta(z_{t-1})) = |\zeta(z_t) - \zeta(z_{t-1})|. \tag{8}$$

Accordingly based on (8), the likelihood of average intensity ratio for combo matching is given as

$$p_r(z_{t-1}|z_t) \propto exp\left(-\frac{D^2(\zeta(z_t), \zeta(z_{t-1}))}{2\sigma^2}\right), \tag{9}$$

2.3   Combo Object Matching

In calculating matching score, we conduct the fusion of multiple features and define the matching score separately for single and combo object.

For single object, we simply combine multiple features introduced above. Then, the likelihood for single object matching can be calculated as

$$p(o_{t-1}|o_t) = \prod_{i=1} p_i(o_{t-1}|o_t), \tag{10}$$

where i is the index of features. In our case, three different features are considered, which follows that $p_i(o_{t-1}|o_t)$ can be the likelihood of intensity distribution, average intensity of interest points or texture feature.

The score for single object matching can be calculated as

$$\delta_o(o_{t-1}, o_t) = \ln p(o_{t-1}|o_t) = \sum_{i=1} \ln p_i(o_{t-1}|o_t). \tag{11}$$

For combo object, we combine multiple features, and compare frame t against $t-1$ detected objects, which include vehicle, boat, and their combined single object. Then, the likelihood for combo object matching can be calculated as

$$p(z_{t-1}|z_t) = \prod_{i=1} p_i(z_{t-1}|z_t), \tag{12}$$

where i is the index of features. In this case, four different features are used, which follows that $p_i(z_{t-1}|z_t)$ can be the likelihood of intensity distribution (7), average intensity of interest points, texture feature and average intensity ratio (9), respectively. Note that each likelihood here involves three single objects (i.e., vehicle, boat and their combined single object) comparison.

The score for combo object matching can be calculated as

$$\delta_z(z_{t-1}, z_t) = \ln p(z_{t-1}|z_t). \tag{13}$$

Table 1. Experimental data

Year	No. of frames (No. of boats)		
	Waitangi	Raglan	Takapuna
2010	77,760 (1772)	79,200 (783)	86,400 (2844)
2011	83,520 (1422)	82,080 (908)	84,960 (2777)
2012	69,120 (1140)	80,640 (724)	77,760 (2174)
Total	230,400 (4334)	241,920 (2415)	249,120 (7795)

3   Experimental Results

The image data we used for our experiments was collected from a network of web cameras overlooking key boat ramps in New Zealand. Table 1 describes the experimental data, which includes 2010–2012 image sequences captured at Waitangi, Takapuna and Raglan boat ramp, and which records the number of frames, and the truth number of objects from manual count for each ramp and each year. The frame size of the video is 720×576 pixels, and the frame rate is 1 frame/minute. The total number of frames tested for Waitangi, Takapuna and Raglan is 230,400, 241,920 and 249,120, respectively.

In our experiment, the proposed tracking approach was compared with the state-of-the-art approaches. The same initializations were set to all algorithms for fair comparison. The parameters of all methods were tuned to achieve the best performance, and the ground truth was manually labeled in advance for comparisons. The proposed system was implemented in Matlab. It was run on a Quad Core 3.4 GHZ CPU with 8 GByte memory. The whole process is fully automatic and requires no manual intervention.

3.1   Performance of Boat Counting

In our boat counting experiments, the proposed method was evaluated and compared with four conventional tracking methods, which include particle filter (PF) [1], and extended mean shift (EMS) [2], cascade particle filter (CPF) [3] and object tracking in stereo videos (STEREO) [4]. For performance evaluation, we calculate the differences between the ground truth (i.e., daily boat number from manual count), and daily boat number provided by different algorithms by NRMSE, normalized root mean squared error

$$NRMSE = \frac{\sqrt{\frac{\sum_{i=1}^{K}(N_i - C_i)^2}{K}}}{N_{max} - N_{min}}, \tag{14}$$

where N denotes the ground truth of boat number from manual count, and C is the number from one tracking method. K is the total number of days for boat counting, and N_{max} and N_{min} represent the maximum and minimum number of boat from manual count, respectively.

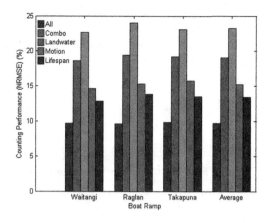

Fig. 2. boat counting performance (in terms of NRMSE) with and without consideration of different factors.

Table 2 presents the NRMSE in percentage for five object tracking methods boat counting at Witangi, Raglan, and Takapuna, respectively. As seen from the table, PF fails to track over 80 % object of interest. EMS, CPF and STEREO perform relatively better than PF, but their errors still above 40 %, which obviously does not satisfy our application requirement for monitoring recreational fishing efforts over time. In contrast, the proposed method gives average less 10 % counting error for all three maritime boat ramps. This demonstrates that our approach successfully overcomes the extremely dynamic background and severe unpredictability problems compared with the state-of-the-art methods.

For the proposed method, we further investigate its four key factors that impact boat counting performance. This include (1) modeling separately land and water scenes (denoted as Landwater factor), (2) modeling boat-vehicle combo tracking (Combo), (3) increasing motion continuity of objects (Motion), and (4) utilizing lifespan templates of launching and retrieving boat to improve counting accuracy (Lifespan), in which factor (3) and (4) are not the focus this paper. A sensitivity experiment on these factors is performed using the same data for evaluating boat counting performance, one factor is unselected each time. In this test, we suppose that object counting with all factors obtains the best performance, and removing one of the factors leads to the decrease of performance. The significance of each unselected factor is then indicated by the decrease of counting performance measured in NRMSE. Figure 2 displays the counting performance without different selected factor as well as that of with all factors. As seen, the Landwater factor has the greatest influence on the counting performance. If without consideration of the Landwater factor, the counting accuracy is expected to reduce 22 % in average. This can be explained that if lack of accurate background estimation, the accuracy of object detection may deduct largely. The second important is the Combo factor, which causes over 15 % counting accuracy loss. This indicates that for fishing trailer tracking,

Table 2. Boat counting performance

Method	PF [5]	EMS [2]	CPF [3]	STEREO [4]	Proposed
Waitangi	87.23	53.61	47.36	45.56	**9.71**
Raglan	84.49	51.58	45.57	42.31	**9.62**
Takapuna	89.62	59.32	51.43	48.47	**9.83**
Average	87.11	54.84	48.12	45.45	**9.72**

correlation between vehicle and boat contributes significantly to the success of boat tracking and counting.

4   Conclusion

Tracking trailered boat at maritime boat ramp involves extra challenges on land-water dynamic scene and unpredictable variations of boat-vehicle combination. Based on a reliable multi-feature object matching mechanism, we propose in this paper a combo object model for fishing trailered boat tracking. The proposed method is capable of simultaneously tracking two combined objects and their combination, monitoring the connection of two objects, and estimating their collective activities. Experimental comparative tests and quantitative performance evaluations on boat counting for three real world boat ramps demonstrate the merits of the proposed approach. Note that severe unpredictability caused by the extremely low 1 frame per minute rate is noticed, but not addressed in this work. Our future work will therefore focus on how to improve appearance continuity in the tracking environment of low frame rate.

References

1. Khan, Z., Balch, T., Dellaert, F.: Mcmc-based particle filtering for tracking a variable number of interacting targets. IEEE Trans. Pattern Anal. Mach. Intell. **27**(11), 1805–1918 (2005)
2. Porikli, F., Tuzel, O.: Object tracking in low-frame-rate video. In: Image and Video Communications and Processing, vol. 72, 6 April 2005
3. Li, Y., Ai, H., Yamashita, T., Lao, S., Kawade, M.: Tracking in low frame rate video: a cascade particle filter with discriminative observers of different life spans. IEEE Trans. Pattern Anal. Mach. Intell. **30**(10), 1728–1740 (2008)
4. Chuang, M., Hwang, J., Williams, K., Towler, R.: Tracking live fish from low-contrast and low-frame-rate stereo videos. IEEE Trans. Circuits Syst. Video Technol. **25**(1), 167–179 (2015)
5. Breitenstein, M., Reichlin, F., Leibe, B., Koller-Meier, E., Van Gool, L.: Online multiperson tracking-by-detection from a single, uncalibrated camera. IEEE Trans. Pattern Anal. Mach. Intell. **33**(9), 1820–1833 (2010)

Erratum to: Towards Robustness to Fluctuated Perceptual Patterns by a Deterministic Predictive Coding Model in a Task of Imitative Synchronization with Human Movement Patterns

Ahmadreza Ahmadi and Jun Tani[✉]

KAIST, Daejeon, South Korea
ar.ahmadi62@gmail.com, tani1216@gmail.com

Erratum to:
Chapter "Towards Robustness to Fluctuated Perceptual
Patterns by a Deterministic Predictive Coding Model in a Task
of Imitative Synchronization with Human Movement
Patterns" in: A. Hirose et al. (Eds.):
Neural Information Processing (Part I), LNCS,
DOI: 10.1007/978-3-319-46687-3_44

In the original version, the name of the first author was mispelled. It should read as "Ahmadreza Ahmadi"

The updated original online version for this chapter can be found at
DOI: 10.1007/978-3-319-46687-3_44

© Springer International Publishing AG 2017
A. Hirose et al. (Eds.): ICONIP 2016, Part I, LNCS 9947, p. E1, 2016.
DOI: 10.1007/978-3-319-46687-3_70

Author Index

Printed in the United States
By Bookmasters